EcoProduction

Environmental Issues in Logistics and Manufacturing

Series editor

Paulina Golinska, Poznan, Poland

For further volumes:
http://www.springer.com/series/10152

About the Series

The EcoProduction Series is a forum for presenting emerging environmental issues in Logistics and Manufacturing. Its main objective is a multidisciplinary approach to link the scientific activities in various manufacturing and logistics fields with the sustainability research. It encompasses topical monographs and selected conference proceedings, authored or edited by leading experts as well as by promising young scientists. The Series aims to provide the impulse for new ideas by reporting on the state-of-the-art and motivating for the future development of sustainable manufacturing systems, environmentally conscious operations management and reverse or closed loop logistics.

It aims to bring together academic, industry and government personnel from various countries to present and discuss the challenges for implementation of sustainable policy in the field of production and logistics.

Paulina Golinska
Editor

Logistics Operations, Supply Chain Management and Sustainability

Springer

Editor
Paulina Golinska
Engineering Management
Poznan University of Technology
Poznan
Poland

ISSN 2193-4614 ISSN 2193-4622 (electronic)
ISBN 978-3-319-07286-9 ISBN 978-3-319-07287-6 (eBook)
DOI 10.1007/978-3-319-07287-6
Springer Cham Heidelberg New York Dordrecht London

Library of Congress Control Number: 2014941519

Printed on acid-free paper

Springer is part of Springer Science+Business Media (www.springer.com)

Preface

Nowadays, logistics and supply chain management are influenced by the sustainable policy challenge. Managers and scientists search for innovative approaches to eco-friendly organization and coordination of the logistics processes and the supply chain configuration, which allow to pollute less.

In order to do this, companies have to fix their common environmental objectives, sharing technical information about products, planning, and processes or starting common programs to reduce adverse impacts on the environment. In order to meet the sustainability goals the following areas have to be taken into consideration:

- green supplier selection,
- collection of used products and their reuse,
- reduction of hazardous substances,
- optimal location of logistics and manufacturing facilities for minimizing unnecessary transport,
- reduction of CO_2 emission.

The book entitled "Logistics Operations, Supply Chain Management and Sustainability" aims to present state-of-the art researches and practical applications. In this book the focus is placed on a multidisciplinary approach. It presents viewpoints of the academic and the industry personnel on the challenges for implementation of sustainable police in logistics. Authors present in the individual chapters the result of the theoretical and empirical research related to the following topics:

- Supply Chain Management and Sustainability,
- Reverse Logistics and Environmental Sustainability,
- Modeling and Optimization of the Manufacturing Operations,
- Optimization of the Location Problems, the Inventory Management, and the Vehicle Routing Problems.

This book includes research contributions of geographically dispersed authors from Europe, North America, Africa, and Asia. It is a clear indication of a growing interest in sustainable development in logistics. The high scientific quality of the chapters was assured by a rigorous blind review process implemented by leading researchers.

This monograph provides a composition of theoretical trends and practical applications. The advantage of this book is presentation of country-specific applications from a number of different countries around the world.

I would like to thank all the authors who responded to the call for chapters and submitted manuscripts to this volume. The International Congress on Logistics and SCM Systems has for 10 years provided an international forum for leading researchers, educators, and practitioners to discuss ideas, exchange experiences on the latest development, and seek opportunities for collaboration on the latest implementation and enhancement of sustainable logistics and SCM systems in the dynamic market. This book presents selected papers from authors who attend the 9th International Congress on Logistics and SCM Systems (ICLS 2014). Although not all of the received chapters appear in this book, the efforts spent and the work done for this book are much appreciated.

I would like to express my gratitude to the Board of the International Federation of Logistics and SCM Systems (IFLS) for the valuable contribution to the volume:

- Honorary Chairman—Prof. Karasawa, Yutaka, Kanagawa University, Japan
- Advisor—Prof. Katayama, Hiroshi, Waseda University, Japan
- Chairman—Prof. Lee, Young Hae, Hanyang University, Korea
- Senior Vice Chairman—Prof. Kachitvichyanukul, Voratas, Asian Institute of Technology, Thailand
- Vice Chairmen—Prof. Lai Kin Keung, City University of Hong Kong; Prof. Rim Suk-Chul, Ajou University, Korea; Prof. Wakabayashi Keizo, Nihon University, Japan and Wu Yen-Chen Jim, National Sun Yat-Sen University, Taiwan.

I would like to thank all reviewers whose names are not listed in the volume due to the confidentiality of the process. Their voluntary service and comments helped the authors to improve the quality of the manuscripts.

Paulina Golinska

Contents

Part I
Supply Chain Management and Sustainability

The Impact of Demographic Changes on Human Resources Management in European Supply Chains-Selected Aspects

Halina Brdulak

Abstract As an analysis of demographic trends in the world, Europe and Poland clearly indicates, the elderly people's share of total population has been steadily expanding, just as the generation born at the turn of the millennium grows up. These two demographics are going to immensely influence the future supply chain. The present chapter analyses the behaviours and attitudes of Generation Y (Millennials)—in their roles as present and future customers and employees—against the background of other selected social groups. The changing economic and business environments are discussed, along with new approaches to interactions within the supply chains, e.g. through coopetition ("cooperative competition"). Using the example of Poland's lead logistics provider (LLP), Lean Management is presented as an approach where intergenerational differences can be tapped for the purpose of efficient supply chain management.

Keywords Demographic changes · Aging of society · Generation Y · Coopetition · Lean management

1 Introduction

The changes going on in the local, regional and global environments over the past decade, coupled with the instability of financial markets, have not been without influence upon the evolving models of supply chain management and configuration. Most importantly, a growing importance has been assumed by risk management—the skill of identifying risks, determining the probability of their occurrence, and assessing their magnitude. The future supply chain will also reflect

H. Brdulak (✉)
Warsaw School of Economics, Warsaw, Poland
e-mail: halina.brdulak@sgh.waw.pl

P. Golinska (ed.), *Logistics Operations, Supply Chain Management and Sustainability*, EcoProduction, DOI: 10.1007/978-3-319-07287-6_1,

changes in human resources management (such as the aging of society and the *acceleration* of Generation Y), in processes (a push to shorten processes and lower their costs, while keeping the quality high) and in technology (technology-based integration of supply chain processes, Big Data, and replacement of simple work by robotics) (Future Value Chain 2020 2012; Brdulak 2014).

In the focus of the present chapter are chances and risks in the field of human resources management, which is seen as key to staying keeping competitive in a turbulent supply chain environment. The article begins with presentation of numerical data on the aging of society (with special attention devoted to Europe and Poland), along with projections of the trend's getting still more pronounced in the future. An added factor that to some extent can influence these developments is the level of migration. The chapter goes on with an analysis of behaviours and values of Generation Y, young people of up to 35 years of age who have already taken up employment, and whose attitudes differ quire perceptibly from those known from the past. The last part discusses the question of which management system utilises most effectively—from the viewpoint of the entire supply chain—the advantages offered by employee's generational and by-gender diversification. Finally, to support the chapter's conclusions, the example is presented of a global-network logistics operator.

The thesis advanced in this chapter is that efficient supply chain management must be rooted in an open communication culture, where problems arising from workforces' age diversity are proactively resolved (Brdulak 2009) and where, most importantly, stakeholders understand the processes taking place in the organisation. The competitive pressures in the future will be for continuous cost reductions and quality improvements—and for solutions to be developed by shared effort throughout the supply chain. Such a culture, as this author argues, can be build by embracing Lean Management.

2 Demographic Changes in the World, Europe and Poland: On a Time Scale of 50–100 Years

According to a 2012 UN analysis (World Population Prospects 2013), the population aged 60 or more in the most developed regions of the globe will grow from 287 million in 2013 to 417 million in 2050 and 440 million in 2100. The respective figures in the developing regions will be 554 million, 1.6 and 2.5 billion, representing a much faster pace (3.7 % a year in 2010–2015), and leading to growth in senior citizens' numbers there—not only in absolute terms, but also as a proportion of the total populace in developing countries.

The population in the less developed regions is still young, with children below 15 years of age (1.7 billion) representing 28 % of total population and young persons aged 15–24 (1.1 billion) accounting for a further 17 %. In the group of the least developed countries, the respective proportions for children and young people are 40 and 20 %.

On the other hand, in the more developed regions children account for 16 % and young people for 12 % of total population. In the future, the numbers for children are expected to change only insignificantly (206 million in 2006, 210 million in 2050 and 202 million in 2010), with young people actually registering a decline (152 million in 2013, 142 million in 2050, 138 million in 2010).

In addition to these demographic changes, the supply chain configuration will also be impacted by the population's regional distribution patterns. As shown in Table 1, markets will be moving towards Asian and African countries, the likely areas of increasing concentrations of the population. And the European population will very probably shrink considerably over the coming years. These changes will be influenced by the expected migration levels, as presented in Table 2.

Most of the global population today live in a few countries. According to data for 2013, China and India accounted between them for 37 % of the total, with a further 22 % inhabiting another eight countries (USA, Indonesia, Brazil, Pakistan, Nigeria, Bangladesh, Russia and Japan). By 2050, this group of the most populous countries will be joined by several entrants from Africa.

In Europe the proportion of those aged 60+, standing at 23 % in 2013, will go up to 34 % in 2050, with the oldest people increasing their share from, respectively, 4.5–9.5 %. But those in the working age will make up only 41 % of the population in 2050, down from 50 % in 2013. This pattern of changes will be similar in Poland, where the 65+ age group contributed 18 % to total population in 2012 (ahead of the up-to-14-year-olds by 2.3 % points), and in 2050 is going to be twice as strong as children (33 % vs 15.6 %, respectively). The total Polish population is expected to dwindle to 30 million by that time (down from 38 million in 2013). These trends are plotted in Fig. 1.

3 Sociological Changes and How They Affect Employee Behaviours

Globally operating companies have been faced with a challenge posed by the diversity of workforces in terms of behavioural patterns, requirements and sets of values. It takes a skilful combination of the previously mentioned elements for such workforces to be able to cooperate and benefit from synergy effects.

Based on the birth-date criterion, sociologists distinguish four major demographic cohorts in the contemporary world.

- Generation Y (born between 1982 and 2002, sometimes referred to as Nexters or Millennials),
- Generation X (1965–1981),
- Baby Boomers (1946–1964),
- Traditionals (pre–1946).

Table 1 Percentage distribution of the world population by development group and major area, estimates and projections according to medium variant, 1950–2100 (World Population Prospects 2013)

Group or area	1980	2013	2050	2100
World	100	100	100	100
More developed regions	24.3	17.5	13.6	11.8
Less developed regions	75.7	82.5	86.4	88.2
- Least developed countries	8.8	12.5	19.0	27.0
- Other less devel.countries	66.8	70.0	67.4	61.2
Africa	10.8	15.5	25.1	38.6
Asia	59.2	60.0	54.1	43.4
Europa	15.6	10.4	7.4	5.9
Latin America and Caribbean	8.2	8.6	8.2	6.8
Northern America	5.7	5.0	4.7	4.7
Oceania	0.5	0.5	0.6	0.6

Table 2 The major countries of net receivers and net emigration of international migrants (2010–2050) (World Population Prospects 2013)

Countries	Number of net receivers/emigration
USA	+1,000,000 (annually)
Canada	+205,000
United Kingdom	+172,500
Australia	+150,000
Italy	+131,250
Russian Federation	+127,500
France	+106,250
Spain	+102,500
Bangladesh	−331,000 (annually)
China	−300,000
India	−284,000
Mexico	−210,000
Pakistan	−170,000
Indonesia	−140,000
Philippines	−92,500

The research focus today is on the youngest demographic, Generation Y, most of whom are not independent financially and without regular (permanent) employment—and even when being so employed, they often draw on parental support (Finansowy portret młodych 2014).

Only 29 % of young people in Poland work under contracts of permanent employment. Overall (for all age groups), permanent employment is most common among those with post-primary vocational education (82 %) and higher education (87 %), followed by those with primary education (55 %) and secondary education (70 %).

The Millennials identify themselves not by the occupation they hold, but rather the activities they perform—for example, computer engineering rather than working for IBM—and consequently they set more store by values and life-work

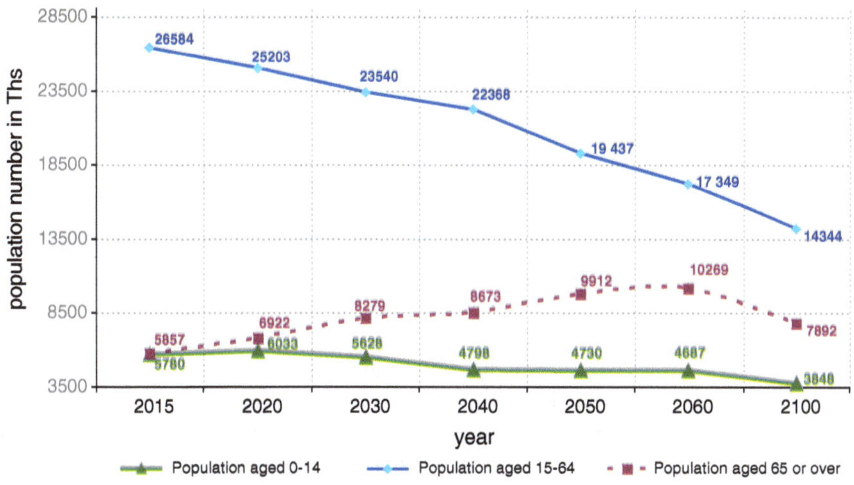

Fig. 1 Demographic Changes in Poland, 2015–2100 (World Population Prospects 2013)

balance. More than one-fifth of the Millennials expect to work for six or more companies in the course of their occupational career, and only 28 % speculate that they will have fewer than three jobs (DeVry University). They value "what they do" much more than "who they work for", and they want to make a difference in the world while ignoring the existing authorities, hierarchies, and the urge to "buy stuff" (in a certain required sequence) (Forbes 2012).

The threshold of adulthood—and its very definition—has been evolving. As early as the late 20th century, adulthood was believed to begin upon completing education, embarking on a career, getting full financial independence and starting a family, but today, at a time of the financial crisis and immense problems with finding a job, this definition no longer corresponds to the reality—and perhaps the reality no longer corresponds to this definition, too.

According to research findings of Achieve Global and Kelly Global Workforce Index 2013,[1] the question about linkage between an employee's age, on the one hand, and their behaviour and attitude to work on the other has no unequivocal answer. The Achieve Global study covered 512 employees, coming from the United States (44 %), Asia (23 %, including 12 % from China and 11 % from Singapore), Europe (33 %, including 13 % from Germany and 21 % from the United Kingdom). In breakdown by age group, Generation Y was the strongest, with a 30 % share, and the Traditionalists brought up the rear (13 %). The study revealed differences in respondents' replies in accordance with their nationality. The opinion that behaviour and attitude towards work are linked to age was expressed much less frequently by employees from Europe and the United States, compared to those from Asia.

[1] http://www.kellyocg.com/uploadedFiles/Content/Knowledge/Kelly_Global_Workforce_Index_Content/Global_Trends_that_Shaped_Job_Choice_Recruitment_and_Workplace_Performance.pdf. dostęp 28.02.2014.

The Kelly Global Workforce research covered more than 1,00,000 employees from all over the world, including Poland. Respondents judged seven qualities of a job—new experience, financial stability, learning and growth, career advancement opportunities, respect, recognition and workplace flexibility—and, overall, the highest score went to respect. Only Generation Y put more value on career development opportunities, providing yet another indication of their drive for self-improvement. At the bottom of the scale came to new job experience, so ranked by all groups. In the opinion of more than 30 % of Polish respondents, generational diversity among employees is conducive to greater efficiency in the workplace.

The research findings indicate that on many questions the views of Generation Y coincide with those of Generation X (both are least inclined to believe in benefits of age diversity in the workplace, and they strongly prefer traditional, financial rewards). On the other hand, the Baby Boomers believe—more frequently than others—that they better understand the multigenerational differences, and they more strongly appreciate non-financial rewards, such as extra days off or access to training. All groups prefer direct contact above online communication in the workplace.

No definitive conclusions can be drawn from these findings. Some researchers also warn against *pigeonholing* and dividing employees into three or four age categories, and then assigning them certain features and behaviours. But based on the research, some weak signals can indeed be picked up, to be possibly used in further observation.

The present author, being an academic worker of Poland's largest economic university and also an advisor to management of a global logistics provider, is inclined to accept the following conclusion: generational differences are indeed visible in the behaviours of, both, students and employees of the large logistics company. There are also differences in the attitude to learning within the younger generation—students must be strongly *nudged* to take interest in a subject, whereas young employees are open to interaction with other people who share their values (no matter their age and regardless of hierarchies). But the key to cooperation among age- and gender-diverse groups is respect and consideration for everyone. These observations will be discussed in greater detail in the chapter's closing part.

4 Supply Chain Trends

A 2012 study from Alcatel Lucent (http://www2.alcatel-lucent.com/knowledge-center/public_files/megatrends/megatrends_n2g_en_market_analysis.pdf) identifies seven megatrends, expected to fundamentally shape the future of the economy. These include "digital native acceleration", *ed-you-cation* and *rejuvenaging*—three areas closely connected with the processes diagnosed earlier in this chapter.

The definition of aging has been evolving and, for example, the retirement age in Poland (and some other countries, too) has been prolonged to 67, with all its implications for employment. People aged 60 now face several years of working, and if they want to use this time actively they should, to some extent, adjust to the new trends (ed-you-cation). They already rate highly physical and intellectual prowess as well as healthy eating. Among EU-sponsored programmes, there are quite a few which specifically target older people's needs. Universities of the third age—and even the fourth age—have been playing an increasing role, and courses are offered in computer and internet skills. Meanwhile, the economy is being entered by members of Generation Y, which in the quoted study is referred to as the Digital Natives.

A turbulent environment, pressure for cost reductions and continued product improvement, ever shorter product lives (sometimes down to several months, especially in the IT sector), coupled with the ongoing demographic changes, have the effect of enforcing the emergence of new management models, too. The University of Tennessee in 2013 (Game Changing Trends In Supply Chain 2013) identified four key areas of competency influencing the supply chain competitiveness. They are presented in Fig. 2.

In order to build a competitive supply chain in a changing environment, the prerequisite is closer collaboration between the parties involved, with interactions increasingly involving a common search for solutions, rather than competition. But such performance is still rare, requiring as it is a change in parties' approach to business. In the University of Tennessee study, supply-chain collaboration averaged between a 2 and 3 on a 10-point scale.

Furthermore, researchers have increasingly been pointing to the importance of intellectual competences, attitudes, behaviours, talent. In this writer's opinion, operational excellence is driven in equal measure by technology and by the performance of people who deal with its deployment (and subsequent improvement). This explains why supply chain coopetition, or collaboration by competitors at the supply chain's particular stages, have increasingly been used as a springboard from which to jump off to a position which individual companies are unable to reach alone—as marked by continued reductions of process costs and time, and rising quality.

The strategy of coopetition is largely based on experiences gained from the pursued negotiating strategies. The methods of principled negotiation (negotiation on the merits) and creative negotiation, involving both parties' collaboration and based on trust, make it possible to reach much better outcomes than either party has expected. But these methods require moving beyond set patterns, understanding the values which underlie the partner's business and, most importantly, they require respecting the partner. On the other hand, under the traditional management model, negotiations are seen as the pursuit of a war strategy. Similarly, war games are frequently deployed to explain the workings of markets.

Very much in the same vein as the coopetition concept is the highly popular management model known as Porter Diamond, which places emphasis on understanding the roles of all players in the contemporary economy. In network economics, the links to the other elements must be strong enough to enable the operation of the whole network, but on the other hand they should be sufficiently weak to make it possible to easily "work your way" into other network elements, thus adjusting to the changing environment. This idea is connected with

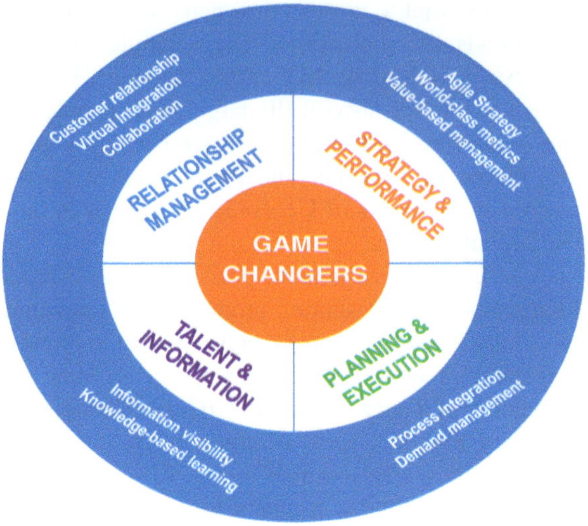

Fig. 2 Game Changing Trends In Supply Chain, University of Tennessee, Spring (2013)

developments such as service customisation and flexible adjustment to business requirements. Gartner research and advisory company says the practice of "working your way" into the partner's value chain will get the upper hand in the future. In the opinion of the present author this in nothing else than a "*networkisation*" of relations, where boundaries between organisations get blurred. One indication that this is indeed the case is the emergence of the prosumers, who complete the last stage of production by themselves, individualising the product and adjusting it to their needs—even if they not always are fully aware of their new role.

5 Lean Thinking as an Approach Combining Intergenerational Competencies

Noting the growing influence exerted on the economy by the Millennials, analyses of future supply chains emphasise the need for motivational and recruitment systems to be suited to the requirements of that generation. *Stick and carrot*, still the leading motivation system at most companies, does not work with Gen Y. Immersed in technology as their default environment, freely surfing throughout the internet and making full use of the social media, they are more oriented to values and to value-based collaboration. Hierarchical structures and authorities are alien to them. They take pleasure in leisure time and can perform their job from various venues, not necessarily having to be gathered under one roof. Their behaviours and needs are fairly different than in the case of those aged 40+ and 50+, who were born in the 1960s and 1970s, and for whom technology change came while they were reaching maturity. The latter's world, based on traditional values, hierarchies and authorities, was more ordered and more stable. Now these two generations should identify the areas where they can meet each other and work together.

A positive response to these challenges is provided by Lean Management, an offshoot of Toyota Production System, which rests on two pillars: respect for people and continuous improvement. Activities are aimed to reduce costs, cut cycle times, and improve quality. Problems are resolved by searching for the ultimate root cause, and growth is based on jointly pursued improvement projects. Teams are so selected as to ensure that different viewpoints are represented (interdisciplinary teams), and that they are age- and gender-diverse, which results in greater creativity. Under this approach, with the passage of time, people learn to cooperate and social capital increases. The subsequent part of the present chapter describes the operation of this system in a service sector company, a logistics operator.

5.1 Case Study: Lead Logistics Provider

The logistics provider in question operates globally, being present in 130 countries around the world, and it earned more than €20 billion in sales in 2012. Polish sales topped 1.4 billion zloty in 2013. Its logistic operations involve various transport segments—the largest being road haulage, followed by ocean-going, air and railway transport—and the company is also involved in warehousing. It has 17 branches around Poland.

The point of departure for the present case study is provided by an analysis of changes in employee age structure in 2011 and 2013, and in individual age groups' contributions to management (at middle level).

The company currently has a workforce of more than 1,800, mostly blue-collar workers. The women-men ratio stood at 36–64 in 2011, changing only slightly in 2013, to 35–65—just as the average employee age (from 35.7 to 36.3 years, respectively). The workforce's distribution pattern by age and by gender in 2013 is presented in Fig. 3. Members of Generation Y (aged 21–30) account for nearly 37 % of total workforce, this share being virtually stable over the three year period under review (2011–2013). The proportion of those aged 51 and above declined only insignificantly (from 12.7 to 11.4 %, respectively).

Bigger changes were seen in management. At the middle management level, the contribution from 21–30-year-olds rose in the period from 17.5 % to more than 19 %, while the share of those aged 51+ decreased by 0.7 % point (from 12.9 to 12.2 %, respectively). The substantial, and slowly increasing, proportion of the youngest managers, coupled with the continuing weight of people aged 50 and more (at above 12 %) necessitates changes in the company's motivation systems and management practices.

Noticing these trends, the company has launched several programmes targeting the youngest generation (21–30 years of age), including School of the Young. Also, a special programme for managers and directors aged 45+ was provided (School of Leaders, first edition), in addition to the Diversity Management

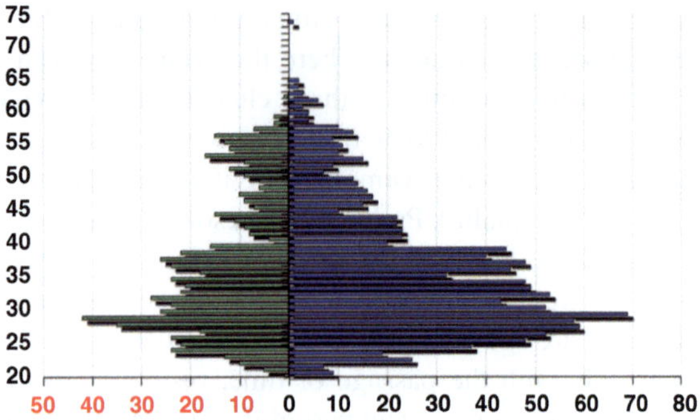

Fig. 3 Age Structure of LLP employees, by gender, 2013

programme for top managers, part- financed by the EU. The company opted for Lean culture back in 2011, when searching for more effective ways to tap employee potential and improve competitiveness. The new work regimen has been introduced in three stages, each involving different groups of company divisions, with completion planned for 2014. Initially run with the help of an outside company, Lean workshops have since 2013 been handled by trained internal coaches (Lean enablers), with only a little help from a third-party trainer. As a result, as early as late 2011, changes could be noticed in employee behaviours: work was being done in teams involving both seasoned managers and younger promoters of change; and with a view to exchange of experience, visits were made to the company's other branches (it should be noted that previously branches were run more like separate entities, in terms of management, although they were linked by a network of road connections). In 2013, as many as around 90 % of all employees were involved in process streamlining via participation in kaizen teams. On average, each team comprised five persons from different units, diversified in terms of age and gender. Those activities were backed by deploying A3 reporting, a Lean tool, for problem-solving purposes.

In this writer's opinion, these experiences demonstrate that they can be put to use in resolving problems related to age diversity and different motivation systems. Lean Management has increasingly been embraced by service and production companies, as a way to overcome problems simultaneously posed by globalisation, regionalisation, increased competition, and pressure on prices. A Lean language has also been developing, which can be spoken by all supply chain parties. Thus the future may also bring increased supply-chain transparency, which all managers regard as a driver of better management.

6 Conclusions

The changes taking place in the world, whether brought about by natural phenomena or factors of economic, demographic or sociological nature, enforce a new approach to supply chain configuration and management. It thus becomes a matter of necessity that there should be tighter collaboration between various parties along this chain, including between competitors (in certain areas), and also tighter intergenerational collaboration (in the context of the aging of society and *acceleration* of Generation Y). Such opportunities are provided by Lean Management, a system originated at Toyota and offering a chance for an Asian culture of teamwork to be transposed into the individualistically oriented European culture. The Lean language is becoming something like the Lingua Franca for contemporary supply chain management. Given that good communication and understanding of the other party's needs are among prerequisites of management, this author argues that Lean is particularly well placed to help realise these expectations.

References

Brdulak H (2009) Diversity management as a business model. Women Bus 1–4:1–7

Brdulak H (2014) Poszukiwanie nowych paradygmatów w zarządzaniu łańcuchem dostaw w warunkach stagnacji gospodarczej (Searching For New Paradigsm In Supply Chain Management In Terms Of Economic Stagnation) In: Zeszyty Naukowe Uniwersytetu Gdańskiego, Gdańsk

Finansowy portret młodych (2014) Instytut Badawczy Millward Brown. www.biznes.fn.pl. Accessed 28 Feb 2014

Forbes (2012) How millennials are redefining their careers as hustlers. www.forbes.com. From 19 Jul 2012

Future Value Chain 2020 (2012) Building strategies for the new decade. Capgemini 2012. www.futurevaluechain.com

Game Changing Trends in Supply Chain (2013) University of Tennessee, Spring 2013

World Population Prospects (2013) The 2012 revision. United Nations. New York. http://esa.un.org/wpp

Barriers of the Supply Chain Integration Process

Anjali Awasthi and Katarzyna Grzybowska

Abstract Complex systems of supply chains need to be integrated. Such integration is essential in order to achieve sustainable logistics of the system. This chapter presents an approach to this issue based on DEMATEL methodology. This chapter presents an approach to this issue based on identifying the barriers in supply chain integration and understanding their cause effect relationships using the DEcision MAking Trial and Evaluation Laboratory (DEMATEL) methodology. A total of 17 barriers affecting the integration of business entities in the supply chain were identified through a survey addressed to experts from Poland and Canada. The results of the study show Lack of Resource sharing (integration), Lack of Organisational compatibility, Lack of Information sharing, Lack of Responsibility sharing, and Lack of Planning of supply chain activities as top five barriers in supply chain integration. Therefore, organizations should investigate causes behind these barriers and take appropriate measures to resolve them to ensure seamless integration across their supply chains.

Keywords Supply chain · Integration · Barriers · DEMATEL · Cause effect relationship · Impact analysis

A. Awasthi (✉)
Concordia University, CIISE—EV 7.636, 1455 De Maisonneuve Blvd. West, Montreal, QC H3G 1M8, Canada
e-mail: awasthi@ciise.concordia.ca

K. Grzybowska
Poznan University of Technology, Strzelecka 11, 60-965 Poznan, Poland
e-mail: katarzyna.grzybowska@put.poznan.pl

P. Golinska (ed.), *Logistics Operations, Supply Chain Management and Sustainability*, EcoProduction, DOI: 10.1007/978-3-319-07287-6_2,

1 Introduction

Supply chain is metastructure (metasystem), characterised by a dynamic holarchy constituted of holons (business organisations) cooperating with each other. Business entities join in the supply chain by providing a diverse and unique ability or skill; this ability or skill is their characteristics. The more the supply chain expands, the less consistent and intimate the created system becomes; consequently lack of integration occurs. This results in the internal links and relationships becoming less stable; the cooperation between the entities might be then hindered.

The supply chain consists of permanent links, which constitute its core, and dynamically modified licks, for example, appropriate for a specific task. They are referred to as joining links. Following the completion of a given task, the joining links are separated from the core of the supply chain and the co-operation is discontinued (Awasthi et al. 2014; Grzybowska 2010a, b). A supply chain is a network of organisations which are involved in different processes and activities that produce value in the form of products and services in the hands of the ultimate consumer (Christopher 1998). The Supply Chain is a concept designed to manage entire supply chains consisting of numerous participating organisations (Mentzer et al. 2001). Supply chain management (SCM) plays an important role on increasing productivity of any organization and there is an increasing interest on implementing efficient SCM techniques in the competitive environment (Zandin and Maynard 2001). In order to develop supply chains (sustainable supply chains), all the involved organizations should work cohesively and constructively towards the bigger goal of achieving the triple bottom line objectives (economic, environment, social) of sustainability (Awasthi et al. 2014; Bai et al. 2012; Elkington 1994; Seuring 2013; Seuring and Müller 2008).

In 1995, R. Ganeshan, T. P. Harrison and D. Brown, S. Wilson and H. L. Lee, C. Billington defined the supply chain as a network of places (Brown and Wilson 2005; Ganeshan and Harrison 1995; Lee and Billington 1995). In the opinion of B. M. Lambert, J. R. Stock and L. M. Ellram, the supply chain should be organized so that the enterprises therein involved adjust to the flow and changes the supply chain undergoes (Lambert et al. 1998).

In order to discuss the supply chain, a list of items understood as factors or characteristics which determine the establishment of a supply chain has been created. It is a typical list of constitutive elements also allowing for the identification of significantly different supply chains (Grzybowska 2010a). The first listed element is the size of the supply chain (1). Even just two entities might constitute a basic supply chain as they fulfil all the constitutive elements of the supply chain. The previous research, however, suggests that for the supply chain to be discussed three cooperating cells need to be identified. The entities constituting the chain perform their designated roles (2). These roles are usually determined in accordance with the specific tasks roles. These roles commonly complement each other and reflect the existing relationships between the entities. The entities in the supply chain are also connected by business connections and relationships. These relationships are established during the contacts which occur due to the roles the entities fulfil, as well as the company status (3). The quality of the contact between the entities affects the integrity of the supply chain they constitute (4). Another constitutive element of the supply chain is communication (5), which may be also understood as coordination mechanisms essential to build a supply chain and maintain it. As a result of communication between the entities certain interactions take place (6), understood as the exchange of stimulus and responses to the business partner's activities. Interactions in the supply chain have varying degrees of intensity, complexity and length. Each supply chain is internally organised (7), the operations of business entities within this organised structure are more or less efficiently coordinated.

Between the entities functioning in the supply chain, the network of connections and relationships is established. The strength of the positive relations represents the level of integration (consistency). Integration (consistency) is one of the constitutive elements of the supply chain which depend on the quality of relationships between the entities constituting the supply chain as well as the size of the established system.

The analysis of degrees of consistency and organisation of the supply chain (integration), as well as the relationships result from it, has allowed to create two concepts (strategies) of how the entities are included in the supply chain. Integration is understood here as the degree of unification of the entities or else the stage of separate units becoming a whole. Depending on the selected strategy of cooperation between the enterprises functioning in the supply chain the approach to consistency of the supply chain is modified.

The first concept concerns full integration of the entity in the supply chain. The concept stems from the total commitment of the company in one organisational system of the supply chain and results from the complete symbiosis. Full integration involves close cooperation of enterprises in the supply chain; this cooperation is beneficial for the engaged links. When there is close integration and symbiosis the benefits can be mutual. In some cases, integration is so deep that both sides become dependent; at the same time, however, it ensures the enterprise survival in a changing and dynamic market. In biology such a strategy is called mutualism.

A looser form of still symbiotic cooperation is protocooperation. It's free integration which brings benefits to both parties but unlike mutualism it leaves the entities independent. In protcooperation the entities interact periodically. Both forms, mutualism and protocooperation, are typical examples of business interactions characterised by non-antagonistic relations where the parties remain friendly and non-competing.

The advantages of cooperation based on the concept of full integration may also be one-sided. Should that occur, the benefits of cooperation are enjoyed by one side only, although the other side is not harmed. Thus solution can be referred to as commensalism where the "+/0" interaction can be observed and one of the business partners generates benefits when the other does not make a loss but does not achieve benefits either.

Full integration of the supply chain is linked closely with one system. It may result from extremely specialised production/service activity in the supply chain (e.g., support activity) or be associated with the performed role of roles (e.g., leader or manufacturer). Full integration of the entity with the supply chain can also result from entity's resources being insufficient to join in and become integrated in some other supply chains. However, should the company decide to invest in new resources and increase their abilities, it may be that within the existing supply chain the entire business entity is not needed nor wanted. In this case, the company should get involved in other business arrangements and accept new tasks to ensure continuous development.

The model of full integration of an entity in the supply chain is characterised by a greater integrity. This strategy is more likely to be implemented in the case of small structures and supply chains with an innovative character. The greater the supply chain, the lower level of integration. This is due to the high number of cooperating organisations and low level of coordination of their operations.

The concept of partial inclusion in the supply chain assumes that only fragments of the entities constitute the chain. They are involved in a range of operations in other more or less integrated supply chains, with varied levels of engagement. Given link is involved in several independent/different supply chains. Their relationship, degree of contact intensity, type of bond and degree of integration affects the way they operate and the activities they undertake.

The openness of this type of supply chains makes them susceptible to all sorts of inputs and changes, including replacement of the chain links. In this case, however, restrictions on the entities are put in place, for example, ensuring they refrain from getting involved in a competitive supply chain. Therefore, despite common elements of these systems, certain clear limits of the supply chains exist.

2 Barriers of the Integration

A survey conducted among experts researching the supply chains, identified 17 factors affecting the integration of the supply chain type systems. They are listed in Table 1. In this survey the respondents were asked to indicate the importance of 17 listed enablers on a five-point Likert scale. On this scale, 1 and 5 correspond to 'very low importance' to 'very high importance', respectively. In total, questionnaires were sent to 20 experts in Poland and Canada. All of them were analysed.

Table 1 Factors affecting the integration of the supply chain

No.	Factors
1	Information sharing
2	Coordination
3	Trust
4	Willingness to collaborate
5	Communication
6	Common business goals
7	Responsibility sharing
8	Planning of supply chain activities
9	Flexibility
10	Benefit sharing
11	Joint decision making
12	Organizational culture
13	Organisational compatibility
14	Resource sharing (integration)
15	Top management support
16	Technological readiness
17	Training

3 The DEMATEL Methodology

Decision Making and Trial Evaluation Laboratory (DEMATEL) was developed in the belief that the appropriate use of scientific research methods could improve understanding of the specific problem. The Science and Human Affairs Program of the Battelle Memorial Institute of Geneva developed it between 1972 and 1976 to study and resolve the complicated and intertwined problem group (Tzeng et al. 2007; Wu and Lee 2007). DEMATEL is a sophisticated method for establishing a structural model involving causal relationships among complex factors (Gabus and Fontela 1972, 1973). One of the group decision-making methods is decision-making trial and evaluation laboratory (DEMATEL) method, which uses matrices and diagrams for visualizing the structure of complicated causal relationships (Fontela and Gabus 1976). DEMATEL was applied to solve problems concerning decisions in order to clarify the essential features of the problems and help make countermeasures. Tzeng et al. (2007) and Liou et al. (2007) used the fundamentals of this method to transform the attributes of the application and evaluation into a non-independent multi-criteria evaluation of problems. DEMATEL then determines the interdependent and constraining relations based on the specific features of the subjects. In this way, it reflects the essential features and the evolving trend of the system.

This technique is widely used in solving complex problems (Hori and Shimizu 1999; Huang et al. 2007; Lin and Wu 2008; Lin and Tzeng 2009; Liou et al. 2008; Seyed Hosseini et al. 2006; Tsai and Chou 2009; Tzeng et al. 2007; Wu 2008; Wu and Lee 2007) such as user interface (Hori and Shimizu 1999), e-learning evaluation (Tzeng et al. 2007), developing global managers' competencies (Wu and Lee 2007), reprioritization of failures in analyzing FMEA system (Seyed Hosseini et al. 2006), the innovation policy portfolios for Taiwan's SIP mall Industry (Huang et al. 2007), selection of knowledge management strategy (Wu 2008), causal analytic method for group decision making (Lin and Tzeng 2009), airlines safety measurement (Liou et al. 2008), and finally selection management systems (SMEs) (Tsai and Chou 2009).

DEMATEL is a popularly used method to model the relationship between variables. It is based on digraphs which separate the involved variables into two groups—cause and effect. A basic contextual relation among elements is portrayed where values represent the strength of influence. The various steps of DEMATEL are presented as follows:

1. Generate the direct relation matrix
   The direct relationship matrix represents the aggregate influence scores for various variables over each other obtained from expert ratings on a scale of 0 to 4 where the notations are: 0 (No influence), 1 (somewhat influence), 2 (medium influence), 3 (high influence), and 4 (very high influence).
   Let A represent the $n \times n$ matrix obtained by pairwise comparisons in terms of influences and directions between variables where a_{ij} represents the degree to which variable i affects variable j i.e. $A = \lfloor a_{ij} \rfloor_{n \times n}$
2. Normalize the direct relation matrix
   The normalized direct relation matrix is obtained from direct relation matrix as follows

$$B = [b_{ij}]_{n \times n} = \frac{A}{\max_{1 \le i \le n} \sum_{j=1}^{n} a_{ij}}, \text{ where } 0 \le b_{ij} \le 1 \quad (1)$$

   and the principal diagonal elements of B are all equal to zero.
3. Develop the total relation matrix
   The total relation matrix is obtained from normalized direct relation matrix using the following equation

$$C = [c_{ij}]_{n \times n} = B(I - B)^{-1} \quad (2)$$

   where I is the identity matrix.
4. Produce a causal diagram
   The sum of the rows and the sum of columns are denoted by vectors D and E.

$$D = [d_{ij}]_{n \times 1} = \left[\sum_{j=1}^{n} e_{ij} \right]_{n \times 1} \quad (3)$$

$$E = [e_{ij}]_{1 \times n} = \left[\sum_{i=1}^{n} e_{ij} \right]_{1 \times n} \quad (4)$$

Table 2 List of Barriers

C1	Lack of trust
C2	Lack of coordination
C3	Lack of communication
C4	Lack of information sharing
C5	Lack of planning of supply chain activities
C6	Lack of top management support
C7	Lack of organisational compatibility
C8	Lack of flexibility
C9	Lack of benefit sharing
C10	Lack of joint decision making
C11	Lack of resource sharing (integration)
C12	Lack of responsibility sharing
C13	Lack of technological readiness
C14	Lack of common business goals
C15	Lack of willingness to collaborate
C16	Lack of organizational culture
C17	Lack of training

The horizontal axis (D + E) represents the importance of the variables whereas the vertical axis (D − E) shows the cause and effect relationships. The variables with positive (D − E) values are the cause factors whereas those with negative are effect factors.

5. Depict structural relation between variables
The structural relation amongst variables is shown through an inner dependence matrix by retaining only those variables whose effect in the matrix C is greater than the threshold value. The threshold value δ can be given by the experts, based on literature review or obtained by averaging the values of C matrix elements.

4 Empirical Analysis: Discussion

In this section, we present the application of DEMATEL technique to identify the relationship between various barriers (Table 2) of supply chain integration considered in this study.

To assess the degree of influence and relationship of various barriers with each other, we performed literature review and discussed with experts from academia. Table 3 presents the direct relation matrix containing influence and relationship of various barriers with each other. The ratings are provided on a scale of 0 to 4 where the notations are 0 (No influence), 1 (somewhat influence), 2 (medium influence), 3 (high influence), and 4 (very high influence).

Table 4 shows the normalized relation matrix for the barriers obtained using Eq. 1.

Table 3 Direct relation matrix for barriers

	C1	C2	C3	C4	C5	C6	C7	C8	C9	C10	C11	C12	C13	C14	C15	C16	C17
C1	0	2	0	3	0	3	4	0	2	4	4	0	0	0	2	2	1
C2	0	0	4	2	4	2	4	4	0	0	3	4	1	0	3	2	2
C3	1	3	0	2	4	2	4	1	2	3	4	3	4	0	2	2	4
C4	4	4	4	0	3	2	4	0	0	3	4	4	2	2	3	2	4
C5	1	3	3	3	0	2	4	4	0	3	3	2	2	2	3	2	1
C6	2	2	2	2	2	0	1	1	2	2	2	2	1	1	2	1	2
C7	2	3	2	2	3	2	0	2	2	3	4	2	4	2	1	1	0
C8	1	2	2	2	2	1	4	0	0	0	2	2	1	1	1	1	2
C9	1	2	2	2	2	2	4	2	0	4	3	2	0	1	2	1	0
C10	1	1	2	2	2	2	4	1	1	0	4	2	0	2	1	1	2
C11	3	3	3	3	3	2	2	2	2	3	0	3	4	2	3	3	3
C12	2	3	2	3	3	4	4	2	4	4	3	0	2	2	3	2	2
C13	2	2	2	2	2	1	1	2	2	2	2	2	0	2	1	2	0
C14	2	2	2	2	2	1	4	0	0	2	1	4	0	0	1	1	2
C15	2	3	3	3	4	1	4	0	4	4	2	2	0	3	0	1	2
C16	2	2	2	2	2	0	4	1	2	2	2	2	0	2	1	0	2
C17	2	1	1	2	2	1	1	1	2	2	1	1	1	1	1	1	0

Table 5 depicts the total relation matrix obtained using Eq. 2.

Table 6 shows the D and E vectors obtained using Eqs. 3–4 to develop the causal diagram.

Table 7 shows the various barriers in decreasing order of their impacts (D + E). It can be seen that the top 5 barriers are Lack of Resource sharing (integration), Lack of Organisational compatibility, Lack of Information sharing, Lack of Responsibility sharing and Lack of Planning of supply chain activities. Organizations should focus on eliminating these barriers in particular to achieve integration. The barrier with least impact is lack of flexibility.

Table 8 shows the barriers organized in terms of their relationships (D−E). The *+ive* ones are the causes while the *–ive* ones are the effects. It can be seen in Table 7 that Lack of Information sharing, Lack of Willingness to collaborate, Lack of Responsibility sharing, Lack of Benefit sharing, Lack of Communication, Lack of Common business goals, Lack of Organizational culture, Lack of Technological readiness and Lack of Flexibility are the cause variables (barriers) which affect remaining other barriers.

Figure 1 shows the impact relationship map for the 17 barriers based on their impact (D + E) and relationship (D−E) values. It can be seen that C11, C7, C4, C12 and C5 are the barriers with highest impact. The cause variables are present in the upper half of the graph and have D − E value >0. The effect variables are present in the lower half and have D − E value <0.

Since C11 (Lack of resource sharing) is the barrier with most impact, we will develop inner dependency matrix to identify the causes/effects barriers that affect it. The threshold value δ is obtained by averaging the values of Total relation matrix (Table 5) and is equal to 0.1759. The inner dependency matrix is shown in

Table 4 Normalized direct relation matrix for barriers

	C1	C2	C3	C4	C5	C6	C7	C8	C9	C10	C11	C12	C13	C14	C15	C16	C17
C1	0	0.044	0	0.067	0	0.067	0.089	0	0.044	0.089	0.089	0	0	0	0.044	0.044	0.022
C2	0	0	0.089	0.044	0.089	0.044	0.089	0.089	0	0	0.067	0.089	0.022	0	0.067	0.044	0.044
C3	0.022	0.067	0	0.044	0.089	0.044	0.89	0.022	0.044	0.067	0.089	0.067	0.089	0	0.044	0.044	0.089
C4	0.089	0.089	0.089	0	0.067	0.044	0.89	0	0	0.067	0.089	0.089	0.044	0.044	0.067	0.044	0.089
C5	0.022	0.067	0.067	0.067	0	0.044	0.89	0.089	0	0.067	0.067	0.044	0.044	0.044	0.067	0.044	0.022
C6	0.044	0.044	0.044	0.044	0.044	0	0.022	0.022	0.044	0.044	0.044	0.044	0.022	0.022	0.044	0.022	0.044
C7	0.044	0.0677	0.044	0.044	0.067	0.044	0	0.044	0.044	0.067	0.089	0.044	0.089	0.044	0.022	0.022	0
C8	0.022	0.044	0.044	0.044	0.044	0.022	0.089	0	0	0	0.044	0.044	0.022	0.022	0.022	0.022	0.044
C9	0.022	0.044	0.044	0.044	0.044	0.044	0.089	0.044	0	0.089	0.067	0.044	0	0.022	0.044	0.022	0
C10	0.022	0.022	0.044	0.044	0.044	0.044	0.089	0.022	0.022	0	0.089	0.044	0	0.044	0.022	0.022	0.044
C11	0.067	0.067	0.67	0.067	0.67	0.044	0.044	0.044	0.044	0.067	0	0.067	0.089	0.044	0.067	0.067	0.067
C12	0.044	0.067	0.044	0.067	0.067	0.089	0.089	0.044	0.089	0.089	0.067	0	0.044	0.044	0.067	0.044	0.044
C13	0.044	0.044	0.044	0.044	0.044	0.022	0.022	0.044	0.044	0.044	0.044	0.044	0	0.044	0.022	0.044	0
C14	0.044	0.044	0.044	0.044	0.044	0.022	0.089	0	0	0.044	0.022	0.089	0	0	0.022	0.022	0.044
C15	0.044	0.067	0.067	0.067	0.089	0.022	0.089	0	0.089	0.089	0.044	0.044	0	0.067	0	0.022	0.044
C16	0.044	0.044	0.044	0.044	0.044	0	0.089	0.022	0.044	0.044	0.044	0.044	0	0.044	0.022	0	0.044
C17	0.044	0.022	0.022	0.044	0.044	0.022	0.022	0.022	0.044	0.044	0.022	0.022	0.022	0.022	0.022	0.022	0

Table 5 Total relation matrix for barriers

	C1	C2	C3	C4	C5	C6	C7	C8	C9	C10	C11	C12	C13	C14	C15	C16	C17
C1	0.096	0.169	0.124	0.184	0.136	0.159	0.251	0.080	0.126	0.220	0.235	0.124	0.085	0.081	0.145	0.127	0.119
C2	0.124	0.173	0.247	0.205	0.267	0.170	0.308	0.196	0.114	0.180	0.259	0.246	0.142	0.106	0.200	0.156	0.174
C3	0.160	0.251	0.181	0.224	0.284	0.186	0.330	0.148	0.168	0.263	0.303	0.244	0.212	0.119	0.195	0.170	0.225
C4	0.236	0.292	0.281	0.201	0.285	0.203	0.360	0.135	0.142	0.287	0.328	0.283	0.184	0.170	0.232	0.184	0.244
C5	0.151	0.241	0.234	0.232	0.190	0.175	0.320	0.197	0.115	0.247	0.270	0.216	0.163	0.152	0.204	0.161	0.160
C6	0.135	0.169	0.163	0.164	0.177	0.097	0.192	0.103	0.126	0.179	0.191	0.164	0.104	0.099	0.146	0.107	0.140
C7	0.161	0.226	0.200	0.199	0.235	0.167	0.217	0.150	0.146	0.234	0.274	0.202	0.193	0.143	0.154	0.133	0.125
C8	0.107	0.160	0.153	0.153	0.167	0.110	0.236	0.076	0.076	0.123	0.178	0.154	0.104	0.092	0.115	0.099	0.131
C9	0.126	0.187	0.181	0.180	0.196	0.153	0.277	0.135	0.093	0.236	0.234	0.182	0.099	0.111	0.158	0.117	0.111
C10	0.122	0.157	0.170	0.171	0.185	0.145	0.260	0.107	0.109	0.143	0.240	0.172	0.095	0.125	0.130	0.111	0.146
C11	0.208	0.261	0.252	0.253	0.273	0.193	0.309	0.169	0.173	0.274	0.232	0.254	0.212	0.165	0.222	0.196	0.215
C12	0.191	0.267	0.238	0.258	0.279	0.239	0.356	0.174	0.217	0.301	0.302	0.197	0.176	0.169	0.227	0.178	0.197
C13	0.134	0.170	0.164	0.164	0.176	0.118	0.195	0.124	0.123	0.177	0.191	0.166	0.082	0.120	0.124	0.128	0.099
C14	0.135	0.169	0.161	0.163	0.176	0.121	0.252	0.082	0.085	0.178	0.171	0.204	0.087	0.077	0.124	0.106	0.139
C15	0.171	0.242	0.234	0.233	0.273	0.159	0.325	0.117	0.197	0.276	0.254	0.216	0.120	0.172	0.144	0.140	0.178
C16	0.139	0.175	0.168	0.169	0.182	0.102	0.262	0.107	0.128	0.184	0.199	0.170	0.090	0.123	0.128	0.088	0.142
C17	0.114	0.119	0.114	0.136	0.144	0.096	0.153	0.083	0.105	0.147674	0.137258	0.115184	0.08416	0.081605	0.100643	0.086896	0.074

Table 6 Total cause and effect for barriers

	C1	C2	C3	C4	C5	C6	C7	C8	C9	C10	C11	C12	C13	C14	C15	C16	C17
D	2.470	3.275	3.671	4.058	3.435	2.463	3.166	2.241	2.784	2.596	3.872	3.974	2.463	2.438	3.460	2.562	1.895
E	2.518	3.437	3.273	3.295	3.633	2.602	4.610	2.191	2.251	3.657	4.005	3.318	2.242	2.112	2.757	2.294	2.626
D + E	4.988	6.712	6.945	7.354	7.069	5.065	7.776	4.433	5.035	6.253	7.877	7.292	4.705	4.550	6.218	4.857	4.521
D − E	−0.048	−0.162	0.398	0.762	−0.198	−0.138	−1.44	0.050	0.533	−1.061	−0.133	0.655	0.220	0.325	0.702	0.267	−0.731

Table 7 Impact table

Barrier	ID	D + E
Lack of resource sharing (integration)	C11	7.877
Lack of organisational compatibility	C7	7.776
Lack of information sharing	C4	7.354
Lack of responsibility sharing	C12	7.292
Lack of planning of supply chain activities	C5	7.069
Lack of communication	C3	6.945
Lack of coordination	C2	6.712
Lack of joint decision making	C10	6.253
Lack of willingness to collaborate	C15	6.218
Lack of top management support	C6	5.065
Lack of benefit sharing	C9	5.035
Lack of trust	C1	4.988
Lack of organizational culture	C16	4.857
Lack of technological readiness	C13	4.705
Lack of common business goals	C14	4.550
Lack of training	C17	4.521
Lack of flexibility	C8	4.433

Table 8 Relationship table

Barrier	ID	D–E
Lack of information sharing	C4	0.762
Lack of willingness to collaborate	C15	0.702
Lack of responsibility sharing	C12	0.655
Lack of benefit sharing	C9	0.533
Lack of communication	C3	0.398
Lack of common business goals	C14	0.325
Lack of organizational culture	C16	0.267
Lack of technological readiness	C13	0.220
Lack of flexibility	C8	0.050
Lack of trust	C1	−0.048
Lack of resource sharing (integration)	C11	−0.133
Lack of top management support	C6	−0.138
Lack of coordination	C2	−0.162
Lack of planning of supply chain activities	C5	−0.198
Lack of training	C17	−0.731
Lack of joint decision making	C10	−1.061
Lack of organisational compatibility	C7	−1.443

Table 9. It contains Table 5 elements whose value exceeds δ. Looking at C11, we find that it is impacted by all barriers but except C14 and C17. Therefore, these barriers act as control barriers for C11 and should be carefully monitored. The impact values is highest for C4 (Lack of Information sharing), C3 (Lack of Communication) and C12 (Lack of Responsibility sharing).

Table 9 Inner dependence matrix

	C1	C2	C3	C4	C5	C6	C7	C8	C9	C10	C11	C12	C13	C14	C15	C16	C17
C1				0.184			0.251			0.220	0.235						
C2			0.247	0.205	0.267		0.308	0.196		0.180	0.259	0.246			0.200		
C3		0.251	0.181	0.224	0.284	0.186	0.33			0.263	0.303	0.244	0.212		0.195		0.225
C4	0.236	0.292	0.281	0.201	0.285	0.203	0.360			0.287	0.328	0.283	0.184		0.232	0.184	
C5		0.241	0.234	0.232	0.190		0.320	0.197		0.247	0.270	0.216			0.204		
C6					0.177		0.192			0.179	0.191						
C7		0.226	0.200	0.199	0.235		0.217			0.234	0.274	0.202	0.193				
C8							0.236				0.178						
C9		0.187	0.181	0.180	0.196		0.277			0.236	0.234	0.182					
C10					0.185		0.260				0.240						
C11	0.208	0.261	0.252	0.253	0.273	0.193	0.309			0.274	0.232	0.254	0.212		0.222	0.196	0.215
C12		0.267	0.238	0.258	0.279	0.239	0.356		0.217	0.301	0.302	0.197	0.176		0.227	0.178	0.197
C13					0.176		0.195			0.177	0.191						
C14					0.176		0.252			0.178		0.204					
C15		0.242	0.234	0.233	0.273		0.325		0.197	0.276	0.254	0.216					0.178
C16					0.182		0.262			0.184	0.199						
C17																	

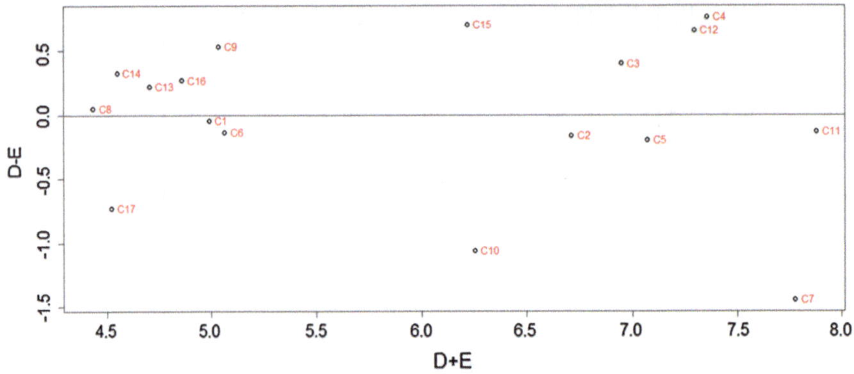

Fig. 1 Impact-relationship map

5 Managerial Implications

The results of our DEMATEL study show Lack of Resource sharing (integration), Lack of Organisational compatibility, Lack of Information sharing, Lack of Responsibility sharing, and Lack of Planning of supply chain activities as top five barriers in supply chain integration. Therefore, managers of interested organizations can look into causes behind these barriers and take appropriate measures to resolve them to ensure seamless integration across their supply chains.

6 Conclusions

In this chapter, we presented the barriers in supply chain integration and investigated their importance and causal relationships using Decision Making and Trial Evaluation Laboratory. The results of our DEMATEL study show Lack of Resource sharing (integration), Lack of Organisational compatibility, Lack of Information sharing, Lack of Responsibility sharing, and Lack of Planning of supply chain activities as top five barriers in supply chain integration.

The next step of our work involves identifying and evaluating alternatives using multicriteria decision making approaches to address the proposed barriers for efficient supply chain integration.

References

Awasthi A, Grzybowska K, Chauhan SS, Goyal SK (2014) Investigating Organizational Characteristics for Sustainable Supply Chain Planning Under Fuzziness. In: Kahraman C, Öztayşi B (eds), Supply Chain Management Under Fuzziness, Studies in Fuzziness and Soft Computing 313. Springer, Berlin, pp 81–100

Bai C, Joseph S, Wei X, Koh L (2012) Evaluating ecological sustainable performance measures for supply chain management. Supply Chain Management Int J 17(1):78–92

Brown D, Wilson S (2005) The black book of outsourcing hoboken. Wiley, New Jersey

Christopher M (1998) Logistics and supply chain management. Financial Times/Pitman Publishing, London

Elkington J (1994) Towards the sustainable corporation: win–win-win business strategies for sustainable development. Calif Manag Rev 36(2):90–100

Fontela E, Gabus A (1976) The DEMATEL observer, DEMATEL 1976 Report. Battelle Geneva Research Center, Geneva

Gabus A, Fontela E (1972) World problems, an invitation to further thought within the framework of DEMATEL. Battelle Geneva Research Centre, Geneva

Gabus A, Fontela E (1973) Perceptions of the world problematique: communication procedure, communicating with those bearing collective responsibility (DEMATEL Report no. 1). Battelle Geneva Research Centre, Geneva

Ganeshan R, Harrison TP (1995) Am introduction to supply chain management, department of management sciences and information systems, 303 beam business building. Penn State University, University Park

Grzybowska K (2010a) Constitutive elements of the supply chain, operations and logistics management, E Pawłowski (red.). Publishing House of Poznan University of Technology. p 77–91

Grzybowska K (2010b) Spójny łańcuch dostaw—sposób na wzrost efektywności metastruktur i przetrwanie w okresie kryzysu, Zmiana warunkiem sukcesu. Odnowa przedsiębiorstwa—czego nauczył nas kryzys?, Prace naukowe UE nr 128, Wydawnictwo Uniwersytetu Ekonomicznego we Wrocławius. 319–326

Hori S, Shimizu Y (1999) Designing methods of human interface for supervisory control systems. Control Eng Pract 7(11):1413–1419

Huang CY, Shyu JZ, Tzeng GH (2007) Reconfiguring the innovation policy portfolios for Taiwan's SIP Mall industry. Technovation 27(12):744–765

Lambert BM, Stock JR, Ellram LM (1998) Fundamentals of logistics management. Irwin/MacGraw-Hill, Boston

Lee HL, Billington C (1995) The evolution of supply-chain-management models and practice at Hewlett-Packard. Interfaces 25(5):42–63

Lin CJ, Wu WW (2008) A causal analytical method for group decision making under fuzzy environment. Expert Syst Appl 34(1):205–213

Lin CL, Tzeng GH (2009) A value-created system of science (technology) park by using DEMATEL. Expert Syst Appl 36:9683–9697

Liou JH, Tzeng GH, Chang HC (2007) Airline safety measurement using a novel hybrid model. J Air Transp Manage 13(4):243–249

Liou JJH, Yen L, Tzeng GH (2008) Building an effective safety management system for airlines. J Air Transp Manage 14(1):20–26

Mentzer JT, DeWitt W, Keebler JK, Min S, Nix NW, Smith CD, Zacharia ZG (2001) Defining supply chain management. J Bus Logistics 22(2):1–25

Seuring S (2013) A review of modeling approaches for sustainable supply chain management. Decis Support Syst 54(4):1513–1520

Seuring S, Müller M (2008) From a literature review to a conceptual framework for sustainable supply chain management. J Clean Prod 16(15):1699–1710

Seyed Hosseini SM, Safaei N, Asgharpour MJ (2006) Reprioritization of failures in a system failure mode and effects analysis by decision making trial and evaluation laboratory technique. Reliab Eng Syst Saf 91(8):872–881

Tsai WH, Chou WC (2009) Selecting management systems for sustainable development in SMEs: A novel hybrid model based on DEMATEL, ANP, and ZOGP. Expert Syst Appl 36(2):1444–1458

Tzeng GH, Chiang CH, Li CW (2007) Evaluating intertwined effects in e-learning programs: a novel hybrid MCDM model based on factor analysis and DEMATEL. Expert Syst Appl 32(4):1028–1044

Wu WW (2008) Choosing knowledge management strategies by using a combined ANP and DEMATEL approach. Expert Syst Appl 35(3):828–835

Wu WW, Lee YT (2007) Developing global managers' competencies using the fuzzy DEMATEL method. Expert Syst Appl 32(2):499–507

Zandin KB, Maynard HB (2001) Maynard's industrial engineering handbook. McGraw-Hill, New York

The Essence of Integration in Supply Chains and Reverse Supply Chains: Similarities and Differences

Martyna Kupczyk, Łukasz Hadaś, Piotr Cyplik and Żaneta Pruska

Abstract The purpose of this chapter is to introduce the issue of integration in supply chains and in reverse supply chains. The Authors specify the factors key to integration and examine their impact on the process. The first part addresses the issues affecting forward flows and the benefits offered by a close cooperation between the links in the supply chain. In the second part the Authors discuss the factors which may contribute to a more seamless operation of backward flows. The third part presents the outcome of a survey into the measures taken by companies with a view to developing closer relationships with business partners in individual chains. These results provided the grounds for selecting the most crucial integration factors and highlighted the similarities and differences in integration-oriented activities taken in supply and reverse supply chains. In the last part of the chapter the Authors investigate the issue of barriers to building close relationships between supply chain partners.

Keywords Integration · Supply chain · Reverse supply chain

1 Introduction

In broad terms, integration is defined as a process of bringing individual elements into one whole. From the economic point of view, integration denotes an economic process consisting in consolidating companies, branches and industries (Pfohl HCh 1998). In other words, it refers to creating, or to the process of creating a large unit composed of a number of small units. Keeping in mind that the concept of logistics imposes a systemic way of thinking, Pfohl holds that an integration process is the centre of performing logistics tasks (Pfohl 1998).

M. Kupczyk (✉) · Ł. Hadaś · P. Cyplik · Ż. Pruska
Poznan School of Logistics, ul. Estkowskiego 6, 61-755 Poznan, Poland
e-mail: martyna.kupczyk@wsl.com.pl

P. Golinska (ed.), *Logistics Operations, Supply Chain Management and Sustainability*, EcoProduction, DOI: 10.1007/978-3-319-07287-6_3,

Depending on the level of integration we may classify integration as occurring: internally—across departments or functions; integration above departments externally—involving entire companies (Pfohl HCh 1998; Yu et al. 2013). External integration plays key role in supply chains, namely in the groups of enterprises, collaborating with a view to ensuring a smooth flow of goods within the streams in the organization. Logistics experts stress that the supply chain integration is an indispensable ingredient of the company's success, because it determines its survival and prosperity in a competitive environment. Competitiveness is on an increase, fuelling the pressure for cost-cutting and driving the efficiency of companies and the productivity of their assets. A. Harrison and R. van Hoek believe that the closer the integration, the greater its impact on the performance of the supply chain. Expanding the scope of integration reduces the level of uncertainty inherent in the flow of goods in supply chains, which adds to enhancing the company's efficiency and shortening production times (Harrison and van Hoek 2010).

The essence of integration in managing a supply chains is presented by SCM house authored by H. Stadtler below (Fig. 1).

Along with coordination, the integration of the networks of enterprises one of the pillars on which the roof of the house rests. The house depicts two major objectives of supply chain management—competitiveness and customer service (Jain and Dubey 2005).

2 Factors at Play in Supply Chain Integration

Supply chains are integrated by way of various instruments and tools. A number of publications on this issue feature different integration factors—operating at various levels of detail, e.g. one of the authors points to the exchange of information (Fechner 2007), another emphasizes the exchange of information between the partners across specific areas such as communicating production plans or inventory structure and volume (Harrison and van Hoek 2010). A review of the literature on the subject identifies the following supply chain integration factors: standardization, unification of solutions, the exchange and the application of information, cutting-edge IT technologies, operational strategy, management concept, trust, partnership, centralization of inventories, modern methods of inventory management, joint planning, joint product designing, joint ventures, inter-organizational decision making with supply chain members (Alfalla-Luque et al. 2013; Fechner 2007; Jayaram et al. 2010; Rutkowski 2000; Szudrowicz 2002; Szymonik 2011). According to the Authors, the greatest importance is carried by partnership based on trust, cooperation in standardization, unification and planning as well as the exchange of information, supported by state-of-the-art IT tools. Each factor will be described in more detail and its impact on the

Fig. 1 House of SCM (Stadtler 2005)

integration and the functioning of the supply chain will provide a justification for their selection.

Standardization means setting standards, introducing norms or unifying production. A standard, on the other hand, is a commonly established criterion, which refers to the most desirable properties of a thing, for example of an item that is being manufactured. Standardization can be viewed as an integration factor, because it facilitates communication on a number of activities which influence the manufacturing and transporting of goods. Unification uses a similar mechanism, it is an activity intended to bring something to one form or in line with a norm, namely unification. Unifying solutions and activities foster a better understanding between companies, helping to ensure a streamlined and unobstructed performance of operations and to avoid potential conflicts due to misunderstanding and diversity in action (Sobol 1997). Standardization may apply to various areas of the supply chain's operation, yet the most important and susceptible to standardization areas include (Szudrowicz 2002):

- information, namely information identifiers and carriers, which enable automatic identification and acquisition of data,
- communication—electronic messages, which come in place of chapter documentation and traditional forms of communication and information exchange,
- transporting goods, where the standard is the distribution chain, enabling the use and adaptation of packaging to storage and transport facilities.

The use of standards in the exchange of information helps to prevent conflicts and reduces the time that companies devote to communicating with business partners, as well as bring down the error rate in documentation and in logistics operations as well as time reductions. The standardization of solutions applied in the process of transporting products contributes to removing barriers which arise due to a mismatch between individual links in the supply chain (Szudrowicz 2002).

The functioning of the supply chain is to a large extent determined by the form, quality and availability of the information useful in facing both in-chain competitive struggles, and for integrating business partners. The supply chain will respond efficiently to market demand and maintain a high customer service level with relatively low costs provided that we ensure a multi-directional exchange of information between all of the supply chain links and a constant access to the information. There should be a flow of information between the supply chain links regarding, among others: sales, inventory at suppliers and recipients, production plans, orders and their status, invoices as well as promotions and modifications in the in progress. The exchange of this information within the supply chain gives an extra competitive edge and a boost for market leadership (Szudrowicz 2002). We could even go as far as to say that the success of collaborating enterprises is predicated on the exchange of information between them (Rutkowski 2000). The understanding of the needs of the partners and the partners' openness to share information essential to achieve a common goal assist in making decisions on manufacturing products and to their flow in logistics chains (Fechner 2007). The information quality comes to the forefront as well, among others its: relevance, accuracy, validity, completeness, coherence, appropriateness, availability, reliability and conformity (Fechner 2007). The value of the information is of key importance to decision-makers who operate under stress and uncertainty gripping the market. The information helps to assess the circumstances or the situation accurately, to forecast future economic conditions, evaluate the factors at play in the immediate and general environment, having impact on an efficient and effective functioning of the company and, most importantly, enables better and more accurate decisions. What is more, good information supported by cutting-edge information and communication technologies enhances the efficiency and the speed of flows in the supply chain (Fechner 2007). Hence information integration is often accompanied by electronic integration. The oldest solution in the field of a rapid transmission of information regarding orders and deliveries is Electronic Data Interchange system (EDI). Technologies using the internet are becoming more and more common. The rapid development of the internet coupled with an increasing computing power provide cost-efficient opportunities for integrating information systems, whereas a faster and broader exchange of information promotes the streamlining and tightening of cooperation between suppliers, manufacturers and distributors. Apart from the internet and EDI, advantage can be also taken of the solutions such as electronic methods of performing transactions, namely automated methods of order placement, invoicing, planned order notifications and dispatched deliveries, payments and sharing a common information system.

Joint planning at all three levels: strategic, tactical and operational enable partners to make a concerted analysis of the situation to assess a future demand more accurately and to develop plans that will duly address this demand. The cooperation in this area may include joint planning of new products, forecasting demand, preparing stock replenishment schedules and harmonizing individual plans within the organization (Harrison and van Hoek 2010).

According to A. Szymonik and K. Rutkowski, the key tools and instruments which integrate enterprises into supply chains area trust and partnership (Rutkowski 2000; Szymonik 2011).

Trust means that the other party shares our values and norms that it will act to our benefit, never detriment. Trust is related to (Szymonik 2011):

- kindness, namely concern for and the motivation to act in the best interest of the other party,
- integrity—entering into agreements in good faith, being honest and keeping promises,
- competencies, namely the skills and/or capability to perform tasks as assigned,
- predictability of the actions that other parties will take, based on which future situations can be anticipated.

The forms of running business over the internet require a great deal of mutual trust in all of the relations across the supply chain. Communication comes in handy in building trust, because it means sharing key information between business partners. A high degree of trust is generated by frequent and top quality communication—when the information is current, relevant, reliable and clear (Szymonik 2011).

Partnership is defined in a number of ways: as a cooperation of at least two entities having one common objective; as a mature form of relations and collaboration with others or as the allocation of tasks and resources aimed at optimizing the use of the partner's capability with a view to sharing profits, losses and obligations. The key to establishing partnership is the involvement of a number of entities, which contribute to achieving one common goal in a conscious and deliberate way. The partnership between the links in the supply chain is based on trust and sharing of both rewards and risks. It may bring benefits in the form of additional synergy effects, competitive advantage, reduced uncertainty, enhanced flexibility and speed of operations, easier access to skills and capabilities being in short supply, easier acquisition of information as well as structuring and maintenance of the pool of resources and warehouse or transportation facilities, the application of more advanced technology, an increased production flexibility, greater opportunities for reducing costs and deploying creating innovative solutions. It should be emphasized that partner relations between the participants of the supply chain do not imply that every partner will have an equal share. They mean that neither party should feel at a disadvantage. One should also remember that competition or rivalry and conflicts of interests are unavoidable in partner relationships. What is more, partnership requires the transformation of individual chain link structures and shifts in competence as regards the role and impact of individual elements of the system, namely transport and procurement. Implemented changes should be geared towards creating planning and designing sustainable structures, developing horizontal communication, removing bottlenecks and root causes of being closed off to others, all with a view to creating the conditions which will foster a more frequent exchange of information and a more formalized information flow systems. As the environment in which supply chains are operating is extremely dynamic and demanding, partnership must be continuously improved (Szymonik 2011).

3 Integration Factors and Managing Recyclable Materials

In the case of supply and maintenance chains, products flow from the point of production to the point of consumption. It does not provide a comprehensive picture of the cooperation between companies in a supply chain. What should also be taken into consideration is the reverse supply chain, flowing in the opposite direction. A reverse supply chain is related to withdrawing from the market all those products which fail to attract customers or which should be disposed of after they have been used, or which have defects, making them dangerous for users—e.g. as a result of a hidden defect visible in operation, a defect due to contamination, improper workmanship or wrong composition (Fechner 2007).

Reverse logistics has implications on the entire supply chain, resulting in the need for :

- analysing the life cycle of finished products in order to demonstrate the impact of environmental factors on their production, usage and disposal,
- designing products to make them easy to dismantle and reuse,
- collaboration with the suppliers and sub-suppliers of materials and ingredients which are fit for reuse.

The cooperation within the reverse supply chain may bring benefits such as an enhanced flow of waste and reduced costs operations in individual links. On top of the above, it also helps to measure efficiency, effectiveness and the costs of the functioning of individual links and identify and expose weak points, bottlenecks and inefficient solutions in place in the supply chain (Szołtysek 2009).

As is the case with forward flows, the supply chains with backward flows aimed at enabling the reuse of recyclable materials, the key factors of integration include partnership, trust and the exchange of information between the chain participants.

Recyclable materials differ largely from the materials derived from original sources—they can't be ordered in fixed quantities or scheduled for delivery on a fixed date. The inflow of such materials is determined by final users, the activities related to waste sorting and the life cycles of products, which are the sources of these materials. It is, thus, hard to forecast the quantity and quality of recyclable materials. What is more, the acquisition of such materials implies much more effort, because they are usually acquired in small quantities and from multiple, unrelated sources, namely from final users. All of the above puts the exchange of information on the locations, quantities and types of resources to the forefront of attention.

To enable the operations pertaining to the flow of products from manufacturer to the consumer, the supply chain must be first supplied with the information on the locations of products intended for withdrawal, then put their flow to a halt and protect them against further sales or consumption, whereas the reverse chain (backward flow) must withdraw product for repair or disposal purposes ensuring that no item has been left out on the market. The use of information standards, product coding, automatic identification, databases, IT systems and electronic data exchange systems ensures the efficiency of backward flows in supply chains, including those related to the use of recyclable materials (Fechner 2007). The waste supply chain will be ineffective without efficient information sharing. All the data related to the points of origin of waste, their quantity and type, collection and storage locations, manufacturers and recipients should be brought together into one information system, which provide the information to all the interested parties and enable its application in waste management, monitoring and in operating activities (Szołtysek 2009) (Fig. 2).

The use of recyclable materials requires a lot of cooperation between various links in the supply chain. As is the case with products of full value sent to the market, also recyclable materials require creating and putting in place a system for collection, warehousing and transportation of the goods. The collaboration in this respect may involve, for example, introducing reusable containers or settling the issue of transport. Backward flows require more than one participant and the supply chain links must agree on: what and when should be reversed? How it should be collected and packaged? Who will be in charge of transportation? The performance of these processes is impossible without mutual trust, especially if indirect links are reluctant to participate in backward flows, all the more if nothing is to be gained from them. The partners must be sure that their efforts related to the acquisition and redirection of recyclable materials will not go down the drain and that the manufacturer will collect recyclable materials.

In closed production cycle models much attention is given to providing maintenance services, which entails the use of spare parts and subassemblies supplied by the manufacturer, as well as the acquisition of used parts of equipment that is under repair (Szołtysek 2009). The incorporation of service points in the

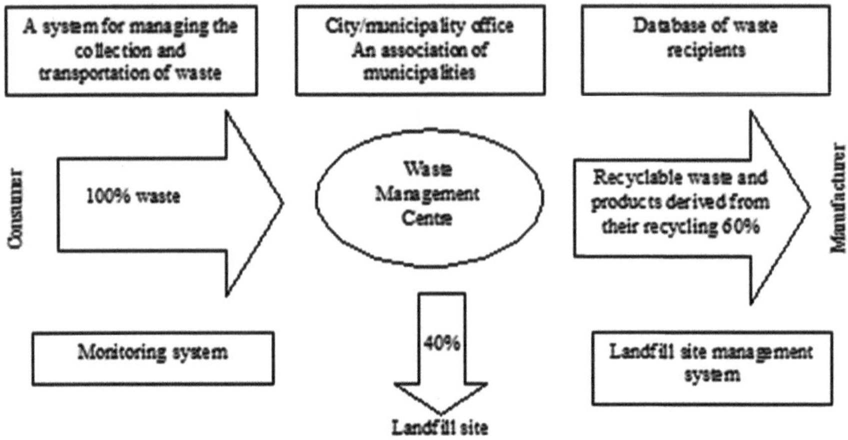

Fig. 2 Integrated information flow in a logistics system for managing municipal solid waste (Szołtysek 2009)

supply chain requires integration with manufacturers. The efficiency of this collaboration will improve greatly thanks to (Szołtysek 2009):

- effective inventory level control—*high-tech* industry 10 % of waste are highly rotational, 10 % out of which is involved in daily repair procedures—which is why this inventory group should be given special attention,
- the implementation of advanced forecasting methods as regards the demand for spare parts,
- the implementation of bar codes and RFID technologies to improve the accuracy of the information spare part inventory levels,
- the enhancement of the quality of repair and regeneration works,
- continuous monitoring of supply chain parameters and benchmarking them with the competition and customer satisfaction.

Close relations and trust between the supply chain participants are instrumental to this end.

4 Evaluation of Supply Chain Relationships: Survey Results

The framework of the project entitled: "Investigating multi-aspect conditions for integration in forward and backward logistics chains with respect to integrated production planning in the context of using recyclable materials", financed and supervised by the National Science Centre, involved carrying out a survey among the Polish enterprises, aimed at evaluating the relationships fostered in supply chains and in reverse supply chains 372 production and service providing enterprises from various industries participated in the survey. They were both SME companies (micro—15.3 %, small—26.6 %, medium—24.2 %) and large companies (33.9 %). The survey questions helped to assess if the companies participate in backward flows, realized from the customer to the manufacturer, if they attempt to build closer relationships with business partners within supply and reverse supply chains, which activities are taken to meet this objective and what prevents them from forming close relationships.

The results of the survey show that 82 % of respondents participate in backward flows, namely the flows of goods occurring from the consumer to the manufacturer 82 % respondents gave their opinions on building close relationships both in supply chains and in reverse supply chains. The involvement of the remaining 18 % of enterprises was limited to supply chains only.

The identification of integration factors among the enterprises under review was initiated by the following question: do you try to build closer relations within particular supply chains? The survey revealed that most of companies aim at fostering closer relations with business partners both in the supply chain and in the reverse supply chain. 17.5% of answers referred to supply chains, namely the chains occurring from the manufacturer to the market (Fig. 3).

The chart indicates that 80.9 % of respondents strive for developing closer relationships with business partners in supply chains. The respective figure for reverse supply chains is 63.4 %. The discrepancies in these results may be due to the fact that forward and backward chains are realized through separate channels and because companies are more engaged in and give higher priority to the operations related to forward flows.

Integration factors investigated in the survey included: joint planning, joint stock replenishment, efficient communication and exchange, the use of electronic systems of communication/data interchange, joint financial investments or sharing financial resources, joint research, sharing risks and profits, standardization (e.g. in as regards the information exchange), unification (of processes, products, etc.) and the implementation of electronic codes for products.

The comparison of integration factors conducive to fostering closer relationships with business partners in the supply chain and in the reverse supply chain show relatively minor differences. The factors of key importance in both chains are, in order of priority: efficient communication and data exchange, the use of electronic communication/data interchange systems and standardization (e.g. regarding information exchange). In supply chains these indications were slightly higher, between 3.8 % and 15.4 % (Fig. 4). It can be therefore inferred that communication, the exchange of information and the establishment of common standards play key role in integrating and streamlining collaboration.

The following two places in the ranking for both the chains are implementing electronic codes for products and joint planning, with one difference; in the case of

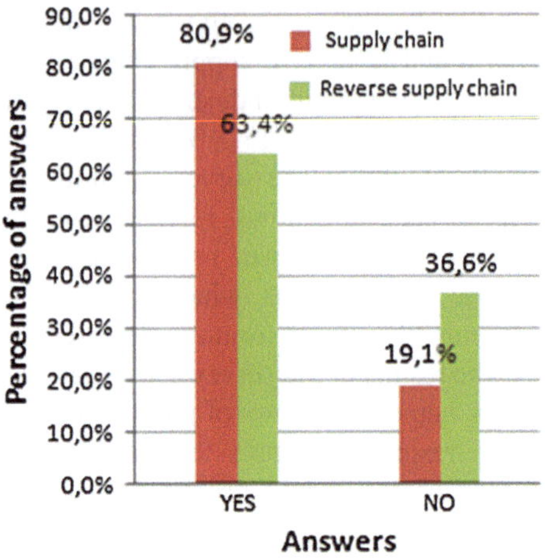

Fig. 3 Willingness to develop closer relationships in supply chains and reverse supply chains—comparison

supply chains joint planning occupies the fourth position, whereas code implementation the fifth; as regards the reverse supply chain these positions are swapped. Higher percentage of indications was recorded in the supply chain, which means that these factors have a much more substantial impact on the flows realized toward the market. Every fifth enterprise on average points to joint research, joint stock replenishment, unification and sharing risks and benefits as relationship-building activities in both chain types (Fig. 4). According to the respondents, joint research is more essential for supply chains rather than for reverse supply chains. It may well mean that many enterprises fail to notice the potential offered by joint research as regards the use of recyclable materials and managing backward flows. In the area of unification, the sharing of risks and rewards and joint financial investments tend to be indicated more often in the case of reverse supply chains. These factors may play primary role in implementing a closed loop for packaging. It can be related to investments in the purchasing of a sufficient amount of packaging to be circulating in the loop, creating work stations, at which the packaging will be cleaned and repaired as well as adapting the entire supply chain for reusing the packaging of the same type and establishing one common standard in this respect. Both in the supply chain and the reverse supply chain all of the possible answers were given by respondents, none of suggested activities aimed at building closer relationships were left unanswered. Each answer had a percentage of indications no lower than 10 %. Answers were also given in the category: *Other*.

The research also addressed the issue of barriers to building closer relationships with business partners. Supply chains and reverse supply chains were presented as separate categories, as was the case with the integration factors. Obtained results were similar in both the chain types. The order of priority of answers given in both supply chains and reverse supply chains is as follows:

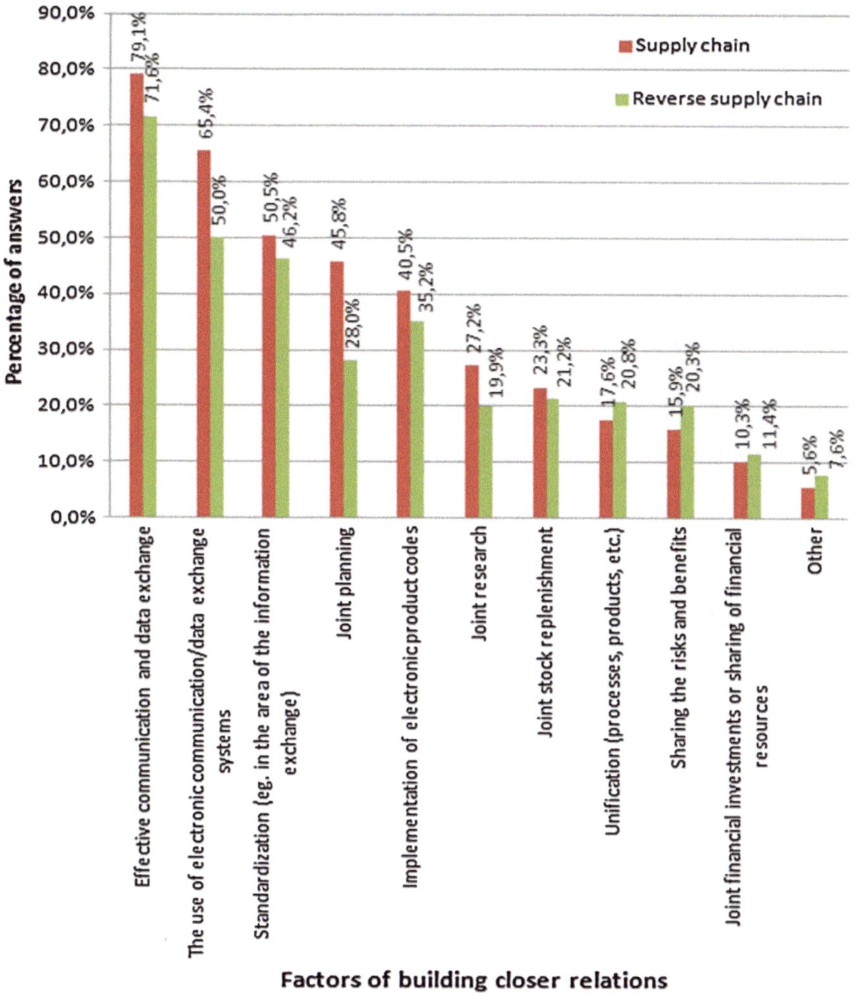

Fig. 4 Factors conducive to building closer relationships with business partners in supply chains and reverse supply chains—comparison

- high frequency of business partner changes,
- each links is focused on its objectives and activities,
- lack of reliable data (no data registration),
- no understanding of the needs of other participants.

Percentage differences in figures pertaining to both chains were minor and no higher than 3.3 %. The sequence of answers arranged in descending order shows

Table 1 Barriers to building closer relationships with business partners in supply chains and reverse supply chains—comparison

Barriers to building closer relationships	Percentage of indications for supply chain (%)	Percentage of indications for reverse supply chain (%)	Percentage differences (%)
High frequency of business partner change	35.8	32.5	3.2
Each link is focused on its objectives and activities	33.6	32.0	1.6
Lack of reliable information (no data registration)	32.5	32.0	0.5
Lack of understanding of the needs of other users	29.6	30.9	−1.3
Fear of information sharing	28.2	24.7	3.5
Various IT systems in individual links and problems with integration	27.7	24.5	3.2
General lack of confidence	26.9	28.2	−1.3
Negative leader (a link in the chain), imposing behaviours which are convenient for him	25.8	19.6	6.2
Differences in manufacturing technology, handling and/or storage	21.5	16.1	5.4
Inability to notice potential benefits for the entire chain	21.0	23.9	−3.0
Fear of joint investments	18.8	16.7	2.2
No funds for pursuing joint investments	18.5	19.1	−0.5
Cultural differences	12.1	9.4	2.7
Other	10.2	18.0	−7.8

certain differences for both supply chains and reverse supply chains. These discrepancies in indications regarding individual barriers to building closer relationships are so negligible that seeking a pattern of dependence is pointless. All of the barriers were indicated by every fifth or every forth company on average in both chains (Table 1).

What immediately grabs attention is a high percentage (10.2 and 18 % in individual chains) of indications in the category of *Other*, which demonstrates that there are other significant barriers to fostering closer relationships not identified in the literature review and, thus not addressed in the survey.

5 Summary

The survey carried out among Polish enterprises indicates that integration factors pertaining to both the supply chain and the reverse supply chain are: communication and exchange of information, establishing common standards (with particular emphasis on implementing electronic codes for products) and joint planning, which has been confirmed by the literature on the subject. Despite the declarations of taking actions regarding the chain integration, Polish enterprises are still faced with many barriers emerging in this field. The barrier most often referred to by respondents is the frequency of business partners which is a symptom of a problem with maintaining long-term relationships with the same business partners. It is a major obstacle, which impedes the development of close relationships and should be eliminated in the first place. In the past sourcing from one supplier was associated with overdependence. However, in the case of relationships based solely on price negotiations it is quality that suffers most often. Focused on cost reduction, a supplier tends to satisfy the most basic requirements and leave other ignored. It means that the recipient incurs extra costs related to necessary inspections or the risk of sub-standard quality of goods if they are given low priority by the supplier. Enterprises must see an opportunity and a room for growth in establishing mutually beneficial and long-term relationships with suppliers, such as enhanced quality, sharing innovations, cost reductions and integrated production and supply schedules. From the supplier perspective, such a partnership may pose a sustainable barrier to entry for the competition.

Other common barriers to building close relationships in supply chains such as the focus on meeting individual needs and objectives as well as inability to understand the needs of other chain participants show that many enterprises fail to perceive themselves as the participants in the supply chain. Most of them are focused on pursuing their internal goals, turning a blind eye to the needs of the supply chain partners. Companies must come to the realization that more and more often competitive advantage is a matter of the effectiveness and efficiency of the entire supply chain and that the closer the cooperation between the supplier and the recipient in key areas of operation, the more sustainable competitive advantage can be delivered. The table below presents selected barriers to building close relationships with business partners along with the tools useful in overcoming these barriers (Table 2).

The survey shows that supply chains and reverse supply chains vary only slightly in fostering close relationship and the integration in general. In both the chains the same factors come into play and the same problems are being faced. Enterprises are more engaged in fostering closer relationships within the supply chains, which means that forward flows are assigned a higher priority and that backward flows are still underdeveloped in Poland.

Table 2 Barrier to building close relationships and the tools for overcoming these barriers

Barriers to building close relationships	Tools helping to overcome barriers
High frequency of business partner change	- Selecting partners having an established market position (if the reason for changing partners is incompetence or bankruptcy of partners) - Assessing the suppliers - Establishing supplier selection criteria (specifying the main points of interest—key to smooth functioning) and finding the most suitable supplier, well-matched to the criteria - Considering not only the price, but also product quality and service when selecting the supplier - Negotiating mutually beneficial terms of collaboration, which will provide benefits in the long run - Having more trust and respect in the supplier
Each link is focused on its own objectives and activities	- Encouraging the partners to collaborate closely for mutual benefit - Assuming the role of a leader in the chain - Making partners aware of potential benefits offered by close cooperation in the context of competition (other supply chains)
Lack of reliable information	- Creating databases - Encouraging employees to pay greater attention to the quality of collected data - Dlear definition of the scope of responsibilities and rights as regards access to the information - Creating operational procedures as regards capturing and recording data - Updating data wherever it is used - Implementing cutting-edge information and communication technologies (e.g. EDI, Internet) - Implementing information exchange standards
Lack of efficient communication and data sharing	- Implementing state-of-the-art information and communication standards (e.g. EDI, Internet) - Implementing information exchange standards - The use of identifiers and carriers enabling an automatic identification of and acquisition of data
Fear of information sharing	- Signing confidentiality clause with partners - Mutual exchange of information (reducing the risk and fears—each party gave and receive access to and data from partners) - Establishing detailed terms and conditions for the exchange of information and the scope of the information (each partners identifies which and how detailed information he or she requires, and for what purpose)

(continued)

Table 2 (continued)

Barriers to building close relationships	Tools helping to overcome barriers
Various systems in individual links and integration problems	- Customization of systems to meet individual needs of the partners - The use of Application Programming Interface, API - Investment in the system to be used by all of the partners
No cooperation in planning and stock replenishment	- Communicating production plans and inventory structure and volume - Advising partners in advance of interruptions in deliveries receipts of products and changes to the production plan - Willingness to help the partner in crisis and emergency situations - Organizing regular meetings of the partners' planning managers to figure out win-win solutions

Acknowledgments This paper has been the result of the study conducted within the project entitled "Investigating multi-aspect determinants of integration in backward and forward logistics chains in respect of integrated production planning in the context of using recyclable materials" pursued at the Poznan School of Logistics in Poznan. The project was financed by The National Science Center based on the decision number DEC-2011/03/B/HS4/03419.

References

Alfalla-Luque R, Medina-Lopez C, Dey PK (2013) Supply chain integration framework using literature review. Prod Planning Control 24(8–9): 800–817

Fechner I (2007) Zarządzanie łańcuchem dostaw. Wyższa Szkoła Logistyki, Poznań

Harrison A, van Hoek R (2010) Zarządzanie logistyką. Polskie Wydawnictwo Ekonomiczne, Warszawa

Jain K, Dubey A (2005) Supply chain collaboration: a governance perspective. Supply Chain Forum: Int J 6(2):50–57

Jayaram J, Tan K-C, Nachiappan SP (2010) Examining the interrelationships between supply chain integration scope and supply chain management efforts. Int J Prod Res 48(22): 6837–6857

Pfohl HCh (1998) Zarządzanie logistyką. Funkcje i instrumenty. Biblioteka Logistyka, Poznań

Rutkowski K (2000) Teoria i praktyka zarządzania zintegrowanym łańcuchem dostaw. In: Rutkowski K (ed) Zintegrowany łańcuch dostaw. Doświadczenia globalne i polskie. Oficyna Wydawnicza Szkoły Głównej Handlowej, Warszawa

Sobol E (ed) (1997) Słownik wyrazów obcych. Wydawnictwo Naukowe PWN, Warszawa

Stadtler H (2005) Supply chain management and advanced planning—basics, overview and challenges. Eur J Oper Res 163(3):575–588

Szołtysek J (2009) Logistyka zwrotna. Biblioteka Logistyka, Poznań

Szudrowicz A (2002) Problemy badania relacji w łańcuchu logistycznym. In: Chaberka M (ed) Modelowanie procesów i systemów logistycznych. Część II. Wydawnictwo Uniwersytetu Gdańskiego, Gdańsk

Szymonik A (2011) Logistyka i zarządzanie łańcuchem dostaw. Część 2. Difin, Warszawa

Yu W, Jacobs MA, Salisbury WD, Enns H (2013) The effects of supply chain integration on customer satisfaction and financial performance: an organizational learning perspective. Int J Prod Econ 146(1):346–358

Supply Chain Integration in View of Secondary Raw Materials

Żaneta Pruska, Łukasz Hadaś, Piotr Cyplik and Martyna Kupczyk

Abstract Environmental awareness keeps growing and comes to be viewed as an integral part of the management process in companies. We should keep in mind that the waste generated by a company (one of the participants in the supply chain) is part of the total waste generated by the entire chain. Moreover, environmental awareness of individual companies does not mean that the principles of sustainable development are adhered to. What is needed is an integration of companies across the entire chain. Their management should be therefore approached from a holistic perspective. The main goal of this article is to identify integration factors that take into account process and structural changes in the entire supply chain in terms of secondary raw materials. Each of the factors comes with a detailed characteristics of individual levels of integration. The universal character of factors enables the study of chains in various sectors of business practice. In order to achieve the goal of this article a survey was conducted among companies operating on the Polish market. On the basis of these studies the key barriers to the supply of secondary raw materials have been identified. The chapter contains an overview of these barriers, taking the selected channels of supply into account. It should be noted that the focus of individual elements on ecology affects the evaluation criteria associated with the level of integration of the entire chain.

Keywords Waste management · The level of integration · Barriers to supply of secondary raw materials

Ż. Pruska (✉) · Ł. Hadaś · P. Cyplik · M. Kupczyk
School of Logistics, ul. Estkowskiego 6, 61-755 Poznan, Poland
e-mail: zaneta.pruska@wsl.com.pl

P. Golinska (ed.), *Logistics Operations, Supply Chain Management and Sustainability*, EcoProduction, DOI: 10.1007/978-3-319-07287-6_4,

1 Introduction

In the face of the overall degradation of the environment, the initiatives aimed at neutralizing pollution seem to be wholly insufficient. Nowadays efforts should be focused on looking for solutions which will enable the running of business in a manner least detrimental to the environment not only at the stage of manufacturing, but also during production planning or product design (Bojar et al. 2012; Kumar et al. 2012). A supply chain must be perceived from a broader perspective, not only as a selection of individual organization blocks through which products and information flows (Palewicz and Baran 2012). Literature of the subject tends to perceive environmental protection as an integral part of the management process. Strategic actions, which respect the objective of "not overburdening" the environment include, among others: trading in secondary resources and reduced consumption of energy and resources (Adamczyk and Nitkiewicz 2007). Enterprises acting in accordance with the "3R" principle (Reduce, Reuse and Recycle) ought to participate in backward flows and use secondary raw materials as input materials for production processes (Corrêa and Xavier 2013; Malek et al. 2007; Tam and Tam 2006). This new approach taking into account also backward material flows paved the way for the development of ecologistics and reverse logistics, which involve solutions related to the management of various kinds of waste (Kruczek 2012).

Nowadays enterprises start to see the social responsibility of business as the object of a purposeful operation—not only in theory, but also in practice. One can suppose that such companies win the support and trust of their customers, suppliers as well as employees and shareholders. It translates into increased competitiveness and greater chances for further development (Sokołowska 2009). The European Commission's definition emphasizes that social responsibility of business means a voluntary incorporation of social interests, environmental protection issues and relations with such interest groups as e.g. customers, subcontractors or workers into the organization's strategy (Arendarski 2006). The perspective of sustainable development has integrated ecological issues in the economy to a greater extent than in previous years. The reorientation of the world economy has raised many questions regarding the development and evolution of the concept, among others, towards solutions having more relevance to business practice (Sadowski 2008).

The supply of secondary raw materials, which contributes to a decrease in the quantity of stored waste at waste disposal plants, requires including the reverse flow channels in material flows. It should be emphasized that a reverse chain exhibits a much higher level of complexity than a traditional chain due to the uncertainty of supplies as well as frequent mismatches between the supply and demand (Akcalı and Cetinkaya 2011; Grabara 2004). The acquisition of waste from the market (industrial or consumption waste) is not easy for enterprises. Reverse logistics is characterized, among others, by the fact that the material flow takes place from many points to one point, the quality of obtained products is not heterogeneous and stock management is incoherent (Harrison and van Hoek 2010 for: Reverse Logistics Executive Council).

The idea of sourcing secondary raw materials is aligned with the principle of sustainable development and corporate social responsibility. Considering the organizational complexity of the process, many links must cooperate in the chain, which will increase the opportunities for managing the waste generated in logistic chains and also decrease the amount of waste discharged to the environment.

2 The Characteristics of the Supply of Secondary Raw Materials in View of a Survey

The survey included 199 companies which are supplied with secondary raw materials. The researched sample involved 37.2 % large businesses employing over 250 employees, 22.1 % medium-sized enterprises (between 50 and 249 employees), 26.6 % small enterprises employing between 10 and 49 employees and 14.1 % micro businesses (between 1 and 9 employees).

Secondary raw materials can be sourced through the following channels: a regular cooperation with a business partner or taking advantage of an offer of an agent (Pruska et al. 2011). As stated above, this kind of supply is a very specific type of securing resources for production purposes. Many companies are anxious about organizational aspects of logistics and are faced with various barriers, which often hinder such a process. The surveys show that the sourcing of resources is inextricably linked with various obstructions. The respondents were asked the following question: “What logistics difficulties did the company encounter while sourcing secondary raw materials?” The outcomes of the survey have been presented in Fig. 1.

From amongst companies which source secondary raw materials, as many as 39 % declare that they use one selected supply channel. The responses were matched to questions on the choice of a supply source as well as logistic difficulties faced by an enterprise, helping to come up with specific features of selected channels. The comparison of individual supply channels has been presented in Fig. 2.

Enterprises relying on a regular cooperation with a business partner encountered such logistics difficulties as: the need to organize own transport, high price volatility of resources and waste storage costs. Organizations relying on a regular cooperation with an agent faced such difficulties as: untimely deliveries and waste storage costs. High price volatility was not marked by the respondents. The cooperation with a larger number of secondary raw material suppliers presents a different range of logistics difficulties, with the most common being the need to organize own means transport. According to the respondents, other inconveniences of equal importance are: untimely deliveries and the need for organizing specialized transport equipment. High price volatility was not selected by the respondents. It can be inferred from the above that waste supply agents do not change prices in short time periods, which facilitates making cost estimates by waste recipients.

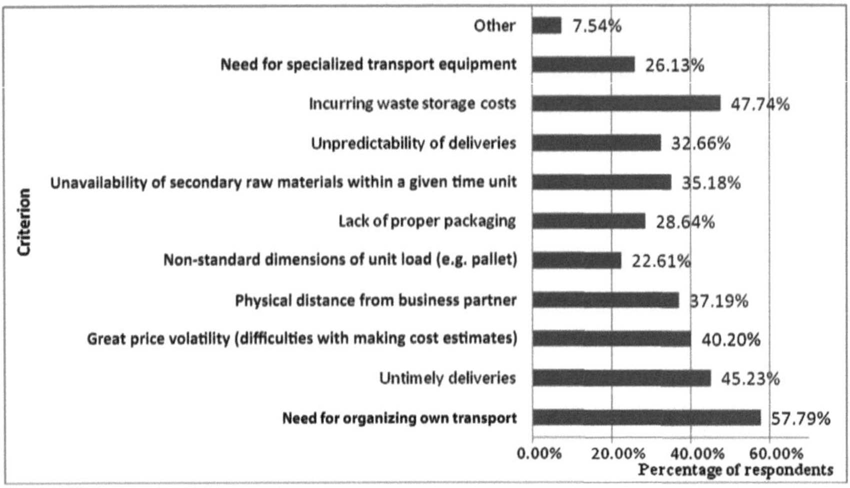

Fig. 1 Logistics difficulties resulting from the organization of the process of supply with secondary raw materials

The survey outcomes help to identify logistics barriers faced by an enterprise in the course of sourcing secondary raw materials. The barriers are encountered not only prior to making a decision to make ecological purchases, but also in the course of the process, and include, for example, high price volatility of resources or untimely deliveries. At a higher level of chain integration some of these barriers can be reduced.

3 Forward and Backward Logistics Chain

According to the definition of M. Jacyna, a model is a representation of reality and "it is treated as a simplified representation of reality, i.e. of its basic features crucial with respect to the intended aim of the study" (Jacyna 2009). This chapter depicts a model developed by the author, in which materials flow between individual participants of the supply chain. The model was designed for the purposes of investigating simulation in modelling reverse chains. Products flow forward, to the customer, and the backward flow of waste is taken into account. The model does not include financial flows, because logistics aspects of material streams' flow were the main focus of study, not payment management by individual links of the chain (Fig. 3).

The model presented above includes not only the flows of resources and products, but also of waste between individual links of the supply chain. The area outside of the participants enables combining various supply chains and constitutes the supply chain's environment.

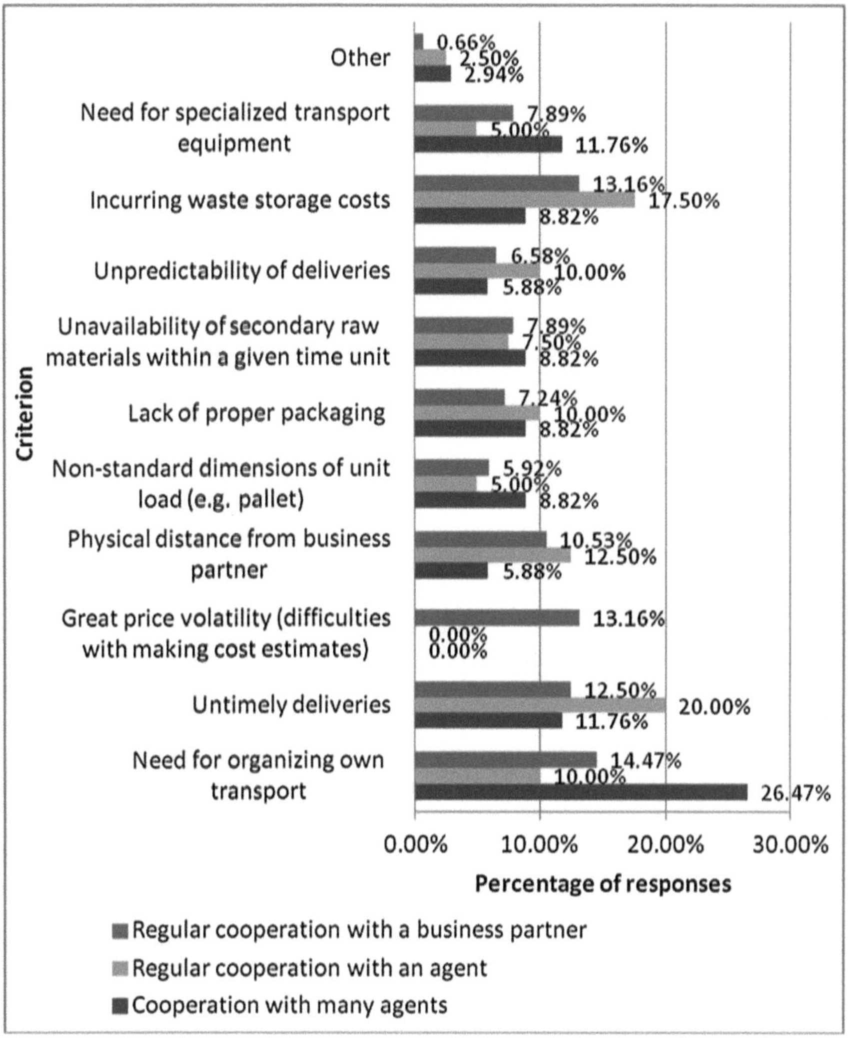

Fig. 2 Logistics problems due to the organization of secondary raw materials supply versus selected supply channels

In the model under review the supplier is a link "preceding" the production floor—namely any entity being a supplier, irrespective of the form of business operation. A production enterprise deals with production ("make" according to SCOR model (Zaitsev 2012)) and sends its products to the market through a distribution network. An end recipient, depending on the enterprise's industry, may be a consumer or another company. An agent, on the other hand, is associated with a company dealing with acquisition/purchase of waste from the market.

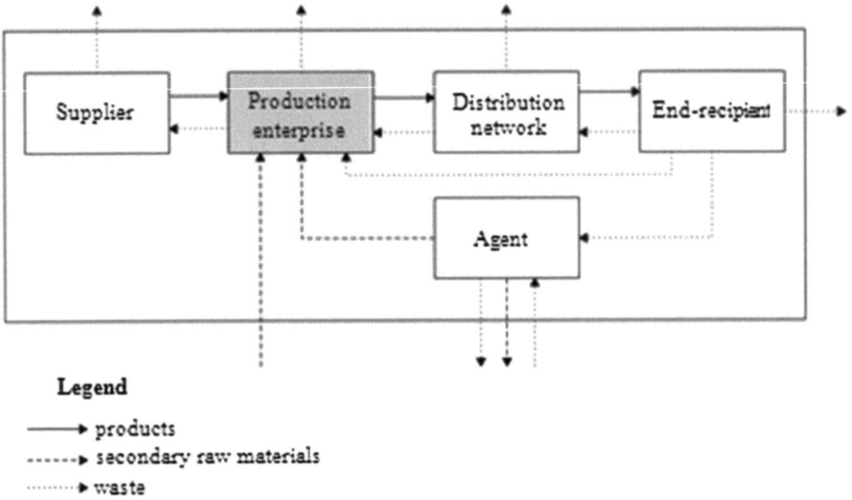

Fig. 3 Forward and backward chain model

The agent does not process the acquired waste, as production operations and the processing of waste into a finished product move the waste forwards the chain. An agent performs such actions as refinement, completing, packing, which means providing waste-related services for recipients.

As has been stressed before, the model in question captures only material flows, leaving out information flows (which are reflected in the model management mechanism) and therefore, an internet-based waste exchange, where waste is traded, is not taken into account as one of its elements. The exchange is a virtual meeting venue for business partners and serves as a platform for conducting transactions involving the purchase/sale of secondary raw waste/materials. Upon establishing the contact, resources are being transferred between the two parties to the transaction. It means that no physical flows are being triggered by the transaction.

The area outside of the supply chain is a platform for making strategic choices and taking various actions (Adamczyk and Nitkiewicz 2007). A given logistics chain can collect waste materials generated by other chains and provide redundant waste it produces to other chains, which will put it to good use in its production processes. The presented model is multiplicative, because it may be combined it into several chains. It makes the model all the more universal and viable in business reality. Such a model can be used for simulating any configuration of supply chains occurring in business practice. The model's multiplicative structure enables a relatively quick mapping of the system in the simulation application and the modelling of variant integration options.

4 Supply Chain Integration Factors in View of the Use of Secondary Raw Materials

In the literature of the subject, we can discern many ways and methods of measuring performance efficiency of logistics chains. They raise a number of barriers such as (Nowakowski 2011):

- *"lack of coherence among logistics indicators of individual logistic areas,*
- *prime focus on using economic rather than operational indicators covering, among others, customer service or product quality,*
- *lack of relationship between logistics metrics in operation and the logistics strategy realized in the logistics chain,*
- *measuring the performance of individual participants rather than the efficiency of the whole chain."*

In view of the above authors formulated the following objective: developing a system for evaluating the integration of forward and backward supply chains. The integration factors presented in this chapter apply to the sourcing of secondary raw materials at each of four integration levels (A—level of the greatest integration, D—level of the lowest integration). The features of integration factors are described in more detail.

The first factor is "Reduction of material waste", and its individual levels are presented in Table 1. It should be noted in view of a holistic approach the entire chain generates waste constituting material losses which when poorly managed. An integrated supply chain will be driven to minimize the losses by respecting the principles of sustainable development. Exchanging the information with other links and getting involved in a joint study will enable a better management of waste generated during production processes and the reduction of the amount of waste sent to the environment. Ecological awareness of individual companies is not enough to perform such actions. It takes the integration of the entire chain. The higher the supply chain's integration level, the greater the spending on such study. Moreover, the study can be based on a greater amount of data coming from the entire chain, which translates into better chances of managing this waste. It requires an exchange of information between the participants—not only on the amount and quality of generated waste, but also technical data connected with production processes. The flow of such is predicated on deep trust between business partners. It is internal data, having a decisive impact on the strength of a link, yet in the case of an integrated chain, unit-oriented thinking is replaced by a holistic approach, which puts the entire chain in the perspective. Excessive generation of waste within an enterprise is an evidence of poor waste management and applies to the entire chain.

The availability of secondary raw materials on the market, namely the possibility to generate supply by individual links, impacts the level of the integration of the supply chain in question. Low or occasional supply compels the use of available resources, namely establishing interactions with alternative supply links,

Table 1 Reduction of material losses factor—performance characteristics at integration levels

Integration level	Performance characteristics
A	Undertaking joint investment activities related to study into the ways of minimizing generated waste and the possibilities of its use along with the information exchange on production processes, including the amount and quality of generated waste
B	Undertaking joint investment activities related to study into the ways of minimizing generated waste and the possibilities of its use or the exchange of information on production processes, including the amount and quality of generated waste
C	Study into the ways of minimizing generated waste and the possibilities of its use in production processes are run independently by individual links. No data exchange regarding production processes from other links
D	Study into the ways of minimizing generated waste and the possibilities of its use in production processes are occasional and run independently by individual links. No data exchange regarding production processes from other links

which may bring down the chain's integration level. This factor is characterized by availability of resources perceived in quantity, time and quality terms. The material's supply, viewed as the capability to meet the needs arising in fixed and stable chain links, or else a continuous change of links, are also reflected by the integration metrics. In other words, such actions impact the quality and cost of the process, chain response time and flexibility as well as cash flow. A continuous change of links can also trigger the need for adapting secondary raw materials to the requirements of production processes. It is due to the fact that various links generate various kinds of waste (secondary raw materials) in the course of various production processes. For instance, the same waste type from different links may have different dimensions or a different level of contamination. In view of the above the link using such resources must tailor its material processing procedure to the requirements of its manufacturing process. The activities pertaining to respective levels of integration for this factor have been presented in Table 2.

The purchase of secondary raw materials instead of primary raw materials is a very specific type of sourcing. The supply of waste is determined by the production processes at the potential supplier. In this respect, the selection of such resources, that is ecological orientation of a chain's production links, has an influence on ratios connected with the integration level of the entire chain. Although the material will be a redundant waste for a single link, charging the link which uses waste materials with transport and storage costs will adversely affect the profitability and cash flow of the entire chain. The higher the integration level, the greater the chance for negotiating favourable terms of such a transaction and ecological benefits. A low level of chain integration, when one of the business partners has a greater bargaining power, entails the risk of:

- "eliminating" secondary raw materials from circulation in a given supply chain, which means directing it to another chain or to a waste storage plant

Table 2 Availability of secondary raw materials—performance characteristics at integration levels

Integration level	Performance characteristics
A	Sustainability of the supply of secondary raw materials from the perspective of time and quantity and no need for changing the material adaptation process so that it would meet the requirements of the production process
B	Disrupted sustainability of the supply of secondary raw materials (storage is required), the quality of resources does not disturb the process of adapting the resources to the production process
C	Low sustainability of supply (need for seeking external links—short-term cooperation), the quality of resources may affect the process of adapting the resources to the production process
D	"Occasional" supply, the quality of resources may affect the process of adapting the resources to the production process

Table 3 Supply and storage of secondary raw materials—performance characteristics at integration levels

Integration level	Performance characteristics
A	Negotiating the most favourable terms and conditions of purchase/sale of secondary raw materials/waste and sharing storage costs (win-win)
B	Negotiating the most favourable terms and conditions of purchase/sale of secondary raw materials/waste or sharing storage costs
C	The recipient has a greater bargaining power. The supplier is in charge of transport (performs or commissions the performance of the services) and incurs storage costs until a given resource is in demand by the recipient
D	The supplier has a greater bargaining power. The recipient is in charge of transport (performs or commissions the performance of the service) and at the time and place set by the supplier, and incurs storage costs

(costs related with transport and storage incurred by a potential supplier will not exceed potential revenues from the transaction),

- purchasing secondary raw materials from a different link or using primary raw materials.

Performance characteristics of the "Supply realization and storage of secondary raw materials" factor has been delineated in Table 3.

Another factor refers to the ability to handle returns from the market and information flows on secondary raw materials. Its distinctive features are presented in Table 4. Products withdrawn or returned from the market are a source of supply of secondary raw materials. Forecasting return quantities is a derivative, among other things, of the product's lifecycle, but it may also involve the materials generated by other producers. The ability to accept returns is predicated on the amount of free space available in the warehouse, and is subject to forecasting.

The exchange of information regarding the projected availability of secondary raw materials between business partners is likely only if the integration level of the

Table 4 Ability to handle returns from the market and information flow on secondary raw materials—performance characteristics at integration levels

Integration level	Performance characteristics
A	Forecasting returns from the market and cooperation with business partners regarding waste transport. The exchange of information on projected availability of secondary raw materials between the supply chain partners (long-term plans)
B	Forecasting returns from the market without the cooperation with business partners regarding waste transport. The exchange of information on projected availability of secondary raw materials between some of the supply chain partners (time and amount of resources). Joint planning, short-term plans
C	No return forecasts (amounts), preparing the ground for cooperation regarding waste transport between the chain partners. Occasional exchange of information between some partners with regard to projected availability of secondary raw materials
D	No return forecasts (amounts), no cooperation with business partners regarding waste transport. Suppliers are sought when the need arises (very costly order placement procedure)

chain is high. Poor information flow affects the costs generated not only by a costly order placement procedure (looking for suppliers each and every time), but also by the storage and the organization of the entire process. Such an organization is reflected in the chain's ability to respond to the market's requirements and, in consequence, to the chain's reliability.

The presented integration factors are universal, which enables investigating supply chains from various industries.

5 Conclusion

Sourcing secondary raw materials poses a logistics problem for enterprises. The focus on ecology of individual links is not enough to run a business in keeping with the principle of sustainable development. The integration across a logistics chain is necessary to bring the economy closer to the society and social responsibility, especially in the context of intergenerational justice, which implies using natural resources in a moderate and responsible way. What is more, corporate managers should keep in mind that being involved in material flows triggered by a given chain means taking responsibility—also for the operations of one's partners. Any materials and substances discharged as in the form of waste outside of the chain by one of the participants combine into the waste generated by the entire chain.

The forward and backward logistics chain model presented in this chapter shows how closely business partners are connected and that there are alternative sources of secondary raw materials. Integration factors such as:

- reduction of material losses,
- secondary raw material availability,
- supply and storage of secondary raw materials,
- ability to handle returns from the market and the flow of information on secondary raw materials

lie at the heart of evaluating the integration of supply chains in the context of the supply of secondary raw materials. The factors recommended above are dedicated to all industries and show how supply chains can enhance their integration level and reduce the amount of generated waste and manage it effectively. Not isolated units, but entire supply chains are responsible for the condition of the environment.

Further study will be conducted based on the division of supply chain's integration factors into simulated and non-simulated factors. Simulation experiments will be performed for each integration level. The supply chain model that has been developed will be subject to a number of simulations with a view to verifying its integration ratio. Simulation results will provide the grounds for reviewing the activities assigned to respective integration levels (A, B, C or D) in terms of their impact on the integration of the supply chain.

Acknowledgments This chapter has been the result of the study conducted within the project entitled "Investigating multi-aspect determinants of integration in backward and forward logistics chains in respect of integrated production planning in the context of using recyclable materials" pursued at the Poznan School of Logistics in Poznan. The project was financed by The National Science Center based on the decision number DEC-2011/03/B/HS4/03419.

References

Adamczyk J, Nitkiewicz T (2007) Programowanie zrównoważonego rozwoju przedsiębiorstw. Polskie Wydawnictwo Ekonomiczne, Warszawa

Akcalı E, Cetinkaya S (2011) Review. Quantitative models for inventory and production planning in closed-loop supply chains. Int J Prod Res 49(8):2373–2407

Arendarski A (2006) W obronie przyszłych pokoleń, E-mentor 2006, vol 1(13), pp 4–7

Bojar E, Bojar M, Żelazna-Blicharz A, Blicharz P (2012) Eco-management in polish companies. Problemy Ekorozwoju 7(2):107–113

Corrêa HL, Xavier LH (2013) Concepts, design and implementation of reverse logistics systems for sustainable supply chains in Brazil. J Oper Supply Chain Manag 6(1):1–25

Grabara J (2004) Zasady zamkniętych pętli łańcuchów dostaw. In: Hejduk IK (ed) Teoria i praktyka modelowania systemów logistycznych. Wydawnictwo Uczelniane Politechniki Koszalińskiej, Koszalin

Harrison A, van Hoek R (2010) Zarządzanie logistyką. Polskie Wydawnictwo Ekonomiczne, Warszawa

Jacyna M (2009) Modelowanie i ocena systemów transportowych. Oficyna Wydawnicza Politechniki Warszawskiej, Warszawa

Kruczek M (2012) Model łańcucha logistyki odwrotnej zużytego sprzętu elektrycznego i elektronicznego, Zeszyty Naukowe Politechniki Śląskiej 2012, Series: Organizacja i Zarządzanie z. 60, pp 165–178

Kumar S, Teichman S, Timpernagel T (2012) A green supply chain is a requirement for profitability. Int J Prod Res 50(5):1278–1296

Malek B, Iqbal M, Ibrahim A (2007) Use of selected waste materials in concrete mixes. Waste Manag 27(12):1870–1876

Nowakowski T (2011) Niezawodność systemów logistycznych. Oficyna Wydawnicza Politechniki Wrocławskiej, Wrocław

Palewicz J, Baran J (2012) Organizacja łańcucha dostaw w branży odzieżowej. Logistyka 2012:531–544

Pruska Ż, Malak M, Stachowiak A, Adamczak M (2011) Determinanty wyboru strategii zaopatrzenia w ramach zrównoważonego rozwoju przedsiębiorstw. Logistyka 5:1223–1230

Sadowski A (2008) Zrównoważony rozwój z perspektywy logistyki zwrotnej. Problemy Ekorozwoju 3(2):129–132

Sokołowska A (2009) Społeczna odpowiedzialność małego przedsiębiorstwa — przejawy i dylematy, E-mentor, 2009, No 5(32), pp 31–37

Tam VWY, Tam CM (2006) A review on the viable technology for construction waste recycling. Res Conserv Recycl 47(3):209–221

Zaitsev E (2012) Modelowanie wiarygodności łańcucha dostaw. LogForum 1:61–69

Integration Level Measurement System in Modeling Forward and Backward Supply Chains

Łukasz Hadaś, Piotr Cyplik and Michał Adamczak

Abstract This article presents a system for measuring the level of integration of a forward and backward supply chain and its use in modeling the structure and processes occurring in the management of the flow of goods. A supply chain involves both forward and backward flows and is related to S&OP model. The paper features also the methodology for developing a system for evaluating and validating the measurement system along with its application guidelines. It also presents selected results of a study into the barriers to integration and operational metrics to evaluate the performance of the supply chain and the specificity of their use in the simulation model. The main part of the article contains a description of integration categories. A comprehensive list of 19 integration categories subject to evaluation is supplemented with the description of the integration levels pertaining to selected categories. The last section discusses principles for evaluation and determining the level of integration of the supply chain and guidelines for its transformation in order to improve its efficiency.

Keywords Supply chain management · Measurement of integration assessment · Forward and backward supply chain

Ł. Hadaś (✉) · P. Cyplik · M. Adamczak
Poznan School of Logistics, Estkowskiego 6, 61-755 Poznan, Poland
e-mail: lukasz.hadas@wsl.com.pl

P. Cyplik
e-mail: piotr.cyplik@wsl.com.pl

M. Adamczak
e-mail: michal.adamczak@wsl.com.pl

P. Golinska (ed.), *Logistics Operations, Supply Chain Management and Sustainability*, EcoProduction, DOI: 10.1007/978-3-319-07287-6_5,

1 Introduction

The integration of the supply chain is an evolutionary process, individual chain participants become integrated gradually. This integration can be evaluated by way of determining its level. The authors and advocates of the concept of the supply chain management agree that the road towards achieving a complete integration is long, challenging and involves many stages. Literature on the subject is not consistent on detailed procedures aimed at driving higher levels of cooperation. There are plenty of such concepts which apply to integrated supply chains. The most interesting concepts, from the point of view of the authors, are:

- SCOR—supply chain reference model (2010),
- 5-tier Poirier model (Poirier 1999),
- 4-tier compass model (Simichi-Levi et al. 2000),
- 4-tier Stevens model,
- 3-tier A.T. Kearney model.

Each of presented models is focused on the selected aspects of integration or on IT tools supporting this integration. They are various takes on the subject matter in question, starting with the models based on a set of reference recommendations up to project management models. Hammer (2001) take notice that it is in the integration of business processes across firms in the supply chain where the real "gold" can be found. Better managing business processes through process integration within and across participants of the supply chain can make the transactions and relationship structures in the supply chain more efficient and effective (Lambert 2004). Therefore, Stock (2002) imply that integration of processes within and between firms in the supply chain is the key to SCM success. Internal integration can contribute to achieving reductions in costs, stock-outs, and lead time; and successful external integration is a necessity to achieve a strong competitive position (Gimenez and Ventura 2003). In this way, a firm is likely to obtain superior performance when achieving high levels of integration both internally and externally. The Authors put forward a comprehensive methodology for evaluating the level of the supply chain integration, drawing on their vast experience in diagnosing logistics systems at enterprises (Cyplik and Hadaś 2011), creating dedicated hybrid solutions (Hadaś and Cyplik 2012) and an in-depth study of barriers to integration in industry. To achieve a high level of usefulness, the model's validation was based on diagnosing integration barriers in market conditions and modelling supply chains in simulation applications.

2 Survey into the Barriers to the Integration of the Supply Chain Processes

The system for measuring the integration level has been developed in line with the methodology of works and is based on:

- The analysis of the results of survey (general and detailed),
- The analysis of selected models of internal and external integration.

The survey aimed at identifying the barriers to the integration in the supply chain had 392 respondents representing various enterprises and management levels. After a preliminary selection (rejecting incorrectly completed surveys) as many as 372 questionnaires were duly analysed. The survey results included:

- A standard structure for enterprise employees (higher level management ca. 5 %, medium level management ca. 38 %, employees ca. 57 % of the respondents),
- The most important departments and functions (stock management/warehouse ca. 22 %, production/assembly 18 %, sales ca. 17 %, distribution/transport ca. 17 %, finance ca. 6 %),
- Small and medium-sized enterprises (189 respondents) and large enterprises (126 respondents).

The issues and detailed questions were arranged into 17 groups and allowed for identifying the categories of internal and external problems, which pose barriers to enhancing the cooperation between the enterprises.

The survey findings were not analysed in depth (due to paper length restrictions), it should be noted, however, that provided answers fall into the following categories:

- Supply chain structure (quantity and the type of links),
- The type of integration measures undertaken by enterprises,
- Barriers which hamper the enhancement of the integration level (see Fig. 1.)
- The characteristics of reverse flows,
- Barriers to increasing the amount of recyclable materials,
- Problems, which are perennial in planning production and logistics processes.

The findings of this research provided an important input to the categories of activities evaluated in the integration process.

3 Supply Chain Integration Factors

The Authors' system for measuring the integration level in the supply chain is based on the analysis of the integration level of individual factors that have been previously identified. Such a solution enables a comprehensive analysis of the supply chain integration levels.

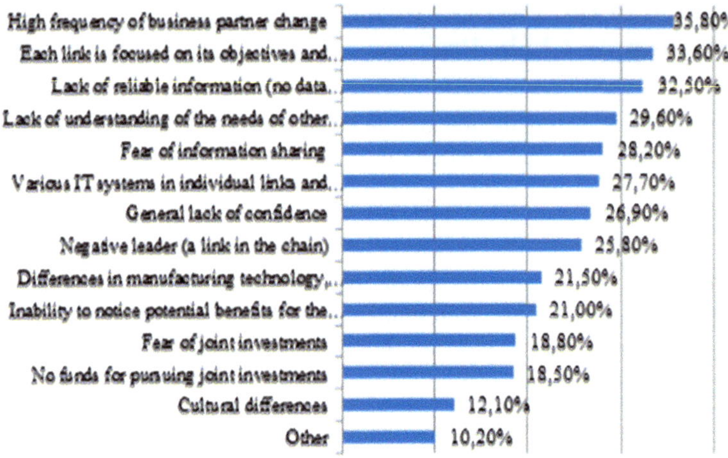

Fig. 1 Barriers to building closer relationships in supply chain. Own study

Four levels—grades have been identified for each of the 19 integration factors. Each level (level A B C D, where A means the highest level) comes with a clearly defined characteristics, based on which the supply chain activities can be measured (by assignment to an appropriate category). When characterising the integration levels, the Authors kept in focus:

- manifestations of integration,
- or the symptoms of their absence,

which facilitates the evaluation based on the typical characteristics of the functioning of a given supply chain, obtained during, e.g. interviews with managers.

Below you will find integration activities selected (due to paper length restrictions) from among 19 categories. In view of the restrictions of article 19 integration category of investment activities is presented without a detailed description.

3.1 Integration Category: R&D Works

Being successful on the market nowadays is often predicated on offering innovative products, tailored to meet customer needs. Conducting research development works is a costly and risky venture. When acting in isolation, companies do not have sufficient funds at their disposal and in case of the research failure the risk of falling into financial problems is quite high. Undertaking R&D works within the framework of the supply chain considerably reduces the risk of failure and fosters the economies of scale, which brings down unit costs of developed solutions. A detailed description of particular integration levels is presented in the Table 1.

Table 1 Research & development works—performance characteristics at particular integration levels

Integration level	Performance characteristics
A	A centralised system for planning and coordinating R&D works (related to the issues consistent with the adopted competing strategy), taking into account the strengths of each partner in the supply chain; the reverse chain and related material feeds are taken into account in the process of designing products and processes
B	Research projects are run within supply chains, they are aimed at delivering innovations of key importance for the chain's competitiveness, more than two partners are engaged in the project
C	Incidental research projects initiated between two partners in the supply chain
D	R&D works conducted in isolated links are based on individual know-how demand forecasts

Own study

Table 2 Investment activities—performance characteristics at particular integration levels

Integration level	Performance characteristics
A	Centralised decision making related to investment activities, of key importance for the chain's competitiveness, the scope of the investment project applies to production facilities, warehouses and distribution channels
B	In the supply chain we can distinguish the investments undertaken jointly by more than two partners, which are aimed at enhancing chain's competitiveness, decisions taken at the level of the agreements between the chain's links, the scope of the investment project involves warehouses and distribution channels
C	Isolated joint investment projects (aimed at increasing competitiveness) undertaken by two partners from the supply chain, the scope of the investment project involves distribution channels
D	No joint investment activities

3.2 *Integration Category: Investment Activities*

Investment activities related to building production facilities, warehouses and to the development of the distribution network determine the expansion of enterprises and supply chains. Running investment projects requires collecting a sufficient amount of funds. Acting in isolation is not conducive to attracting and raising capital. Making a concerted effort not only facilitates the acquisition of capital (greater assets), but also allows for a more effective use of new investments, which reduces ROI. Investment activities for particular integration levels are presented in the Table 2.

Table 3 Demand forecasting—performance characteristics at particular integration levels

Integration level	Performance characteristics
A	Forecasting at the level of the entire supply chain with the time horizon covering cumulative lead time and the absolute forecast error per each link being no less than 90 %.
B	Forecasting at the level of the entire supply chain on condition that the time horizon covers cumulative lead time and with the absolute forecast error per each link being no less than 90 %.
C	Irrespective of the forecasts for individual links in the supply chain (no exchange of the information, trend-based, etc.
D	No formalized forecasting procedure

Own study

3.3 Integration Category: Demand Forecasting

Forecasting demand is a key information input in the supply chain. Errors in forecasts have their consequences for the material flow, inventory volume and the level of customer service. All of the above mentioned consequences trigger extra costs of the supply chain's functioning. It can be therefore concluded that forecasts with lower error rates will have a beneficial impact on the supply chain return. One of the conditions of precise forecasting is the origin of the information source and the length of the forecast period. If the data is derived from a source that is close to the customer and the final end user, then the error rate is lower. On top of the above, if the data is processed (forecast preparation) by fewer entities, then the cumulative forecast error rate is lower. This is what makes the cooperation between the supply chain partners so vital. A detailed description of the activities undertaken on the four identified integration levels is presented in the Table 3.

3.4 Packaging Unification

Packaging performs various functions. In view of the integration of the supply chain, the focus is on the logistics functions. The primary purpose of the packaging from the point of view of the supply chain is protecting the product from any losses (damage) and reducing packaging handling costs in storage (all phases) and in transit. Packaging unification is intended to bring to a minimum any operations involving packages, in particular at the stage of warehouse receipt and issuing. Standardising packages in the supply chain may effectively rule out the need for repacking or changing the shapes of logistics units. Such a unification is also likely to bring down inventory levels in the supply chain, which may lead, in consequence, to further operational cost cuts. The measures taken at integration level are presented in the Table 4.

Table 4 Packaging unification—performance characteristics at particular integration levels

Integration level	Performance characteristics
A	Packaging unified across the entire reverse chain, matched to the transportation fleet and the material movement technology. Returnable packaging in use to a large extent (over 75 % in total for the chain) with closed loop, no repacking except for packaging purposes
B	Packaging unification and the application of closed loop of reusable packaging between more than two partners including the supply chain leader
C	A section unification of selected types of packaging between the closest partners in the supply chain
D	Internal (one link) packaging unification

Own study

Table 5 Integration of the information flow—performance characteristics at particular integration levels

Integration level	Performance characteristics
A	Information available in real time across the entire supply chain
B	Information available in real time between selected supply chain partners, including the chain leader
C	Information available in real time between two selected supply chain partners
D	No tools which reduce the information transfer time

Own study

3.5 Integration Category: Integration of the Information Flow

Information availability mentioned in the chapter on forecasting as a factor behind lower forecast error rates is not the only that an uninterrupted flow brings. If we assume that inventory in logistics may take a material or a time-related form, then the more time is lost in the course of the information flow in the supply chain, the greater inventory levels it requires. It can be inferred, therefore, that accelerating the information flow may lead to reducing the operational costs of the supply chain. Potential benefits go beyond the flow time. Up-to-date information usually has a lower error rate and enables a more precise identification of further measures to take. A detailed description of activities in the supply chain at each of the 4 integration levels is presented in the Table 5.

3.6 Sales Support

Sales support pertains not only to promotional activities, targeted directly at the customer, but to all of the operations in the supply chain intended to communicate

Table 6 Sales support—description of activities at individual integration levels

Integration level	Performance characteristics
A	Activities supporting sales (supplying increased amounts of goods to the market, promotional activities in the sales network) coordinated across the entire chain, in which at least several supply chain partners are engaged, and whose frequency is equal or greater than one activity per quarter, the coordination of sales support activities for all assortment groups
B	Joint (engaging at least several supply chain partners) marketing activities, aimed at promoting all product groups, irregular and incidental by nature
C	Occasional (less often than once a year) collaboration between the links in the supply chain in supporting sales—and only with regard to the assortment groups which contributes to the highest material flow
D	No activities sales support activities performed by several supply chain partners. Sales support activities undertaken by isolated supply chain links.

Own study

Table 7 Transport activities—description of activities at individual integration levels

Integration level	Performance characteristics
A	Optimization of transport processes across the entire supply chain, taking into account reverse flows
B	Optimization of transport processes across the entire supply chain
C	Optimization of transport processes between two closest supply chain partners
D	Transport between supply chain links controlled autonomously

Own study

these promotional activities to the customer and to meet the demand increased in response to these activities (Table 6).

3.7 Transport Activities

Transport activities are one of the pillars of the supply chain (Table 7).

3.8 Optimization of Material Flow

Material flow is instrumental for the functioning of the supply chain. It is a primary flow, triggering: financial flows, document flows and, indirectly, also information flows. The optimization of material flow plays key role in bringing down the costs of functioning of the supply chain and is a major activity leading to the improvement of the chain (Table 8).

Table 8 Optimization of the material flow—description of activities at integration levels

Integration level	Performance characteristics
A	Adapting the volume of material flow for all the links in the supply chain to current demand (including backward flows)
B	Optimization of the Economic Order Quantity (EOQ) between all of the links in the supply chain
C	Optimization of the Economic Order Quantity (EOQ) between selected partners in the supply chain
D	No optimization, flows realized with the application of FOQ aligned to logistics minimum

Own study

Table 9 Financing the operation of the supply chain—description of activities at individual integration levels

Integration level	Performance characteristics
A	Central financing of the supply chain, possibility to grant loans on preferential terms between all of the links in the supply chain, or extending payment dates if necessary
B	Key partners in the supply chain (with the highest turnover) may grant loans for operational activity or extend payment dates between one another, if necessary
C	Selected supply chain partners grant loans for operational activity or extend payment dates between one another, if necessary
D	Autonomous financing of each link in the supply chain

Own study

3.9 Financing the Operation of the Supply Chain

Financing the functioning of the supply chain is a decisive factor for its competitiveness. Cash flows between the links can be uneven, which might trigger the loss of liquidity of some of them (Table 9).

3.10 Stock Management in the Supply Chain

Collaboration in the supply chain in respect of stock management provides ample room for reducing inventory level and related costs. Along with growing integration, the focus of the functioning of the supply chain is shifted from management issues in individual companies onto the allocation of stocks that would be optimum from the point of view of the entire chain (Table 10).

Table 10 Stock management in the supply chain—description of activities at individual integration levels

Integration level	Performance characteristics
A	Centralised stock, real-time flow of information regarding demand
B	Centralised stock, real-time flow of information regarding demand to the closest link from the customer in the supply chain
C	Centralised stock, the supplier is responsible for the stock, no flow of information regarding demand
D	Individual stock replenishment without collaboration with other links

Own study

Table 11 Quality, accuracy and information standards—description of activities at individual levels of integration

Integration level	Performance characteristics
A	Accurate and up-to-date information, continuous and open access to the partner's data, data made available and transferred by way of the information network, information network—configuration of access (who has access), scope (who has access to what data) and frequency as regards data updating, established and observed information standards (form)
B	Accurate and up-to-date information, made available on a regular basis at a higher frequency than in group C and at the request of the partner, established and observed information standard (form)—fixed mode of information transfer (e.g. e-mail, fax), fixed document layout and content, data transferred with the use of modern technological solutions, regularly updated data—fixed data updating/capturing frequency
C	Accurate, but not up-to-date data, made available on a regular basis at the request of the partner, fixed information standard (form), data supplemented irregularly—no fixed data updating/capturing frequency
D	Inaccurate and not up-to-date information, made available at the request of the partner, no set information standard (form)—transferred data can have any form, data updated occasionally—no fixed data updating/capturing frequency

Own study

3.11 Quality, Accuracy and Information Standards

The functioning of the supply chain depends, to a large extent, on the form, quality and availability of the information used for integrating business partners (Table 11).

3.12 Key Areas of Collaboration

The level of integration is affected by the areas in which companies constituting the supply chain cooperate with one another. Undoubtedly the cooperation within a greater number of areas of operation and areas related to the execution of the

Table 12 Key areas of collaboration—description of activities at individual level of integration

Integration level	Performance characteristics
A	Sphere expanded by a developing a joint market strategy as well as R&D
B	Sphere expanded by the execution of processes
C	Sphere expanded by collaboration in stock replenishment
D	Basic sphere involving purchasing and procurement

Own study

Table 13 Collaboration regarding recycling—description of activities at individual integration levels

Integration level	Performance characteristics
A	Joint (for many supply chain links) designing of the product, taking into account the possibility of recycling, swapping environmentally harmful materials to less harmful, namely producing less harmful substances during the reclaiming processes, adapting production processes to such a use of secondary raw materials that would enable the closing of the loop and the execution of recycling within the supply chain, without the involvement of additional supply chain links (e.g. enterprises dealing in the processing of secondary raw materials).
B	Unification of goods and materials, implementing structural changes in goods to facilitate their disassembly.
C	Regular cooperation regarding backward flows related to the use of secondary raw materials of minimum two supply chain links, cooperation between the links on the organization of backward flows, collaboration with suppliers regarding the use materials fit for recycling.
D	Incidental use of secondary raw materials or no use at all.

Own study

company's and the supply chain's strategy is an evidence of a higher level of integration, because it takes a higher level of cooperation, trust and involvement (Table 12).

3.13 Collaboration Regarding Recycling

In the context of the management of secondary resources, reverse supply chains (with backward flows) play major role. They handle the flow of goods and information crucial for the acquisition of products which had been used, packaging or waste generated during manufacturing processes as well as their delivery to the facilities, where they can be reused, processed, subject to recycling or safely disposed of (Table 13).

Table 14 Response to external disruptions of the supply chain—description of activities at individual levels of integration

Integration level	Performance characteristics
A	Joint undertaking of corrective measures aimed at restoring the stability of the supply chain and sharing the responsibility together by all the partners. Common contingency plans keep the supply chain in operation—i.e. help to maintain core business activity—in spite of disruptions. An emergency replacement of a given link with its competitor is possible, although such an action does not result in the change of the chain participants.
B	Disruptions do not spread beyond two links. Contingency plans are in operation, yet they do not cover the full catalogue of disruptions. Increased probability of eliminating the link which was the focal point of disruption.
C	Disruptions do not cover more than half of the supply chain. No procedures for responding to selected disruptions.
D	Emerging disruptions affect the entire supply chain. In the event of disruptions individual links must take corrective measures themselves and there is a risk of being eliminated from the chain can't be ignored.

Own study

3.14 Response to External Disruptions of the Supply Chain

The supply chain's stability in the face of disruptions depends on its level of integration. The higher the integration level, the higher the degree of trust towards business partners and the more equal sharing of risk regarding the provision of a product to the final recipient (Table 14).

3.15 Reduction in Material Losses

An integrated supply chain will minimize these losses, which means it will follow ecodevelopment-oriented ways of operation. The exchange of information with other links and conducting joint research helps to manage and reduce the amount of the waste generated in the course of the production process (Table 15).

3.16 Delivery Performance and the Storage of Secondary Raw Materials

In this context, the selection of the right materials, namely pro-ecological orientation of supply chain links impacts the metrics measuring the degree of integration of the entire chain. Although a given material may be treated as redundant waste for one link, carrying all of the transport and storage-related costs onto the

Table 15 Reduction of material losses—description of activities at individual integration levels

Integration level	Performance characteristics
A	Join investment decision-making regarding research into the ways of minimizing the amount of generated waste and the possibilities of their use, as well as the exchange of information regarding production processes taking into account the quantity and quality of the waste.
B	Join investment decision-making regarding research into the ways of minimizing the amount of generated waste and the possibilities of their use or the exchange of information regarding production processes taking into account the quantity and quality of the waste.
C	Study into the ways of minimizing the amount of generated waste and the possibility of their use in production processes, conducted independently by individual links. No exchange of information regarding production processes from other links.
D	Study into the ways of minimizing the amount of generated waste and the possibilities and the possibilities of their use in production processes are conducted occasionally and independently by individual links, or no such activities are taken at all. Lack of the exchange of data regarding production processes from other links.

Own study

Table 16 Delivery performance and the storage of secondary raw materials—description of activities at individual integration levels

Integration level	Performance characteristics
A	Negotiating the most favourable for the parties (business partners) delivery terms of purchase/sales transaction of raw materials/waste and sharing storage costs (win-win)
B	Negotiating the most favourable for the parties (business partners) delivery terms of purchase/sales transaction of raw materials/waste or sharing storage costs
C	The recipient has a greater bargaining power. The supplier is in charge of transport (performs or commissions the performance of the services) and incurs storage costs until a given resource is in demand by the recipient.
D	The supplier has a greater bargaining power. The recipient is in charge of transport (performs or commissions the performance of the service) and at the time and place set by the supplier, and incurs storage costs.

Own study

link sourcing the secondary raw materials will adversely affect the profitability and the cash flow of the entire family. The higher the integration level, the greater the possibility of negotiating favourable conditions for both parties, which translates into the "*ecologicality*" of the entire chain (Table 16).

Table 17 Description of availability of a secondary raw material—description of activities at individual

Integration level	Performance characteristics
A	Sustainability of the supply of secondary raw materials from the perspective of time and quantity and no need for changing the material adaptation process so that it would meet the requirements of the production process
B	Disrupted sustainability of the supply of secondary raw materials (storage is required), the quality of resources does not disturb the process of adapting the resources to the production process
C	Low sustainability of supply (need for seeking external links—short-term cooperation), the quality of resources may affect the process of adapting the resources to the production process
D	"Occasional" supply, the quality of resources may affect the process of adapting the resources to the production process

Own study

3.17 Description of Availability of Secondary Raw Material

The supply of secondary raw material, namely the ability to satisfy one's own needs within fixed and stable links in the supply chain or an ongoing change of the supply chain links are reflected by the values of integration metrics (Table 17).

3.18 Ability to Handle Returns from the Market and Information Flow on Secondary Raw Materials

Table 18

3.19 Monitoring the Results of the Supply Chain Links

Table 19

Table 18 Ability to handle returns from the market and information flow on secondary raw materials—description of activities at individual integration levels

Integration level	Performance characteristics
A	Forecasting returns from the market and cooperation with business partners regarding waste transport. The exchange of information on projected availability of secondary raw materials between the supply chain partners (long-term plans).
B	Forecasting returns from the market without the cooperation with business partners regarding waste transport. The exchange of information on projected availability of secondary raw materials between some of the supply chain partners (time and amount of resources). Joint planning, short-term plans.
C	No return forecasts (amounts), preparing the ground for cooperation regarding waste transport between the chain partners. Occasional exchange of information between some partners with regard to projected availability of secondary raw materials.
D	No return forecasts (amounts), no cooperation with business partners regarding waste transport. Suppliers are sought when the need arises (very costly order placement procedure).

Own study

Table 19 Monitoring the performance of the supply chain links—description of activities at individual integration levels

Integration level	Performance characteristics
A	An integrated system for monitoring the results of supply chain, compatibility (compliance) of metrics and indicators at the strategic and tactical level and matching them to the character of a given entity's activity at the operational level.
	Fixed standards of exchanging the information on monitored parameters.
	The practice of consulting the managers of key supply chain links about the matters related to the monitoring of the supply chain results.
	Established key procedures regarding co-management and changing the norms and standards.
B	Compatibility of indicators/metrics (in terms of quality and logic) at the tactical and operational level of functioning for key links in the supply chain. Tactical and operational activity of the managers of particular links in the supply chain does not distort the results of the partners or such a situation is not reported by the measurement system.
C	A logically coherent metric pyramid in the system monitoring individual companies.
	Compliance in respect of selected metrics and indicators among the partners in the supply chain.
D	Random metrics at the operational, tactical and strategic level. The lack of relationship between the systems for monitoring the supply chain results. A tendency towards numerous local improvements without the analysis of its impact on the entire supply chain.

Own study

4 Methodology for Determining the Integration Factor for the Supply Chain Under Review

A full integration of all of the links in the supply chain is not possible due to:

- its natural dynamics (variability),
- adopted strategies for realizing particular functions (e.g. short-term agreements regarding the distribution and sales or purchasing according to the "multi-sourcing" strategy).

What is more, it has been assumed in the works on the evaluation model that integration is not an end, but a tool for driving the company's performance metrics. In view of the above one should:

- evaluate the level of integration of the supply chain for a given group of key entities (links),
- **relate the selection of key links in the supply chain with the status of supply item** (category) and the selected strategy for choosing the sources for a single supply item,
- adopted metric for evaluating the importance of a given supply chain link should take into account the status (managerial decision) and value (measured in currency units) of the supply stream,
- evaluation of the level of integration should not determine, but result from the sourcing strategy; it should also provide the information required for taking action at the tactical and operational level.

The methodology for determining integration factor for the supply chain under study involves 4 basic stages: preselection (step 0), analysis (step 1), selection (step 2), evaluation (step 3) (Figs. 2 and 3).

The first stage (step 0) involves preselecting the assortment, which will be taken into account in the process of measuring the supply chain integration. The aim of preselection is to review the position occupied by an item in the classification from the point of view of its relevance (adequacy) in the context of the current situation on the supplier market as well as in terms of its consistency with functional strategies of the enterprise (mainly: B&R, production, logistics and finance). Secondly, items sourced according to the so-called "*multi sourcing*" strategy are being rejected, because enhancing integration with their suppliers is pointless and financially unjustified. Based on the initial preselection of assortment we identify the links providing supply items, whose management strategy implies intensifying the cooperation. In the subsequent step one should determine the volume of the value stream in monetary terms for each represented period of time (period of time depending on market demand characteristics). Nonetheless to measure stream value we should assume the amount of time **longer than stock replenishment cycle in the supply chain—to avoid a situation, in which** a given supply chain link weighs more at the measurement moment due to a block in the flow or a speculative stock accumulation. It is a view taken from the perspective of the value of the material

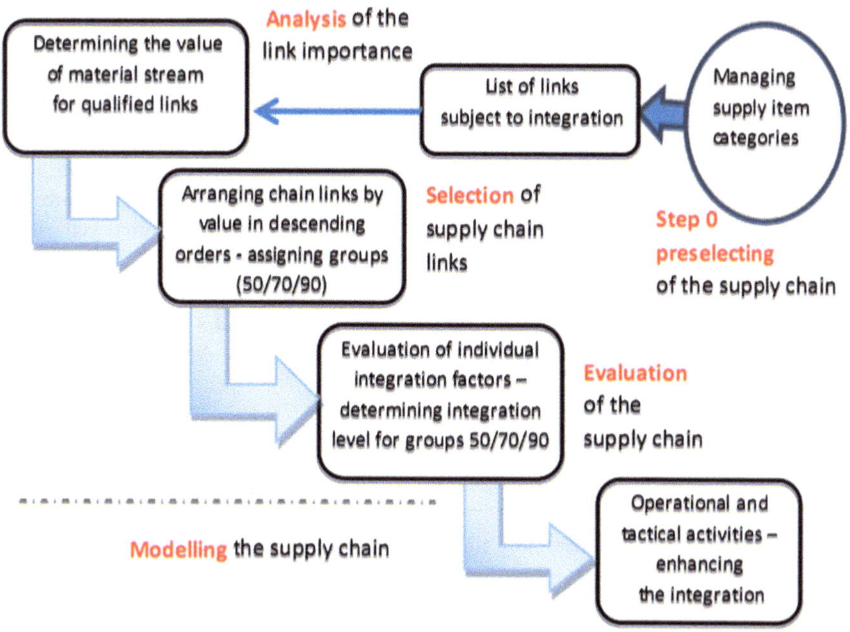

Fig. 2 Framework methodology for determining integration factor in the supply chain under study

stream, which flows through individual links. **Hence, to meet the demand for individual groups, at least the following must be subject to analysis**:

- 50 % of value (for group C) and its representative links,
- 70 % of value (for group B) and its representative links,
- 90 % of value (for group A) and its representative links,
- **in the material flow in the supply chain which had been qualified in the preselection stage**.

Such a measurement method also provides a focus for integration activities. As the value of material streams increases downstream (by adding values), the integration occurring closer to the customer produces better results and is more economically justified (high inventory costs, diversified products) than in the upper part of the stream (unprocessed raw materials).

The next step involves the selection of a group of links 50/70/90, where 50 means 50 % of the value of the stream and its links (arrangement in descending order). In a similar fashion, group 70 consists of the first 50 plus another 20, etc. (Table 20).

The level of integration is scored for each of 50/70/90 categories (see table Integration level matrix—integration value). The assignment to the supply chain integration level depends on the worst result (the so-called "Achilles' heel") for all the factors under study. Thus, an enterprise may reach a score of C70, which

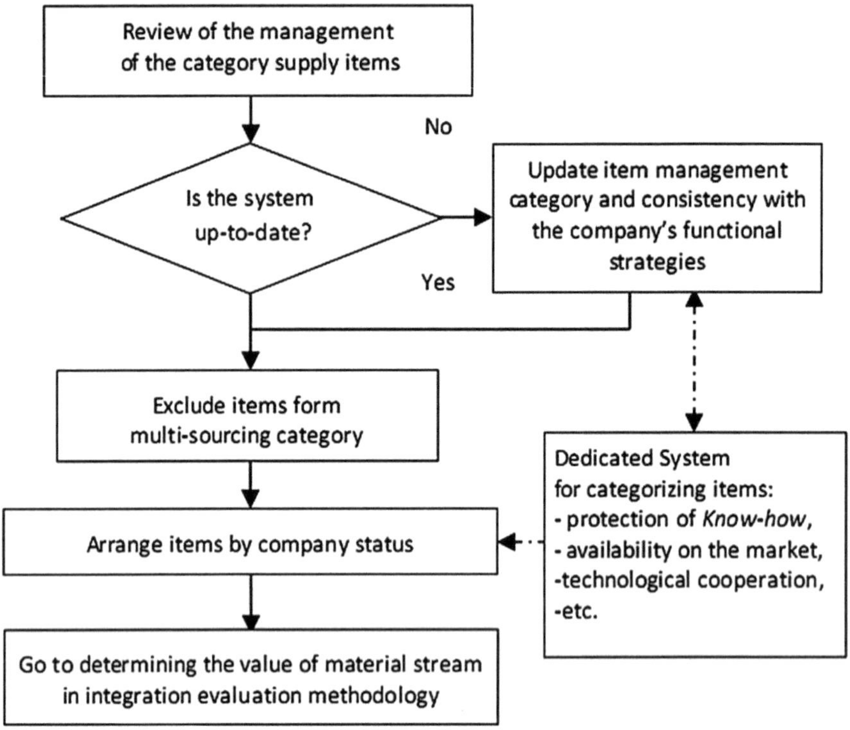

Fig. 3 Algorithm for preselecting supply chain links for the purpose of measuring the level of integration

Table 20 Integration level matrix—integration value

Integration scope	Integration factors							
Stream value	Integration levels	1	2	3	4	5	6 ….	19
A (>90 %)	A			√				√
B (>70 %)	B	√	√		√		√	
C (>50 %)	C					√		
D (<50 %)	D							

Own study

means that at the value level of 70 % the integration level C was achieved, but may equally receive a score of B50 meaning that for the key value of 50 % the integration level was B. It must be, however, the first 50 % from the point of view of the stream value, not any. What is more, the achievement of class A—"*best class*" is **only possible** provided that the group of links under study contains **all** the links with the top item categorization ratio.

5 Conclusion

The methodology for evaluating the level of integration presented in this paper is in the first place a useful tool for supply chain management purposes. The analysis results are applied for modelling, namely process and structural changes of its constituent components. The sequence of activities taken by managers focuses the matrix evaluating the integration score (ABCD) for each category (Sect. 3). In order to improve the operational performance of the supply chain, measures are taken aimed at driving the integration level in the critical (receiving the lowest scores) of 19 categories. The entire method combines into a coherent improvement tool. In the works to follow the Authors intend to:

- Specify the procedure for preselecting supply items and supply chain links based on a multi-criteria analysis (among others functional strategies, the importance of the company's position for technological development and competitiveness, the position value and its market availability),
- Verify simulation study in practice—case studies,
- Expand the scope of integration activities along with the evaluation of their usefulness under specific conditions.

The model developed by the Authors offers a huge potential and opportunities. The validation of the solution as a reference solution in business practice takes time and a broad range of activities—ranging from simulation experiments up to promoting the solution among business practitioners.

Acknowledgments This paper has been the result of the study conducted within the project entitled "Investigating multi-aspect determinants of integration in backward and forward logistics chains in respect of integrated production planning in the context of using recyclable materials" pursued at the Poznan School of Logistics in Poznan. The project was financed by The National Science Center based on the decision number DEC-2011/03/B/HS4/03419.

References

Cyplik P, Hadaś Ł (2011) Production system virus analysis tool (PSVA)—problem identification and analysis framework—case study. LogForum 7(1):1

Gimenez C, Ventura E (2003) Supply chain management as a competitive advantage in the Spanish grocery sector. Int J Logistics Manage 14(1):77–88

Hadaś Ł, Cyplik P (2012) Practice of building production planning system of company with a wide range of products—case study. LogForum 8(3):191–200

Hammer Michael (2001) The superefficient company. Harvard Business Rev 79(8):82–91

Lambert Douglas M (2004) Supply chain management: processes, partnerships, performance. Supply Chain Management Institute, Sarasota

Poirier Ch C (1999) Advanced supply chain management. Berrett-Koehler Publishers, San Francisco, p 23–41

Simichi-Levi D, Kaminsky P, Samichi-Levi E (2000) Designing and managing the supply chain. concepts, strategies and case studies. McGraw-Hill/Irwin, Boston 2000, p 243–244

Stock, James R (2002) Marketing myopia revisited: lessons for logistics. Int J Phys Distrib Logistics Manage 32(½): 12–21

Supply Chain Operations Reference (SCOR®) Model (2010) Overview—Version 10.0, Supply Chain Council. http://supply-chain.org/f/SCOR-Overview-Web.pdf [04.2013]

Green Supplier Selection Criteria: From a Literature Review to a Flexible Framework for Determination of Suitable Criteria

Izabela Ewa Nielsen, Narges Banaeian, Paulina Golińska, Hossein Mobli and Mahmoud Omid

Abstract Green supplier selection (GSS) criteria arise from an organization inclination to respond to any existing trends in environmental issues related to business management and processes, so GSS is integrating environmental thinking into conventional supplier selection. This research is designed to determine prevalent general and environmental supplier selection criteria and develop a framework which can help decision makers to determine and prioritize suitable green supplier selection criteria (general and environmental). In this research we considered several parameters (evaluation objectives) to establish suitable criteria for GSS such as their production type, requirements, policy and objectives instead of applying common criteria. At first a comprehensive and deep review on prevalent and green supplier selection literatures performed. Then several evaluation objectives defined to assess the green supplier selection criteria include: frequency, compatibility, quantifiable, easy to understand and assessment. By developed framework suitable criteria can be selected using multi attribute decision making methods. The main contribution of this research is developing a framework to help managers for creating their green supplier criteria list.

Keywords Green supplier selection criteria · General and environmental criteria · Evaluation objectives · Framework

I. E. Nielsen (✉)
Department of Mechanical and Manufacturing Engineering, Aalborg University, Aalborg, Denmark
e-mail: izabela@m-tech.aau.dk

N. Banaeian · H. Mobli · M. Omid
Department of Agricultural Machinery, University of Tehran, Tehran, Iran

P. Golińska
Department of Production Management and Logistics, Poznań University of Technology, Poznań, Poland

P. Golinska (ed.), *Logistics Operations, Supply Chain Management and Sustainability*, EcoProduction, DOI: 10.1007/978-3-319-07287-6_6,

1 Introduction

Supply chain management (SCM) practices have flourished since the 1990s and are one of the most important competitive strategies used by modern enterprises. The main aim of supply chain management is to integrate various suppliers to satisfy market demand. Meanwhile, supplier selection and evaluation plays an important role in establishing an effective supply chain (Chen 2011). So manufacturers in the late 1980's went beyond what was required in the legislation and went for a greener approach in their operations systems (Testa and Iraldo 2010). Green Supply Chain Management (GSCM) has emerged in the last few years, integrates environmental issues and covers all phases of product's life cycle from design, production and distribution phases to the use of products by the end users and its disposal at the end of product's life cycle (Borade and Banson 2007).

One of the important stages in supply chain management which regards all the activities from the purchasing of raw material to final delivery of the product is the supplier selection process. So the most significant business decision faced by purchasing managers in a supply chain is the selection of best suppliers while trying to satisfy multiple criteria (Arikan 2013). Hence Supplier selection by its nature involves the need to trade-off multiple criteria, as well as the presence of both quantitative and qualitative data.

Most review papers on supplier selection are concentrated on methods. As the last studies, Chai et al. (2013) provided a systematic literature review on articles published from 2008 to 2012 on the application of decision making techniques for supplier selection. They showed that Analytic Hierarchy Process (AHP), Analytic Network process (ANP), Linear programming (LP) and Data Envelopment Analysis (DEA) are the most popular technique for solving supplier selection problems. Also Govindan et al. (2013) reviewed 33 papers from appearance to 2011 on green supplier selection, studied types of criteria and techniques, finally introduced the most popular. The most popular approach in green supplier selection was AHP followed by ANP, DEA too Govindan et al. (2013).

A common area of research in the field of supply chain was establishing the criteria for supplier selection and evaluation since the 1960s. A glance at the history of common criterion shows three primary categories of emphasis: in the late 1970s and early 1980s, cost was the main focus; then in the early 1990s, cycle time and customer responsiveness were considered, and, finally, in the late 1990s, the focus shifted to flexibility. Now, environmental factors are a key issue which gives rise to the new paradigm of focusing on green supply chains (Haung and Keskar 2007).

As Govindan et al. (2013) noted, despite the large number of literature to supplier selection, there is narrow literature on green supplier selection and especially criteria definition. So it is obvious that plenary review about green criteria will be helpful in guiding future researchers. It is an important note that instead of applying common criteria for GSS, better way is to establish suitable criteria which are fitted to kind of production, company policy and other objectives of decision makers, consumers and government.

The main goal of this chapter is developing a framework which can help to select suitable criteria for GSS. Common environmental and general supplier selection criteria presented and classified. After that to select suitable criteria evaluation objectives clarified. At the end of the study, our set of supplier selection criteria including environmental, general and new-defined, assessed by evaluation objectives. Purchase managers can select their appropriate green criteria by developed framework.

2 Green and Conventional Supply Chain

A definition of a supply chain can be proposed as: the global network of players involved in executing the production and distribution of a predefined good from raw materials through to delivery to satisfy a consumer demand (Dornfeld 2013). GSC is similar while managers simultaneously consider environmental impacts and work to reduce those impacts over time. GSCM takes considerations to ecology as well as economy as an objective, while Conventional SCM is usually concentrated on economy as a single objective (Luthra et al. 2011). Some of characteristic and differences between SCM and GSCM are presented in Table 1.

3 Framework for Green Supplier Selection Criteria

First step of SCM implementation includes the purchase, movement and storage of raw materials (Harland 1996). Decision making on supplier selection is based on criteria, so definition and selection of criteria has an important role in your selection. Standard supplier selection criteria (SSC) generally intend to cover issues such as quality, capacity in terms of finance, services, and equipment, quantity, responsiveness, and others. Green supplier selection criteria (GSSC) arise from an organization inclination to respond to any existing trends in environmental issues related to business management and processes. This chapter helps decision maker to determine and prioritize supplier selection criteria in right direction.

Before starting procedure of supplier selection, managers should pay attention to criteria which decision making is based on it. One form of criteria definition could be to ask managers to indicate the criteria they use for supplier selection and makes the suppliers different or to indicate in how far they agree with the above developed criteria.

Often using relevant criteria is not enough for supplier selection. Nowadays competitive and high technology market has reduced differences between suppliers in several aspects like price, quality and etc. so importance of some criteria is going to be unnoticeable.

As noted before, a large number of criteria have overlapping meaning so a summarized set of relevant general and environmental criteria for GSS should be

Table 1 Characteristics of conventional and green supply chain management (Luthra et al. 2011)

S. no.	Characteristics	Green supply chain management	Conventional supply chain management
1	Objectives	Ecological and economic	Economic
2	Ecological optimization	High ecological impacts	Integrated approach Low ecological impacts
3	Supplier selection criteria	Eco logical aspects long term relationship	Price switching suppliers Quickly short term relationship
4	Cost pressure	High	Low
5	Flexibility	Low	High
6	Speed	Low	High

collected. Also sometimes based on new condition of market and kind of products, managers need to define new criteria. So for more precise supplier selection, it is better to omit some of traditional criteria and add new ones which have fitted to market structure and kind of production. Therefore it is necessary to design a framework to select and prioritize updatable and suitable criteria.

Supplier selection process is shown in Fig. 1. Because of adequate literature on selection methods, the detailed focus is on supplier selection criteria process (main parts of step 1 and 2). This chapter proposed a framework to provide exact and suitable criteria for green supplier selection, explanation about step 1 and 2 are presented comprehensively.

Step 1

3.1 Literature on General Criteria

Many previous studies on supplier selection and evaluation defined numerous evaluation criteria and selection frameworks for supplier selection. In consideration of the criteria for supplier selection, the pathfinder work by Dickson (1966) has been one of the most cited studies.

Weber et al. (1991) re-examined Dickson's work by reviewing published articles between 1966 and 1990. Twenty three distinct criteria are identified in various supplier selection problems in Dickson (1966) and Weber et al. (1991). Among the 23 criteria identified for vendor selection, the product quality was ranked as most important, it was followed by on-time delivery, performance history of supplier and warranties and claimed policies, and so on.

In a later work, Cheraghi et al. (2004) continued to extend these key players initial work to obtain the current perspective of supplier selection by analyzing articles published between 1990 and 2001. They (Cheraghi et al. 2004) also provided an update of Dickson's seminal work with 13 more criteria. Next Thiruchelvam and Tookey (2011) reviewed supplier selection studies and comprise the selection criteria between two period of 1966–2001 and 2001–2010.

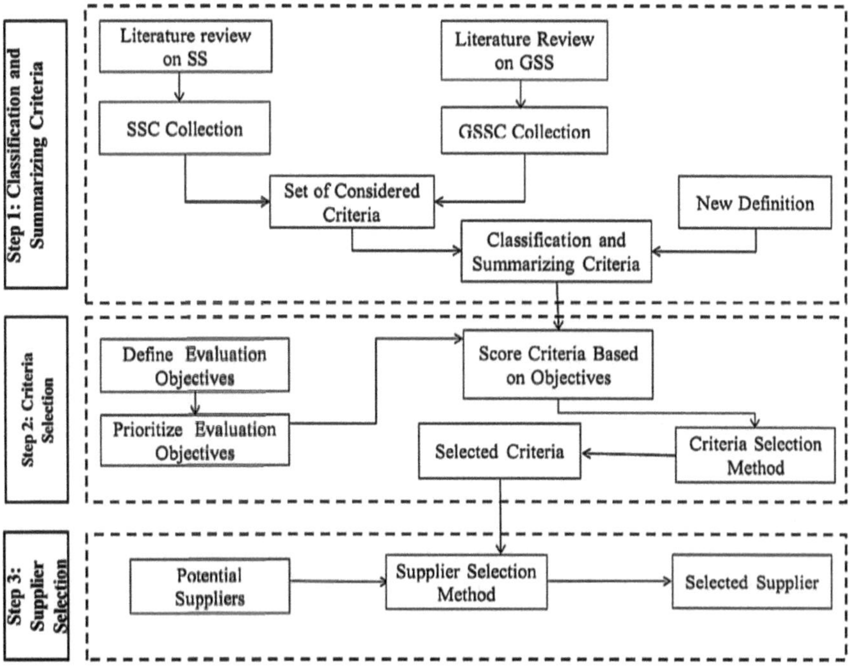

Fig. 1 Framework to determine green supplier selection criteria

Appendix 1 shows summary results of review chapter on most important general criteria on supplier selection.

Chang et al. (2011) used DEMATEL to find influential factors in selecting SCM suppliers and found that "technology ability", "stable delivery of goods", "lead-time", and "production capability" criteria are more influential than other evaluation criteria. They also suggested extending the scope of this study and exploring the addition of a green supply chain in future studies. Often each of these important criteria contains several sub criteria which is described completely in past literature (Kannan et al. 2013).

3.2 Literature on Environmental Criteria

Purchasing function in relation with other functions has a greater role to play in environmental management performance of an organization. Zhu et al. (2010) discussed that purchasers are key personnel for ensuring environmental preferable decisions in supplier selection and that they are best placed and best qualified to adopt a more environmental friendly purchasing practice.

Having seen the significance of collaborative supply chain and centralized purchasing in environmental management performance it is now obvious that deciding which suppliers to collaborate with and how to select suppliers is a very crucial decision for the organization performance.

The literature is comprehensively collected from chapters associated with the descriptors “Green/Environmental supplier selection”, “vendor selection”, from academic databases including Science Direct, Emerald, Springer-Link Journals, IEEE Xplore, Academic Search Premier, and World Scientific Net. After a methodological decision analysis of all collected articles, 57 international journal articles published up to now (Appendix 2) are reviewed. This review of literature provides the basis for identification of supplier selection criteria. Figure 2 shows the growth of green supplier selection form the appearance to 2013.

Based on reviewing chapters about 90 criteria were purposed (Appendix 2). A very large set of criteria, some of which overlap, are utilize in various works (Govindan et al. 2013). Sometimes authors used the same criteria but in a different statement; these kind of dispersion in criteria, make researcher ambiguous to select criteria. Unfortunately lack of concordant studies about basic green criteria caused widespread number of green criteria in each study. Nonetheless evolving criteria have emerged due to competitive market as number of suppliers has increased over the years. Therefore, inception of new criteria is essential to narrow and qualify not only deserving but exceptional suppliers. Under the current rapid GSCM transformation edge, reliability and flexibility of each supplier are considered as key contributing factors.

Environmental supplier selection criteria may be developed with intent of focusing on meeting government regulations, focusing on nature of product, focusing on process improvement, and focusing on buying company’s environmental policy. Therefore in real condition offering an exact framework for supplier selection is impossible. Tradeoff between kinds of criteria is commonplace in decision making that need careful integration. Some of authors classified the Environmental criteria, which can help in the better perception of them. Supplier selection decision making involves multi-criteria comprising both qualitative and quantitative criteria, several authors (Humphreys et al. 2003b; Kannan et al. 2014; Zhu et al. 2010) categorize the environmental criteria into two groups of Qualitative and Quantitative Criteria. Depending on whether an organization is using a reactive or proactive environmental management strategy, one or both groups of criteria may be used at the same time.

A potential supplier may incur costs investing in environmental management of its processes or it may be a source of environmental costs because of its destructive processes. Sometimes manager consider waste rate, non-appropriate by-products, emissions and natural resource depletion obtained by an inventory life cycle of suppliers. Often quantitative criteria are based on the cost in monetary (economic) terms (A), waste or by-product (gas emission, liquid and solid waste) (B) and energy consumption (C). More details about quantitative environmental criteria are represented in Table 2.

A big part of green criteria is qualitative which focuses on more intangible (e.g. company image and reputation) have been used for supplier selection. Qualitative criteria are more subjective and their application depends on the weight given to each one depending on its importance to the organization or industry and total points score obtained on the bases of the measured parameters. Also, the

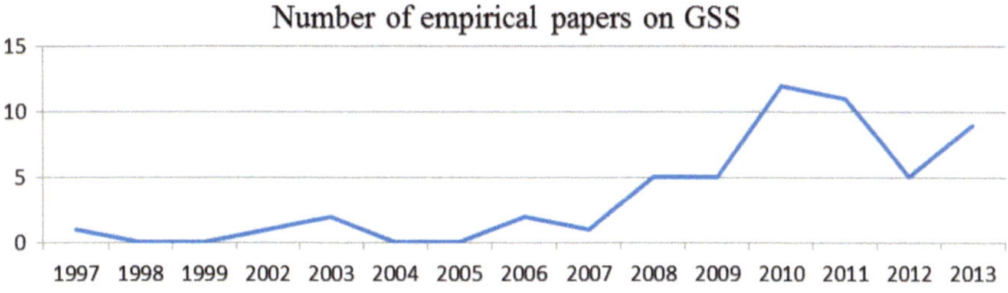

Fig. 2 Trend of green supplier selection chapters

weightings of supplier evaluating criteria depend on business priorities and strategies. Some qualitative criteria are:

- Management competences: Senior management support, Environmental partners, Training, Information exchange, reputation
- Green image: Customer's purchasing retention, Green market share, Stakeholders relationship
- Design for Environment: Recycle, Reuse, Remanufacture, Disassembly, Disposal
- Environment Management Systems: Environmental certification (ISO 14000, Eco-label and etc.), Implement and Operation, Environmental planning, Environmental policies
- Environment competencies: Clean technology availability, Use of environmental friendly material, Pollution reduction capability, Returns handling capability

Also Wong et al. (2012) categorize the green criteria into three groups of design, production and end of life management phases (Fig. 3).

3.3 Summary of Literature Review

Now Based on literature review (Appendix 1 and 2), presented criteria is determined and relevant supplier selection criteria are demonstrated in Table 3. Accordingly on general criteria, most of authors emphasized on quality, delivery and price; although service and management are important as well. To summarize from reviewing 57 GSS articles, the most important environmental criteria are: Environmental Management System (EMS) (20 frequencies), green image (8 frequencies), environmental competences (6 frequencies), design for environment (5 frequencies), environmental improvement costs (5 frequencies) and environmental performance (5 frequencies). 35.08 % of authors used EMS as a criterion for supplier selection process.

Green supplier selection is considering environmental issues in conventional supplier selection process, so decision makers can use a collection of general and environmental criteria and add new criteria based on company policies, finally set summarized potential criteria. Final criteria will be selected based on evaluation objectives.

Table 2 Different type of quantitative environmental criteria

Quantitative environmental criteria		
A Monetary (economic) terms	Pollutant costs/effects	Environmental costs caused by a potential supplier
	Improvement cost	The degree of commitment the supplier has in environmental management
B Waste or by-product	Greenhouse gases emission (carbon footprint)	CO_2 & CH_4, NO, HFCs, PFCs, and SF_6 which are unified into CO_2 equivalent
	Air pollution	NO_x, SO_x, and VOCs
	Waterborne pollution	COD, BOD, ammonia, sulfur dioxide, and nitrogen
	Solid waste	Represent the waste that is not appropriate to be recycled and disposal is the only treatment method
C Natural resource depletion	Land	
	Abiotic	Minerals, fossil fuels, water
	Biotic	Plant, fish, game

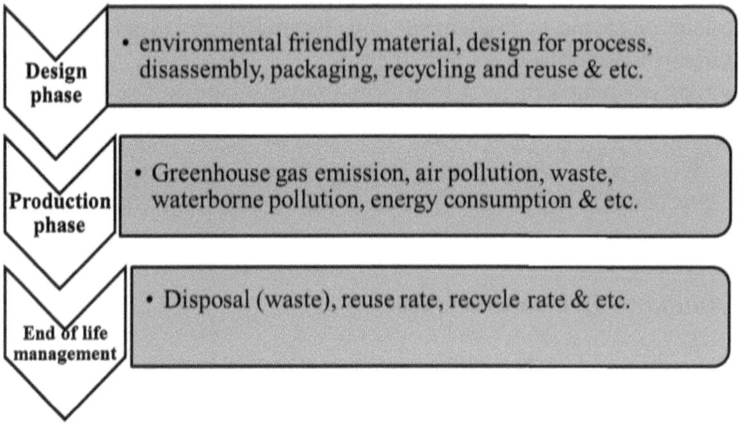

Fig. 3 Classification based on different phase of production

Table 3 Summary of common general and environmental supplier selection criteria

Relevant supplier selection criteria	
General	Quality, delivery, price, production facilities (manufacturing or technical capability) and capacity, service (repair service), management and organization, financial position, geographical location
Environmental	EMS, green image, environmental (green) competences, design for environment, environmental improvement costs, environmental performance, measure of carbon management, environment protection, hazardous substance management, pollution control

Step 2

3.4 Definition of Evaluation Objectives and Criteria Selection

Reviewing literature showed that most of supplier selection criteria are selected based on relevancy (Govindan et al. 2013; Weber et al. 1991) which is important but not adequate, because variation of products, industries and etc. will effect on some criteria. For example often all suppliers offer similar price, this implies that despite of price criterion relevancy, it is no longer relevant to be used as a basis in selecting suppliers (Magdalena 2012).

As noted before, some concepts of GSCM are ambiguous and haven't clear definition. Also most of environmental criteria are termed as qualitative and they required subjective decisions to be made during their evaluation. When decision makers are asked to identify their methods of selecting a supplier, their responses are linguistic and inexact. Due to the nature of this unquantifiable and/or incomplete information, degree of precision is rather low. Hence, recently integrated models by Fuzzy set theory, Gray Relation Analysis and etc. are common (Shen et al. 2013; Lu et al. 2007; Grisi et al. 2010; Yang and Wu 2008; Buyukozkan and Cifci 2011; Chen 2010). The strength of these kind of models are that despite the vagueness of experts' opinions in the evaluation process, the models are easy to apply (Govindan et al. 2013). So it is obvious that using easily understanding and quantitative criteria are more tangible for supplier selection and lead to obtaining high accuracy results.

Sometimes professional characteristic based on production type, changing consumer vision, special social obligation, sanctions and political limitations and many other reasons can effect on criteria selection. Considering these situations, some criteria are easily understand and quantifiable, but the measurement process is difficult, takes time or need huge amount of cost which make it rather impossible to consider them. Therefore it's better to consider criteria which are easily assessed. So researchers have to ask themselves these questions:

1. How many is the frequency of criterion in literatures? (Number of references to criterion and frequency)
2. How compatible is criterion with production type?
3. How compatible is criterion with external policies (producer, consumer and government)?
4. How much intelligible and easily understanding is criterion concept?
5. Is criterion quantifiable or measureable?
6. How much easily assessment is criterion?

Generally manager prefers to apply relevant, compatible, intelligible, quantifiable and easily assessment criteria for decision making. Also evaluation objectives can prioritize, if there are significant differences between evaluation objectives. After definition of evaluation objectives, criteria can be assessed and the best ones selected using evaluation objectives.

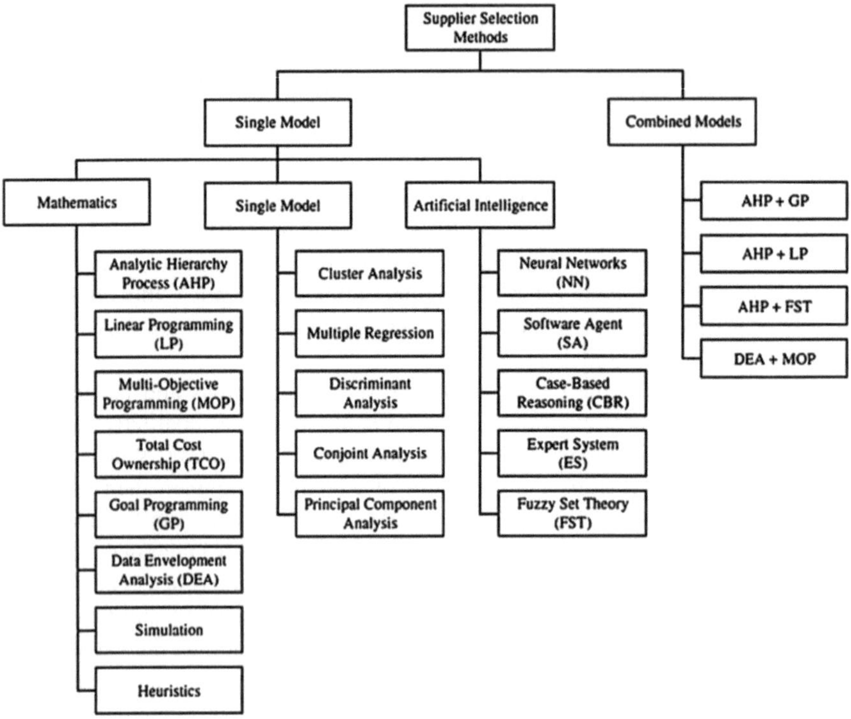

Fig. 4 Existing analytical methods for supplier selection (adopted from Chen 2011)

Step 3

3.5 Supplier Selection

Fundamentally supplier selection is divided in two kind of single sourcing and multiple sourcing. In single sourcing, one supplier can satisfy all the buyer's requirements but in multiple sourcing company need to select a collection of selected suppliers and their allocations. Ranking methods are usually applied to single source models, but in multiple source models mathematical programming models are developed.

Also supplier selection methods can divide to single and combined models as illustrated in Fig. 4 (Chen 2011). Recently because of high flexibility in combined models, they are more developed. Govindan et al. (2013) studied decision making methodologies on GSS in two groups of individual and integrated methodology approach. Chai et al. (2013) reviewed decision making techniques of supplier selection in three main group of: (1) Multi-criteria decision making (MCDM) techniques (2) Mathematical programming (MP) techniques, and (3) Artificial intelligence (AI) techniques.

Typically, when organizations seek to develop or choose a supplier evaluation and selection method, the organization's specific requirements are introduced. Therefore a choice from a range of different selection methods with different applications is needed and based on company requirements, decision makers (purchase managers) decide to use which kind of supplier selection and methods.

4 Conclusion

The green supplier selection problem and suitable criteria selection are very important for operation of all companies. Solving this problem can directly affect the prosperity of them. From a systematic perspective, study on green supplier selection problem can be separated into two parts, i.e. criteria and methods. However many chapters have been conducted to determine what criteria and methods should be used to evaluate supplier, but there is an absence of information for giving a systematic framework to the managers and help them to specify criteria for green supplier selection, in fluctuated condition or product. This chapter focused on presenting a step by step framework to determine suitable criteria for green supplier selection.

A comprehensive literature review on general and environmental supplier selection performed , relevant criteria introduced and merged together. Evaluation objectives defined to assess criteria and select the best ones. Condition of company, suppliers and competitive market affects the policy purchase, and eventually will affect the evaluation objectives. For example more intelligible, compatible, quantifiable and easily assessment criteria are better for green supplier selection.

Several methods based on problem nature (number of evaluation objectives, number of criteria, kind of data …) can be applied in criteria selection. Simplest way to prioritize evaluation objectives and criteria is using an expert team, Delphi method or pairwise comparison. These methods can determine the weight of each evaluation objectives. Linear programming can be used in criteria selection process. Finally the preferences of studied company can determine the criteria selection method.

Analysis of criteria for finding specific green supplier selection criteria in different industries is an important area. Therefore in future studies the proposed framework can be applied either in special products or different industries.

Appendix 1: Frequency of general supplier selection criteria based on review chapters

No.	Criterion	Dickson (1966)	Weber et al. (1991) (1966–1990)	Cheraghi et al. (2004) (1966–2001)	Ho et al. (2010) (2000–2008)	Thiruchelvam and Tookey (2011) (1966–2010)
		Mean rating	Number of articles	Number of articles	Number of articles	Number of articles
1	Quality	3.508	40	71	68	108
2	Delivery	3.417	44	75	64	111
3	Performance history	2.998	7	11	NA	21
4	Warranties and claim policies	2.849	0	1	NA	6
5	Production facilities and capacity	2.775	23	35	NA	55
6	Price	2.758	61	81	63	118
7	Technical capability	2.545	15	30	25	54
8	Financial position	2.514	7	15	23	32
9	Procedural compliance	2.488	2	4	NA	4
10	Communication system	2.426	2	7	NA	14
11	Reputation and its position in industry	2.412	8	10	3	18
12	Desire for business	2.256	1	2	NA	4
13	Management and organization	2.216	10	17	25	39
14	Operating controls	2.211	3	5	NA	5
15	Repair service	2.187	7	18	NA	29
16	Attitude	2.120	6	14	NA	20
17	Impression	2.054	2	6	NA	10
18	Packaging ability	2.009	3	5	NA	9
19	Labor relation record	2.003	2	4	NA	10
20	Geographical location	1.872	16	17	NA	29

(continued)

(continued)

No.	Criterion	Dickson (1966)	Weber et al. (1991) (1966–1990)	Cheraghi et al. (2004) (1966–2001)	Ho et al. (2010) (2000–2008)	Thiruchelvam and Tookey (2011) (1966–2010)
		Mean rating	Number of articles	Number of articles	Number of articles	Number of articles
21	Amount of past business	1.597	1	1	NA	3
22	Training aids	1.537	2	3	NA	3
23	Reciprocal arrangements	0.610	2	5	NA	5
24	Manufacturing capability	NA	NA	NA	39	NA
25	Service	NA	NA	NA	35	NA
26	Research and development	NA	NA	NA	24	NA
27	Flexibility	NA	NA	NA	18	19
28	Risk	NA	NA	NA	3	NA
29	Safety and environment	NA	NA	NA	3	NA
30	Reliability	NA	NA	NA	NA	11
31	Process improvement	NA	NA	NA	NA	12
32	Product development	NA	NA	NA	NA	19
33	Environmental and social responsibility	NA	NA	NA	NA	9
34	Occupational safety and health	NA	NA	NA	NA	4
35	Integrity	NA	NA	NA	NA	5
36	Professionalism	NA	NA	NA	NA	4
37	JIT	NA	NA	NA	NA	5
38	Commitment	NA	NA	NA	NA	9
39	Economy situation	NA	NA	NA	NA	1
40	Long-term relationship	NA	NA	NA	NA	4
41	Political situation	NA	NA	NA	NA	2

Appendix 2: Frequency of environmental criteria based on reviewing GSS chapters

No.	Environmental criterion	Govindan et al. (2013) (1996–2011)	Frequency (current research)	References
1	Adherence to environmental policies	1	1	Awasthi et al. (2010)
2	Air emission	NA	1	Chiou et al. (2011)
3	Air pollution treatment cost	1	1	Yeh and Chuang (2011)
4	Alternative green supply chain systems, project, practices, etc.	2	2	Buyukozkan and Cifci (2011, 2012)
5	Cleaner production	1	1	Chen et al. (2010)
6	CO_2 management competencies	NA	1	Theiben and Spinler (2014)
7	Carbon risk assessment	NA	2	(Hsu and Hu 2009; Hsu et al. 2011)
8	Carbon policy	NA	2	(Hashemi et al. 2013; Hsu and Hu 2009)
9	Carbon accounting and inventory	NA	3	(Hashemi et al. 2013; Hsu and Hu 2009; Hsu et al. 2011)
10	Carbon reduction targets	NA	2	(Hsu and Hu 2009; Hsu et al. 2011)
11	Carbon governance	NA	3	(Hashemi et al. 2013; Hsu and Hu 2009; Hsu et al. 2011)
12	Carbon disclosure and report	NA	2	(Hsu and Hu 2009; Hsu et al. 2011)
13	Carbon emission verification	NA	3	(Hashemi et al. 2013; Hsu and Hu 2009; Zhang et al. 2013)
14	Carbon policy	NA	1	Hsu et al. (2011)
15	Carbon footprint	1	2	(Kumar and Jain 2010; Kumar et al. 2014)
16	Commitment of GSCM from managers	NA	1	Shen et al. (2013)
17	Current environmental efficiency	2	2	(Noci 1997; Grisi et al. 2010)
18	Design for environment	3	5	(Awasthi et al. 2010; Humphreys et al. 2006; Humphreys et al. 2003a, b; Torng and Tseng 2013)
19	Eco-design	1	2	(Hong-Jun and Bin 2010; Shen et al. 2013)

(continued)

(continued)

No.	Environmental criterion	Govindan et al. (2013) (1996–2011)	Frequency (current research)	References
20	Environmental and legislative management	NA	1	Tuzkaya et al. (2009)
21	Environmental management system	11	20	(Amindoust et al. 2012; Azadnia et al. 2012; Chen et al. 2010; Chiou et al. 2008, 2011; Grisi et al. 2010; Hsu et al. 2013; Humphreys et al. 2006, 2003a, b; Kannan et al. 2013; Kuo et al. 2010; Lee et al. 2009; Li and Zhao 2009; Shen et al. 2013; Thangchattu and Siripokapirom 2010; Torng and Tseng 2013; Wittstruck and Teueberg 2011; Yan 2009; Zhang et al. 2003)
22	Environmental policies	1	1	Awasthi et al. (2010)
23	Environment protection	1	4	(Hsu et al. 2011; Kannan et al. 2013; Ma and Liu 2011; Yang and Wu 2008)
24	Environmental authentication	2	2	(Awasthi et al. 2010; Hong-Jun and Bin 2010)
25	Environmental commitment	1	1	Large and Thomsen (2011)
26	Environmental competences	3	6	(Amindoust et al. 2012; Grisi et al. 2010; Humphreys et al. 2006, 2003b; Kannan et al. 2014)
27	Environmental criteria	1	1	Lu et al. (2007)
28	Environmental impacts	NA	1	Boosothonsatit et al. (2012)
29	Environmental issues	1	1	Kuo et al. (2010)
30	Environmental performance	3	5	(Chiou et al. 2008, 2011; Feyziogelu and Büyüközkan 2010; Large and Thomsen 2011; Theiben and Spinler 2014)
31	Environmental Programs at the supplier's facilities	1	1	Handfield et al. (2002)
32	Environmental friendly product design	NA	1	Azadina et al. (2012)
33	Environmental permits and reporting	NA	1	Chiou et al. (2011)
34	Environmental friendship	NA	1	Ying-tuo Yang (2011)
35	Environmental improvement costs	2	5	(Govindan et al. 2013; Humphreys et al. 2006, 2003a, b; Tuzkaya et al. 2009)

(continued)

(continued)

No.	Environmental criterion	Govindan et al. (2013) (1996–2011)	Frequency (current research)	References
36	Greenhouse gas emissions	NA	1	Shaw et al. (2012)
37	Green product performance	1	1	Wen and Chi (2010)
38	Green collaboration with suppliers	1	1	Large and Thomsen (2011)
39	Green competencies	3	4	(Büyüközkan and Cifci 2010; Chiou et al. 2008; Lee et al. 2009; Noci 1997)
40	Green design	1	3	(Chen et al. 2010; Hsu and Hu 2008; Lee et al. 2009)
41	Green image	4	8	(Cheraghi et al. 2004; Grisi et al. 2010; Humphreys et al. 2006, 2003a; Kannan et al. 2013; Lee et al. 2009; Shen et al. 2013; Tuzkaya et al. 2009; Wen and Chi 2010)
42	Green knowledge transfer	1	1	Bai and Sarkis (2010)
43	Green logistics dimension	2	3	Buyukozkan and Cifci (2011, 2012, 2010)
44	Green management system	1	1	Wen and Chi (2010)
45	Green managerial innovation	1	2	(Awasthi et al. 2010; Chiou et al. 2011)
46	Green process management	NA	1	Tuzkaya et al. (2009)
47	Green market share	1	1	Awasthi et al. (2010)
48	Green process innovation	NA	1	Chiou et al. (2011)
49	Green organizational activities dimension	2	3	Buyukozkan and Cifci (2011, 2012, 2010)
50	Green product	1	2	(Kannan et al. 2013; Lee et al. 2009)
51	Green innovation	NA	1	Kannan et al. (2013)
52	Green product innovation	1	1	Chiou et al. (2011)
53	Green projects partnership	1	1	Vachon and Klassen (2006)
54	Green purchasing	1	1	Chen et al. (2010)
55	Green R&D projects	1	1	Awasthi et al. (2010)
56	Green supplier assessment	1	1	Large and Thomsen (2011)
57	Greening the supplier	1	1	Chiou et al. (2011)

(continued)

(continued)

No.	Environmental criterion	Govindan et al. (2013) (1996–2011)	Frequency (current research)	References
58	Hazardous substance management	NA	4	(Chiou et al. 2011; Kannan et al. 2013; Torng and Tseng 2013; Zhang et al. 2013)
59	Internal green production plan	1	1	Chen et al. (2010)
60	Involvement in initiatives for carbon management	NA	2	(Hsu and Hu 2009; Hsu et al. 2011)
61	ISO 14000	1	2	(Chen et al. 2010; Thangchattu and Siripokapirom 2010)
62	Labeling/certification	1	1	Handfield et al. (2002)
63	Life cycle assessment	1	1	Chen et al. (2010)
64	Life cycle management	NA	2	(Hsu and Hu 2009; Hsu et al. 2011)
65	Management systems of carbon information	NA	1	Hsu and Hu (2009)
66	Measure of carbon management	NA	4	(Chiou et al. 2011; Hashemi et al. 2013; Hsu and Hu 2009; Hsu et al. 2011)
67	Outside environmental management	1	1	Hong-Jun and Bin (2010)
68	Packaging/reverse logistic	1	1	Handfield et al. (2002)
69	Partnership with green organization	1	1	Awasthi et al. (2010)
70	Pollution production	NA	1	(Shen et al. 2013)
71	Pollution	NA	1	Azadina et al. (2012)
72	Pollution preventation	NA	1	Chiou et al. (2011)
73	Pollution control	1	4	(Kannan et al. 2013; Lee et al. 2009; Tuzkaya et al. 2009; Yan 2009)
74	Pollution Control initiatives	1	1	Awasthi et al. (2010)
75	Purchasing's environmental capabilities	1	1	Large and Thomsen (2011)
76	R&D green product	1	1	Chen et al. (2010)
77	Resource reduction	NA	2	(Chiou et al. 2011; Torng and Tseng 2013)
78	Resource consumption	NA	2	(Shen et al. 2013; Torng and Tseng 2013)
79	Recycle	1	2	(Bala et al. 2008; Wittstruck and Teueberg 2011)
80	Reuse	1	1	Yan (2009)
81	Social responsibility and environmental competencies	1	2	Buyukozkan and Cifci (2011, 2012)

(continued)

(continued)

No.	Environmental criterion	Govindan et al. (2013) (1996–2011)	Frequency (current research)	References
82	Staff environmental training	1	2	(Awasthi et al. 2010; Shen et al. 2013)
83	Supplier's green image	2	2	(Noci 1997; Wen and Chi 2010)
84	Training related to carbon management	NA	2	(Hsu and Hu 2009; Hsu et al. 2011)
85	Total product life cycle cost	NA	1	Lee et al. (2009)
86	Distance	NA	1	Kumar et al. (2014)
87	Transportation time	NA	1	Yeh and Chuang (2011)
88	Use of environmental friendly material	NA	2	(Awasthi et al. 2010; Shen et al. 2013)
89	Use environment friendly technology	NA	2	(Awasthi et al. 2010; Shen et al. 2013)
90	Waste management	NA	1	Handfield et al. (2002)
91	Waste water and solid waste	NA	1	Chiou et al. (2011)
92	Waste recovery disposal	NA	1	Yan (2009)

References

Amindoust A et al (2012) Sustainable supplier selection: a ranking model based on fuzzy inference. Appl Soft Comput 12:1668–1677

Arikan F (2013) An interactive solution approach for multiple objective supplier selection problem with fuzzy parameters. J Intell Manuf. doi: 10.1007/s10845-013-0782-6

Awasthi A et al (2010) A fuzzy multicriteria approach for evaluating environmental performance of suppliers. Int J Prod Econ 126:370–378

Azadnia AH et al (2012) Sustainable supplier selection based on self-organizing mapneural network and multi criteria decision making. In: international congress on interdisciplinary business and social science 2012 (ICIBSoS 2012). Procedia Soc Behav Sci (65): 879–884

Bai C, Sarkis J (2010) Green supplier development: analytical evaluation using rough set theory. J Clean Prod 18:1200–1210

Bala A et al (2008) Experiences with greening suppliers. The Universitat Autònoma de Barcelona. J Clean Prod 16(15):1610–1619

Boosothonsatit K et al (2012) A generic simulation model for green supplier selection. In: Leveraging technology for a sustainable world: proceedings of the 19th CIRP 23–35 May, pp 587–592

Borade AB, Banson SV (2007) Domain of supply chain management-a state of art. J Technol Manage Innov 12:109–121

Buyukozkan G, Cifci G (2011) A novel fuzzy multi-criteria decision framework for sustainable supplier selection with incomplete information. Comput Ind 62:164–174

Buyukozkan G, Cifci G (2012) A novel hybrid MCDM approach based on fuzzy DEMATEL, fuzzy ANP and fuzzy TOPSIS to evaluate green suppliers. Expert Syst Appl 39:3000–3011

Büyüközkan G, Cifci G (2010) Evaluation of the green supply chain management practices: a fuzzy ANP approach. Prod Plann Control iFirst 1–14

Chai J et al (2013) Application of decision-making techniques in supplier selection: a systematic review of literature. Expert Syst Appl 40:3872–3885
Chang B et al (2011) Fuzzy DEMATEL method for developing supplier selection criteria. Expert Syst Appl 38(3):1850–1858
Chen CC et al (2010) Implementation of green supply chain management in uncertainty. In: IEEE international conference on IEEM 7–10 Dec, pp 260–264
Chen Y-J (2011) Structured methodology for supplier selection and evaluation in a supply chain. Inf Sci 181:1651–1670
Cheraghi SH et al (2004) Critical success factors for supplier selection: an update. J Appl Bus Res 20(2):91–108
Chiou CY et al (2008) Comparative investigation on green supplier selection of the American, Japanese and Taiwanese electronics industry in China. In: IEEE international conference on IE&EM 8–11 Dec, pp 1909–1914
Chiou TY et al (2011) The influence of greening the suppliers and green innovation on environmental performance and competitive advantage in Taiwan. Transp Res Part E 47: 822–836
Dickson GW (1966) An analysis of vendor selection system and decisions. J Purchasing 2(1):5–17
Dornfeld DA (2013) Green manufacturing: fundamentals and applications. pp 86
Feyziogelu O, Büyüközkan G (2010) Evaluation of green suppliers considering decision criteria dependencies. In: Lecture notes in economics and mathematical systems 1, 634. Multiple criteria decision making for sustainable energy and transportation systems, part 2: pp 145–154
Govindan K et al (2013). Multi criteria decision making approaches for green supplier evaluation and selection: a literature review. J Clean Prod XXX: 1–18
Grisi RM et al (2010). Supplier performance evaluation for green supply chain management. Bus Perform Meas Manage Part 4: 149–163
Handfield R et al (2002). Applying environmental criteria to supplier assessment: a study in the application of the analytical hierarchy process. Eur J Oper Res 141:70–87
Harland CM (1996) Supply chain management, purchasing and supply management, logistics, vertical integration, materials management and supply chain dynamics. In: Slack N (ed) Blackwell encyclopedic dictionary of operations management. Blackwell, UK
Hashemi SH et al (2013) A grey-based carbon management model for green supplier selection. In: IEEE international conference 15–17 Nov, pp 402–405
Haung SH, Keskar H (2007) Comprehensive and configurable metrics for supplier selection. Int J Prod Econ 105:510–523
Ho W et al (2010) Multi-criteria decision making approaches for supplier evaluation and selection: a literature review. Eur J Oper Res 202(1):16–24
Hong-Jun L, Bin L (2010) A research on supplier assessment indices system of green purchasing. In: IEEE international conference on ICEE 13–14 March, pp 314–317
Hsu C-W et al (2013) Using DEMATEL to develop a carbon management model of supplier selection in green supply chain management. J Clean Prod 56:164–172
Hsu CW, Hu AH (2008) Green supply chain management in the electronic industry. Int J Environ Sci Technol 5(2):205–216
Hsu CW Hu AH (2009) Applying hazardous substance management to supplier selection using analytic network process. J Clean Prod 17:255–264
Hsu CW et al (2011) Using DEMATEL to develop a carbon management model of supplier selection in green supply chain management. J Clean Prod 56:164–172
Humphreys P et al (2006) Employing dynamic fuzzy membership functions to assess environmental performance in the supplier selection process. Int J Prod Res 44(12):2379–2419
Humphreys P et al (2003a) Using case-based reasoning to evaluate supplier environmental management performance. Expert Syst Appl 25:141–153
Humphreys PK et al (2003b) Integrating environmental criteria into the supplier selection process. J Mater Process Technol 138(1–3):349–356

Kannan D et al (2014) Fuzzy axiomatic design approach based green supplier selection: a case study from Singapore. J Clean Prod. http://dx.doi.org/10.1016/j.jclepro.2013.12.076
Kannan D et al (2013) Integrated fuzzy multi criteria decision making method and multi-objective programming approach for supplier selection and order allocation in a green supply chain. J Clean Prod 47:355–367
Kumar A, Jain V (2010) Supplier selection: a green approach with carbon footprint monitoring. In: IEEE international conference on SCMIS pp 1–9, 6–9 Oct 2010
Kumar A et al (2014) A comprehensive environment friendly approach for supplier selection. Omega 42:109–123
Kuo RJ et al (2010) Integration of artificial neural network and MADA methods for green supplier selection. J Clean Prod 18(12):1161–1170
Large RO, Thomsen CG (2011) Drivers of green supply management performance: evidence from Germany. J Purchasing Supply Manage 17(3):176–184
Lee AHI et al (2009) A green supplier selection model for high-tech industry. Expert Syst Appl 36:7917–7927
Li X, Zhao C (2009) Selection of suppliers of vehicle components based on green supply chain. In: IEEE international conference on IE&EM 21–23 Oct, pp 1588–1591
Lu LYY et al (2007) Environmental principles applicable to green supplier evaluation by using multi-objective decision analysis. Int J Prod Res 45(18–19):4317–4331
Luthra S et al (2011) Barriers to implement green supply chain management in automobile industry using interpretive structural modeling technique: an Indian perspective. J Ind Eng Manag 4(2):231–257
Ma X, Liu T (2011) Supplier selection analysis under the green supply chain. Int Conf Automot Logistics 5:205–209
Magdalena R (2012) Supplier selection for food industry: a combination of taguchi loss function and fuzzy analytical hierarchy process. Asian J Technol Manage 5(1):13–22
Noci G (1997) Designing green vendor rating systems for the assessment of a suppliers environmental performance. Eur J Purchasing Supply Manage 3(2):103–114
Shaw K et al (2012) Supplier selection using fuzzy AHP and fuzzy multi-objective linear programming for developing low carbon supply chain. Expert Syst Appl 39:8182–8192
Shen L et al (2013) A fuzzy multi criteria approach for evaluating green supplier's performance in green supply chain with linguistic preferences. Resour Conserv Recycl 74:170–179
Testa F, Iraldo F (2010) Shadows and lights of GSCM (green supply chain management): determinants and effects of these practices based on a multi-national study. J Clean Prod 18(10–11):953–962
Thangchattu C, Siripokapirom S (2010) Green supplier selection consensus by neural network. In: IEEE international conference on ICMEE 1–3 Aug, pp 313–316
Theiben S, Spinler S (2014) Strategic analysis of manufacturer-supplier partnerships: An ANP model for collaborative CO2 reduction management. Eur J Oper Res 233:383–397
Thiruchelvam S, Tookey JE (2011) Evolving trends of supplier selection criteria and methods. Int J Automot Mech Eng (IJAME) 4:437–454
Torng C, Tseng KW (2013) Using fuzzy analytic hierarchy process to construct green suppliers assessment criteria and inspection exemption guidelines. HCI 28:729–732
Tuzkaya G et al (2009) Environmental performance evaluation of suppliers: a hybrid fuzzy multi-criteria decision approach. Int J Environ Sci Technol 6:477–490
Vachon S, Klassen RD (2006) Green project partnership in the supply chain: the case of package printing industry. J Clean Prod 14(6):661–671
Weber CA et al (1991) Vendor selection criteria and methods. Eur J Oper Res 50(1):2–18
Wen UP, Chi JM (2010) Developing green supplier selection procedure: a DEA approach. In: IEEE international conference on IE&EM, pp 79–74, 29–31 Oct 2010
Wittstruck D, Teueberg F (2011) Development and simulation of a balanced scorecard for sustainable supply chain management: a system dynamics approach. In: Writschaftinformatik proceedings 2011, paper 86. http://aisel.aisnet.org/wi2011/86

Wong TN et al (2012) CSR and environmental criteria in supplier selection. In: Kachitvichyanukul V, Luong HT, Pitakaso R (eds.) Proceedings of the Asia pacific industrial engineering & management systems conference (APIEMS) 2–5 Dec, pp 74–84
Yan G (2009) Research on green suppliers evaluation based on AHP & genetic algorithm. In: IEEE international conference on SPS 15–17 May, pp 615–619
Yang Y, Wu L (2008) Extension method for green supplier selection. In: IEEE international conference on WiCom 12–14 Oct, pp 1–4
Yeh WC, Chuang MC (2011) Using multi-objective genetic algorithm for partner selection in green supply chain problems. Expert Syst Appl 38:4244–4253
Ying-tuo P Yang C (2011) Iron and steel companies green suppliers' selection model based on vague sets group decision-making method. In: International conference on electronics, communications and control (ICECC) 9–11 Sept, pp 2702–2705
Zhang HC et al (2003) Using fuzzy multi-agent decision-making in environmentally conscious supplier management. CIRP Ann Manuf Technol 52(1):385–388
Zhang Y et al (2013) Green partner selection in virtual enterprise based on pareto genetic algorithms. Int J Adv Manuf Technol 67:2109–2125
Zhu Q et al (2010) A portfolio-based analysis for green supplier management using the analytical network process. Supply Chain Manag Int J 15(4):306–319

A Model for Optimizing Traceability of Product in a Supply Chain Based on Batch Dispersion

Muhammad Saad Memon, Young Hae Lee and Sonia Irshad Mari

Abstract Product recall appears as a nightmare for manufactures as it may lead to bankruptcy. Recent product recalls stimulated manufacturers to enhance product safety in their supply chain. Implementation of traceability system now becomes necessary part of product safety strategies. Product recall crisis dramatically affect the revenue of firms which may lead to serious outcomes. Large recall induces great recall cost such as cost of notification and destroying or repairing recall products. Effective design of traceability system may turn out to be significant tool for managing recall crisis and product safety issues in production system. Traceability systems are widely used to minimize the probability and quantity of recalls however literature shows that manufacturers still not getting substantial results. This chapter develops a model to optimize the traceability based on batch dispersion methodology in order to minimize the expected recall cost and other operational cost aiming to increase shareholder profit. The simulation analysis is also carried out to test the various production strategies and its impact on profit. The analysed result show that expected shareholders profit reduces with reduction in level of traceability. It is also concluded that manufacturers should not only depended on operational costs for batch size optimization but shareholders profit should also be considered during recall crisis.

Keywords Traceability optimization · Product recalls · Batch dispersion · Shareholder profit

M. S. Memon · Y. H. Lee (✉) · S. I. Mari
Department of Industrial and Management Engineering, Hanyang University, Seoul, South Korea
e-mail: yhlee@hanyang.ac.kr

M. S. Memon
e-mail: msmemon@hanyang.ac.kr

S. I. Mari
e-mail: sonia@hanyang.ac.kr

P. Golinska (ed.), *Logistics Operations, Supply Chain Management and Sustainability*, EcoProduction, DOI: 10.1007/978-3-319-07287-6_7,

1 Introduction

Manufacturers are facing immense problem regarding their product safety which leads to huge product recalls. United Sates department of agriculture reported 82 number of recalls which contains around 3,475,115 pounds of meat in year 2012 (USDA 2012). Food Standards Australia reported around 60 recalls of food products in 2012 which is more than average of past years (FSANZ 2013). Other than food the higher recalls results shown in consumer products. According to US Consumer Product Safety Commission 5095 products have been recall in year 2013 (CPSC 2013).

These statistics shows that producers are still facing recall crisis. Product recall crisis dramatically affect the revenue of firms which may lead to serious outcomes. Large recall induces great recall cost such as cost of notification and destroying or repairing recall products. Manufacturers always faced problems in recalling the defected products because of poor traceability system.

Effective design of traceability system may turn out to be significant tool for managing recall crisis and product safety issues in production system. Traceability system is widely used to minimize the quantity of recalls however literature shows that manufacturers still not getting substantial results. Recall statistics shown that microbial contamination is leading reason of total recalls followed by contamination of foreign material in finished products.

Practitioners always try to trace the raw material batches contaminated by microbial contamination in food industry and defected component batches in assembly lines but they failed because of non-optimized traceability system. Also it is very costly for companies to refund the customer's money or replacement against defected products. But situation some time is going to be worse when defected products lead to customer health problems or injuries.

US Consumer Product Safety Commission (CPSC) indicates that total recall cost in US is more than $700 billion each year just for consumer products (Langhorst 2007). Most companies now take insurance to cover defined cost involved in the recall. But companies face hard time when insurance cannot be claimed due to the severity of product recall specially recall of unsafe and contaminated products. According to US Food and Drug Authority (USFDA), the company's stock price has declined by 22 % within two weeks after recall announcement. The average cost of recall in US companies is around $10 million.

Keeping in view the current challenging situation regarding product safety, this chapter develop traceability optimization model for minimizing the impact of recall on shareholders profit. Also simulation analysis has been carried out to test the various levels of traceability by changing batch dispersion.

2 Related Work

In today's global market, the supply chain presents significant risk management challenges. Having an appropriate traceability and recall plan in place is critical to managing supply chain risks. Ability to track products through all stages of the supply chain has always been important for companies, but in the event of a product recall, having an efficient system in place is critical (Kumar and Schmitz 2011).

Traceability will help to build trust and the establishment of long-term relationships among supply chain partners and consumers (Alfaro and Rabade 2009). Advances in track and trace technologies such as RFID is giving companies more visibility to the supply chain and can be beneficial in a product recall (Philips et al. 2007; Zang and Fan 2007). Improved product and manufacturing process designs will minimize likelihood of quality issues and product recalls (Kumar and Schmitz 2011). Recalls have several costs associated with them, both direct and indirect. Table 1 lists direct and indirect cost drivers adapted from (Kumar and Schmitz 2011).

Many researchers focus on traceability system in order to increase the product safety. Large part of literature consist of traceability technology while very less literature is available on manufacturing methodologies needed for efficiently utilization of traceability system. In best of our knowledge Dupuy et al. (2005) proposed batch dispersion methodology to optimize the traceability system in order to reduce the impact of recall. Wang et al. (2009) used batch dispersion methodology to optimize the batch size considering traceability factor. Rong and Grunow (2010) used same methodology to optimize the chain dispersion in food distribution system. Memon et al. (2013) also used batch dispersion methodology considering more robust recall cost management in three stage multi-procured bill of material (BOM).

All of these researches focus on optimizing batch size such that traceability system should be optimized considering minimum total cost. However it is also very important to analyze the impact of traceability optimization on shareholders profit. This chapter developed the model for minimizing the direct cost of recall using batch dispersion methodology and analyzed its impact on shareholders profit.

2.1 Traceability Optimization

In order to evaluate the accuracy of the traceability in the production process, Dupuy et al. (2005) introduce new measures: downward dispersion, upward dispersion and batch dispersion in food production as shown in Fig. 1. However these measures also applicable to all type of batch production systems. The *downward dispersion* of a raw material batch is the number of finished product batches which contain parts of this raw material batch. For e.g. raw material batch is used in four batches of finished product (see Fig. 1a). The *upward dispersion* of a finished product batch is the number of different raw material batches used to produce this batch. For e.g. finished product batch consist of three batches of raw materials

Table 1 Direct and indirect recall costs

Direct recall costs	Indirect recall costs
Notification costs	Loss of market share
Loss of sales	Subsequent loss of sales
Cost to maintain business interruption	Negative impact to brand image
Inventory losses	Cost to rehabilitate image
Cost of refund/compensation	Collapse of organization
Logistics costs	Negative impact to morale
Fines/lawsuits	Cost to rehabilitate reputation

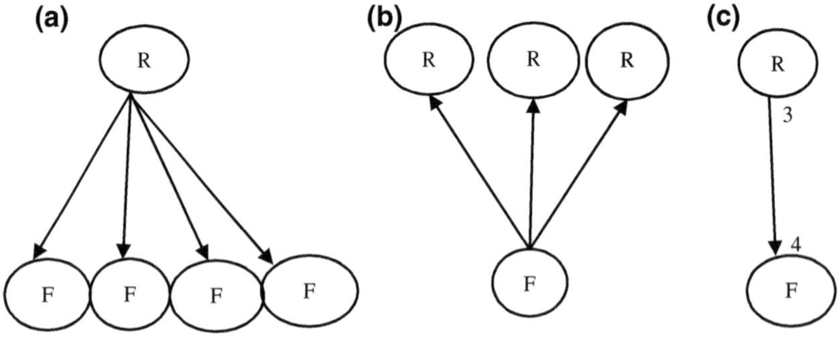

Fig. 1 Simple batch dispersion concept: **a** downward dispersion, **b** upward dispersion, **c** total dispersion

(see Fig. 1b). The *total dispersion* is the summation of all downward and upward dispersions and is called as Batch dispersion (see Fig. 1c), where three batches of raw material used in four batches of finished product. In practice, batch numbers assigned at the start of the manufacturing processes accompany products as identifications throughout a supply chain. If a safety problem comes from a raw material batch, all the finished products containing this raw material have to be identified and recalled. For example if three raw material batches used in four finished products batches (see Fig. 1c), and only one batch of raw material found to be tainted then all four batches of finished products will be recalled, hence increased recall cost.

In many types of industries where batch production process is carried out, raw material batches from different suppliers with different prices and quality attributes are often mixed together. This is usually known as a batch dispersion problem which concerns relevant disassembling and assembling processes in the production (Dupuy et al. 2005). The above discussion of batch dispersion shows that; higher batch dispersion means lesser level of traceability because when batch dispersion increases, it will be difficult to identify the source of contamination. Hence it is clear that traceability of tainted product is directly influenced by batch dispersion. Optimization of traceability means, how effectively one can identify the source of product? This means that if batch dispersion is high then manufacturers will have to face difficulty in identifying source of problem, and they may have to recall large number of finished products.

3 Proposed Model for Traceability Optimization

The proposed model is designed to minimize the expected recall cost with integration of traceability factor. In order to increase the product safety or reduce the impact of recall, manufacturer has to incorporate traceability in their process and by minimizing the batch dispersion. Probability of recall will be increased with increase in batch size because larger batch size requires more number of raw material batches which increase the chances of contamination. It is interesting to know that, increase in level of traceability will also have influence on other operational cost, therefore we also analyze the impact of traceability level on other operational costs. Notations used in model are shown in Table 2.

Typical economic production quantity (EPQ) models assume fixed costs for each production setup. The EPQ balances set up and inventory costs in order to minimize total costs. Here, the production setup cost in a planning period is described in Eq. 1.

$$C^{Setup} = \frac{D}{Q} SC \tag{1}$$

A larger manufacturing batch size reduces the setup cost component in the overall product cost. Many types of manufacturers often face the fact that retailers want products to be delivered in small quantities at fixed time intervals to minimize their inventory holding cost. This may result in the products produced in one batch being allocated in several deliveries. In many cases, for quality assurance purposes, products cannot be delivered until the whole batch is produced (Wang et al. 2009). This model follows same methodology of Wang et al. (2010) to calculate the holding cost of finished product, the accumulated inventory during the production uptime cannot be used for delivery until the whole batch is completed. The next batch production resumes and completes at the time when the inventory from previous batch is exhausted and is ready for next delivery (see Fig. 2). In the model, we only consider the holding cost of finished product inventory accumulated during the production uptime (the triangular area in Fig. 2) and the inventory ready for delivery (staircase area in Fig. 2). The average product inventory level is given in Eq. 2.

$$I_{Avg} = \frac{Q^2}{2P.nt} + \frac{(n-1)x}{2} \tag{2}$$

Table 2 Notations of traceability model

Indices	
i	Raw material batches
j	Component batches
f	Finished product batches
Parameters	
D	Demand rate of a product, units/period
P	Production rate for a product, units/period (here $P > D$)
SC	Unit set-up cost in production, $/set-up
H	Inventory holding cost of finished products, $/unit/period
x	Shipment quantity of products at a regular interval (units/shipment)
t	Time interval between successive shipments
$B_{RC}(K)$	Proportion of the *k*th type of raw material used in a component batches
$P_{i,k}$	Unit price of batch i of the *k*th type of raw materials
P_F	Unit transaction price of product.
C_S	Shipment charges of recalled product ($/unit/period)
C_C	Notification charges of recalled product ($/period)
C_{RH}	Cost for holding and dispose of recalled products, $/unit/period
Q	Quantity of batch of *K*th type of raw material
K	Type of raw material
Decision variables	
Q	Quantity of a finished product batch
S	Integer variable which is quantity of *K*th type of raw material batches used in finished product batch f
x_{RC}	Binary variable equal to 1 if the type of raw material batch i is used in the component batch j and 0 otherwise
x_{CF}	Binary variable equal to 1 if the component batch j is used in the finished product batch f and 0 otherwise
n	Integer variable which is the frequency of product shipments, number/period

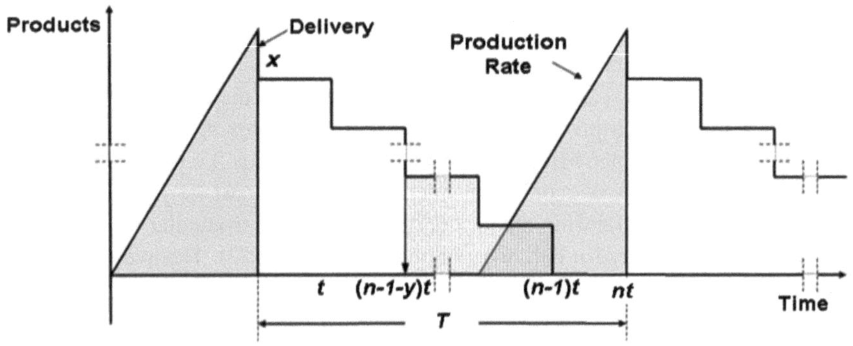

Fig. 2 Product inventory level with respect to time (Adopted from Wang et al. 2010)

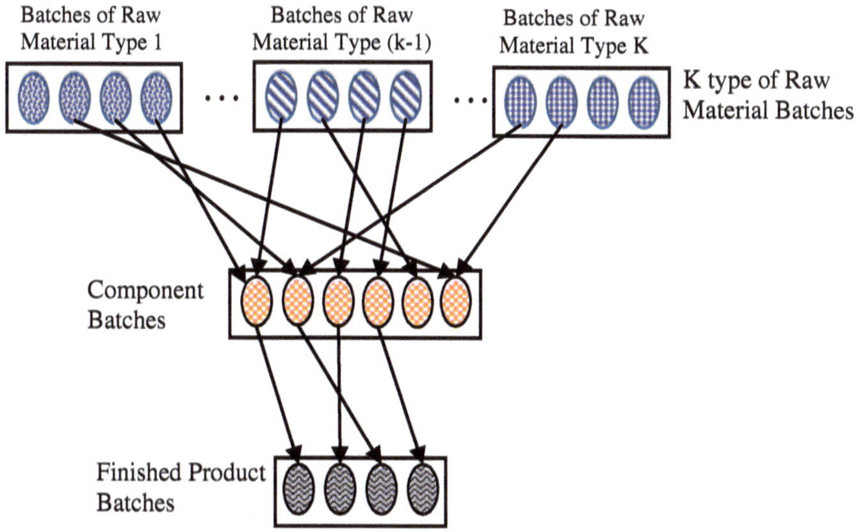

Fig. 3 Graphical model of 3-level batch dispersion

Equation 2 can be obtained by calculating the relevant area of one batch production cycle in Fig. 2 and dividing it by the cycle time period nt. The area consists of one triangle area with the value of $Q^2/2P$, and $(n^2-n)/2$ rectangle areas with the value of tx. The integer variable n can be replaced by Q/x. The inventory holding cost equation can be therefore given as Eq. 3.

$$C^{Holding} = \frac{QxH}{2Pt} + \frac{Q}{H} - \frac{xH}{2} \quad (3)$$

The major aim of this chapter is to incorporate traceability system in existing batch production system, which would create additional costs for producers. Probability of recall will be increased with increase in batch size because larger batch size requires more number of raw material batches which increase the chances of contamination. Traceability system effectiveness can be measured by how much decrease in frequency of product recall and amount of product to be withdrawn in recall crisis. We consider the two upmost important recall costs in this model, which include recall notification cost and product retrieval cost together with value of product. Product retrieval cost depends on the quantity of recalled products. Whereas product notification cost is fixed cost incurred in recall crises. In this model, we only consider the recall shipment cost and holding cost of recalled item as product retrieval cost, the expected recall cost for a demand period in three level of BOM (see Fig. 3) can be described in Eq. 4.

$$C^{\text{Recall}} = \sum_{k=1}^{M}\sum_{i=1}^{S}\sum_{j=1}^{N} X_{RC} \cdot \sum_{j=1}^{N}\sum_{f=1}^{O} X_{CF} \cdot \{D \cdot (P_F + C_{RH} + C_S) + C_N\} \quad (4)$$

where

$$S = \frac{Q \times B_{RC}(k)}{Q_k} \tag{5}$$

The Integer variable S in Eq. 5 depends on the finished product batch size. If batch size increases, the more number of Kth type of raw material batches will be required to fulfill the finished product batch size. Naturally if more numbers of raw material batches used in each batch of finished product then it will be very difficult to trace the source of problem in final finished product batch. This type of problem widely occurs in food and pharmaceutical industries. In consumer product industries like toys, automobiles, etc., manufacturers randomly assembled components from different suppliers in same finished product batch which again lead to same problem of increased complexity in traceability and will lead to high expected recall cost. This problem can easily solved using batch dispersion methodology by minimizing the expected recall cost as shown in Eq. 4.

The total cost of production also includes raw material cost. It is necessary to calculate the raw material cost or all procured materials cost, as individual procured material batches may have different price and product safety risk attributes because of different types of procured materials which may come from different suppliers. These batches are often mixed together in bill of materials for balancing between cost and quality. The cost of procured material is the sum of the cost of all raw material batches. The cost of individual procured material batches can be calculated by multiplying the unit price with batch quantity. Therefore, the total material cost for a production cycle can be given by Eq. 6.

$$C^{Material} = \frac{D}{Q}\left(\sum_k \sum_i \sum_j X_{RC} \times \sum_j \sum_f X_{CF} \times S \times P_{i,k}\right) \tag{6}$$

3.1 Shareholder Profit

The shareholder profit depends on firm's performance, if the performance is ineffective then there will be more probability of product recall. During recall crisis firms face two types of costs: direct cost and indirect cost but here we only consider the direct costs as these costs affect shareholders profit. It is hopeful that our results will be useful for practitioners to assess the impact of recall on profit. In order to find out the impact of recall on shareholders profit we assume that there will be recall or no recall in next month. If recall will occur then expected recall cost C^{recall} will be imposed to shareholders. Therefore firm's stock price or the value of a stock at the end of the month will be as shown in Eq. 7 if no recall occur and as shown in Eq. 8 if recall occur.

$$S_e^{NR} = PV \text{ (If no recall occurs)} \quad (7)$$

$$S_e^{R} = PV - C^{\text{Recall}} \text{ (If recall occurs)} \quad (8)$$

where,

$$PV = \textit{Present value of stock} \times \textit{Number of Shares Outstanding}$$

The value of a stock at the month beginning is the present value of future profit. We assumed that there is probability p for occurring recall in the next month. The probability p that a product recall occurs must be assessed and included in the decision-making process (Saltini and Akkerman 2012). Therefore stock price at the start of month will be as shown in Eq. (9).

$$S_b = p(PV - C^{\text{Recall}}) + (1 - p)PV \quad (9)$$

Thus if recall occur in the next month, then shareholder profit will be as in Eq. 10. It can be seen that shareholder will loss the amount equals to cost incurred on product recall. If recall is totally unexpected (i.e. $p = 0$) then Eq. 10 will be equal to C^{recall} means shareholder will gain the extra profit equal to amount of expected C^{recall}.

$$S_e^{R} - S_b = -(1 - p)C^{\text{Recall}} \quad (10)$$

And if no recall occurs in the next month then, shareholder will gain profit as shown in Eq. 11.

$$S_e^{NR} - S_b = pC^{\text{Recall}} \quad (11)$$

4 Simulation Experiment

For analysing the optimization model and its impact on shareholder profit gain, three cases as example are simulated using spreadsheet simulation to illustrate the different manufacturing strategies. For case one all material batches are assumed to be mixed in final finished product batch. Case two considers the medium level of dispersion in which very few raw material batches are used in finished product batch. The last case considers least possible utilization of raw material batches in finished product batch. It is obvious that the probability of recall will be higher under high batch dispersion as many raw material batches used mixed which result poor traceability (Dupuy et al. 2005; Wang et al. 2009). On the other hand low batch dispersion results less probability of recall (see Table 3 for description). Table 4 shows the input parameters of the model used for simulation analysis.

Table 3 Description of simulation setup

Case	Description	Traceability level	Probability of recall (%)
Case I	All raw material batches mixed used	Lower	15
Case II	Few raw material batches mixed used	Medium	10
Case III	Least raw material batches mixed used	Higher	5

Table 4 Input parameters of traceability model

Demand	6000 units			
Production rate	9000 units			
Setup cost	$300/setup			
Holding cost	$1/unit			
Shipment quantity	150 units			
Type of raw material batches	Type K1	Type K2	Type K3	Type K4
Size of raw material batches	1000 units	500 units	500 units	300 units
Price of raw material batches	$10	$7	$8	$6
Price of finished product batch	$1			
Cost of holding the recalled product	15 % of Product price (Resende-Filho and Buhr 2010)			
Cost of shipping the recalled product	10 % of Product price (Resende-Filho and Buhr 2010)			
Notification cost of recall	4 % of Product price (Resende-Filho and Buhr 2010)			

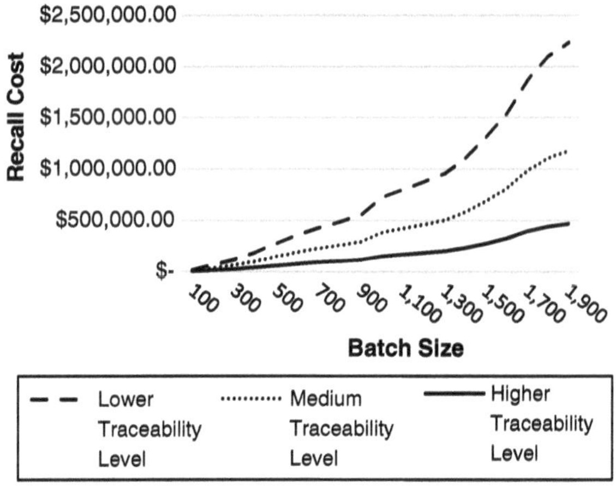

Fig. 4 Recall cost under different cases

Simulation result shows that recall cost can be reduced by minimizing the batch dispersion. Figure 4 also reveals that recall cost increases with size of batch, this is because many direct recall costs depends on size of batch (such as logistics and holding costs). This shows that firms need to be cautioned when selecting the optimum batch size because larger batch size may difficult to handle in recall crisis. On the other hand it is obvious that increment in batch size also increase the

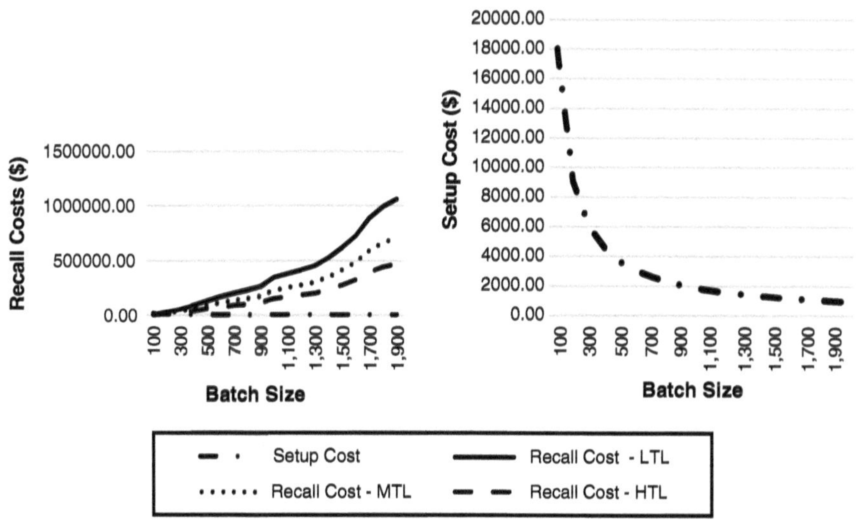

*HTL=High traceability level, MTL=Medium traceability level, LTL=Low traceability level

Fig. 5 Effect on setup cost and recall costs under different batch sizes

Fig. 6 Reduction in stock price of the firm after recall announcement (where PV = $100,000)

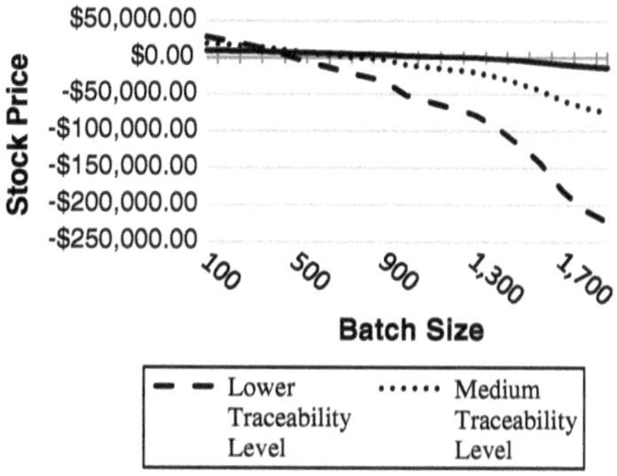

setup cost, which means reduction in production efficiency. The setup cost analysis shown in Fig. 5, this analysis will be important for practitioners because production managers always think that reduction in batch size will increase setup cost on large scale, whereas analysed results show that setup cost is much lower than recall cost. It can be seen that even if production system is working on low traceability level, the setup cost is very low as compared to recall cost under low traceability level.

It is also analysed that stock price of firm reduces after announcement of recall as shown in Fig. 6. Analysis shows that stock price affected less under higher traceability level whereas if firms working on lower traceability level its stock price reduces dramatically. It also reveals that if batch size is too large then under lower traceability firm may not survive during recall crisis and it is also possible

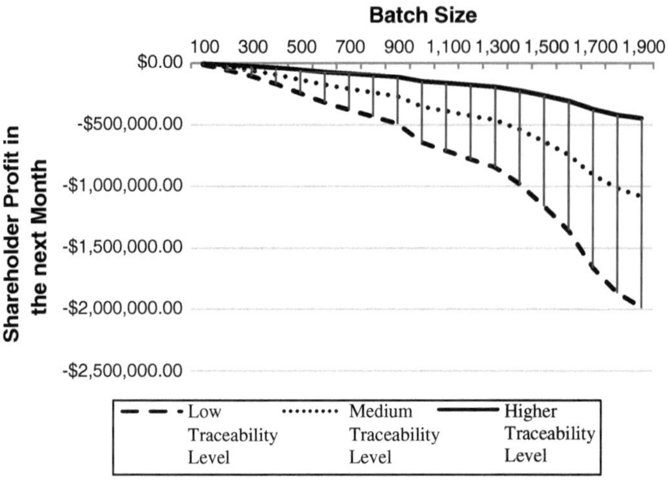

Fig. 7 Shareholders profit reduction after recall announcement

that firm may go bankrupt if recall size is too large to handle. Therefore firms should analyse the stock price when optimizing the batch sizes because analysis results shows that larger batch sizes reduce the stock prices dramatically which leads to complete failure of organizations.

Ultimately the recall crisis affects the shareholder wealth. Figure 7 shows that profit of shareholders always reduces after recall announcement. However the profit reduction can be minimized by adopting the higher traceability methodology. As can be seen from Fig. 7 that the difference between lower traceability and higher traceability is too large therefore firm should analyse the batch size and its capabilities for profit sharing among shareholders. If batch size is too large then during recall crisis firms may not bear the loss.

This analysis shows that firm not only consider the operational costs for batch size optimization but should also incorporate loss of expected profit during recall crisis. The firms may also select the level of traceability depend on their available resources and its impact on financial loss during recall crisis. We can see that profit loss is much less under higher traceability level than lower traceability level. Implementation of traceability system also depends on other operational costs such as setup costs and inventory holding cost because higher traceability level on one hand reduces the profit loss during recall crisis while on the other hand increases the setup cost. Therefore it is also necessary to analyse operational costs impact with respect to level of traceability. Hence correlation analysis (see Table 5) of various operational costs and recall costs is also added to further analyse the impact of traceability system on firm's performance and shareholder profit.

Correlation analysis shows that increase in level of traceability will also increase in setup cost hence operational performance may be effected. Material cost and recall cost are directly correlated, this shows that increase in batch size will result in wastage in raw material in case of recall crisis, similar case with the

Table 5 Correlation analysis of operational costs and recall cost

	Material cost	Holding cost	Setup cost	Recall cost
Material cost	1.00			
Holding cost	0.97	1.00		
Setup cost	−0.75	−0.72	1.00	
Recall cost	0.97	0.96	−0.59	1.00

holding cost. At this stage it is clear that the only factor is to be taken in account when deciding for level of traceability is setup cost. The example data in this chapter reveals that setup cost is much lower than the recall cost (as shown in Fig. 5) and it is also possible in real case scenario, therefore decision makers should analyse recall cost, setup cost and probability of recall when deciding the optimum batch size.

5 Conclusion

This chapter gives insights about impact of recall under various different traceability methodologies on firm's shareholders profit where the batch sizes and level of traceability on shareholder profit are analysed. The results are very helpful to practitioners for analysing their batch sizes and recall cost management. The chapter also developed traceability optimization model based on batch dispersion methodology for minimizing the recall cost in three stage BOM. It is very helpful for managers to understand the importance of recall cost minimization in order to withstand in recall crises. Analysis shows that increase in level of traceability will also reduce the operational efficiency, while expected recall cost can be minimized. This may cease the implementation of batch dispersion methodology for minimizing the expected recall cost, because practitioners will perceive too much reduction in production efficiency. For this reason, we also added correlation analysis of various operational costs and recall cost, and result shows that setup cost and recall costs are negatively correlated. This means that reduction in level of traceability (increment in batch dispersion) will reduce the production costs just because of less setup cost, but on the other hand it will increase expected recall cost.

The simulation experiment in this chapter shows that setup cost is much lower than expected recall costs, which is normally true in all type of industries. This shows that practitioners should only focus on expected recall costs and probability of recall, when designing production system and deciding optimum batch size. It is also seen from experiments that batch size is directly proportional to expected recall cost and larger batch size means more customers will be affected from tainted products. For future work, this chapter can be extended to find optimum batch size such that total loss to shareholders should be minimum. The other work subjected to future study is to develop a multi-objective optimization model for making decision on batch size while minimizing recall and operational cost with maximization of shareholders profit.

Acknowledgments This research was supported by Basic Science Research Program through the National Research Foundation of Korea (NRF) funded by the Ministry of Education, Science and Technology (2012R1A1B3000631).

References

Alfaro JA, Rabade LA (2009) Traceability as a strategic tool to improve inventory management: a case study in the food industry. Int J Prod Econ 118:104–110

CPSC (2013) Product recall statistics. http://www.statisticbrain.com/product-recall-statistics/

Dupuy C, Botta-Genoulaz V, Guinet A (2005) Batch dispersion model to optimise traceability in food industry. J Food Eng 70:333–339

FSANZ (2013) Food recall statistics. http://www.foodstandards.gov.au/industry/foodrecalls/recallstats/Pages/default.aspx

Kumar S, Schmitz S (2011) Managing recalls in a consumer product supply chain—root cause analysis and measures to mitigate risks. Int J Prod Res 49:235–253

Langhorst P (2007, 2013). Product recall statistics—an FAQ for manufacturers. http://ezinearticles.com/?Product-Recall-Statistics—An-FAQ-for-Manufacturers&id=900367

Memon MS, Lee YH, Mari SI, Cho SY (2013) Traceability optimization for product safety in supply chain management. In: 8th international congress on logistics and SCM systems (ICLS2013). Tokyo, Japan, pp 55–64

Philips J (2007) Consumer product recalls prompt product visibility questions. Text World 157

Resende-Filho M, Buhr B (2010) Economics of traceability for mitigation of food recall costs. Available at SSRN 995335

Rong AY, Grunow M (2010) A methodology for controlling dispersion in food production and distribution. OR Spectrum 32:957–978

Saltini R, Akkerman R (2012) Testing improvements in the chocolate traceability system: impact on product recalls and production efficiency. Food Control 23:221–226

USDA (2012) Recall summary for calendar year 2012. http://www.fsis.usda.gov/wps/portal/fsis/topics/recalls-and-public-health-alerts/recall-summaries

Wang X, Li D, O'brien C (2009) Optimisation of traceability and operations planning: an integrated model for perishable food production. Int J Prod Res 47:2865–2886

Wang X, Li D, O'Brien C, Li Y (2010) A production planning model to reduce risk and improve operations management. Int J Prod Econ 124:463–474

Zang C, Fan Y (2007) Complex event processing in enterprise information systems based on RFID. Enterp Inf Syst 1:3–23

Investigating the Readiness of the Grocery Retail Chains for Virtual Supply Chain Technology in Egypt

Sama Gad, Khaled Hanafy and Sara Elzarka

Abstract The advancement in information technology has significantly changed the operations of supply chain management, and promoted efficiency and effectiveness which directly had a positive impact on market competitiveness. The grocery retail sector is probably one of the largest business sectors in many countries and it is one of the sectors that witnessed the applications of evolving supply chain technologies over the past few years such as radio frequency identifications (RFID), quick response codes (QR codes) and more. Therefore, this paper will provide a critical analysis and emphasis on the idea of adopting the virtual supply chain (VSC) concept by the large grocery retail stores in Egypt in order to investigate the grocery retail sector readiness towards applying the VSC. Research approach: The research is based on the qualitative analysis using a case study approach applied on the top two grocery retail chains in Egypt which are; Carrefour and Fathallah. Semi structured interviews were conducted with Fathallah's IT manager, and the receiving manager of Carrefour Alexandria and with subject matter experts (SMEs) to collect the necessary information for a SWOT analysis that would emphasize the strengths and weaknesses of both chains towards applying VSC in addition to the opportunities and threats that may encourage or hinder the application of VSC. Findings and Originality: According to the literature reviewed, it was found that there is a significant gap in literature regarding the application of technologies in supply chain in emerging and developing countries in general, which renders this research important to the discipline. This research is considered the first in investigating the application of

S. Gad (✉) · K. Hanafy · S. Elzarka
College of International Transport & Logistics - Arab Academy for Science, Technology and Maritime Transport, Alexandria, Egypt
e-mail: sama.hossam@gmail.com

K. Hanafy
e-mail: kmhanafy@aast.edu

S. Elzarka
e-mail: sara_elzarka@aast.edu

P. Golinska (ed.), *Logistics Operations, Supply Chain Management and Sustainability*, EcoProduction, DOI: 10.1007/978-3-319-07287-6_8,

the VSC in the Middle East, especially in Egypt. The findings of this research showed that the grocery retail chains are rather ready for the application of the VSC, and it was shown that such application can support them in improving and developing their supply chain processes. Research impact: This research will bring the topic to the attention of researchers to further investigate and exploit the area of technology in supply chain in Egypt and it will contribute to close the existing gap in the literature. Practical impact: Applying the concept of VSC will increase the retailers' efficiency, responsiveness, market coverage and accessibility and consequently its profit which will impact the economy as well. Additionally, customers will also benefit from the convenience and ease of access.

Keywords Virtual supply chain · Grocery retail sector · Supply chain management · Egypt

1 Introduction

The changing life-style of Egyptians towards the use of advanced electronic devices and mobile phones and their rising familiarity of technology lead to opening new ideas to the current businesses to take advantage of, in order to enhance their businesses and to make it easy and convenient to the customers as well. The consumer survey conducted in 2010 reveals that the Egyptians retail purchasing habits recorded that Egypt is ranked the 13th on the most attractive retail market worldwide and that it is one of the promising and fast expanding markets in its region (Ramzy and James 2012). Nevertheless, the expanding number of shopping malls and grocery chains indicate the extent to which the market is lucrative which encourages investment, whether local or foreign.

This research examines the utilization of one of the recently emerged concepts in the retail industry which is the virtual supply chain (VSC), i.e. the coherent and intensive utilization of technology between the supply chain members. This research investigates the effectiveness of the "high tech" route to helping customers feel informed about products in Egypt. Real online retail systems are chosen to assess and develop the extent to which a person feels informed about an item. In addition, this research reviews the quick response (QR) code which is a key tool to effectively manage the operation process of the VSC with the involvement of different software. The virtual architecture in the supply chain is also discussed based on the application of the VSC concept by the international retailer Tesco.

Moreover, this research highlights the main challenges that the organization might face in the application of virtual shopping concept by interviewing SMEs of two grocery retail stores in Egypt.

2 Literature Review

Based on the analysis and review of literature in VSC, this research is mainly concerned with several key enabled technologies and management systems supporting the application of the VSC concept within the enterprise.

2.1 Virtual Supply Chain Basic Concepts

The term "*supply information*" is defined as information that originates from the upstream of a supply chain and has an impact on the price, quality or delivery time of material flows in the downstream of the supply chain (Li et al. 2006). Any change occurring at an upstream stage of a supply chain can affect performances of firms at downstream stages through the change of the price, quality and delivery time of material flows directly or indirectly. On the other hand, supply chain integration in a virtual environment was defined as the process by which suppliers, partners, and customers within a shared market space collaboratively plan, implement and manage the flow of information, services and products along the supply chain in a way that improves business operations in terms of speed, agility, real time control, or customer response (Li et al. 2006).

In virtual environments systems, integration means to integrate disparate data sources, to present real-time content and information to all virtual network participants and finally to create an environment where every participant and every activity in the process is integrated (Scholz 2000). Virtual environment involves limited human interaction (seamless integration) as manual processes are too slow and error-prone to support the new business model. The goal of a structured data exchange is to move data from one organization to another with near-zero tolerance for the loss of the actual content or meaning of it.

A virtually enabled supply chain network is a series of value-added processes/stages owned by one or more enterprises, starting with material/information suppliers and ending with consumers (Hwang and Seruga 2011). Another definition denotes that the virtually enabled supply chain network is an organization that is subject to constant changes, demonstrating a specific potential when required, and overcoming time and space barriers (Kisperska-Moroń 2010). Virtual supply chains could be described by such attributes like temporary character, focus on customers, geographical dispersion, and intensive support of IT systems, network structure and an extensive use of key competencies of their members.

Virtualization implies the use of IT and communication technology by organizations in managing their interactions and key business operations with customers, suppliers and employees (Zarour et al. 2005). It uses the concept of hardware abstract layer (HAL) which is presented to the operating system in each virtual server as its usual hardware platform. Therefore, a virtual machine can be moved to and from different physical servers with different physical configurations and vendors. By simplifying the hardware and hence reducing the cost of new system implementation, it is then possible to increase the redundancy of the system or reduce power consumption with less physical hosts. Virtualization of supply chain systems involves IT as well as organizational changes (Liu et al. 2005).

A number of technology advocates have claimed that virtual reality and other advanced technology will encourage playful interaction with the product and thereby make online shopping a more engaging and enjoyable (Smith et al. n.d).

In particular, QR code is used in both commercial tracking application and convenience-oriented application; it is developed by a Japanese corporation Denso-Wave in 1994 and has been widely used in Japan since then and also known by mobile tagging (Lorchirachoonkul and Mo 2010).

The authors listed some desirable features of the QR code comparable to Radio Frequency Identification System (RFID):

- High-capacity data encoding. It stores up to 7,089 bytes of information.
- High area density. Coding in both vertical and horizontal direction and in smaller area.
- Error correction. Built-in error correction capability allowing up to 30 % code corruption.
- Omni-directional and multi-scan. Can be scanned 360° at high speed with multiple tags.
- Structured append feature. Up to 16 smaller QR code can be embedded in a single QR code.

In a supply chain environment, information such as product code, serial number and production date can be encoded using QR code. QR code is essentially a picture containing encoded data, which can be decoded by filtering the pixel of the image. Therefore, it can be easily captured as an image by a digital camera and can be quickly decoded on virtually any computing device that connects to the Internet. At the moment, most mobile phones have built-in cameras that can read the QR code, decode it with free software uploaded on the phone and then either link them to a web site, or store the entire business information somewhere in a server (Lorchirachoonkul and Mo 2010).

2.2 Actual VSC Practice: Tesco

Tesco is a British, international grocery retailer that operates in 14 markets across Europe and it has grown to be the third largest retailer in the world. Tesco follows a strategy that focuses on innovation, growing retail services and market share, and acting as a responsible member of the communities they serve (Lorchirachoonkul and Mo 2010). In South Korea, Tesco Home Plus had 113 branches operating and, prior to the virtual shopping campaign, Tesco Home Plus was striving to be number one among the competitors. With fewer retail locations and customers, Tesco Home Plus needed an effective marketing campaign to break into the market and take the spot as the number one grocery retailer in Korea.

In order to gain market control in Korea, Tesco Home Plus collaborated with an innovative agency specializing in digital and mobile marketing. According to their research, Koreans are the second hardest working people in the world, and due to their busy lives, commuting to and from work, leave little time for grocery shopping. Moreover, the research showed that the number of smart phone subscribers in Korea had surpassed 10 million and was on the rise. They found that Koreans were increasingly reliant on smart phones and other mobile devices for their daily tasks. Acknowledging this lifestyle and the openness to innovative technology, Tesco Home Plus tailored its mobile campaign (Nielsen 2004). The campaign was rather simple. Tesco strategically located virtual grocery stores i.e. posters in bus stops and underground stations displaying the most frequently demanded products by Koreans. These posters showed the products' images, prices, offers and QR codes. Commuters would simply scan the QR codes of the desired products while waiting for their bus/underground, and have their products delivered right to their doors. Tesco Home Plus reported a 130 % increase in online sales as a result following the campaign. Furthermore, over 10,000 customers have used the virtual grocery stores around Korea. This campaign shows the extent to which mobile platforms can be used to engage consumers and create and maintain brand value for a company (Nielsen 2004).

3 Research Methodology

The review of literature in SCM showed the significant growth of empirical research in the discipline in an attempt to investigate the applications of SCM theories in different settings. However, there is a major gap in empirical research investigating the VSC, its applications and challenges. Therefore, this research would follow an exploratory approach due to the limited empirical research and due to the ambiguity of current available literature base in warranting specific hypotheses. Moreover, the exploratory approach would allow the researchers to thoroughly understand the role that VCS can play in the groceries retail sector in Egypt. Semi-structured interviews were conducted with Fathallah's IT manager and the receiving manager of Carrefour Alexandria. The interviews duration ranged between 60 and 90 min. The interviews aimed at identifying the potentials of VSC application in Egypt in terms of the elements required for application, namely, IT infrastructure, logistics and supply chain.

4 Data Analysis and Findings

4.1 *The Respondents' Profiles*

4.1.1 Fathallah

Fathallah is the first family-based wholesale grocery chain in Alexandria since the eighties. Fathallah has two types of stores; Hypermarket and mini-market. Fathallah has seventeen branches of hypermarkets and an expanding number of mini-markets to serve the surrounded neighborhoods. Mini-markets are the only branches that provide delivery to customers, where these mini-markets replenish its stocks from the nearest Fathallah store or warehouse. Fathallah has four main warehouses which are located in four different suburban areas to serve the network of stores in an efficient and effective manner.

The interview was conducted with Mr. Amr Fathallah who is the IT Specialist for Fathallah since 2007. He is responsible for managing seven branches of Fathallah, in addition to, managing the IT systems, and developing new software and hardware solutions for Fathallah's daily operations.

4.1.2 Carrefour

Carrefour is one of the most popular retail stores around the world. Carrefour, a French based retailer, started in Egypt by the end of 2002 and since then, Carrefour has been the most popular retailer in Egypt. Carrefour is characterized by its large selling area with over 55,000 different product, a free parking space, and very attractive offers. Majid AL Futtaim Retail Group manages Majid Al Futtaim Hypermarkets, a joint venture company with the world's second largest retail Carrefour, and offers shoppers the same quality, variety and value-for-money that have made the brand a household name to millions over the world (Carrefour Co. n.d.).

Carrefour is expanding across the Middle East region in the last few years and still planning to open new stores in the coming future. Its' reputation is based on the quality and freshness of the products, customer service and competitive prices compared with the other local retailers. Carrefour has more than 13,000 employees in eleven countries.

4.2 *IT Infrastructure*

The purpose of this section is to provide detailed information about the current IT infrastructure and applications in both stores, the methods by which they manage their operations and their IT readiness to adopt VSC.

Mr. Fathallah reflects Fathallah's corporation way of thinking and he works on developing and implementing new technologies that may add to their business.

He strongly supports e-shopping and classify its advantages for both sides; the customers and the company. First for the customer: it provides him a full coverage of the products available, easy access to the items needed with minimal time and keeps the customer updated automatically with the latest promotion and offers. As for the company, Fathallah would benefit from e-shopping as it is considered as an efficient method for marketing that requires less shelf space, less labor costs and insurance expenses, less security measures and consequently minimum total running/carrying cost. Fathallah targets the different customers' segments but when adopting the virtual shopping concept, it will target mainly category/class (A) of customers who own smart phones and are aware of using the internet.

Fathallah works on a cross functional basis where all its departments are connected with internal network that ensures the information visibility within the company. Fathallah uses a reliable information system called Fujitsu Retail Power—an off-the-shelf software—that delivers in-store transaction and customer relationship functionality and enables companies to minimize the time, cost and risk of creating real-time interoperability between disparate applications and devices. On the other hand, Carrefour manages its functional operations separately and it is all linked by using the Oracle Platform Security Services (OPSS) also off-the-shelf software, in order to ensure the security and audit for the products. Based on the official website of OPSS, it provides enterprise product development teams, systems integrators (SIs), and independent software vendors (ISVs) with a standards-based, portable, integrated, enterprise-grade security framework for Java Standard Edition (Java SE) and Java Enterprise Edition (Java EE) applications.

Fathallah may face some IT infrastructure barriers, e.g. Abees warehouse in Alexandria, which is located in the suburb area of the city, is not connected by the internet so they rely on the 3G coverage which is unstable and costs more than the normal connection as the companies need more speed to share the large amount of data.

Technology is involved in Fathallah's daily operations as it has fifteen inches price checkers and it aims to apply TESCO's advanced price checker (NB. TESCO's price checkers can compare products attributes and prices). And also Fathallah's technology needs to be improved based on the availability of money and the IT specified budget and also according to the business needs. Fathallah aims at adopting new software that can include an electronic shelf label that automatically updates the prices on shelf.

Fathallah and Carrefour both use Personal Digital Assistant (PDAs) in their warehouses for receiving and dispatching products, as it replaces the manual data entry by scanning the products received which minimize the time consumed in checking and receiving the ordered products and also reduces the paper work which saves stationary and materials expenses.

Fathallah takes in consideration minimizing the human interaction in the future in order to minimize the errors as well i.e. more automation means more accuracy of information, faster exchange, better performance, easily audit and control and less threat to employees' injury.

Fathallah and Carrefour both believe that the VSC concept is a promising concept to be applied in Egypt, as it makes the demand forecasting faster and easier and also enhances the availability (or the high response rate) of products with faster updated and more accurate data exchange. In addition, it minimizes the percentage of errors in warehouses and in stock merchandize. To apply the VSC concept both companies will need a strong online application, Wireless access point, switch to run the application, expend more to secure the online users against hackers through firewalls and securing the customers' bank accounts and customers personal information.

4.3 Supply Chain Activities

This section examines the current logistics and supply chain practices of the respondents in terms of suppliers' management, suppliers' relationship management (SRM), distribution, delivery and transportation, which are all vital elements for the application of VSC.

In terms of suppliers' management and SRM, this part highlights the number of suppliers of each retail chains with a detailed focus on the top suppliers and the contractual nature between them.

Fathallah has a flexible policy with around 2,500 supplier, the top suppliers bond with a long-term contracts like; PepsiCo, Procter and Gamble (P&G), Unilever, Nestle', Henkel and Juhayna. They have information visibility along the process as they quarterly share their sales and promotions, develop programs, train employees, and marketing. Annual meetings are conducted at the end of each year to check whether goals have been achieved, and to discuss new development plans for the following year.

Carrefour uses a mix of local vendors and customized product offerings to cater to local tastes, which is why Carrefour is ranked number one in the Egyptian retailers' market based on the amount of stock keeping units (SKUs) available in stores and the high speed of service provided. Moreover, the number of suppliers that Carrefour deals with differs according to the population and the geographical area it serves. For example, Carrefour Alexandria-City Center branch deals with an average of 190 suppliers whereas Carrefour Cairo-Maadi, the Head Quarter in Egypt, deals with nearly 300 suppliers.

In terms of logistics operations, based on Fathallah's working SC plan, the VSC concept can cut from the operations cost if the logistics activities are outsourced. Also applying the VSC concept is reflected in less human involvement as Fathallah tries to replace employees with automated machines in order to decrease human errors in the process. Also, Carrefour Egypt was setting a plan to improve the technological side of their operations, which means less paper work and replace the manual steps of receiving the goods with the PDAs. But due to the political unrest in Egypt, Carrefour is postponing the development of new strategies until the market environment becomes more stable.

In terms of distribution and delivery, applying the VSC concept will affect Fathallah's Return on Investments (ROI) within one year based on the already available infrastructure with good customer awareness campaign. VSC will speed up the process of delivering orders by benefiting from the use of software that provides real time information sharing between warehouses/distribution centers (DC) and coverage. A SC network restructure would be needed to include more DCs or warehouses. The situation differs for Carrefour, as its service yards are divided into partitions to hold the products in and it works as a receiving point or a transit stage between the receiving of products and shelf display. Carrefour lacks the automatic connection and transferability between the yards and the shelves, as it requires reliable software, so Carrefour still rely more on the human interaction rather than on technology. Carrefour does not currently provide home delivery service, so it may represent a starting point to fulfill more customers' preferences.

Regarding the transportation plan, Fathallah own its fleet that work between Fathalla's branches only, as suppliers are responsible for the delivery of their products to Fathallah's warehouses. In applying VSC, it would be better to use the services of a third party logistics company with good strategy; as the trucks will be fully loaded with planned schedule and the delivery will be from the nearest mini market or distribution center not from the hyper markets in order to minimize the transportation time. Also the VSC concept will provide a full utilization of available assets of warehouses, DCs, branches and that means less inventory and carrying cost, while Carrefour Egypt outsources its transportation and logistics activities by the international company Agility logistics in order to provide Carrefour the needed services to minimize the overall supply chain process cost. Carrefour is working according to a future logistics plan on a monthly basis, includes all branches with tones or amounts of products ordered all clearly classified on a scheduled calendar, with a frequent checkup if they are working on schedule. Carrefour City-Center branch receives orders from more than 298 companies on average per day.

Mr. Fathallah mentioned that Egyptians awareness of using internet increased recently so companies have great potentials in expanding in the market adopting technology and internet which will be reflected in more profit to the company and larger market share as well.

5 Findings Comparison

The following table summarizes the main challenges of implementing the VSC concept in both grocery retail stores in a comparison form (Table 1).

6 Conclusions and Recommendations

In summary of research results, this research discussed the VSC concept with all its' dimensions in the literature part and also applied on two of the most successful groceries retail stores in Egypt; Carrefour and Fathallah with a critical emphasis on

Table 1 Findings comparison

Challenges of implementation		Fathallah	Carrefour
Technical challenges	Network infrastructure	Well established network infrastructure connects all branches and the warehouses	All internal departments are connected through the ERP software and the employees use their PDAs for data entry
	Operation systems	Fujitsu retail power	OPSS
	Tracking systems	Use bar coding system and may need to improve to the RFID technology or the QR code	Use bar coding system and may need to improve to the RFID technology or the QR code
		Fathallah work on a cross-functional basis so it is possible to apply a tracking system	Each department of Carrefour work as a separate entity connected through the ERP system
Cultural challenges		Support the VSC concept and ready to raise the hiring requirements for new staff	Take the VSC concept into consideration and will establish a separate SC department
Partners relation		Strong partners relation with high level of information visibility	Good partners relation but needs to be supported by up-to-date information sharing
Marketing		Applied between the SC partners	Applied with their partnered companies and also for the neighborhood customers
Partners trust		Fathallah has a strong partners trust with its suppliers with risk sharing and also development programs	Carrefour has a good relationship with its wide range of suppliers who are mainly relying on the trust linking between them
Logistics activities		Performed in house	Outsourced to agility logistics
Orders delivery cost		Fathallah's own fleet delivers orders between its branches only	Carrefour outsource their transportation to agility logistics
		Suppliers use their own fleet to send the ordered goods to Fathallah	Suppliers use their own fleet to send the ordered goods to carrefour
		Fathallah's mini-markets deliver orders to customers	Carrefour does not deliver groceries to customers with the exception of home appliances

the implementation challenges, and the findings of the interviews were encouraging those businesses to adopt a new technology as it will add to the Egyptian groceries retail sector.

The contribution of this research relies on applying the VSC concept on the Egyptian groceries retail industry with taking in consideration the time frame that it could be applied within and linking it to the preferences and the daily activities of the Egyptian customers. In addition to the analysis of retailers current working SC activities and plans and discussing their future short and long term plans to develop and expand in the Egyptian market in a new innovative way.

Both companies are performing on a satisfactory level based on their current real working plans, but here indicated some suggested points that the two companies may take it in consideration to support their existing supply chain and business activities.

The increasing demand from the Egyptian customers on the large retail stores represents a real opportunity to those retailers and provides them a promising profit margin and gives a competitive advantage to them in the Egyptian retail industry and the global market as well.

6.1 Recommendation for Fathallah

- Specifying more budgets for the IT department to be able to use advanced software and servers.
- Develop IT training programs for the current employees.
- Make contractual partnership with Internet Service Provider (ISP) to cover the warehouses which located in urban areas.
- Build online mobile application for different mobile devices operating systems like; IOS, Android, Symbian OS, etc.
- Well trained Fathallah employees work as representatives in all stores and the highly populated areas to spread the new idea and technology and increase the awareness of customers of using it.
- Periodical check and analysis of their current achievement to control their current situation and check for improvements and following up plans.

6.2 Recommendation for Carrefour

- Link with the outsourced companies (i.e. Agility logistics) by networks using software to enable the online visibility of information all day around.
- Involve more automation in their logistics activities and intense the use of PDAs and make it connected wireless with the warehousing and cashiers software.
- Build a strict contract with the suppliers to clearly indicate the amount of products ordered and a detailed description of it and also indicate the accurate delivery time.

- Replace the manual entry of scheduled plans to being electronic in order to allow and facilitate the following up of progress plans and up-to-date checkup of approaching the adoption of new technology.

7 Research Limitations and Future Research

VSC is a newly developed concept that is still not well recognized and very few retail grocers abroad are using it. Undertaking this research on the Egyptian retail grocery especially in Alexandria city, will not be sufficient due to the limited number of medium and large groceries retail stores available. Therefore, the research population was limited to; Carrefour, Metro, Fathallah and Zahran.

Another limitation of this research is concerning the sample. The sample was only chosen from Alexandria in Egypt. However, it is considered as the second largest city of Egypt after Cairo the capital city; nonetheless, it may be better if the research covered other areas and cities.

One of the study's limitations was the difficulty of data collection. This research is based on the evaluation of two of the previously mentioned grocery retail stores in Alexandria, Egypt regarding the VSC concept those retailers are; Fathallah and Carrefour City Center. This is due to the limited accessibility to specialized managers in Cairo retailers' in addition to the cultural differences between the different cities in Egypt based on; the educational level and the income level that may accept the usage of new advanced technologies. The lack of reliable statistics on the Egyptian population usage of electronics in their shopping also represents an obstacle on carrying this research.

The major limitation of the study was the generalization of the data collected from the interviews conducted as the samples used to evaluate and apply this concept are restricted only on two of the previously mentioned grocery retail stores in Alexandria to explore their readiness for applying the VSC technology, this is due to the limited accessibility to contacts in other retailers.

References

Carrefour Co. (n.d) About us: Carrefour Co. http://www.carrefour.com.eg/aboutus.aspx. Accessed 13 July 2012

Hwang HJ, Seruga J (2011) An intelligent supply chain management system to enhance collaboration in textile industry. Int J u- e- Serv 4(4)

Kisperska-Moroń D (2010) Evolution of competencies of logistics and supply chain managers. Log forum. http://www.logforum.net/vol6/issue3/no3. Accessed 9 April 2012

Li G, Lin Y, Wang S, Yan H (2006) Enhancing agility by timely sharing of supply information. Supply Chain Manag: Int J 11(5): 425–427

Liu S-H, Liao H-L, Peng C-J (2005) Applying the technology acceptance model and follow theory to online E-learning users' acceptance behavior. Inf Syst 6(2): 176–178

Lorchirachoonkul W, Mo JPT (2010) RFID implementation with virtual infrastructures. Bus Process Manage J 16(6): 917–924

Nielsen T (2004) TESCO: a case study in supermarket excellence. Coriolis Res 4–7

Ramzy O, James RO (2012) The egyptian retailing industry. J Int Bus Econ 11(3): 1, 2, 5

Scholz C (2000) The virtual corporation: empirical evidence to a three dimensional model. The academy of management conference. Toronto. http://orga.uni-sb.de/bibliothek/artikel/aom_toronto.pdf. Accessed 2 Jan 2014

Smith SP, Johnston RB, Howard S (n.d) Virtual models in online shopping: do they help or hinder customers? http://is2.lse.ac.uk/asp/aspecis/20060120.pdf. Accessed 10 April 2012

Zarour N, Boufaida M, Seinturier L, Estraillier P (2005) Supporting virtual enterprise systems using agent coordination. Knowl Inf Syst 8(3):330–349

Fuzzy TOPSIS/SCOR-Based Approach in Assessment of RFID Technology (ART) for Logistics of Manufacturing Companies

Bartłomiej Gładysz and Krzysztof Santarek

Abstract It is not clear how to evaluate radio frequency identification (RFID) potential for use in production companies supply chains (internal and external). We propose Assessment of RFID Technology (ART) modular method (strategy, tactics, operations, evaluation) and elaborate tactical module (ART-T). ART-T application in a manufacturing company is illustrated. We apply TOPSIS (technique for order of preference by similarity to ideal solution), SCOR (supply chain operations reference) model, linguistic scales, fuzzy sets and questionnaires to answer the question "*which processes should be chosen for RFID-based reengineering and/or improvement*". Research output shows ART-T flexibility and ease of use at factory level.

Keywords RFID · Assessment of RFID technology (ART) · Supply chain · Manufacturing · Fuzzy TOPSIS · SCOR · Linguistic scales

1 Introduction

1.1 Radio Frequency Identification

For the purpose of this chapter we define RFID as following: every time when radio frequencies (RF) are used to identify and/or locate tagged object, we say about RFID technology. RFID technology is each technology based on RF and enabling the process of identification and/or location. For detailed description of physics and applications authors refer to Dobkin (2012), Finkenzeller (2010),

B. Gładysz (✉) · K. Santarek
Faculty of Production Engineering, Institute of Organization of Production Systems, Warsaw University of Technology, ul. Narbutta 86, 02-524 Warsaw, Poland
e-mail: b.gladysz@wip.pw.edu.pl

P. Golinska (ed.), *Logistics Operations, Supply Chain Management and Sustainability*, EcoProduction, DOI: 10.1007/978-3-319-07287-6_9,

Sweeney (2010), Yan et al. (2008). RFID is the technology that offers wide spread of applications for manufacturing companies. Its applications are not limited to predefined use cases and RFID is not only next autoidentification technology, better and faster than barcodes. RFID also enables dramatic changes in business processes that could be performed in a way previously impossible due to technology limitations. In the chapter we consider applications in logistics processes, both external and internal, of manufacturing companies. Kevin Ashton, a co-founder of the Auto-ID Center at Massachusetts Institute of Technology said: "*Calling RFID a radio bar code is like calling a car a motorized horse*".

Numerous authors report successful RFID deployments and benefits in many areas e.g. production logistics (Zelbst et al. 2012), warehousing (Osyk et al. 2012), external supply chains (Sarac et al. 2010). At the same time authors report difficulties in return on investment calculations and high costs. It means that RFID implementation needs to be preceded by detailed analysis.

1.1.1 Assessment of RFID Technology (ART) for Logistics of Manufacturing Companies

For the purpose of this chapter authors follow relabeling perspective on logistics versus supply chain (Larson and Halldorsson 2004). Table 1 shows list of selected papers on design, implementation and evaluation of RFID solutions for logistics processes. There were no papers found that treat about synthetic method indicating: if RFID should be considered, which and how processes should be improved, what will be the effect of changes.

Analyzed works allow evaluation of specific solution, classification of benefits in different application areas (especially in supply chains), indication of relations of qualitative benefits and measurable factors, evaluation in terms of characteristics of specific economies, branches and markets. All the authors (Bottani and Rizzi 2008; Bottani and Volpi 2006; Lee and Lee 2010; Sarac et al. 2010; Savino et al. 2012; Ugazio and Pigni 2010) conclude that RFID is promising technology that eliminates numerous dysfunctions of communication processes in supply chains and is implemented successfully for logistics processes in many branches. At the same time authors (Bottani and Rizzi 2008; Bottani and Volpi 2006; Lee and Lee 2010; Sarac et al. 2010; Savino et al. 2012; Ugazio and Pigni 2010) emphasize the need of detailed pre-implementation analysis and calculations of investment profitability. Starting point for our research was the output of Bottani and Volpi (2006), Ugazio and Pigni (2010), as those papers treat about preliminary evaluations of RFID implementation, but are limited and lack of:

- indication which processes should be improved, how to improve/reengineer processes, what will be the effect of changes,
- emphasis on manufacturing processes—they focus on sourcing and delivery processes, so they are not suited for manufacturing companies,
- issues related to business process modeling, reengineering,

Table 1 Selected papers on design, implementation and evaluation of RFID solutions for logistics processes

No.	Main problems and characteristic
Savino et al. (2010)	Designing business processes for supply chain using RFID—support for implementation of supply chain management information system
Wei et al. (2010)	Designing and evaluating RFID system, SCOR-based integration of infrastructure with manufacturing system
Sobottka et al. (2012)	Designing sophisticated and innovative RFID solutions
Leung et al. (2013)	Designing RFID systems as IT innovations—based on general framework for "mindful IT innovation"
Li et al. (2010)	Analyzing benefits and barriers of RFID implementation based on literature review and questionnaire surveys
Lutton et al. (2008)	Justifying RFID implementation—very high generalization level
Becker et al. (2010)	Evaluating RFID impact on processes based on authors reference model of processes and benefits
Bottani and Rizzi (2008), Ha et al. (2014), Wang et al. (2008)	Evaluating RFID implementation benefits in supply chain of selected market and branch
Hardgrave et al. (2005), Hardgrave et al. (2006)	Evaluating RFID implementation benefits in supply chain—case study
Sarac et al. (2010)	Evaluating RFID implementation benefits in supply chain—literature review of case studies (ex post evaluation) and methods for justification (ex ante)
Lee and Lee (2010), Ustundag and Tanyas (2009)	Evaluating RFID benefits based on simulation model
Dai and Tseng (2012)	Evaluating RFID benefits based on analytical models
Attaran (2012)	Evaluating RFID benefits based on literature review, case studies and critical success factors
Bottani and Volpi (2006)	Evaluating RFID potential impact based on SCOR
Ugazio and Pigni (2010)	Evaluating RFID implementation profitability

Source Gładysz Ph.D. research—in progress

- (Ugazio and Pigni 2010); considerations on how to choose process for improvement—it enables ex post evaluation,
- (Bottani and Volpi 2006); details and indication how to transpose from strategic to tactical level.

Literature lacks of research and synthetic, factory-level methods to solve problems related to design of RFID solutions on consecutive levels of strategy, tactics, operations and evaluation. Authors propose 4-steps ART (Assessment of RFID Technology) method to analyze RFID potential applications in logistics processes of manufacturing company. ART is described in Table 2.

ART application area is manufacturing company that has no RFID in its portfolio and is at initial phase of analyzing RFID potential for improvement of logistics processes. Details of ART on strategy level (ART-S) were discussed in Santarek and Gładysz (2014). In this chapter we focus on ART-T details and its application.

Table 2 ART modules

Module	Research questions
1. Strategy, ART-S	s1/"*if to consider RFID implementation?*"
2. Tactics, ART-T	t1/"*which processes to choose for improvement?*"
	t2/"*which processes could be RFID-supported?*"
3. Operations, ART-O	o1/"*how to reengineer processes on a basis of RFID support?*"
	o2/"*how to improve existing processes?*"
	o3/"*which tasks should be RFID-supported?*"
4. Evaluation, ART-E	e1/"*how to evaluate effectiveness of RFID-supported processes and its impact on a whole system?*"
	e2/"*what is an impact of effectiveness increase on financial performance?*"

Source Gładysz Ph.D. research—in progress

2 ART-T: Evaluation of RFID Applications from Tactical Perspective

2.1 Algorithm, Applied Techniques, Tools and Methods (SCOR, Linguistic Scales, Fuzzy Sets, TOPSIS)

An algorithm of ART tactic (ART-T) module is shown on Fig. 1. Its starting point is positive output of ART-S (strategy) module (see Santarek and Gładysz 2014 for details). It consists of 5 consecutive tasks and is flexible in terms of techniques, methods, models used to perform task. Reference models, tailored and company-specific models, expert evaluation, group decision making and direct questionnaires are used.

We validated ART-T module and applied:

- Supply Chain Operations Reference (SCOR) model for as-is modeling and a base for defining possible applications,
- linguistic scales for expert evaluations,
- fuzzy sets for mathematical representation of linguistic scales,
- fuzzy extension of Technique for Order of Preference by Similarity to Ideal Solution (TOPSIS) for multiple attribute group decision making (MADM) based on expert linguistic evaluations of alternatives (possible RFID applications for logistics processes of a company).

SCOR is the reference model maintained by Supply Chain Council (SCC 2014). There are many works reporting usefulness of SCOR for business process modeling and also for designing and evaluation of RFID solutions (Savino et al. 2010; Wang et al. 2008; Wei et al. 2010; Zelbst 2012). For more detailed information on SCOR we refer to Bolstoff and Rosenbaum (2011), SCC (2014).

Due to the objective alternatives uncertainty, human thought fuzziness and complexity of problem linguistics scales are used for qualitative expert evaluation.

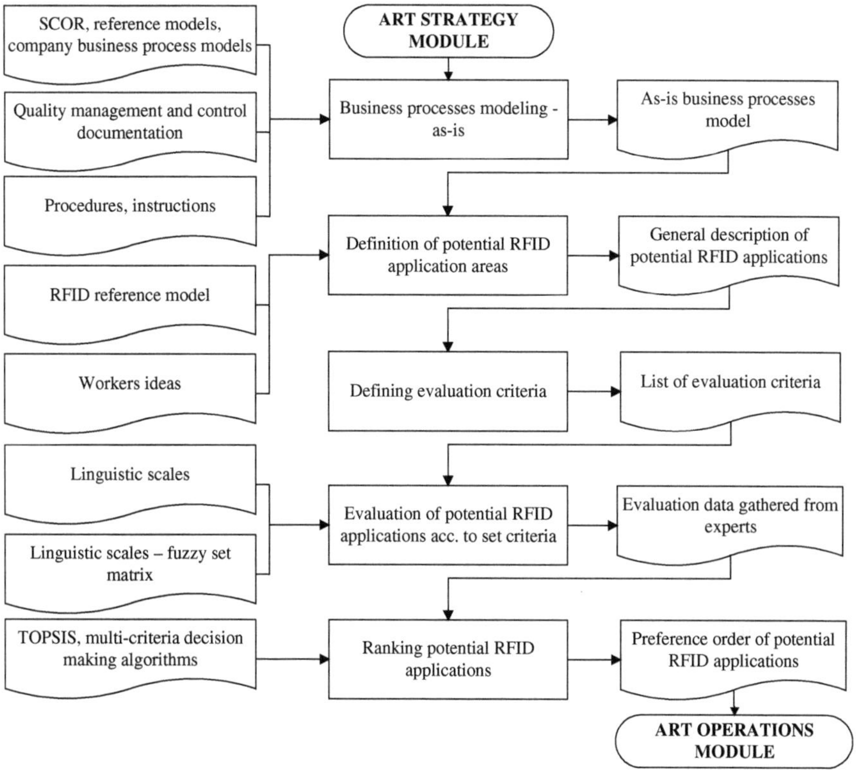

Fig. 1 ART-T steps (*source* Gładysz Ph.D. research—in progress)

Decision environment is impossible to be clearly defined, especially at the early phases of alternatives evaluation. At this stage information is unquantifiable and approximate value is tolerable so according to Xu et al. (2012) linguistics scales are proper mean to be used. For more detailed information on linguistic scales we refer to Chen and Hwang (1992).

Linguistic evaluations are uncertain and fuzzy, so fuzzy sets were used as representation of linguistic scales. For more detailed information on fuzzy sets, its algebraic properties, and applications for technology evaluations we refer to Kosiński and Prokopowicz (2004), Zadeh (1965) and Zimmermann (1991).

We use multiple attribute decision making (MADM) that is "*making preference decisions (such as evaluation, prioritization, selection) over the available alternatives that are characterized by multiple, usually conflicting, attributes*" (Hwang and Yoon 1981). MADM is used to select process to be promoted to next ART modules. We apply fuzzy TOPSIS for MADM and we use translation of linguistic scales into fuzzy sets. For more detailed information on TOPSIS, its fuzzy extension and its applications we refer to Chen and Hwang (1992), Hwang and Yoon (1981) and Mehrjerdi (2013). Table 3 provides linguistic scale and corresponding triangular fuzzy numbers. Linguistic scale and fuzzy numbers were

Table 3 Linguistic scale and fuzzy sets

Linguistic evaluation of potential RFID application	Triangular fuzzy number (l_i; m_i; u_i)
Very low—VL	(0; 0; 0.25)
Low—L	(0; 0.25; 0.5)
Medium—M	(0.25; 0.5; 0.75)
High—H	(0.5; 0.75; 1)
Very high—VH	(0.75; 1; 1)

Source Gładysz Ph.D. thesis research—in progress

chosen according to expert method due to the lack of data to use more sophisticated method (see Hryniewicz 2010) and are consensus of authors' discussion with RFID practitioners.

2.2 Defining Applications

The first step of ART-T (tactics) module is as-is modeling of business processes. For this purpose we used SCOR model that allows quick modeling, when proprietary company-specific models are not available. Definitions of possible RFID applications are based on as-is SCOR-based business process model. Each of SCOR level 1 processes is examined in terms of possibility of RFID application and expected RFID-related benefits. Table 4 shows the template for the definition of possible RFID application. Each described alternative is named according to the name of supported SCOR process (e.g. M for make). If there are more than one alternative related to the same process then number is added (e.g. M1 and M2).

2.3 Evaluation Criteria

Exemplary descriptions of criteria k1–k2 for evaluation of possible RFID solutions are shown in Table 5. Analogically questions and detailed descriptions were prepared for each criterion to guide expert through evaluation process of previously defined RFID applications (see Table 4). Questions were stated in a language natural for decision makers. Criteria k1–k4 are positive (maximization of criterion is expected) and k5–k7 are negative (minimization is expected). Scores for criteria are input of TOPSIS decision matrix. Full list of criteria is given in Table 6.

2.4 Collecting Data for Group Decision Making

Direct questionnaires were chosen for collecting data for group decision making. Each decision maker (expert, medium or high level manager related to potentially RFID-supported processes) receive questionnaire template and discuss with RFID expert (author) following issues:

Table 4 SCOR-based definition of possible RFID applications

Characteristic	Description
SCOR level 1 process	Source/make/deliver/return
Tagged objects	
Reading points	
Tag application point	
Short process description	General description of consecutive steps of new (improved) RFID-based process
Main expected benefits	In terms of cycles, stocks, transparency, information accessibility
Main costs	In terms of hardware and software (including integration with existing systems)

Source Gładysz Ph.D. research—in progress

Table 5 Evaluation criteria k1–k2

No.	Short description	Full description and questions
k1	Potential of labor effectiveness improvement	The need of acceleration of identification/localization tasks for raw materials/ork in progress/intermediates/finished goods or other assets of the company 1. Are identification tasks cycles significant part of total cycle time? 2. Are localization tasks cycles significant part of total cycle time? 3. Would automation and decrease of identification tasks cycles imply decrease of total cycle time? 4. Would automation and decrease of localization tasks cycles imply decrease of total cycle time? 5. Are cycle times too long? VH should be designated if identification is highly time-consuming
k2	Inventory levels	The need of decrease of inventory levels, improvement of inventory transparency and reliability of inventory data 1. Are inventory data needed in real time? 2. Is automation of replenishment task needed? 3. Are inventory data accessible on each necessary level (e.g. work in progress, finished good, warehouse locations etc.)? 4. Would better inventory transparency imply significant decrease of inventory levels and rotation cycles? VH should be designated if inventory levels are high, inventory transparency is crucial and inventory data should be accessible in real time

Source Gładysz Ph.D. research—in progress

- RFID basics, rules, potential, capability, strengths and weaknesses,
- possible RFID applications for logistics of surveyed manufacturer (see Table 4),
- evaluation criteria (see Table 5),
- linguistic scale (see Table 3).

Table 6 Scores for possible RFID applications

Criteria		Alternative 1	…	Alternative n	Criterion weight
Potential of labor effectiveness improvement	k1				
Inventory levels	k2				
Problems with accuracy	k3				
Process weight	k4				
Organizational barriers	k5				
Estimated costs	k6				
Technological barriers	k7				

Source Gładysz Ph.D. research—in progress

The questionnaire outputs are data collected according to the template shown in Table 6. The template is filled in for each expert separately. Experts score each possible application (alternative) according to criteria set (Table 5) and linguistic scale (Table 3), what is input for TOPSIS decision matrix. Scores of criteria weights are input for TOPSIS weight vector and are compatible with the same linguistic scale (Table 3).

3 Illustrative Example

ART tactic module was validated and processed in a manufacturer of medical goods that are produced in batches. Production process must meet rigorous standards of EN-ISO 14644-1:1999 (Class 7 for some departments). Source, deliver, make to order (MTO) and make to stock (MTS) processes (see SCC 2014) are in place.

Data were collected through direct interviews with employees responsible for surveyed areas (areas of possible RFID applications). Those employees were literally: logistics manager, production manager, process engineer and external RFID business solutions expert. Table 6 was filled in four times, each time by another employee.

Table 7 describes example of possible RFID application for manufacturer of medical goods. Three possible RFID solutions were analyzed: related to source (S), make (M) and deliver (D) process [see SCOR (SCC 2014)].

Table 8 shows collected data, decision matrix and weight vector for fuzzy TOPSIS for one expert. Abbreviations are compatible with the scale given in Table 3. Analogically data were collected for all surveyed experts.

Figure 2 shows fuzzy TOPSIS calculations for one expert. Analogically calculations were performed for all experts. ART calculations were automated and Excel spreadsheet was designed. Data collected from all experts (see Table 8) are

Table 7 Description of possible RFID application in area of make

Characteristic	Description
Process name	MAKE
Tagged objects	Production order paper (POP) or returnable transport item (RTI)
Reading points	Stations, inter-department gates automated identification by fixed reader when POP/RTI reaches designed chokepoint
Tag application point	Printing POP, papers with RFID inlays or inlays attached by first worker, coding POP number in RFID tag or coding tag of RTI at first station
Short process description	1. Attaching RFID labels to POP (possible closed-loop use case if labels inserted in POP cover)
	2. Coding RFID labels on first station
	3. Automated order identification when it reaches specific station or enters/leaves inter-department gates
	3a. Informing planning department about work in progress (WIP)
	3b. Informing purchasing department about materials used
	4. Automated registration of WIP status
	5. Automated data collection (e.g. timestamps) for documentation of production order
Main expected benefits	1. Acceleration of WIP identification and decrease of delay of data accessibility for planning department
	2. Planning based on real time data
	3. Decrease of mistakes in documentation
	4. Decrease of lead times
	5. Decrease of WIP and intermediate products stocks
	6. Real time control of production order status
	7. Less make to stock, more make to order
Main costs	1. ca. 1,000 PLN/fixed reader
	2. ca. 0.5–0.75 PLN/RFID inlay (non-reusable) ca. 1–2 PLN/RFID label (reusable)
	3. integration with MES and ERP
	4. ca. 6,000–8,000 PLN—mobile reader and/or
	5. ca. 8,000–10,000 PLN—RFID printer

Source Gładysz Ph.D. research—in progress

input for spreadsheet and all the calculations are executed automatically. Spreadsheet can be used by any other company. Figures 2 and 3 show screens of spreadsheet for ART-T module.

ART-T spreadsheet final results reached through group decision making with fuzzy TOPSIS are shown in Fig. 3. Three alternatives (S—source, M—make, D—deliver) were considered by four experts.

Possible RFID application for area of "*make*" got the highest score of relative closeness for all considered alternatives in fuzzy TOPSIS, so it is the solution that is promoted to the next ART module (ART-O). Alternative "*deliver*" has also high relative closeness parameter and could be also investigated after "*make*". Alternative "*source*" has definitely the worst score of relative closeness, so it will not be analyzed in ART-O.

Table 8 ART-T—expert 1, fuzzy TOPSIS decision matrix D^1 and weight vector w^1

Criterion		k1	k2	k3	k4	k5	k6	k7
Decision matrix D^1	Alternative source	M	M	VL	H	M	H	L
	Alternative make	VH	VH	L	VH	VL	M	M
	Alternative deliver	L	L	VL	M	H	H	L
Criteria weight vector w^1		VH	VH	VL	M	VH	VH	M

Source Gładysz Ph.D. research—in progress

DECISION MATRIX $D = [x_{ij}]_{m \times n}$

	k1			k2			k3			k4			k5			k6			k7		
	k1	k1	k1	k2	k2	k2	k3	k3	k3	k4	k4	k4	k5	k5	k5	k6	k6	k6	k7	k7	k7
S	0,25	0,5	0,75	0,25	0,5	0,75	0	0	0,25	0,5	0,75	1	0,25	0,5	0,75	0,5	0,75	1	0	0,25	0,5
M	0,75	1	1	0,75	1	1	0	0,25	0,5	0,75	1	1	0	0	0,25	0,25	0,5	0,75	0,25	0,5	0,75
D	0	0,25	0,5	0	0,25	0,5	0	0	0,25	0,25	0,5	0,75	0,5	0,75	1	0,5	0,75	1	0	0,25	0,5

WEIGHTED VECTOR $w = [w_j]$

0,75	1	1	0,75	1	1	0	0	0,25	0,25	0,5	0,75	0,75	1	1	0,75	1	1	0,25	0,5	0,75

NORMALISED DECISION MATRIX $R = [r_{ij}]_{m \times n}$

	k1			k2			k3			k4			k5			k6			k7		
S	0,25	0,5	0,75	0,25	0,5	0,75	0	0	0,5	0,5	0,75	1	0	0	0	0,25	0,33	0,5	0	0	0
M	0,75	1	1	0,75	1	1	0	0,5	1	0,75	1	1	0	0	0	0,33	0,5	1	0	0	0
D	0	0,25	0,5	0	0,25	0,5	0	0	0,5	0,25	0,5	0,75	0	0	0	0,25	0,33	0,5	0	0	0

u_i* lub l_i^-

1	1	1	1	1	1	0,5	0,5	0,5	1	1	1	0	0	0	0,25	0,25	0,25	0	0	0

a	0,75	0,5	0,25	0,75	0,5	0,25	0,25	0	0	1	0,75	0,5	0,75	0,5	0,25	1	0,75	0,5	0,5	0,25	0
	1	1	0,75	1	1	0,75	0,5	0,25	0	1	1	0,75	0,25	0	0	0,75	0,5	0,25	0,75	0,5	0,25
	0,5	0,25	0	0,5	0,25	0	0,25	0	0	0,75	0,5	0,25	1	0,75	0,5	1	0,75	0,5	0,5	0,25	0

NORMALISED WEIGHTED DECISION MATRIX $V = [v_{ij}]_{m \times n}$ $v_{ij} = (v_{ij1}; v_{ij2}; v_{ij3})$

	k1			k2			k3			k4			k5			k6			k7		
S	0,19	0,5	0,75	0,19	0,5	0,75	0	0	0,13	0,13	0,38	0,75	0	0	0	0,19	0,33	0,5	0	0	0
M	0,56	1	1	0,56	1	1	0	0	0,25	0,19	0,5	0,75	0	0	0	0,25	0,5	1	0	0	0
D	0	0,25	0,5	0	0,25	0,5	0	0	0,13	0,06	0,25	0,56	0	0	0	0,19	0,33	0,5	0	0	0

IDEAL POSITIVE SOLUTION v_j^+

v_{1+} = 1	v_{2+} = 1	v_{3+} = 0,25	v_{4+} = 0,75	v_{5+} = 0	v_{6+} = 1	v_{7+} = 0

IDEAL NEGATIVE SOLUTION v_j^-

v_{1-} = 0	v_{2-} = 0	v_{3-} = 0	v_{4-} = 0,06	v_{5-} = 0	v_{6-} = 0,19	v_{7-} = 0

DISTANCE FROM IDEAL POSITIVE SOLUTION $d(v_{ij}^k; v_j^+)$

S	0,569	0,569	0,217	0,421	0,000	0,672	0,000
M	0,253	0,253	0,204	0,355	0,000	0,520	0,000
D	0,777	0,777	0,217	0,503	0,000	0,672	0,000

DISTANCE FROM IDEAL NEGATIVE SOLUTION $d(v_{ij}^k; v_{j-})$

S	0,532	0,532	0,072	0,438	0,000	0,199	0,000
M	0,879	0,879	0,144	0,476	0,000	0,504	0,000
D	0,323	0,323	0,072	0,308	0,000	0,199	0,000

	$d_i^{\rightarrow}$	d_i^-
S	2,374	1,811
M	1,759	2,494
D	2,764	1,350

TOTAL DISTANCE FROM IDEAL POSITIVE SOLUTION ($d_i^{\rightarrow}$); TOTAL DISTANCE FROM IDEAL NEGATIVE SOLUTION (d_i^-)

Fig. 2 Fuzzy TOPSIS results and calculations for one expert (*source* Gładysz Ph.D. research—in progress)

TOTAL DISTANCE FROM IDEAL POSITIVE SOLUTION		d_i^{+}	d_i^{-}	TOTAL DISTANCE FROM IDEAL NEGATIVE SOLUTION
	S	2,374	1,811	
	M	1,759	2,494	
	D	2,764	1,350	

RELATIVE CLOSENESS C_i^-		RANKING
S	0,433	2
M	0,587	1
D	0,328	3

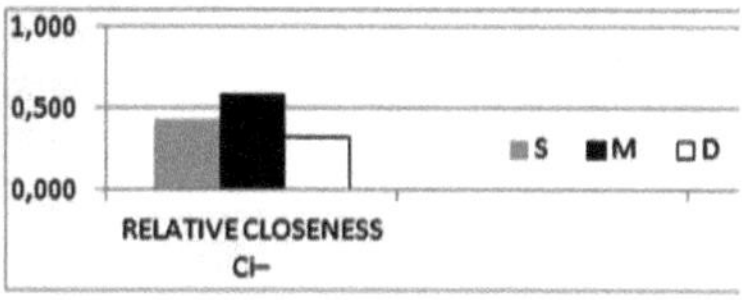

Fig. 3 Final results of ART-T (*source* Gładysz Ph.D. research—in progress)

4 Conclusion

Proposed Assessment of RFID Technology (ART) method was applied to environment of manufacturer of medical goods, but it is designed in a way to be applicable in any manufacturing company. ART details were designed for the purpose of RFID evaluation for logistics of manufacturing company, but the method itself could be customized and applied also for other innovative technologies and early phases of decision making. ART-S (see Santarek and Gładysz 2014) and ART-T applications were successful and show its big potential to be used at early phases of decision making process.

ART-T as a module of more holistic ART method is flexible in terms of techniques and tools used for solving problems and answering formulated questions. Authors chose and showed that SCOR, fuzzy sets, linguistic scales and TOPSIS are proper tools to solve problems related to formulated research questions. Those questions were namely: t1/"which processes to choose for improvement?" and t2/ "which processes could be RFID-supported?" (see Table 2). SCOR was used as a starting point for as-is modeling of business processes and defining potential RFID applications. Linguistic scales and fuzzy sets were used for scoring previously defined alternatives and criteria. Criteria set with detailed descriptions are provided to score each alternative. Fuzzy TOPSIS was used to find alternative that will be promoted to the next ART module (ART-O, see Table 3, Fig. 1).

Although ART-T utilizes several tools, it is easy in application and factory-level what was proved by good understanding and cooperation with surveyed experts. ART is supported by Excel spreadsheet, what enables easy correction of weights and matrixes to check impact of such changes on fuzzy TOPSIS results.

Next steps of research will be extension of already validated modules (strategy and tactics) and application of operations and evaluation modules for the same company, execution of the whole ART for other manufacturing companies, comparison of different tools that can be used in ART (e.g. replacing TOPSIS with other MADM method).

References

Attaran M (2012) Critical success factors and challenges of Implementing RFID in supply chain management. J Supply Chain Oper Manage 10(1):144–164

Becker J, Vilkov LWB, Winkelmann A (2010) A model based approach for calculating the process driven business value of RFID investments. Int J Prod Econ 127(2):358–371

Bolstoff P, Rosenbaum R (2011) Supply chain excellence. A handbook for dramatic improvement using SCOR model. AMACOM, Saranac Lake

Bottani E, Rizzi A (2008) Economical assessment of the impact of RFID technology and EPC system on the fast-moving consumer goods supply chain. Int J Prod Econ 112(2):548–569

Bottani E, Volpi A (2006) A SCOR-model based methodology to assess potential impact of RFID technology on logistic processes. In: Proceedings of RIRL, Pontremoli, Italy, pp 118–129, Sep 2006

Chen SJ, Hwang CL (1992) Fuzzy multiple attribute decision making: methods and applications. Springer, Berlin

Dai H, Tseng MM (2012) The impacts of RFID implementation on reducing inventory inaccuracy in a multi-stage supply chain. Int J Prod Econ 139(2):634–641

Dobkin DM (2012) The RF in RFID: UHF RFID in practice, 2nd edn. Newnes, Oxford

Finkenzeller K (2010) RFID handbook: fundamentals and applications in contactless smart cards and identification. Wiley, New York

Ha O-K, Song Y-S, Chung K-Y, Lee K-D, Park D (2014) Relation model describing the effects of introducing RFID in the supply chain: evidence from the food and beverage industry in South Korea. Pers Ubiquit Comput 18(3):553–561

Hardgrave BC, Waller M, Miller R (2005) Does RFID reduce out of stocks? A preliminary analysis. ITRI UARK, Fayetteville (ARK)

Hardgrave BC, Waller M, Miller R (2006) RFID's impact on out of stocks: a sales velocity analysis. ITRI UARK, Fayetteville (ARK)

Hryniewicz O, (2010) Fuzzy set membership function - design and interpretation methods" (in Polish: "Funkcja przynależności zbioru rozmytego - metody konstrukcji i interpretacji"), in Studies of Polish Association of Knowledge Management (in Polish: Studia i materiały PTZW), vol. 31, no. 31, pp 8–17

Hwang CL, Yoon K (1981) Multiple attribute decision making: methods and applications. Springer, New York

Kosiński W, Prokopowicz P (2007) "Fuzzy sets algebra" (in Polish: "Algebra liczb rozmytych"). Appl Math (in Polish: Matematyka Stosowana) 46(5):37–63

Larson PD, Halldorsson A (2004) Logistics versus supply chain: an international survey. Int J Logistics Res Appl 7(1):17–31

Lee I, Lee B-C (2010) An investment evaluation of supply chain RFID technologies: a normative modeling approach. Int J Prod Econ 125(2):313–323

Leung J, Chu S-C, Cheung W (2013) Design research guidelines for mindful IT innovations: the case of RFID innovation in supply chain management. In: 46th international conference systems science (HICSS), Hawaii

Li S, Godon D, Visich JK (2010) An exploratory study of RFID implementation in the supply chain. Manage Res Rev 33(10):1005–1015

Lutton E, Skinner G, Regan B (2008) RFID rationale and deployment methodology. In: 8th WSEAS international conference application computer science, Venice

Mehrjerdi YZ (2013) Group decision making process for RFID-based system selection using fuzzy TOPSIS approach. Artif Intell Res 2(3):1–15

Osyk BA, Vijayaraman BS, Srinivasan M, Dey A (2012) RFID adoption and implementation in warehousing. Manage Res Rev 35(10):904–926

Santarek K, Gładysz B (2014) "Strategic evaluation of RFID implementation effects" (in Polish: "Ocena strategiczna efektów wdrożenia RFID"). In: Knosala R (ed) Innovations in management and production engineering (in Polish: Innowacje w Zarządzaniu i Inżynierii Produkcji), vol 1. OW PTZP, Opole, pp 191–202

Sarac A, Absi N, Dauzère-Pérès S (2010) A literature review on the impact of RFID technologies on supply chain management. Int J Prod Econ 128(1):77–95

Savino M, Holimchayachotikul P, Brun A (2010) SCOR model based on RFID enable supply chain management module on ERP. In: SKIMA—4th international conference software, knowledge, information management and applications, Paro, Bhutan

SCC (2014) SCOR I Supply Chain Council. https://supply-chain.org/scor. Accessed 11 Feb 2014

Sobottka T, Leitner R, Sihn W (2012) Methodology for the development of RFID value added services to improve supply chain operations. Trans FAMENA 36(2):67–77

Sweeney PJ II (2010) RFID for dummies. Wiley Publishing, Indianapolis

Ugazio E, Pigni F (2010) Measuring RFID benefits in the supply chain. In: Information systems: people, organizations, institutions, and technologies. Springer Physica-Verlag, Berlin, pp 469–476

Ustundag A, Tanyas M (2009) The impacts of radio frequency identification (RFID) technology on supply chain costs. In: Transportation research part E, pp 29–38

Wang S-J, Liu S-F, Wang W-L (2008) The simulated impact of RFID-enabled supply chain on pull-based inventory replenishment in TFT-LCD industry. Int J Prod Econ 112(2):570–586

Wei KQ, Zheng L, Xiang Q, Chen X (2010) Applications of RFID in a SCOR-model driven enterprise production system. In: IEEE 17th IE&EM, Xiamen

Xu Z (2012) Linguistic decision making. Theory and methods. Springer, Berlin

Yan L, Zhang Y, Yang LT, Ning H (2008) The internet of things: from RFID to the next-generation pervasive networked systems. Auerbach Publications, Boca Ranton

Zadeh L (1965) Fuzzy sets. Inf Control 8(3):338–353

Zelbst PJ, Green KW, Sower VE, Reyes PM (2012) Impact of RFID on manufacturing effectiveness and efficiency. Int J Oper Prod Manage 32(3):329–350

Zimmermann H (1991) Fuzzy sets theory and its applications, 2nd edn. Kluwer Academic Publishers, Boston

Comarch EDI Platform Case Study: The Advanced Electronic Data Interchange Hub as a Supply-Chain Performance Booster

Piotr Reichert

Abstract Effective supply-chain management is an ever greater requirement in modern enterprises. In addition to optimising operating costs, companies are looking for ways to be more flexible and to cut time-to-market in a new and rapidly changing economic environment. Enterprises that have numerous business partners are involved in many different supply chains. This means using a number of different sets of technologies grouped around highly sophisticated services. These services can provide access to real-time document exchange and can be articulated with financial institutions' internal ERP systems and other marketplaces. This case study is largely about the Comarch Electronic Data Interchange (EDI) platform, which has been providing an effective solution to meet all B2B needs for over 10 years. The main purpose to meet in this document is to describe potential services, which can be developed after reaching some critical mass of documents turnover and customers. Those services can influence boosts in performance in supply-chain management between suppliers and large multiple retailers and show how a common environment (technology and business) could create added value for enterprises.

Keywords EDI · Supplier · Multiple retailer · Chain management · Corporate finance

1 Introduction

The present very competitive business environment is compelling companies of all sizes to use all of the available tools to become more cost effective, faster or more innovative. At its outset around 2003 Comarch EDI was innovative in that it

P. Reichert (✉)
Comarch SA, Al. Jana Pawła 39A, 31-864 Kraków, Poland
e-mail: piotr.reichert@comarch.com

P. Golinska (ed.), *Logistics Operations, Supply Chain Management and Sustainability*, EcoProduction, DOI: 10.1007/978-3-319-07287-6_10,

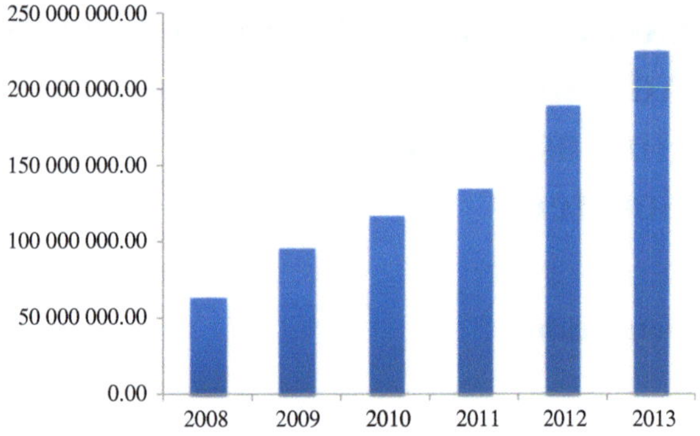

Fig. 1 Number of total documents exchanged 2009–2013 (own *source*)

offered a communication channel from large multiple-retailers to producers or distributors. A strong effort was made to persuade small companies to join this newly-created hub and to bear the additional cost of doing so. At first, companies treated adoption of the hub as a lesser evil when compared to ceasing to do business with retailers.

Invoices were the very first messages to be exchanged. For this there were two major alternatives in the early years of the last decade: IBM AS400 and Comarch EDI. The latter was then the challenger in this very specific market but, after 10 years' development, the Comarch solution was able to integrate 60 messages and was still being extended. Initially it concentrated on attracting as many customers as possible—both multiple retailers, an example of which would be a major supermarket chain, and end customers, such as the entities that supply these chains.

Comarch EDI after 10 years' development:

- Serving 19,000 end customers
- Operating in 33 countries
- Supporting 12 languages
- Exchanging more than 200 million documents and constant growth beginning from 2008 (Fig. 1)
- Integrating more than 60 document types
- Using all market standards (EDIFACT, TRADACOM, XML, AS2, RVA)
- Running projects according to proven procedures for more than a decade

Number of documents processed by Comarch EDI.

Based on the facts already set out, and on a survey prepared by Forrester and presented at Fig. 2, we may conclude that the main needs of end customers focus on:

- Reducing operating costs, which are continuing to rise
- Exchanging electronic documents with partners.

There is still little awareness of employing EDI messages to create new services and of the significant boost this can provide to efficiency and liquidity when running an operational business (Vollmer et al. 2011).

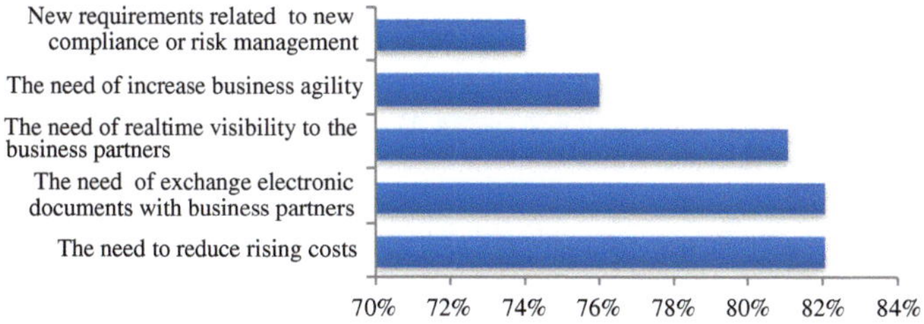

Fig. 2 Key business issues driving the need for B2B improvement

Before 2009 the primary focus was on the customer, but in the recession the attitude shifted to 'cash is king'. Enterprises—especially the more agile among them—made good use of the crisis to implement new services and to take a more open approach to other ways to improve liquidity. They were trying to find other financial partners to join the system that had been created.

Now, with the regulations that will structure Master Management Data beginning from 2014, new services can be expected to become available. This will create new opportunities for logistics projects such as in validation of fiscal data, inventory management and freight.

Moreover, supporting further integration with logistics partners (especially transport) and monitoring delivery to optimize empty runs will present a great challenge to the European and Polish markets. In the passages that follow we will look at how the Comarch EDI hub organises the processes and consider the coming challenges that will influence the direction and substance of the road map.

This platform is very innovative and comprehensive, there is not much bibliography on market about similar solutions. Companies serving incomplete services treat this data as confidential one.

Currently on Polish market there are four significant EDI operators: Comarch SA, Edison, Xtrade and Ininite.[1] Definitely those companies are focused on basic services but only Comarch has created added services, which allows adding new business partners and significant boosting of turnovers in the future.

2 Classical EDI Onboarding

The onboarding process has now become very standardized. However back in Comarch EDI's early days there was no experience in Poland of how to accomplish this professionally and—most importantly—fast. Yet Comarch EDI proved to have the resourcefulness and vision to develop a set of tools, along with a trained and experienced team, to make onboarding highly cost and time efficient.

[1] www.ecr.pl

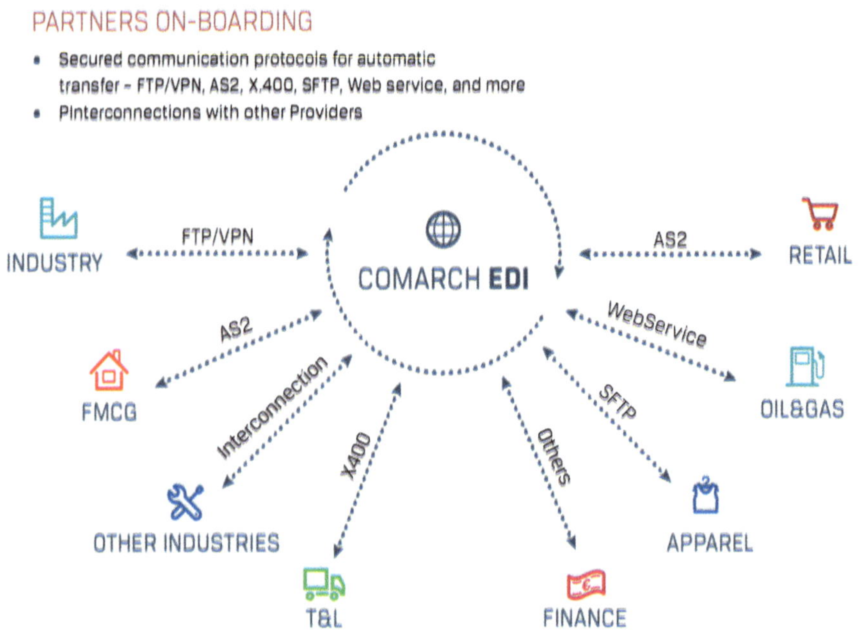

Fig. 3 Partners onboarding

The process of plugging the customer into the hub involves many complex steps. These include setting up connections with multiple retailers, establishing standards and formats and preparing special tools for supporting end customers so that they too can join the system. The complexity and the business players are presented at Fig. 3.

The standard processes are based on the following modules:

- Onboarding module (support in 12 languages),
- Set onboarding scripts,
- Interconnect integration,
- Integration module,
- EDI Web (this module is dedicated to the daily operational work of customers).

Depending on the enterprise's capabilities there are a set of communication channels that can provide secure protocols, such as FTP/VPN, AS2, X400, SFTP, Web services and others. Comarch EDI was able to prepare the first stage of partner onboarding based on these channels. Then, using the dedicated modules, consultants began a campaign aimed at end customers. Launching the first partner went very slowly, but roll-out became smoother and swifter once a certain number of end customers had been passed. This is because each customer typically does business with several retailers and other partners. The key success factor in the early stages was reaching a breakthrough number of customers as soon as possible. In this way onboarding could be repeated more easily with subsequent partners: if end customers are trained properly at the first stage adding another partner is significantly more efficient and cost driven. The table below depicts the procedure for onboarding customers.

In the case of Comarch EDI, the stiffest challenge was to onboard the first 500 customers, which took approximately 18 months beginning from the early years of the last decade. After that the process gathered pace so that 12,000 customers had been onboarded by 2009.

In the first 5–8 years of onboarding the dominant motivation was to reduce operational costs. There was no thought then of using message exchange to boost and optimize the supply chain and to secure financial cash flow—and nor was there the technology needed to achieve it. The 1.2 billion documents that have been processed and stored can be used to write algorithms and prepare a behaviour scheme to formulate additional procedures.

3 Extended EDI Process

3.1 Comarch EDI in the Internal Organizational Systems of Enterprises

One of the many interesting uses of Comarch EDI is the incorporation of a platform for an internal logistics solution that optimises the speed and flow of document processing. The project launched at PepsiCo, whose initial aim was to reengineer 28 business processes from semi-automated to fully automated document exchange, is an excellent example. The desired outcome was to limit the number of Logistics Service Platforms (LSP) and by so doing to unlock further growth opportunities. Once achieved, this leveraging of operating services would deliver cost efficiencies by reducing duplicated operations, documents and resources.

The most rigorous challenge was embedding Comarch EDI into the middle layer, which connects the LSPs with the internal PepsiCo systems (SAP, Roadnet, CGK), and integrating it with the newly-designed processes and organizational structure.

Messages exchanged during the processes:

- INSEDES—instruction to dispatch,
- HANMOV—cargo/goods handling and movement,
- OSTRPT—order status report,
- IFCSUM—multiple consignment transport instruction,
- IFTMAN—arrival notice,
- IFTMIN—transport instruction,
- IFTSTA—transport status,
- IFTMBC—booking confirmation,
- IFTMBF—firm booking (Fig. 4).

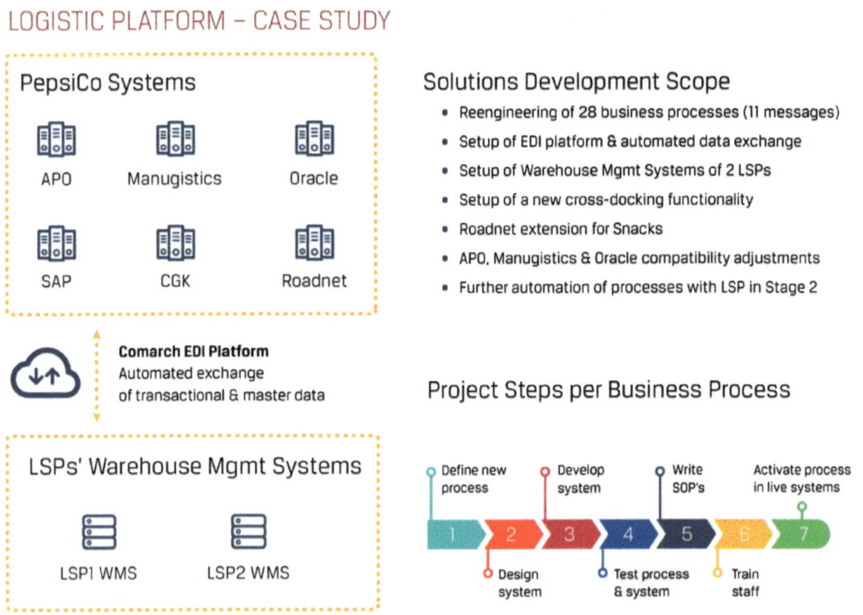

Fig. 4 Logistics platform case study

The pre-implementation PepsiCo structure had:

- 120,000 m^2,
- 68 DCs (including 30 providers),
- 13 Cross docks,
- 48 Bins.

The implementation created new cost and operational savings:

- 62,000 m^2,
- 9 Regional DCs,
- 48 Cross docks,
- 46 Bins.

The structured messages used in the new processes provide excellent base data for further optimization and load balancing. Gathering these data makes it possible to run analytical tools and to prepare new logistics rules for the following:

- Optimizing goods storage in certain DCs,
- Optimizing transport between units,
- Prediction in seasonal storage,
- Optimizing production.

A potential new challenge for PepsiCo could involve exchanging logistics data with other similar companies. The foremost reason for this would be the further pursuit of excellence in freight forwarding. One aspect of this is optimizing delivery within internal structures so that processes can be performed only up to a certain level. The use, for example, of a wider range of DCs coming from other companies will significantly increase transport efficiency. Of course this must be done through independent institutions, which can work only for optimizing purposes. Returning to the overall picture, we can see that PepsiCo took the first step in implementing Comarch EDI for internal processes and is now in a position where it could decide to use the documents and messages for further analysis and, finally, as part of a global logistics system.

3.2 e-Invoicing

e-Invoicing in the Comarch EDI platform purely involves the exchange of electronic data without any document scanning. The general process begins with the electronic issuing of invoices through the following channels: EDI, Web EDI, EDI to PDF and e-mail PDF (Fig. 5).

The e-invoice is then internally processed by the platform beginning with transformation from the issuer's format to a standard format or to the format demanded by a partner. The transformed data are validated in line with business and legacy rules (in accordance with EU, RU and UA regulations) and signed by certified signature or by a signature that is commonly accepted by the partners (depending on the country). They are then delivered to the partners, which mainly involves frequently used channels such as AS2, WebService, FTP/VPN, Web EDI, uploading from portal sites and e-mail.

The process of e-invoicing is still growing—especially in Europe where it is now responsible for transferring approximately 13 % of total e-invoicing in the consumer segment and for 20 % in the business segment. This amount is increasing annually by 20 and 29 % respectively.

The main drivers of growth in electronic invoicing are now operational cost and time efficiency, but there is potential for introducing new services to the market after gathering data for a few years. All efforts are now being directed at regularising the process and at creating a European invoice standard which can be implemented in ERP and in IT invoicing systems.

3.3 EDI and Financing

From the development point of view there was an interesting situation in 2009 when banks expressed interest in B2B financing. It was then that the first ideas were

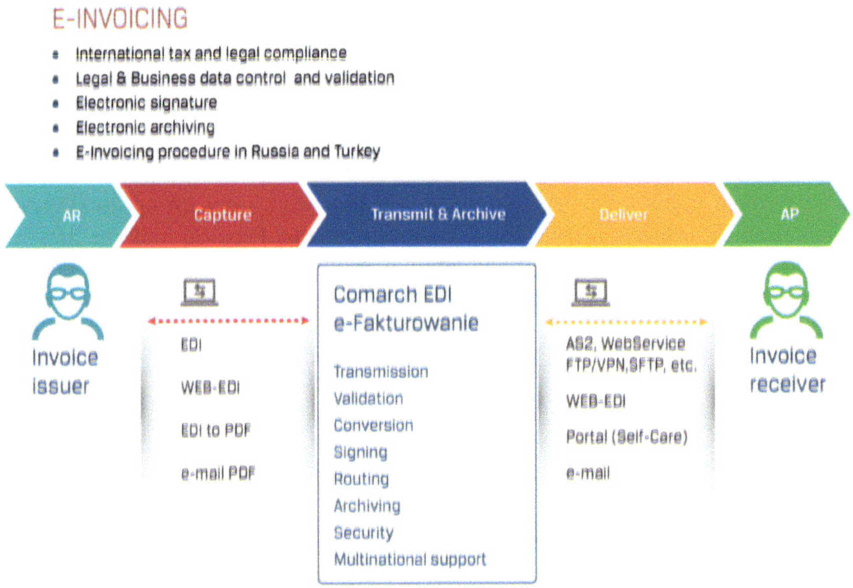

Fig. 5 e-Invoicing process

formulated regarding the use of a huge number of transactions to offer financial services or, in other words, their use to extend corporate financing services.

The banking industry had all the necessary information to rate companies, including that for multiple-retailers and suppliers, at its disposal. This information was very useful in the case of risk assessment. All the information was stored in the banking systems—especially historical transaction data—yet the issue was that it was only payment transactions that were being utilised. The remaining messages, such as invoices, were lying disregarded on the premises of the companies concerned or in the Comarch EDI hub.

From the point of view of financing, this information was crucial for creating new services, which could be offered to enterprises and accepted by multiple-retailers as well. The central challenge was to create a very usable tool and incorporate it into the supply chain. This process began with the issuing of invoices by suppliers to multiple-retailers or to assembly and manufacturing facilities.

To add new players to the message-exchange process involves the sophisticated analysis and alignment of two perspectives and sets of sometimes similar, and sometimes different, information: that of the creditor and debtor.

The bank has all the ratings information based on historical data at its disposal and seeks a common data set:

- Information on the number of clients they have in the banking system is present in the Comarch EDI hub
- Invoices, which can be accepted by the creditor and debtor for financing.

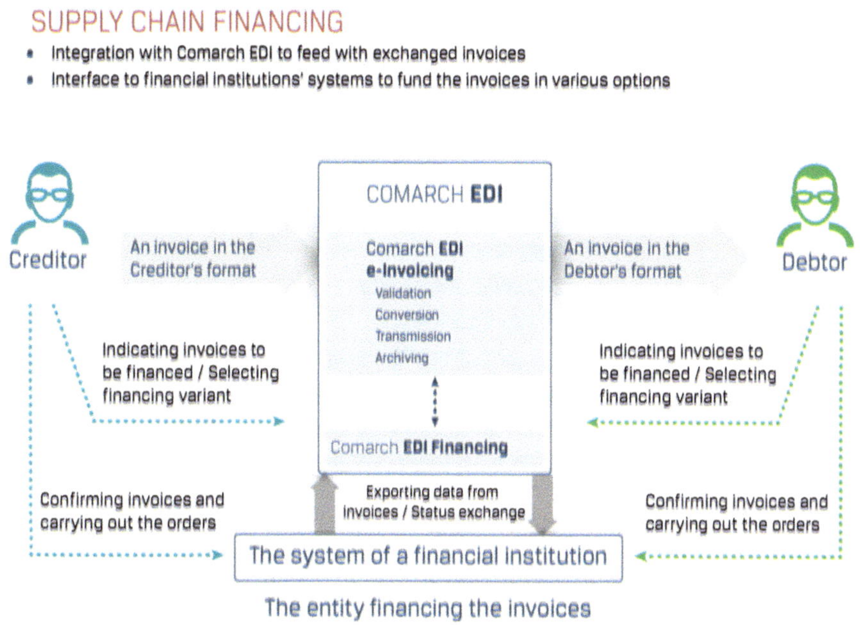

Fig. 6 Supply chain financing process

The creditor can offer invoices for financing through Comarch EDI once the bank has made a preliminary assessment and agreement has been reached concerning the transaction conditions. The creditor can see the debtors whose transactions can be authorised in the system. The debtor has only an acceptance or authorisation role in the system; all rating is done by the Bank. The simplified process is presented at Fig. 6.

There are many variations on these processes that can be applied to the different situations. The most common are:

1. The debtor first accepts invoices for financing
2. The debtor accepts invoices after a proposal by a creditor
3. The debtor proposes invoices for financing and the creditor accepts.

The second process, that in which the debtor accepts invoices following a creditor's proposal, is used in 90 % of transactions on the Polish market. After finalizing the delivery stage, the supplier (creditor) uploads and marks this information in the Comarch EDI system. The multiple-retailer (debtor) has access to its own invoices, which are issued and marked for financing. The debtor has the option to confirm the delivery process or add some remarks such as corrections or deductions.

1. The debtor selects the invoices that can be financed and sends an authorising order

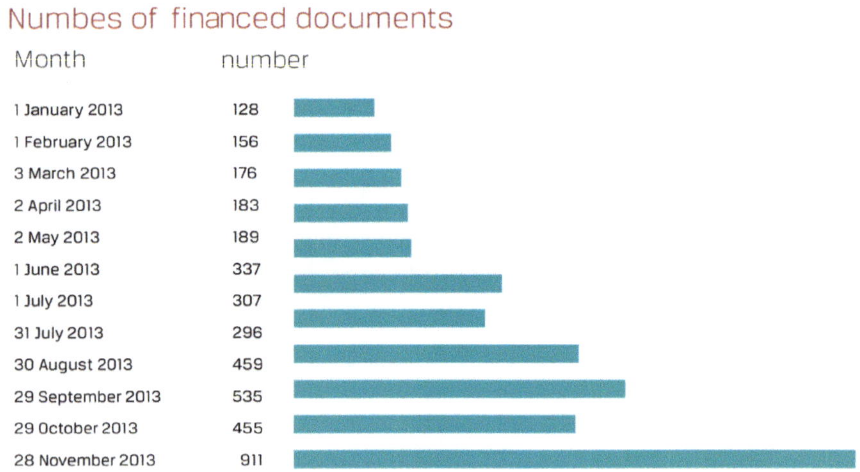

Fig. 7 Numbers of documents financed via Comarch EDI in 2013 (own *source*)

2. Invoices are loaded to PB24; the debtor is automatically signed in from Comarch EDI to PB24 and sees the invoices for acceptance
3. The supplier sees all the invoices proposed for financing by the debtor in Comarch EDI and is automatically signed into PB24. The supplier then accepts the documents to be financed
4. The debtor gives final acceptance to all of the documents from PB24 or Comarch EDI
5. The documents are sent to the TSS system for further financing according to bank procedures. The money is transferred to the creditor
6. PB24 exchanges data with Comarch EDI every day or in real-time
7. The debtor pays back the invoices and PB24 accounts for them.

However complex this process may appear on chapter it can be executed in minutes in real life situations given that other conditions, such as risk assessment and rates agreements, have been satisfied. For the time being there are a number of people on all sides who are responsible for authorising all of these steps, but in future the authorisation stages can be fully automated and authorisation can be performed at very high levels.

This service is very innovative and its use has been growing. Figure 7 shows the numbers of transactions in 2013. Its further successful development depends on marketing campaigns—mostly those conducted by the banks—and on the continued training and education of companies.

Though these numbers at Fig. 7 are not yet impressive, still this extremely innovative service has very considerable potential. It must be borne in mind that it was only launched at the end of 2012 and thus far has been adopted by only one bank: PEKAO SA (Unicredit Group). The optimal situation will arrive when these services are available through larger groups of banks and are operated by Comarch EDI.

4 Data Handling and Data Sharing

The EDI platform has now been on the market for more than 15 years and has added 25 % more documents annually. It is of course a challenge to store these numbers of records, but it is one that also generates new opportunities. Annual growth of approximately 300 million documents involves implementing new technology and a highly refined and customized solution. The platform cannot be built only on one homogeneous database and middle layer. This will not be cost effective because the result will be spending the huge savings from increased use of electronic data on additional data centres, hardware or licenses (Fayard et al. 2012).

Until this data is being used solely for operational purposes, such as opening and closing processes, data archiving or message exchange, there will be little demand for IT technology. This may change, though, when the platform becomes a base for data handling and Big data analysis. This will involve immediate access to data and processing complex queries and algorithms, which may not be feasible with existing infrastructure.

The main operations are now based on Oracle Database and on the Microsoft BizTalk services bus, while the portal and web solutions are made both in .Net and Java base technology. This environment still requires considerable optimization work to be ready in reasonable system-response times. If we are to seriously contemplate using the acquired data for Big data analysis, it will be necessary to separate the data from the operational work.

This new technology should be extended to:

- thick databases as a base for algorithms and rules engines: these databases are now being used to handle and search vast amounts of data in brokerage houses for algorithms trading,
- rules and event engines as a set of tools for creating and searching rules in stored historical data,
- tools for the incorporation of new algorithms to present data and predict some behaviours,
- trained staff for consistent data handling and transforming existing processes.

These measures can lead to significant improvements in efficiency—especially by locating logistics and transportation gaps, looking for seasonal periods and finding new opportunities for building additional services that support day-to-day business.

There is very frequent mention in the analyses from Gartner and Forrester of a strong need for a multi-enterprise application with a common or shared data and process model. The focus is on the E2E community, including multiple tiers of demand channel partners and upstream suppliers (Sterneckert and Griswold 2013). This presents Comarch with a considerable opportunity, though one that will be hard to take. There are a number of competing solutions for this sort of multi-enterprise application but none is dominant. An extension to the Comarch EDI platform could provide the answer. Having all of the data originating from multi retailers and suppliers in one place and creating services based on them could prove both interesting and desirable. Such a service could be used by a freight forwarding operator to optimize deliveries and by distributors for prediction and load balancing in warehouse centres (Titze and Barger 2013).

5 Summary

The EDI platforms on the European market are more or less the same and will follow the same pattern of greater and greater data standardization—where they will diverge is in terms of numbers of documents and customers. The vital ingredient of all platform developments is having a clear business and technology roadmap so that customers receive a precisely defined account of what the services will be like in the future and of the benefits they will generate for them.

The major challenge and opportunity in EDI platform development lies in working on behalf of customers to identify new areas that can yield efficiency gains and greater profits.

The Comarch EDI platform is now focused on technology diversification and has the potential in the near future to become a superb base for using advanced data handling tools. This will naturally create new opportunities and could be a key factor in igniting further growth in the platform's popularity.

Services continue to be offered alongside the EDI platform via the traditional channels of web portals for B2B and B2C customers and mobile applications.

Further integration with logistics operators for delivery tracking and optimizing empty runs represents a further immense opportunity.

Adding new services like invoice financing, dynamic discounting or freight forwarding optimisation create new value on market, which is trailblazing attitude. Those innovations can defiantly boost logistic environment in case of efficiency, time and cost savings.

References

Demirkan H, Delen D (2013) Leveraging the capabilities of service-oriented decision support systems: putting analytics and big data in cloud. Decis Support Syst 55(1):412–421

Fayard D, Lee L, Leitch R, Kettinger W (2012) Effect of internal cost management, information systems integration and absorptive capacity on inter-organizational cost management in supply chains. Accounting, Organizations and Society, 37

Humphrey DB (2010) Retail payments: new contributions, empirical results, and unanswered questions. J Banking Finance 34(8):1729–1737

Moore S (2011) Gartner forecasts global business intelligence market to grow 9.7 percent in 2011. Gartner Research, 02.2011

Schubert P, Legner C (2011) B2B integration in global supply chains: an identification of technical integration scenarios. Accounting, Organizations and Society, 20

Sterneckert K, Griswold M (2013) Six strategies can turn retailers into supply chain leaders. Gartner, 05.2013

The Economist (2010) Data, data everywhere, a special report on managing information, 02. 2010

Titze Ch, Barger Jr R (2013) Use cases for supply chain visibility. Gartner, 10.2013

Vollmer K, Gilpin M, Rose S (2007) B2B integration trends: message formats. Forrester Research, 06.2007

Vollmer K, Gilpin M, Rose S (2011) The future of EDI. Forrester Research, 02.2011

www.comarch.com

www.forrester.com

www.gartner.com

http://ec.europa.eu/internal_market/payments/einvoicing/index_en.htm

www.ecr.pl

Modelling Integration Process Planning in the Supply Chain Using SOP Approach

Michał Adamczak, Łukasz Hadaś, Roman Domański and Piotr Cyplik

Abstract This chapter presents the concept behind a simulation model reflecting the functioning of a supply chain in the context Sales and Operations Planning (SOP) working in business in practice. The methodology assumed by the authors was based on: the analysis of integration models of planning processes, with a particular focus on the concept of Sales and Operations Planning, accompanied by the analysis of the surveys findings. Literature on the subject features SOP mainly as a business tool based on a collective decision making on the tactical level. The solution put forward in this article represents a systemic approach to the planning process integration. It includes the mechanisms of coordination put in place in a wide process planning structure, applied both in companies and within the supply chain. The model serves as a point of departure for further works related to the simulation of the material flow within a supply chain in the context of integrated planning. The final results of whole research will be employed in defining the impact of integrated process planning on the functioning of the supply chain, in particular on the coordination of the material flow and achieved results, as measured by economic and logistics metrics.

Keywords Planning process · SOP planning · Process modelling

1 Introduction

Many enterprises operating on today's market are faced with the challenge of unstable demand, short lead teams and reduced life cycle of goods. One of the ways to solve these problems is to coordinate and integrate logistical processes (including planning processes) within a supply chain. Coordination was defined by

M. Adamczak (✉) · Ł. Hadaś · R. Domański · P. Cyplik
Poznan School of Logistics, Estkowskiego 6, 61-755 Poznań, Poland
e-mail: michal.adamczak@wsl.com.pl

P. Golinska (ed.), *Logistics Operations, Supply Chain Management and Sustainability*, EcoProduction, DOI: 10.1007/978-3-319-07287-6_11,

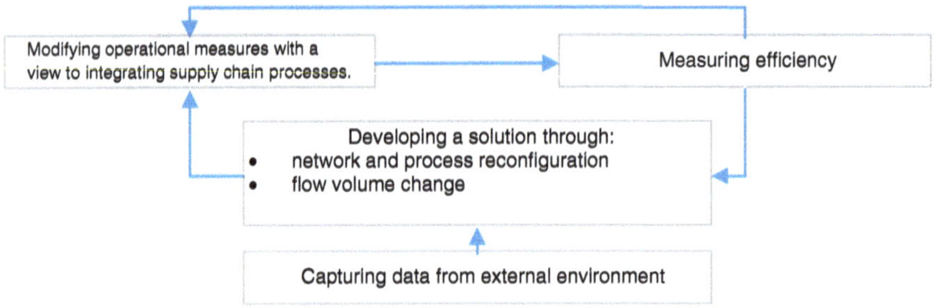

Fig. 1 The model of process integration in a supply chain (Nakanoa et al. 2013)

Malone (1987) as: a "*pattern of decision making and communications among a set of actors who perform tasks to achieve goals*". However, Romano (2003) views integration as "*breaking boundaries between the company's functions and between companies in supply chain*".

The model approach is extremely useful in taking an in-depth and effective examination of logistics processes. As a prime idea of systemic approach, modelling may be perceived as one of the greatest methodological achievements of science, both in the field of fundamental and applied research. Modelling became the only effective cognitive tool for embracing modern economic phenomena (Chaberek 2001).

The modelling of process integration is all the more relevant in the context of this chapter as well as in view of the facts provided therein. Figure 1 shows a selected model of process integration.

The model presented above is based on the continuous improvement loop. It includes two kinds of information inputs: internal, related to the process implementation phase, and external. The focal point of the model is the modification of operational activities and the verification of the efficiency of developed solution. With its dual nature this model fits in very well with the requirements of planning process integration analysis. On one hand it reconfigures processes (possible alignment with different planning variations) and on the other hand it implements operational processes in a modified environment and evaluates their effectiveness. Such an approach enables keeping the efficiency of the prepared plan in check.

Apart from planning there are also other tools that coordinate and integrate processes in a supply chain, including, amongst others: CPFR (collaborative planning, forecasting and replenishment) (Chung and Leung 2005), SCM (supply chain management) (Eng 2005), ERP (enterprise resource planning) (Kelle and Akbulut 2005). They will not be discussed below, because they do not fall within the scope of this chapter.

Based on the presented market conditions, importance of the integration of material and information flows in supply chain authors decided to develop a model of integration planning processes (using SOP approach). The model serves as a point of departure for further works related to the simulation of the material flow within a supply chain in the context of integrated planning.

2 Sales and Operations Planning

Defining the sales and operations plan requires taking a more holistic look at the structure of the planning processes implemented in today's companies. Many authors discuss this issue in their publications (Klimek and 2005; Łopatowska et al. 2007). The planning process structure as presented by Głowacka-Fertsch and Fertsch (2004) is especially interesting according to the Authors of this chapter. Fertsch and Głowacka hold that the structure of planning processes starts with a business plan/company game plan. Such a plan involves long-term quantitative sales forecasts, arranged into groups of goods and markets. After the sales forecasts have been assessed against available resources, a master plan/sales and operations plan is being prepared. The plan usually covers a year and is divided into quarters. The plan basically includes: product range, quantity, sales and production deadlines. Production capacity is balanced against tasks to be performed on a quarterly basis. The plan is complex and usually consists of master plans: financing, sales, production, technical, overhaul and procurement. At lower levels of process planning structure (in a shorter timescale), each of the primary plans is subject to verification in terms of the availability of resources. Master production plan is verified at the rough-cut capacity planning stage. The placement of SOP in the structure of planning processes of a manufacturing enterprise is presented in Fig. 2.

Having identified the location of SOP in the planning process structure, we shall now proceed to the definition. Muzumdar and Fontanella define SOP as: "*a set of business and technological processes allowing the company to match market demand to production and supply capacity in the most effective manner*" (Muzumdar et al. 2007). An expanded definition of SOP planning, including the interconnection of strategic and operational plans, may be found in APICS dictionary (Blackstone and Jonah 2008).

According to Affonso et al. (2008) a SOP plan is of huge significance for the coordination and integration of planning processes in companies, because it brings together many functional areas. The areas that are linked by an SOP plan in an Aberdeen Group report (Aberdeen Group et al. 2008). The coordination and integration of plans is essential for the cooperation of enterprises in the supply chain. SOP combines sales plans, operation plans and the identification of supplier's limitations, and, thus, goes beyond the scope of individual links in the supply chain. The place of a SOP plan in the supply chain is shown in Fig. 3.

The objectives of SOP are as follows (Gray et al. 2007):

- institutionalizes management decision-making and communication, especially in the areas related to customer demand and manufacturing volumes;
- balances demand and supply in a way that meets the needs of the customer;
- shock absorbers when supply and demand are not equal;
- integrates financial and operating plans;
- links strategic plans with detailed plans and schedules;
- regulates all detailed master schedules and sales plans;

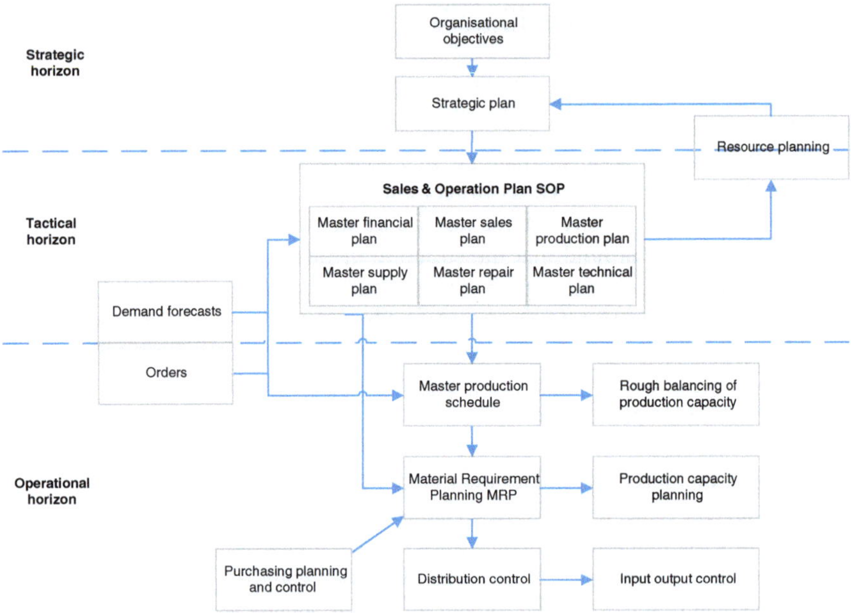

Fig. 2 The placement of SOP in the planning process structure at a production enterprise, own study based on: (Łopatowska et al. 2007)

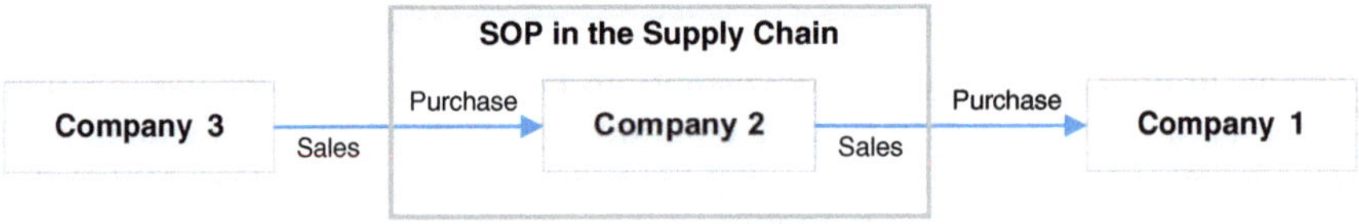

Fig. 3 SOP in the Supply Chain (Affonso et al. 2008)

- provides "long range vision" to other more short sighted processes, especially to resolve potential long-range capacity, material and financial issues.

Many publications describe the idea of SOP (as provided above), the process take on the situation—namely how SOP "should be" developed in terms of the involvement of organisational units, its implementation in the organisation (Lapide 2007; Mellon et al. 2010; Parker et al. 2008; Stahl 2008) and the implementation results (Hitachi Consulting's Chemicals Practice 2011; Ventana Research 2007, 2012). Neither of the publications addresses the issue of modelling integrated planning in terms of SOP.

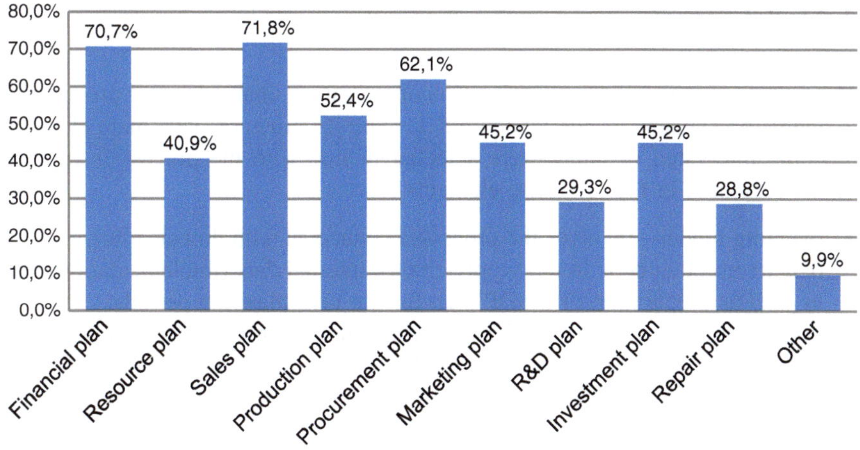

Fig. 4 Detailed planning functions executed in a company

3 Survey Findings

In order to assess and confirm the planning structures identified at the stage of literature review, the Authors performed a survey among companies registered or based in Poland (registered offices or branches). 372 business entities were involved in the survey, both small- and medium-sized enterprises as well as large companies (the outcomes of preliminary studies were presented in Adamczak et al. 2013). Each of the entities was provided with a survey questionnaire, where they had an option to mark the planning functions implemented (each of the entities could mark more than one function). The outcomes of this survey were presented in Fig. 4.

Survey findings indicate that planning most often occurs in sales (72 %), finance (71 %) and procurement (62 % of the answers). Production (52 %), marketing and investment (45 % each) and asset planning (41 %) come second. What follows in the order of importance is the planning of research and development (29 % each). Attention should be given to the fact that 10 % of answers were marked as 'others' in this case. Such a small percentage of functions going beyond the survey list means that the structure of planning processes was successfully identified by the managers, which is also confirmed in literature.

To sum up the survey, key plans which combine into SOP planning approach are indeed put in business practice. A more in-depth research reveals that plans are not integrated in nature and multi-variant solutions with respect to sequencing and planning tactics are not subject to modelling.

4 Model Structure

The first step in modelling the integration of the supply chain planning processes is to define the elements of the model and develop its structure. Two main elements mapping integrated planning were selected in line with the process integration model presented in Fig. 1. These elements are:

- planning model—involves the processes connected with downloading external data (from other cells in the supply chain), internal data (implementation data) as well as preparation of a SOP plan for the next planning periods;
- implementation model—involves the implementation of material flow in line with the prepared SOP plan. It provides actual internal data for the planning model as well as the data for evaluating the plan based on the material flow performance efficiency.

Figure 5 presents the planning model structure.

Description of the notions used on Fig. 5:

$M^*_{a_m}$ - planned impact of marketing activities on assortment *'a'* (or connected with the same positions and requiring the same purchasing materials as assortment 'a') in month *'m'*

$S^*_{a_m}$ - planned total sales figure of assortment *'a'* in month *'m'*

D^*_m - planned maximum level of operational debt in month *'m'* (credit line value)

$Fm^*_{s_m}$ - planned reduction of inventory levels due to maintenance works on the resources dedicated for raw material *'s'* in month *'m'* [quantity of raw material units that may be stored in a warehouse]

$Z_{s_{m max}}$ - maximum total order volume for raw material *'s'* which may be delivered by a supplier in a month *'m'*

$Fp^*_{a_m}$ - planned reduction in production capacity due to the maintenance and repair works of resources used in the production of assortment *'a'* in month *'m'* [working hours]

$Fm^*_{a_m}$ - planned reduction in warehouse levels due to the overhaul and repair works of resources dedicated for assortment *'a'* in month *'m'* [quantity of finished produce units that may be stored in a warehouse]

$l_{a_{m_k}}$ - current inventory level of assortment *'a'* stores at the end of month *'m'*

$l_{s_{m_k}}$ - current inventory level of raw material *'s'* stores at the end of month *'m'*

$P^*_{a_m}$ - planned total volume of production orders for assortment *'a'* performed in month *'m'*

$Z^*_{s_{m-LT}}$ - planned total order volume for raw material *'s'* processed in month *'m-LT'*

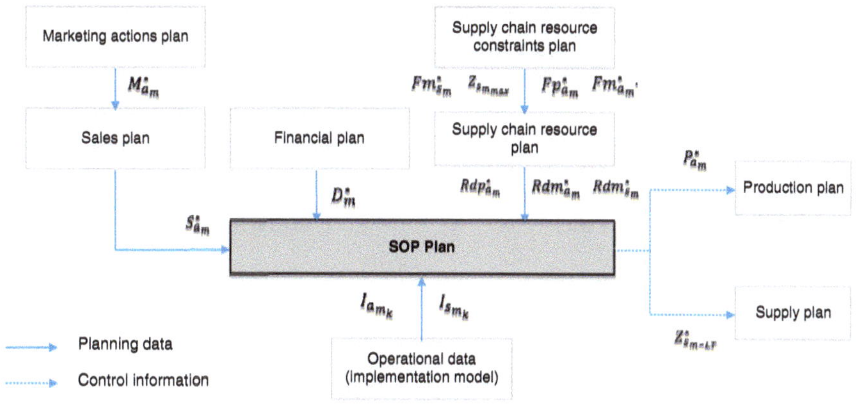

Fig. 5 Integrated planning model structure

The structure of the planning model presented in Fig. 5 indicates the components which impact the final SOP plan. The allocation of these elements in the SOP planning model is justified on two levels:

- Theoretical—resulting from the structures of planning processes and defining sales planning and operational activities by the authors;
- Practical—resulting from the survey on the integration of the forward and backward supply chains.

Production and sales plans are the outcome of the SOP planning process. They are responsible for the material flow, inventory levels (of raw materials and finished products), product availability to end customers and cash flows. These plans are forwarded to the implementation model structure and constitute its main control parameters. The material flow model and its key relationships are shown in Fig. 6.

Material flow implementation model is a simplified version of the production and logistics system in operation in the supply chain. The main flows in the model include:

- material flows—from the supplier through a producer to end customers (a flow taking place within across cooperating supply chain links);
- information flows—can differ in character, hence the division into: controlling flows (information from the SOP model), internal flows (between the model components as well as reporting flows), external flows (to the partners in the supply chain);
- financial flows—reflecting cash flows within one link in the supply chain and also between other links.

This model performs a support function and is primarily intended to provide input data to the planning model as well as the data for the evaluating the efficiency of the production and logistics system operating in line with the plan.

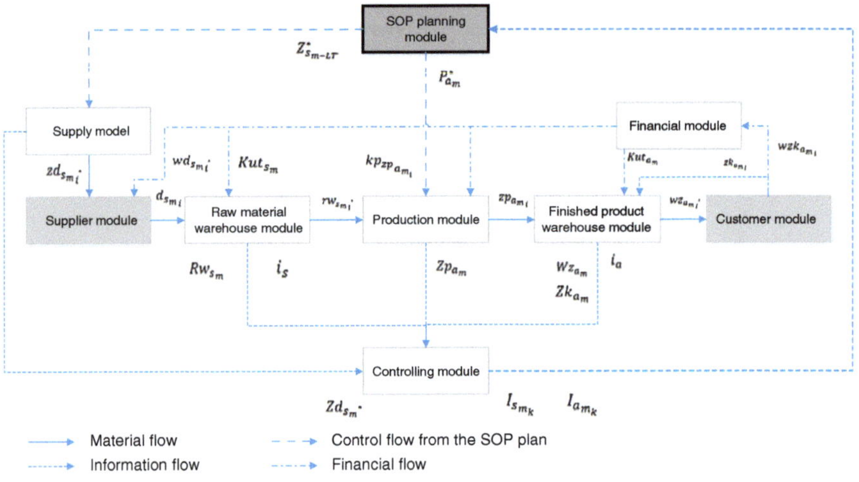

Fig. 6 Structure of material flow implementation model

5 Modelling of the SOP Process

The key element of the planning model is the entity which executes the SOP planning process. Literature on the subject suggests that SOP plan is realized on an annual basis (the value depends on industry characteristics). In the presented model a plan is prepared every month for the following year divided into months. Detailed changes pertain to the closest quarter of the year, the following months are only estimates. The plan diagram in time is presented in Fig. 7.

A SOP plan is formulated at the beginning of each month. As has already been mentioned, input data is supplied from external sources and from internal sources being the products of the material flow and ongoing reporting on operation production. The planning process outcome involves control parameters for material flow for successive months. The SOP plan is developed in line with a strictly defined algorithm. Figure 8 presents the framework SOP implementation algorithm.

Planning and operational data as well as information of limitations fed into the SOP planning algorithm originate from both external sources (sales forecasts, supplier capability) and internal sources (inventory level of finished products, nominal production system capacity). According to the SOP method, the sales plan is in the first place translated into operational activities (sales plan objectives), followed by the verification of the capacity. The frame algorithm presented in Fig. 8 captures only the basic mechanism of data transformation used in for the purpose of generating control information (which governs the material flow in the implementation model), constituting the output of the SOP planning process. More detailed solutions for plan coordination presented in the structure of a planning

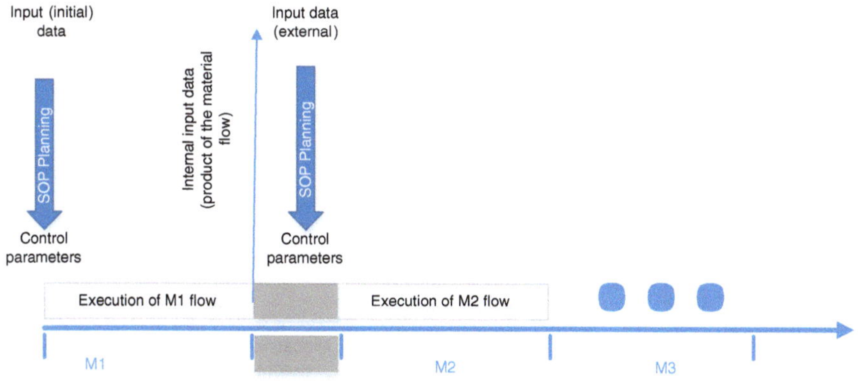

Fig. 7 Placing of SOP planning processes on time axis

model, taking into account the system limitations are the key determinants of the efficiency of an integrated SOP plan. The models enables the evaluation of:

- efficiency of the integrated planning methodology;
- efficiency of corrective procedures;
- feedback between the planning sequence and corrective procedures.

These research problems will be discussed in more detail in the publications to follow.

6 Modelling of the Integration of Planning Processes

The last stage of the modelling process is presenting the activities aimed at developing a model addressing the conditions mapping the integration of planning processes. A set of 7 activities which are distinctive features of integrated planning have been prepared for this purpose. These activities have been presented in Table 1 along with a detailed description of the differences.

The characteristics presented above shall be used as guidelines in modelling planning processes in the proposed model. Their implementation will be treated as the implementation of a given type of planning. Further simulation works will enable the identification of the impact of each of the elements on the efficiency of the integrated plan.

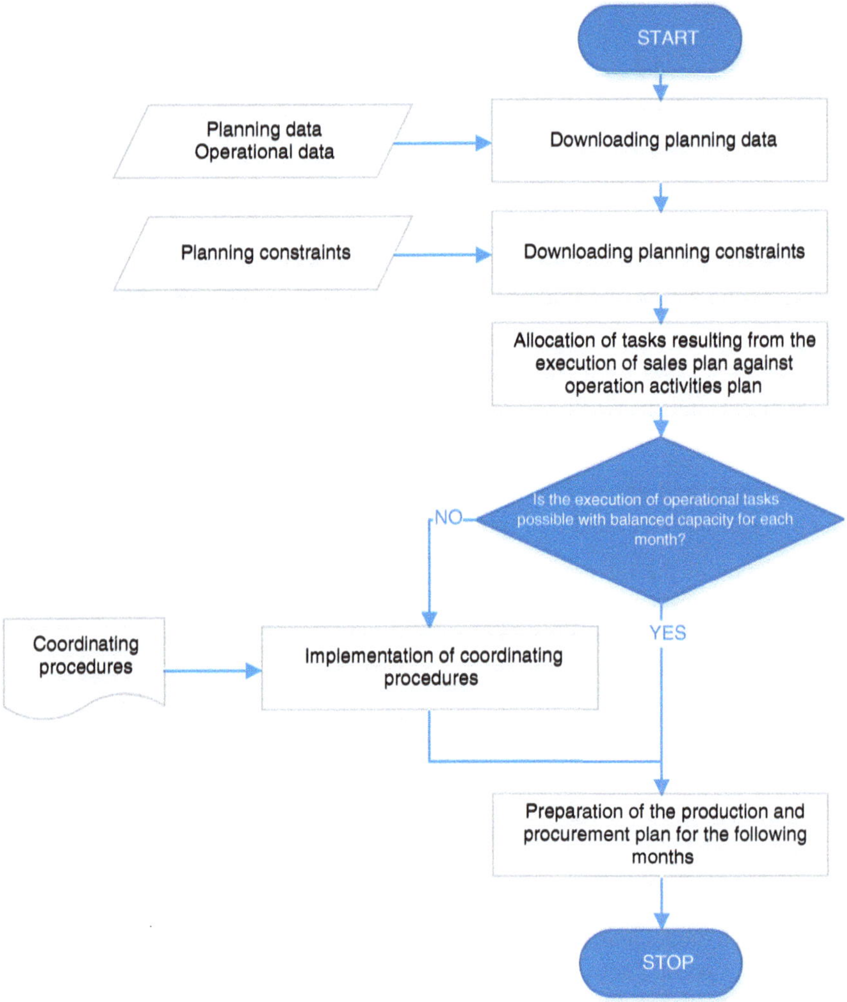

Fig. 8 Framework SOP implementation algorithm

Table 1 Comparison of characteristic features of integrated planning and non-integrated planning

	Planning	Integrated planning
Repair plan	Repair plan in the form of an estimated decline in the availability of resources. Resource availability decline is the same for each of the planning and is an average value based on historical data.	Inspections, maintenance and repairs planned in detail for subsequent planning periods. Capacity reductions for subsequent planning periods reflect the actual stop-overs.
Resources (production capacity, space in warehouse)	Sustainable resource availability over time, described as a percentage of the nominal capacity, estimated based on historical data.	Resource availability variable over time, adjusted to the planned stopovers of machines and devices due to inspections, maintenance and repairs.
Supply plan	Supply capacity seen as unrestricted (not to be included in the preparation of a supply plan), optionally estimated based on order history. The supply plan results from the material resources planning (MRP).	Supplier capacity viewed as the system limitation. The supply plan determined by the production plan as and long term sales plans in case of exceeding the periodical maximum output declared by a supplier and the need for creating buffers.
Sales plan	Sales plan developed for product items. Sales forecasts are subject to a higher error rate (higher values of a standard forecast errors).	A plan developed for product groups. The standard sales forecast error for an assortment group assumes lower values than the analysis of the particular assortment items.
Marketing plan	Marketing plan unrelated to the operations of the supply chain.	Marketing plan involved in the material flow plans (periodical increase in sales figures, launching new products on the market, withdrawing products from the market).
Production planning	Balanced or adjustment production plan.	A mixed production plan including limited production capacity, availability of raw materials, and resource plan (stock levels to build up in order to achieve sales plan targets in the foreseen resources limitations).
Financial plan	Financial plan including costs and incomes, created as a result of the outcome of the sales plan and operations.	A financial plan approached as planning limitation due to the availability of funds (own capital or a credit line), includes not only costs and revenues but also cash flow over time (cash flow).

7 Conclusion

According to the authors modelling integration process planning in the supply chain may be implemented with the use of systemic approach and may consist of the following elements (Fertsch 2006):

- identifying the objective—what will be assessed by the proposed model;
- bringing together the system elements—i.e. its constituent elements;
- establishing a set of interdependencies—interrelations between the system elements;
- transformation mechanism—i.e. the transformation of the process itself and the transformation of activities aimed at the provision of conditions which would distinguish integrated processes from the initial situation.

Activities, which model the integration of planning processes include:

- research based on up-to-date data instead of on estimates made based on historical data—resource planning (production capacity and the available warehouse space) based on nominal data and the repair plans pertaining to subsequent planning periods;
- incorporating other supply chain links into the planning process—approaching production capacity of suppliers as a system limitation, undertaking planning activities aimed at offsetting or ruling out these limitations;
- creating aggregated plans—planning for assortment groups in a monthly timeline and not for particular items;
- incorporating all plans that affect the operations—including marketing activity plan in SOP plan as it impacts operating expenses;
- planning for the purpose of offsetting or eliminating limitations—preparing a production plan matched to the current sales plan and enabling the execution of product issues to end customers also in the periods of time, when resources are insufficient to cover current sales requirements;
- approaching finances as a limitation to operational activity—replacing the revenue and expense plan, resulting from the operations plan, with cash flow plan for the operations plan, being a planning element and its potential limitation at the same time.

The model presented in this publication is a basis for further research aimed at testing various mechanisms coordinating plans within SOP and identifying the impact of the integration of planning processes on the functioning of the supply chain.

Acknowledgments This chapter has been the result of the study conducted within the project entitled "Investigating multi-aspect determinants of integration in backward and forward logistics chains in respect of integrated production planning in the context of using recyclable materials" pursued at the Poznan School of Logistics in Poznan. The project was financed by The National Science Center based on the decision number DEC-2011/03/B/HS4/03419.

References

Aberdeen Group (2008) Sales and operations planning: aligning business goals with supply chain tactics. Aberdeen Group, Inc., a Harte-Hanks Compan, Boston

Adamczak M, Domański R, Cyplik P, Pruska Ż (2013) The tools for evaluating logistics processes. Logforum 9(4):255–263

Affonso R, Marcotte F, Grabot B (2008) Sales and operations planning: the supply chain pillar. Prod Plann Control 19(2):132–141

Blackstone JH, Jonah J (2008) APICS dictionary, 12th edn. University of Georgia, Georgia

Chaberek M (ed) (2001) Modelowanie procesów i systemów logistycznych. Wydawnictwo Uniwersytetu Gdańskiego, Gdańsk

Chung WWC, Leung SWF (2005) Collaborative planning, forecasting and replenishment: a case study in copper laminate industry. Prod Plann Control 16(6):563–574

Eng TY (2005) Mobile supply chain management: challenges for implementation. Technovation 26:682–686

Fertsch M (2006) Podstawy logistyki. Instytut Logistyki i Magazynowania, Poznań

Głowacka-Fertsch D, Fertsch M (2004) Zarządzanie produkcją. Wyższa Szkoła Logistyki, Poznań

Gray CD (2007) Sales and operations planning standard system: with reference software. Trafford Publishing, Bloomington, USA

Hitachi Consulting's (2011) Chemicals Practice, Sales & Operations Planning (S&OP) Survey Review. Hitachi Consulting, Dallas

Kelle P, Akbulut A (2005) The role of ERP tools in the supply chain information sharing, cooperation, and cost optimisation. Int J Prod Econ 93(94):41–52

Klimek A (2005) Miejsce i znaczenie głównego harmonogramu produkcji w zintegrowanych systemach informatycznych opartych na standardzie MRP II/ERP. In: Fertsch M (ed) "Logistyka produkcji", pp 110–122, Biblioteka Logistyka, Poznań

Lapide L (2007) Sales and Operations Planning (S&OP) Mindsets. J Bus Forecast Spring 26(1):18–27

Łopatowska J (2007) Wykorzystanie wnioskowania logicznego w planowaniu i sterowaniu produkcją. In: Fertsch M, Grzybowska K, Stachowiak A (eds), Logistyka i zarządzanie produkcją – nowe wyzwania, odległe granice, pp 252–264, Instytut Inżynierii Zarządzania, Poznań

Malone TW (1987) Modeling co-ordination in organisations and markets. Manag Sci 33(10):1317–1332

Mellon C, Allen B, Prokopets L (2010) Putting S&OP on the fast track. Supply Chain Manag Rev 14(1):41–50

Muzumdar M, Fontanella J (2007) The secrets to S&OP success. Supply Chain Manag Rev 4(1):33–41

Nakanoa M, Akikawab T, Shimazuc M (2013) Process integration mechanisms in internal supply chains: case studies from a dynamic resource-based view. Int J Logistics Res Appl 16(4):328–347

Parker K (2008) S&OP encompassing broader financial and performance parameters. Manufact Bus Technol 26(1):28–39

Romano P (2003) Co-ordination and integration mechanisms to manage logistics processes across supply networks. J Purchasing Supply Manag 9(3):119–134

Stahl RA (2008) Executive S&OP: managing to achieve consensus. Foresight Fall (19):33–41

Ventana Research (2007) Sales and operations planning research study results. Ventana Research, San Francisco

Ventana Research (2012) Optimizing business performance with sales and operations planning the role and importance of demand planning in S&OP. Ventana Research, San Francisco

Problems of Logistic Systems Vulnerability and Resilience Assessment

Tomasz Nowakowski and Sylwia Werbińska-Wojciechowska

Abstract The aim of this chapter is to present and investigate the main concepts of supply chain vulnerability and resilience. Thus, the fundamental differences between vulnerability and resilience definitions are discussed. General aspects of supply chain vulnerability understanding are presented and some problems with vulnerability assessment are underlined. The main issues on vulnerability and resilience assessment are discussed, and the case study of production system resilience estimation is investigated.

Keywords Vulnerability · Resilience · Supply chain performance

1 Introduction

Nowadays, supply chains and logistic systems performance issues are of immediate importance. On the one side, logistic flows in many industries strive to be lean, responsible and agile (Svensson 2000) on the other, the increasing dynamism and uncertainty in the business make environment risk issues become key concerns to the organizations (Singhal et al. 2011). Thus, the need to adapt the operation of logistics systems in such an uncertain and volatile environment has caused the need to formulate new characteristics, like vulnerability and resilience. This process of adding more requirements for the operation of technical systems to meet the needs of its users is shown schematically in Fig. 1.

T. Nowakowski (✉) · S. Werbińska-Wojciechowska
Wroclaw University of Technology, Wybrzeze Wyspianskiego 27,
50-370 Wroclaw, Poland
e-mail: tomasz.nowakowski@pwr.wroc.pl

S. Werbińska-Wojciechowska
e-mail: sylwia.werbinska@pwr.wroc.pl

P. Golinska (ed.), *Logistics Operations, Supply Chain Management and Sustainability*, EcoProduction, DOI: 10.1007/978-3-319-07287-6_12,

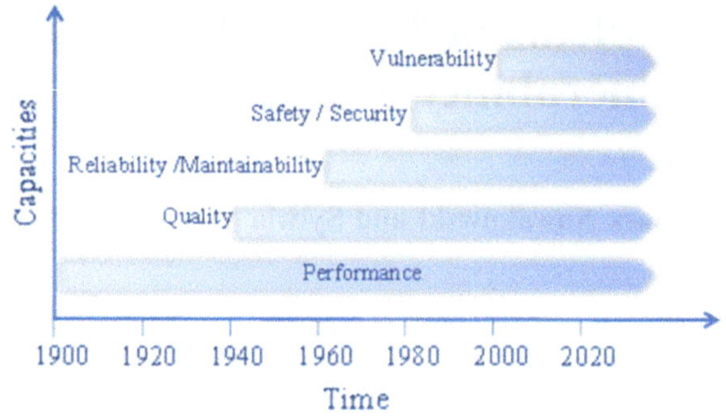

Fig. 1 Scheme of development of technical system capacity requirements (Nowakowski 2013)

The figure shows how within last fifty years the performance requirements of technical objects were supplemented with: customer expectations (quality), abilities to prevent the loss of the object properties in operation time (reliability and maintainability), the effects of undesirable events and protection against them (safety and security) and the ability to restore performance (resilience) [see e.g. (Oke and Golapakrishnan 2009; Supply Chain Resilience 2009, 2010, 2011, 2012, 2013)].

The concept of a supply chain has been defined by many authors, but they are all quite uniform in terms of their content (Folta and Novotny 2013). Thus, following (Christopher 1998) supply chain can be defined as *the network of organisations that are involved, through upstream and downstream linkages, in the different processes and activities that produce value in the form of products and services in the hands of the ultimate consumer*.

To prevent vulnerability, it is essential to manage risks in chains through creating more resilient supply chains that are able to respond to disruptions and adapt themselves to necessary changes (Christopher and Peck 2004). Risk managing in supply chains is the function of supply chain risk management, according to the definition presented e.g. in Christopher (1998), Christopher and Peck (2004). More information can be found e.g. in Singhal et al. (2011), where authors provide a comprehensive review and classification of supply chain risk management literature.

The importance of supply chain risk management issues may be confirmed e.g. by the Supply Chain Resilience Reports in which the challenge of developing resilient supply chains are considered since 2009 (Supply Chain Resilience 2009, 2010, 2011, 2012, 2013). The reports highlight the level, range and cost of disruptions that organizations face, and demonstrate how a disruption in one organization can spread out over the entire supply chain (Supply Chain Resilience 2013). One of the most important key finding results from the last report (Supply Chain Resilience 2013), is that about 75 % of respondents still do not have full visibility of their supply chain disruption levels (survey of 579 respondents from 71 countries).

Following this, the focus of this chapter is to present and discuss the main terms of supply chain vulnerability, and resilience with giving some explanations of the fundamental differences between the mentioned concepts.

2 Supply Chain Vulnerability and Resilience: Literature Overview

Supply chain vulnerability and resilience has become a field of research the past 10 years, and a number of definitions have been made (Berle et al. 2011; Sheffi 2007). To present these concepts more clearly, first the disruption profile should be explained.

Any significant disruption will have a typical profile in terms of its effect on company performance. The performance is measured by sales, production level, profits, customer service or another relevant metric (Sheffi 2007).

The nature of the disruption and the dynamics of the system response can be characterized by the following eight phases (Fig. 2) (Blackhurst et al. 2005; Sheffi 2007):

1. *Preparation.* In some cases, a company can foresee and prepare for disruption, minimizing its effects. In other cases there is little or no warning. For example, there can be maintained higher safety stock in order to avoid an unexpected late delivery.
2. *The Disruptive Event.* The tornado hits, the bomb explodes, a supplier goes out of business or the union begins a wildcat strike. As a result, all organizations performing in the supply chain should be prepared for the next step.
3. *First Response.* Whether there's a physical disruption, a job action or an information technology disruption, first response is aimed at controlling the situation, saving or protecting lives, shutting down affected systems and preventing further damage. In supply chain this step is also connected with proper exchange of information performed between the systems.
4. *Initial Impact.* The full impact of some disruptions is felt immediately. Other disruptions can take time to affect a company, depending on factors such as the magnitude of the disruption, the available redundancy, and the inherent resilience of the organization and its supply chain. During the time between the disruptive event and the full impact, performance usually starts to deteriorate.
5. *Full Impact.* Whether immediate or delayed, once the full impact hits, performance often drops precipitously. As a result, the production level cannot be satisfied, suppliers cannot deliver the proper raw materials, and distributors cannot satisfied the defined customer service level.
6. *Recovery Preparations.* Preparations for recovery typically start in parallel with the first response and sometimes even prior to the disruption, if it has been anticipated. They involve qualifying other suppliers and redirecting suppliers'.
7. *Recovery.* To get back to normal operations levels, many companies make up for lost production by running at higher-than-normal utilization, using overtime as well as suppliers' and customers' resources.
8. *Long-Term Impact.* It typically takes time to recover from disruptions, but if customer relationships are damaged, the impact can be especially long-lasting and difficult to recover from.

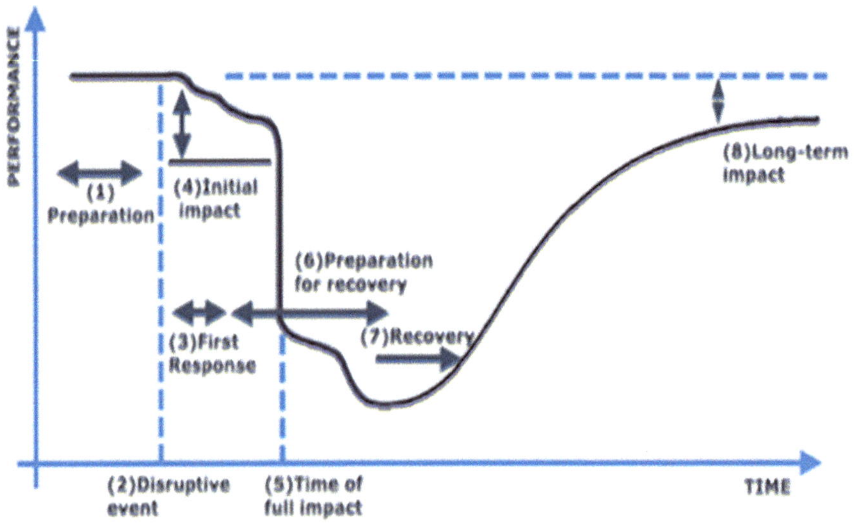

Fig. 2 The disruption profile (Sheffi 2007)

As a result, supply chain vulnerability can be defined as *an exposure to serious disturbance, arising from risks within the supply-chain as well as risks external to the supply-chain* (Chapman et al. 2002).

Svensson, one of the most widely cited author in this field, placed vulnerability and related concepts, such as risk, uncertainty and reliability, within the context of the wider concept of contingency planning. In his work (Svensson 2000), he defined vulnerability as *the existence of random disturbances that lead to deviations in the supply chain of components and materials from normal, expected or planned schedules or activities, all of which cause negative effects or consequences for the involved manufacturer and its sub-contractors*. This definition was the base for developing a model consists of three principal components, namely: source of disturbance, category of disturbance and type of logistics flow. Later, in (Svensson 2002a, c), author constructed the vulnerability consisting of the two components: disturbance and the negative consequence of disturbance. This approach referred to the focal firm's inbound and outbound logistics flows and corresponded to the direct source of disturbance in the model of vulnerability introduced in (Svensson 2000). In the next work (Svensson 2002b), author investigated vulnerability consisted of two components, namely time-dependence and relationship-dependence, which can occur in marketing channels. The higher the dependence the higher is the level of perceived vulnerability. Following this, author in his next work (Svensson 2004) based on the assumption that the gap between perceived dependence and perceived trust influences, and has an impact on, companies' perceived vulnerability in business relationships towards suppliers and customers. The presented research studies are typically survey perceptions of supply chain related risk in an organisation's purchasing department or its first tier supplier base.

The empirical study results of the sources and drivers of supply chain vulnerability were also under investigation of another author in (Peck 2005). In this chapter, given that supply chains link organisations, industries and economies, author suggested that, there should be developed a multi-level framework for analysis of supply chain risk.

The level of vulnerability was making the difference. To date, there is no consensus definition of vulnerability. Researchers from various fields bring their own conceptual models which often address similar problems and processes using different language (Bouchon 2006). A global system property focused on three elements (Albino and Garavelli 1995; Bouchon 2006; Kroger and Zio 2011):

- degree of loss and damage due to the impact of a hazard,
- degree of exposure to the hazards; susceptibility of an element at the risk of suffering loss and damages,
- degree of resilience; the ability of a system to anticipate, cope with, absorb, resist and recover from the impact of a hazard or disaster.

Further studies on supply chain vulnerability reduction regarded e.g. the information sharing issues [see e.g. (Sheffi 2005)], or discrete transport system vulnerability [see e.g. (Walkowiak and Mazurkiewicz 2013)]. In Sapna Isotupa et al. (2014) authors integrated two streams of literature: the supply chain algorithmic models with the insurance-based risk management literature.

The literature on supply chain risks and vulnerability is growing. In (Neil Adger 2006), author reviewed existing knowledge on analytical approaches to vulnerability, taking into account the socio-ecological systems concept. In the same year, author in (Peck 2006) provided a critique and review of the positioning of research in this field. In the mentioned chapter, author explained the relationships between supply chain vulnerability, risk and supply chain management. Issues of risk categorization and mitigation in supply chains were under investigation of authors in (Oke and Golapakrishnan 2009). The problem was continued e.g. in (Longo and Oren 2008). Following this, in work (Wagner and Neshat 2010), authors develop an approach based on graph theory to quantify and mitigate supply chain vulnerability. There are also research studies, which implemented Analytic Network Process approach for the supply chain vulnerability assessment (Fazli and Masoumi 2012).

Moreover, it's should be underlined, that the several scientific research works on supply chain vulnerability were developed considering specific supply chain sectors or technical systems. There were articles on supply chain risk/vulnerability, which dealt with aspects of aerospace manufacturing (see e.g. Haywood and Peck 2003), JIT production systems (see e.g. Albino and Garavelli 1995), telecommunication sector (see e.g. Agrell et al. 2004), maritime transportation system (see e.g. Berle et al. 2011), transportation systems (see e.g. Klibi and Martel 2012; Walkowiak and Mazurkiewicz 2013), power systems (see e.g. Hofmann et al. 2012), or critical infrastructures (see e.g. Kroger 2008; Zio and Kroger 2009).

For more information and briefly literature reviews we recommend reading e.g. (Briano et al. 2009; Nowakowski and Valis 2013; Valis et al. 2012).

Similarly, the number of research studies introduced the concept of supply chain resilience. The *Online Compact Oxford Dictionary* (2014) defined resilience as the (a) *ability to withstand or recover quickly form difficult conditions*; (b) *the ability to recoil or spring back into shape after bending, stretching, or being compressed*. For technical systems, resilience generally means *the ability to recover from some shock, insult, or disturbance, the quality or state of being flexible* (Bouchon 2006). Supply chain resilience may be defined as *the ability of the supply chain to handle a disruption without significant impact on the ability to serve the supply chain mission* (Berle et al. 2011). As reported e.g. in (Creating a Resilient Supply Chains: A practical Guide. Cranfield School of Management 2003), the resilience definitions took into account the following supply chain aspects: its flexibility, agility, velocity, visibility and redundancy. A brief survey of resilience definitions from different disciplinary perspectives was given in (Francis and Bekera 2014). For further information refer to e.g. (Christopher and Rutherford 2004; Sheffi 2005, 2006, 2007).

The mathematical formulation of supply chain risk is provided in e.g. (Bogataj and Bogataj 2007). The proposed model of risk measurement was based on the Input-Output Analysis and Laplace transforms. In another work (Cho 2010), author provided the framework for analysing the cost impact on the supply chain resulting from disruptions in its elements and resources using the concept of influence matrices. In the next year, authors in (Shuai et al. 2011) presented the quantitative measuring method helpful to supply chain resilience assessment. The proposed research method was based on the biological cell elasticity theory.

Another work, in which there is presented a model to help quantify resilience in the supply chain is (Falasca et al. 2008). Authors proposed a simulation-based framework that incorporates concepts of resilience into the supply design process. Another simulation method implementation into supply chain performance is presented in (Carvalho et al. 2012). Authors provide the simulation tool to support the decision making process in supply chain design to create a more resilient supply chain. The resilient supply chains designing issues was later continued in (Bukowski and Feliks 2012; Wicher and Lenort 2013), where authors proposed the frameworks for building a resilient supply chain.

Later, in (Losada et al. 2012), authors presented an optimizing system resilience tool which bases on mixed integer linear program. The supply chain resilience assessment tool, called Supply Chain Resilience Assessment and Management (SCRAM) was investigated in (Pettit et al. 2013).

Moreover, the several scientific research works on supply chain resilience were developed considering specific supply chain sectors or special issues. Following this, the discussion of economic resilience and specification of individual resilience indicators was given in (Rose and Krausmann 2013). The examples of resilience of global companies can be viewed e.g. in (Barac et al. 2011; Wicher et al. 2012), and for automotive supply chains resilience were investigated e.g. in (Carvalho et al. 2012).

The comprehensive literature review on supply chain resilience is presented e.g. in (Briano et al. 2009; Brusset 2013; Carvalho et al. 2012; Christopher and Peck 2004; Haffenen et al. 2012; Longo and Oren 2008; Ponis 2012; Schoon 2005).

To sum up, based on the literature review and Fig. 2, there can be presented the three main definitions of conceptions connected with effective performance of supply chains which are vulnerable to disruptions.

Reliability assessment is focused on the possibility of an unwanted event occurrence. Measures of the reliability should express uncertainty about the appearance of such an event, like failure, fault, error, etc. Thus, reliability (dependability) of a logistic system can be understood (Bukowski and Feliks 2011) as the ability to deliver correct service under normal (ordinary) work conditions in a given time interval.

Safety means absence of critical/dangerous events while security is focused on protecting the system environment against the effects of these damages. Safety is measured generally by risk—two-dimensional combination of probability of an undesirable event and possibility of loss (consequences). Risk assessment consists on process of risk identification related to threat, includes its possibility (likelihood or probability), impact, and consequences.

Resilience takes into consideration not only the discussed issues (reliability and safety) but also the possibilities of restoring the original properties of the system. Thus, resilience means (Bukowski and Feliks 2011) readiness for secure and acceptable service under abnormal (uncommon) work conditions (e.g. disruptions, attacks, accidences, disaster). And the measure of resilience can be understand as time to restore the capabilities of the system [worse than new, as good as new, better than new (Nowakowski 2013)].

3 Supply Chain Vulnerability and Resilience Assessment Issues

> Any threat which could cause an interruption in the flow from raw material to the end user is a supply chain risk and any interruption in the flow of material is a supply chain disruption (Sapna Isotupa et al. 2014).

Disruptions can be divided into three categories to facilitate estimating their likelihood (Sheffi 2007): natural disasters, accidents (faults, failures), intentional attacks and the methods of estimating their likelihood are different.

Natural disasters are relatively frequent, thus statistical models can be used to estimate their likelihood and their magnitude.

Accidents belong to low-probability/high-impact events—statistical models can be used to estimate their likelihood but often there is a lack of certain information. Moreover, accidents involve human factors, and large number of small incidents may foreshadow more significant accidents.

Intentional disruptions follow a different logic—they constitute adaptable threats in which the perpetrators seek both to ensure the success of attack and to maximize the damage. Consequently, "hardening" one potential target against a given mode of attack may increase the likelihood that another target will be attacked or there will be another type of attack. Such attacks are likely to take place at the worst time and in the worst place; when the organization is most unprepared and vulnerable.

The likelihood of intentional threats is more difficult to assess because of these attacks adapt to defensive measures. Despite of it (or because of it) further conceptual and research works on logistics process vulnerability modelling are to be carried on.

Moreover, based on (Chapman et al. 2002; Mascaritolo and Holcomb 2009; Singhal et al. 2011), main supply chain disruptions may arise from such sources as:

- natural disasters, such as earthquakes, cyclones, epidemics,
- terrorist incidents,
- accidents, like fire of supplier's factory, or
- operational difficulties, connected e.g. with variability in supply, demand uncertainties, price variability.

Svensson in his work (Svensson 2000) considered them in terms of analysis inbound logistic vulnerability and divided them into direct and indirect sources of disturbance. Moreover, following (Longo and Oren 2008; Sheffi 2005), terrorist attacks, wars, politic problems or natural disasters should be considered as risks external to the supply chains. Risks related to processes and activities should be considered as internal to the company, and risks coming from the market or from suppliers should be considered as external to the company and internal to the supply chain.

Treating supply chain disruptions as unexpected events occurrence, they can be described as having uncertainty in logistic process realization (Werbińska-Wojciechowska 2013). Uncertainty in the process is connected with definition of perfect operation of logistic system and may be described by 7R formula: Right product, Right quantity, Right quality, Right place, Right time, Right customer, Right price. Thus, different aspects are taken into consideration, such as (Werbińska-Wojciechowska 2013):

- time (in the sense of duration of activity/process, starting/ending moment of activity realization, frequency of activity/demand occurrence),
- quantity (of supply, demand or physical transfer of goods),
- location/place (where activity starts/ends),
- quality (of service/products),
- cost (fluctuation, occurrence).

However, not every disruption occurrence leads to logistic system failure appearance. The critical factor which determines the logistic system failures is time. In a situation, when disruption (connected with e.g. improper delivery quality/quantity, improper location) occurs, there is a necessity to find out if we have enough time to correct the problem. When the spare time let us to remove the disruption—logistic system is not defined as failed. In other words, time redundant system has the ability to tolerate interruptions in their basic function for a specific period of time without having the negative impact on the system task performance.

Typically, the time redundant systems have a defined time resource that is larger than the time needed to perform the system total task (Werbińska-Wojciechowska 2013). Time redundancy is to take additional time to complete the task (in relation to the time necessary for its execution), which can be used to restore the state of the system or improve its technical characteristics. This means that the system with time resource tolerates faults with a short (usually specified) duration.

Following this in the literature there can be found research works dedicated to vulnerability and resilience measurement issues. For example in (Neil Adger 2006), authors stated that, *measurement of vulnerability must reflect social processes as well as material outcomes within systems that appear complicated and with many linkages that are difficult to pin down. Vulnerability is, therefore, not easily reduced to a single metric and is not easily quantifiable*. They proposed a set of vulnerability indicators that incorporate an objective material measure of vulnerability and severity in its distribution. In Fig. 3, there is presented a theoretical framework for supply chain vulnerability indicators, based on the presented above literature review.

The examples of supply chain resilience measurement systems/methods were given e.g. in (Barker et al. 2013), where authors presented two resilience-based component importance measures for networks, in (Cox et al. 2011), where authors investigated the assessment issues of passenger transportation system's resilience, in (Fakoor et al. 2013), where a method for measuring resilience based on fuzzy logic was proposed, and in (Henry and Ramirez-Marquez 2012), where authors developed generic metrics for quantifying system resilience. In work (Francis and Bekera 2014), authors introduced a resilience metric that incorporates the three resilience capabilities (absorptive capacity, adaptive capacity, recovery capacity) and the time to recovery. Following this, in the Fig. 4, there is presented the resilience measurement framework.

Moreover, the complexity of the problem is connected with the necessity of taking into account some factors which might be considered to increase the level of risk, like (Supply chain vulnerability 2002; Wicher et al. 2012):

- focusing on efficiency targets instead of effectiveness issues,
- supply chain globalisation,
- focussing on factories and centralised distribution,
- outsourcing,
- reduction of the supplier base,
- demand variability,
- lack of visibility and control procedures.

This factors are discussed in more depth e.g. in (Singhal et al. 2011; Supply chain vulnerability 2002).

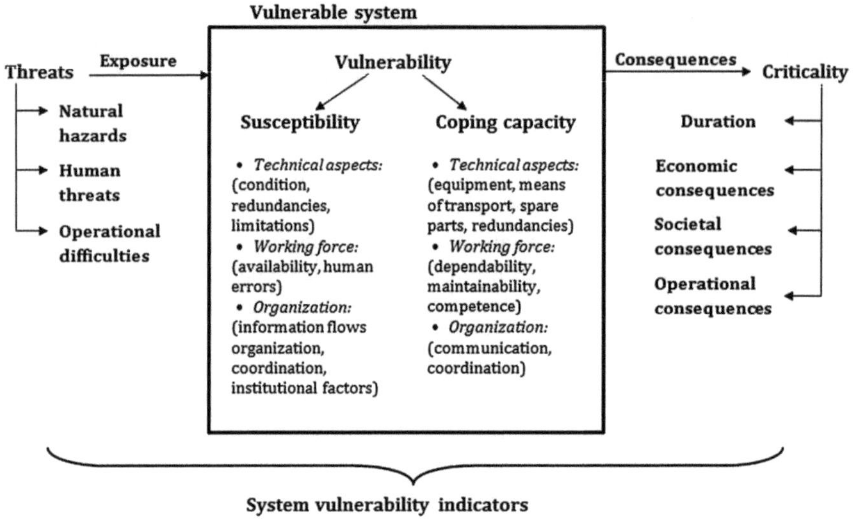

Fig. 3 Framework for supply chains vulnerability indicators

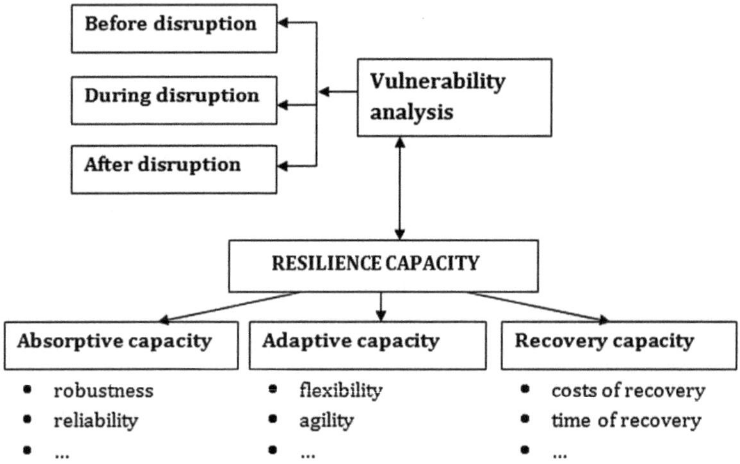

Fig. 4 Resilience framework.Own contribution based on (Francis and Bekera 2014; Sheffi 2007; Werbińska-Wojciechowska 2013)

4 Case Study

The resilience analysis can be connected with the main life cycle phases of any technical (also logistics) systems. Any company may be in one of its life cycle phases—first phase of development, when company increases market share and production levels, phase of stable performance, and the phase of the end of life cycle, when company decreases market share. The system resilience measurement is the most difficult for the first phase of system life cycle. This is mainly connected on the one side with a system recovery capacity definition and on the other, with the necessity to reach the system defined goals in terms e.g. of effectiveness, dependability, or productivity.

To discuss this problem in more depth, the case study is presented. The case company is a production leader in the rail industry and sector related services. The company offers customers e.g. a wide range of passenger and freight rail vehicles, complex transport systems, locomotives, and train control systems (Rysińska 2013).

In 2012, during the summer, there can be observable some organizational disruptions in the company. The symptoms of disruptions may be defined e.g. by such indicators, like On Time Delivery (ODT), which defines the number of deliveries which were received on time in the manufacturing company (Fig. 5).

Late delivery of material to production company results in e.g. the change of the date of delivery of final products to the customer. Thus, this situation affects the dependability of the company and causes even total loss of customer confidence. The causes of late deliveries occurrence are very different. Following the company's analyses, such situations often are connected with suppliers' delays occurrence, or the wrong communication with the customer.

Following this, one of the convenient indicators in the area of supply chain vulnerability may regard to occurred delays of performed processes. This is compatible with vulnerability indicators presented in the Fig. 3 on the one side, and it may be used to define logistic systems reliability on the other.

In the next point of the research analysis, there was stated a question—how to measure the production company resilience? The answer for that question may be given after company's measurement system analysis. One of the possibilities is to measure system resilience in terms of its productivity level—more precisely the indicator of monthly productivity improvement level (Fig. 6)

As we can see in the Fig. 6, in May and June 2012 (when late delivery were noticed), the productivity improvement level did not reach the target levels of 2.5 and 3 %. Moreover, despite that the achieved productivity level increased in next months, the case company did not reach the target level of this indicator. This is mainly connected with the company's phase of life cycle, which is the developmental one. In such a case, the indicator level increases in every month (for the case company 0.5 % in every month to reach level of 5 % in December). Following this, there can be stated such a conclusion, that the resilience time period for companies being in their developmental phase of life cycle is longer, because of the changing indicator target level.

As we can see in the Fig. 6, in August 2012, there can be observable second time period with some disruptions (very low level of analysed indicator). To confirm, if the chosen indicator can be used as a resilience measure, the company's performance measures have been analysed. The disturbances have been noticed in the distribution and production areas (Fig. 7).

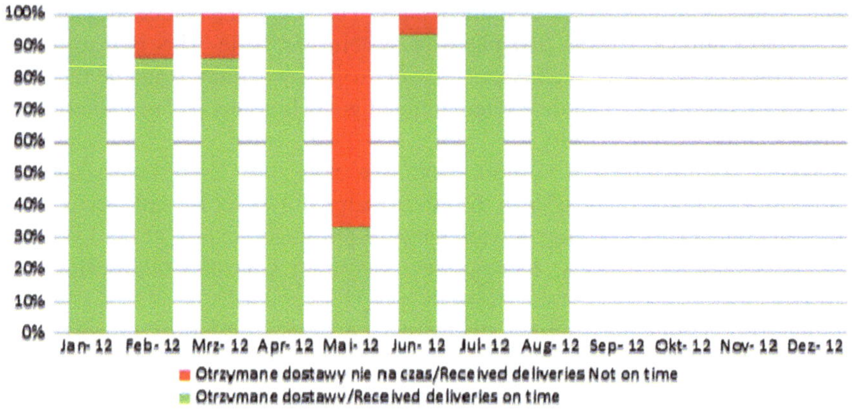

Fig. 5 On Time Delivery indicator in 2012 for the analysed case company (Rysińska 2013)

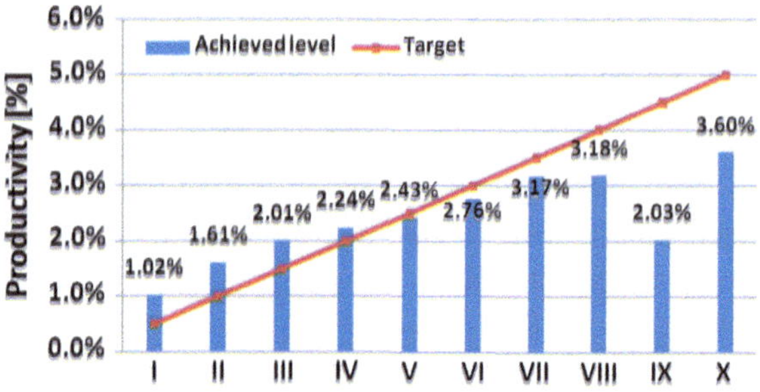

Fig. 6 Case company productivity level in 2012 (Rysińska 2013)

To sum up, based on the presented case study and the resilience framework illustrated in the Fig. 4, the system resilience measure (indicator) can be defined as *the time to restoration of system expected performance level*. Following this, there can be underlined, that the three main concepts: reliability, safety and resilience may be defined in terms of their main measures identification. Thus, system reliability measure is connected with failure/error occurrence probability, system safety is measured by risk ratio, and the resilience is defined by the time period necessary to restore the system to defined performance level.

5 Conclusions

The analysis shows possibilities of resilience assessment for real logistic system. More detailed statistical analysis of the problem is very limited by lack of certain information. Often, the data are regarded as private or secret and results of analysis are impossible to be published.

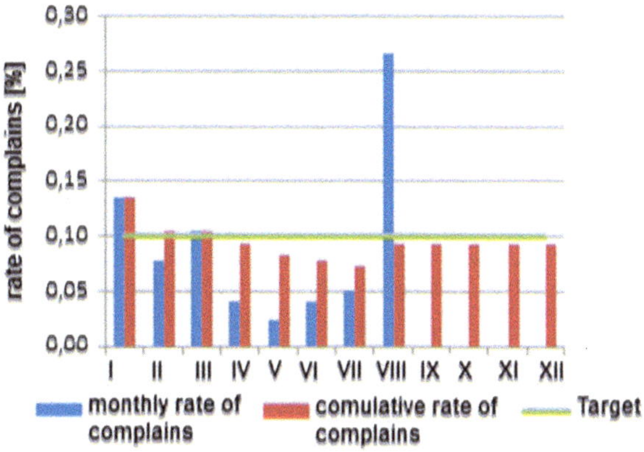

Fig. 7 Case company rate of complains level in 2012 (Rysińska 2013)

On the other side, following the presented in Sect. 2 literature review, the terms of logistic system/supply chain vulnerability and resilience still demand clarification. In the literature, there can be found many different approaches and aspects of vulnerability and resilience, which are under investigation. Thus, this research area still demands examination of real logistic systems performance and development of new complex framework for system vulnerability and resilience measurement.

References

Agrell PJ, Lindorth R, Norrman A (2004) Risk, information and incentives in telecom supply chains. Int J Prod. Econ 90

Albino V, Garavelli AC (1995) A methodology for the vulnerability analysis of just-in-time production systems. Int J Prod Econ 41

Barac N, Milovanovic G, Andjelkovic A (2011) Risk and Resilient supply chain. Facta Univ Series Econ Organ 8(2)

Barker K, Ramirez-Marquez JE, Rocco CM (2013) Resilience-based network component importance measures. Reliab Eng Syst Saf 117

Berle O, Asbjornslett BE, Rice JB (2011) Formal vulnerability assessment of a maritime transportation system. Reliab Eng Syst Saf 96

Blackhurst J, Craighead CW, Elkins D, Handfield RB (2005) An empirically derived agenda of critical research issues for managing supply-chain disruptions. Int J Prod Res 43:19

Bogataj D, Bogataj M (2007) Measuring the supply chain risk and vulnerability in frequency space. Int J Prod Econ 108

Bouchon S (2006) The vulnerability of interdependent critical infrastructures systems: epistemological and conceptual state-of-the art. European Commission Directorate-General Joint Research Centre Institute for the Protection and Security of the Citizen

Briano E, Caballini C, Revetria R (2009) Literature review about supply chain vulnerability and resiliency. Proceedings of the 8th WSEAS international conference on SYSTEM SCIENCE and SIMULATION in ENGINEERING

Brusset X (2013) Supply chains: agile, resilient or flexible? MS&OM conference, 28–30 July 2013, Fontainebleau, France

Bukowski L, Feliks J (2012) Dependability based concept of supply chain resilience. Proceedings of METAL conference, 23–25.05.2012, Brno, Czech Republic
Bukowski L, Feliks J (2011) Evaluation of technical systems dependability with the use of fuzzy logic and experts' knowledge. Proceedings of the 15th world multi-conference on systemics, Cybernetics and Informatics, Orlando (Florida)
Carvalho H, Barroso AP, Machado VH, Azevedo S, Cruz-Machado V (2012) Supply chain redesign for resilience using simulation. Comput Ind Eng 62
Chapman P, Christopher M, Juttner U, Peck H, Wilding R (2002) Identifying and managing Supply-Chain Vulnerability. Logistics Transp Focus 4(4)
Cho S (2010) Modelling of supply chain vulnerability using influence matrices. J Stud Manufact 1
Christopher M (1998) Logistics and supply chain management: strategies for reducing cost and improving service. Financial Times/Prentice Hall, London
Christopher M, Peck H (2004) Building the resilient supply chains. Int J Logistics Manage 15(2)
Christopher M, Rutherford C (2004) Creating a supply chain resilience through agile Six Sigma. Critical Eye Publications LTD
Compact Oxford Dictionary: resilience, http://www.oxforddictionaries.com/definition/english/resilient (11.02.2014)
Cox A, Prager F, Rose A (2011) Transportation security and the role of resilience: a foundation for operational metrics. Transport Policy 18
Creating a Resilient Supply Chains: A practical Guide. Cranfield School of Management, 2003. https://dspace.lib.cranfield.ac.uk/bitstream/1826/4374/1/Creating_resilient_supply_chains.pdf
Fakoor AM, Olfat L, Feizi K, Amiri M (2013) A method for measuring supply chain resilience in the automobile industry. J Basic Appl Sci Res 3(2)
Falasca M, Zobel CW, Cook D (2008) A Decision support framework to assess supply chain resilience. Proceedings of the 5th international ISCRAM conference, Washington, DC, USA, May 2008
Fazli S, Masoumi A (2012) Assessing the vulnerability of supply chain using Analytic Network Process approach. Int Res J Appl Basic Sci 3
Folta M, Novotny P (2013) Vulnerability of industrial supply chain. Proceedings of METAL conference, 15–17.05.2013, Brno, Czech Republic
Francis R, Bekera B (2014) A metric and frameworks for resilience analysis of engineered and infrastructure systems. Reliab Eng Syst Saf 121
Haffenen B, Verner D, Petit F, Bassett G, Phillips J, Carlson L, Collins M, Whitfield R, Folga S, Buehring W (2012) Resilience theory and applications, Technical Report ANL/DIS-12-1, Argonne National Laboratory, Argonne, Ill.: January 2012 http://www.ipd.anl.gov/anlpubs/2012/02/72218.pdf
Haywood M, Peck H (2003) An investigation into the management of supply chain vulnerability in UK aerospace manufacturing. Proceedings of the 8th international symposium on logistics, Sevilla, 6–8th July
Henry D, Ramirez-Marquez JE (2012) Generic metrics and quantitative approaches for system resilience as a function of time Reliab Eng Syst Saf 99
Hofmann M, Kjolle GH, Gjerde O (2012) Development of indicators to monitor vulnerabilities in power systems. http://www.sintef.no/project/Vulnerability%20and%20security/Publications/Papers/PSAM11_ESREL2012_Vulnerability%20indicators.pdf
Klibi W, Martel A (2012) Modeling approaches for the design of resilient supply networks under disruptions. Int J Prod Econ 135
Kroger W, Zio E (2011) Vulnerable systems. Springer, London
Kroger W (2008) Critical infrastructures at risk: a need for a new conceptual approach and extended analytical tools. Reliab Eng Syst Saf 93
Longo F, Oren T (2008) Supply chain vulnerability and resilience: a state of the art overview. Proceedings of the European modeling & simulation symposium, Campora S. Giovanni (CS), Italy
Losada C, Paola Scaparra M, O'Hanley JR (2012) Optimizing system resilience: a facility protection model with recovery time. Eur J Oper Res 217

Mascaritolo J, Holcomb MC (2009) Moving towards a resilient supply chain. J Transp Manage 19(Fall 2008)
Neil Adger W (2006) Vulnerability. Global Environ Change 16
Nowakowski T (2013) Vulnerability vs. dependability of logistic systems. Proceedings of Carpathian logistic congress 2013, Dec 9th–11th 2013, Cracow, Poland
Nowakowski T, Valis D (2013) Selected options of vulnerability assessment. State of Art in Literature Review. Logistics Trans 17(1)
Oke A, Golapakrishnan M (2009) Managing disruptions in supply chains: a case study of a retail supply chain. Int J Prod Econ 118
Peck H (2006) Reconciling supply chain vulnerability, risk and supply chain management. Int J Logistics Res Appl 9(2)
Peck H (2005) Drivers of supply chain vulnerability: an integrated framework. Int J Phys Distrib Logistics Manage 35(4)
Pettit TJ, Croxton KL, Fiksel J (2013) Ensuring supply chain resilience: development and implementation of an assessment tool. J Bus Logistics 34(1)
Ponis ST (2012) Supply chain resilience: definition of concept and its formative elements. J Appl Bus Res 28(5)
Rose A, Krausmann E (2013) An economic framework for the development of a resilience index for business recovery. Int J Disaster Risk Reduction 5
Rysińska D (2013) Analysis of logistics processes' quality on chosen example, (not published), BSc Thesis, Wroclaw University of Technology
Sapna Isotupa KP, Kelly M, Kleffner A (2014) Building resilient supply chains using supply chain and traditional risk management and insurance techniques, http://www.aria.org/drupal/sites/default/files/annual_meeting_docs/aria%20supply%20chain%20isotupa%20kelly%20kleffner%20Final.pdf (11.02.2014)
Schoon M (2005) A short historical overview of the concepts of resilience, vulnerability, and adaptation. Workshop in Political Theory and Policy Analysis Indiana University, Working Paper W05-4. http://www.indiana.edu/~iupolsci/gradcv/schoon/historical_critique.pdf
Sheffi Y (2007) The resilient enterprise, overcoming vulnerability for competitive advantage. The MIT Press, Cambridge
Sheffi Y (2006) Resilience reduces risk. The official Magazine of the Logistics Institute, vol. 12, no. 1
Sheffi Y (2005) Building a resilient supply chain, Harv Bus Rev 1(8)
Shuai Y, Wang X, Zhao L (2011) Research on measuring method of supply chain resilience based on biological cell elasticity theory. IEEE international conference on industrial engineering and engineering management, 6–9 Dec 2011, Singapore
Singhal P, Agarwal G, Lal Mittal M (2011) Supply chain risk management: review, classification and future research directions. Int J Bus Sci Appl Manage 6(3)
Supply Chain Resilience 2013, 5th Annual Survey, Business Continuity Institute. http://www.zurich.com/internet/main/sitecollectiondocuments/reports/supply-chain-resilience-2013-en.pdf
Supply Chain Resilience 2012, 4th Annual Survey, Business Continuity Institute. http://www.zurich.com/internet/main/sitecollectiondocuments/reports/supply-chain-resilience2012.pdf
Supply Chain Resilience 2011, 3rd Annual Survey, Business Continuity Institute. http://www.zurich.com/internet/main/sitecollectiondocuments/insight/supply-chain-survey-2011.pdf
Supply Chain Resilience 2010, 2nd Annual Survey, Business Continuity Institute. http://www.bcipartnership.com/BCISupplyChainResilienceSurvey2010EXECUTIVESUMMARY.pdf
Supply Chain Resilience 2009, 1st Annual Survey, Business Continuity Institute. http://www.bcifiles.com/BCISupplyChainResilienceSurvey2009MainReport2.pdf
Supply chain vulnerability (2002) Executive Report, Cranfield University School of Management. http://www.som.cranfield.ac.uk/som/dinamic-content/research/lscm/downloads/Vulnerability_report.pdf
Svensson G (2004) Vulnerability in business relationships: the gap between dependence and trust. J Bus Ind Mark 19(7)

Svensson G (2002a) A conceptual framework of vulnerability in firms' inbound and outbound logistics flows. Int J Phys Distrib Logistics Manage 32(2)

Svensson G (2002b) Vulnerability scenarios in marketing channels. Supply Chain Manage Int J 7(5)

Svensson G (2002c) Dyadic vulnerability in companies' inbound and outbound logistics flows. Int J Logistics Res Appl Leading J Supply Chain Manage 5(1)

Svensson G (2000) A conceptual framework for the analysis of vulnerability in supply chains. Int J Phys Distrib Logistics Manage 30(9)

Valis D, Vintr Z, Malach J (2012) Selected aspects of physical structures vulnerability—state-of-the-art. Eksploatacja i Niezawodnosc Maintenance Reliab 14(3)

Wagner SM, Neshat N (2010) Assessing the vulnerability of supply chains using graph theory. Int J Prod Econ 126

Walkowiak T, Mazurkiewicz J (2013) Planning of respond actions to emergency situations in discrete transport systems (in Polish), XLI Winter School of Reliability—Critical Infrastructure Dependability, Faculty of Transport, Warsaw Technical University, 2013, Szczyrk, 6–12 January 2013, Szczyrk 2013

Werbińska-Wojciechowska S (2013) Time resource problem in logistic system dependability modelling, Eksploatacja i Niezawodnosc Maintenance Reliab 15(4)

Wicher P, Lenort R (2013) The ways of creating resilient supply chains. Proceedings of Carpathian logistic congress 2013, Dec 9th–11th 2013, Cracow, Poland

Wicher P, Lenort R, Krausova E (2012) Possible applications of resilience concept in metallurgical supply chains. Proceedings of METAL conference, 23–25.05.2012, Brno, Czech Republic

Zio E, Kroger W (2009) Vulnerability assessment of critical infrastructures. IEEE Reliability Society 2009 Annual Technology Report

The Category of Risk Management in a Company with High Level of Customization

Anna K. Stasiuk-Piekarska, Łukasz Hadaś and Magdalena K. Wyrwicka

Abstract The purpose of this chapter is to present selected aspects of organizational risk management in an engineering and manufacturing-to-order enterprise, particularly these focused on the importance of material flow management in a supply chain. The authors reviewed selected well known methodologies of risk management in terms of their usefulness in systems with a high level of customization. They also reviewed norms and standards of 31,000 series in order to distinguish some characteristics of risk. The diagnosis included its division according to the likelihood of being suppressed by the system, when the frequency of its appearance and its influence on the considered system is taken into account. The analysis of companies, focused on production dedicated to the individual customer's order, made it possible to categorize the types of risk connected with the implementation process. Suggesting the different approach to managing risk occurring on various operational level is the final effect of the chapter.

Keywords Risk management · Supply chain · Make to order · Customization

1 Introduction

Risk is an integral part of making a decision, in a private life as well as in a professional one. The more decisions concerning unique situations are taken, the higher degree of uncertainty connected to them (PKN-ISO Guide 73:2012 2012).

A. K. Stasiuk-Piekarska (✉) · Ł. Hadaś · M. K. Wyrwicka
Poznan University of Technology, Strzelecka 11, 60-965 Poznan, Poland
e-mail: anna.stasiuk-piekarska@put.poznan.pl

Ł. Hadaś
e-mail: lukasz.hadas@put.poznan.pl

M. K. Wyrwicka
e-mail: magdalena.wyrwicka@put.poznan.pl

P. Golinska (ed.), *Logistics Operations, Supply Chain Management and Sustainability*, EcoProduction, DOI: 10.1007/978-3-319-07287-6_13,

Counteracting the aftermath of any undesirable action or situation should be planned and well organised, controlled and coordinated, especially in the context of functioning company. A risk management is an attitude which derives from the need of providing safety to a company on a strategic, tactical and operational level. Maintaining the safety of a workplace is an obligation of management team, both as a moral responsibility and as a legal necessity connecting company functions with those of other subjects, e.g. suppliers and customers.

Risk is understood as a ratio of frequency of consequences (time or cost) and probability of occurring an incident which results in a disturbance and its effects on an examined system. According to the standard PKN-ISO Guide 73:2012 (2012) which is a definition of risk management, risk is understood as "*an effect of uncertainty on objectives*"—"*a deviation from the expected—positive and/or negative*", whereas "*an uncertainty*" is defined as "*the state, even partial, of deficiency of information related to, understanding or knowledge of, an event, is consequence, or likelihood*".

The risk management defined above is a set of coordinated actions to direct and control a company with regard to risk (PKN-ISO Guide 73:2012 2012). The standard of risk management outlined by FERMA (Federation of European Risk Management Associations) interpret it as a process which involves methodical dealing with problems being the results of risk connected with company activities. Such handling the risk may and should be beneficial for a company (Raport Specjalny 2014).

Every incident is only on the surface unpredictable and random whereas it is just a visual confirmation of lack of professional knowledge of a person which is concerned (Kaczmarek 2006; Kisperska-Moroń and Krzyżaniak 2009; Stasiuk and Werner 2012; Zsidisin et al. 1999). Taking this into consideration one should perceive risk as a ratio of frequency and probability of occurring an incident involving availability (understood as being ready to use) on an operational level, necessary to produce informative data, resources and objects of work and an executive team of the process which includes a workplace, time, frequency and quantity, all of them can affect the whole production system. It is obvious that one have to notice the necessity of managing operational risk on operational level. Thus, it is essential to update and analyze the stage of utilization of resources since only practical and logical way of using them is advised (Wiśniewska and Janasz 2011). In a case when the process of production is disturbed may appear too low or too high exploitation of available resources.

2 Approach to Risk Management

Every decision made by employees is burdened with the uncertainty, thus it should be transformed into identified risk. Wrong decisions can be the reason that a company won't be able to meet its goals. It is essentials for a company with the high level of customization to outline the role of risk management, especially when the production is controlled, when reacting on deviations is obviously very important. A risk management isn't only insurance of areas exposed to the standard, often emphasized risk. Risk management should be oriented on the proactive approach to prevent any wrong and unscheduled situations in the consistent and earlier planned way. It definitely, shouldn't be a reaction which one may called "*firefighting*", however it is quite often in many enterprises. They often represent the reactive attitude which means reacting in the moment of crisis or incident or even later when the aftermath appears. Such an understanding is reflected in Fig. 1. Authors compared the traditional and modern approach to the risk. In particular, one should pay attention to differences initiating the first phase of the risk management. The reactive approach most often keeps an eye on appearing interferences which should be dealt with. In the modern, proactive approach, the risk management begins with the risk analysis which is associated with the need of identifying the risk and assessing its influence on the whole process.

Moreover, risk management approaches considered in the context of the entire company are distinguished into traditional and integrated in specialist literature, it is shown in the Table 1.

The risk in the traditional approach obstructs achieving business purposes. In the integrated one is supposed to support achieving them. It is possible thanks to the constant risk assessment, updating and assigning responsibilities to a particular person to complete the processes.

3 Overview of Methods of the Risk Management

Risk management, on the basis of the entire enterprise, is called ERM (it is an abbreviation of enterprise risk management). Its first and basic goal is to assign a position of a risk manager. Nowadays, there is also a newer expression of the post, it is a chief risk officer (CRO). With "*the new name*" comes also a wider area of responsibility than traditionally. Obviously, one can't ignore a growth in importance of committees dealing with risk at a workplace in order to provide an integrated and systematic approach in the entire company (Raport Specjalny 2014).

The first attempt of formulating basic components of the corporate monitoring system on the international arena was the document provided by OECD—Principles of Corporate Governance. Its last version was admitted through Member States in 2004. However, in Poland these regulations were accepted only by the Warsaw Stock Exchange under the name "*best practices in public companies in 2005*". In these documents a strong emphasis is being put on the protection of minority shareholders, although it is pointing, both directly and indirectly, at question of risk management.

There have been many guidelines drawn up so far. They are recognizable on the market and include (Daliga 2014):

- An American Enterprise Risk Management—Integrated Framework COSO II 2004;

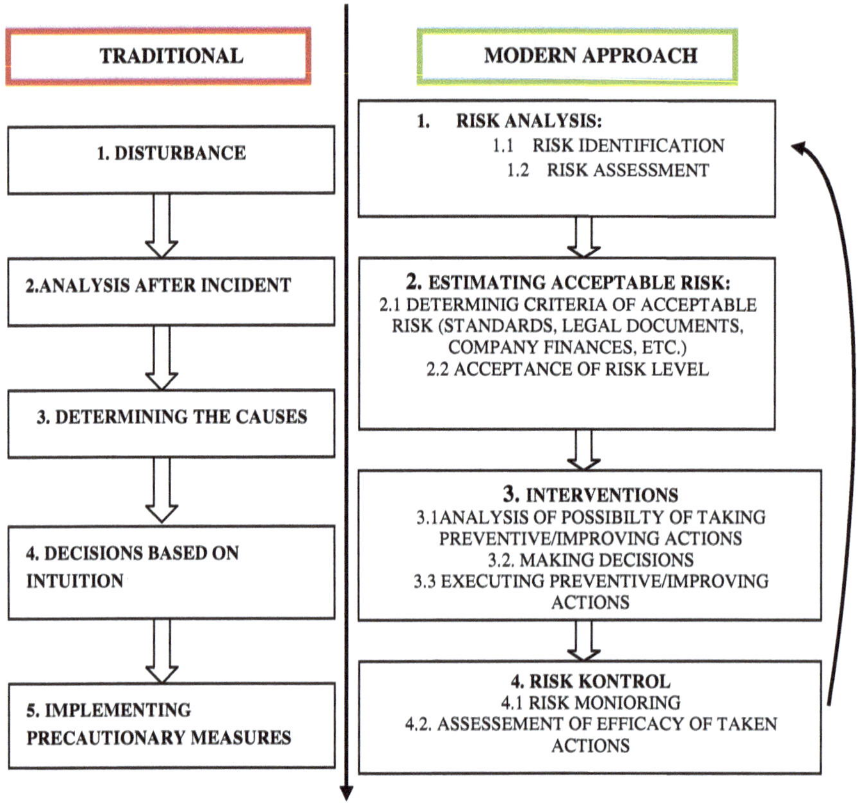

Fig. 1 Comparison of approaches to risk management

Table 1 Comparison of traditional and integrated approaches to risk management

Traditional approach	Integrated approach
Influence on a strategy is limited	Effective support of a company business plan
Reluctance to take a risk	Managing is proactive, it includes analysis, estimation and planning which in turn allows to take the right action in case of a risk situation
Information system is incoherent	Concise, coherent and consolidated communication system
Closed communication	Open communication
Barriers intensification	Integrated approach
Occasional risk assessment (most often when any kind of disturbances appear)	Risk is continuously estimated, updated and dealt with
Lack of assigned responsibilities	Responsibility is assigned to a particular person and to a particular process
Lack of clearly defined roles in the carried out processes	Roles are defined and known

COSO II is one of the most popular practices of risk management and it operates in the area of the organizational order (COSO II—"*integrated frame structure for the corporate risk management*" http://isc.infor.pl/slownik-pojec/haslo,124296.html Accessed 10 Jan 2014;

System components of managing the corporate risk result in an integrated form showing the manner of running the business. They consist of the following http://www.coso.org/documents/COSO_ERM_ExecutiveSummary_Polish.pdf Accessed 10 Jan 2014, http://www.coso.org/-erm.htm Accessed 10 Feb 2014:

- Internal environment
- Setting objectives
- Identification of incidents
- Risk assessment
- Reaction to the first signs of risk
- Controlling actions
- Information and communication
- Monitoring

- British RM Standard AIRMIC/ALARM/IRM 2002
- Australian and New Zealand AS/NZS 4360:2004;
- South African King II.

The most often referred of industry standards are:

- Medical—BRC/IFS (Raport Specjalny 2014);
- Grocery—HACCP ISO 22000, BRC/IOP;
- Sarbanes–Oxley Act 2002 a document regulating activities American listed companies;
- Standard ISO 27001—which regulates managing the safety of information;
- AS/NZS 4360—Australian and New Zealand risk management standard;
- Corporate governance—a code "*Best practices for WSE of listed companies*".

Mentioned above best practices are defined as "*soft*" in terms of their character and voluntary participation. However, there are also "hard" regulation concerning financial markets, also European markets: Basel II for the banking area (Basel Committee on Banking Supervision—International Convergence of Capital Measurement and Capital Standards; A Revised Framework) or Solvency II for the insurance area.

In Poland, there is a document facilitating implementing the risk management, too. Its range defines the norm PKN-ISO 31000: 2012. which also recommends to base the structure of risk management on the following guidelines (PKN-ISO 31000: 2012 2012):

- The risk management is supposed to create and to protect the value (as a possibility pointing and achieving objectives and improving some activities),
- The risk management is supposed to be an integral part of all processes in a company organization (also as an component of managing and planning changes),

- The risk management is supposed to be a part of making decisions (as a facilitation in making conscious decisions),
- The risk management is supposed to refer explicitly to the uncertainty (its character and the way of taking it into consideration),
- The risk management should be systematized, prioritized and on time (its essential in terms of achieving cohesive, credible and compared results),
- The risk management is supposed to use best available information (to confirm that the whole system is as reliable as input data),
- The risk management should be fitted (particularly, to the external and internal context of a company and the profile of its risk),
- The risk management should include cultural and human factors (it is considered as an external and internal possibility and intentions of stakeholders),
- The risk management is supposed to be transparent and comprehensive (it should be based on updated risk management systems which engage stakeholders),
- The risk management should be dynamic, iterative and reacting to changes (in the planning phase monitoring and inspections should be included as necessary part of updating process),
- The risk management should facilitate constant improving of a company (reaching higher and higher level of risk management is supposed to be an effect).

Obviously, the catalogue of the guidelines associated with the structure of the effective risk management system was considerably extended.

4 Diagnosis of Risk in a System

The mass production in the world of global competition, quite often, does not satisfy consumers needs. They demand a wider range of products, available quickly and for a reasonable price (Lubiński 2005; Stasiuk and Werner 2012). To be among the best competitors an enterprise has to create such an offer which will suit individual needs of each customer. This type of dealing with customers is called a customization, it means that a product is produce or modify according to a customer's individual requirements. Analysis of this phenomenon leads into the conclusion that that mass customization is a revolutionary approach to a consumer. Rudnicki defined "*Mass customization as producing goods to respond to demands of relatively wide market including specific needs of an individual customer with expenses similar to those of mass production*". Customization can have different forms in a value chain (Rudnicki 2012; Stasiuk and Werner 2012; Wirkus and Maciągowski 2012). It is shown in (Fig. 2).

The levels of customization shown above one should understands as follows (Rudnicki 2012):

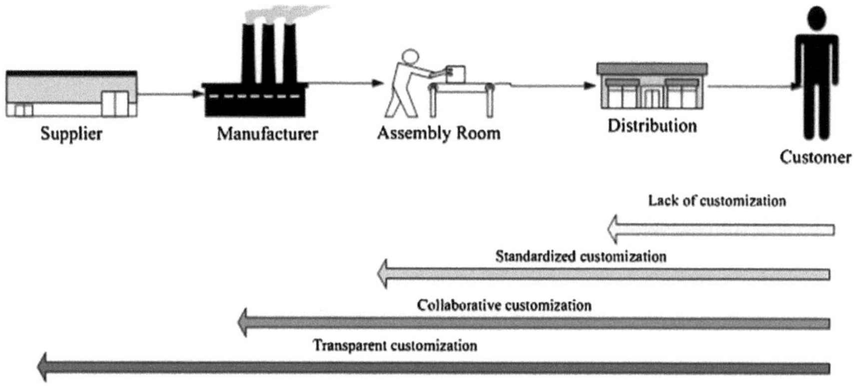

Fig. 2 Levels of customization in a supply chain

- Transparent customization—a company provides individual customers with unique products. In this case there is a need to accurately assess customer needs,
- Collaborative customization—a company talks to individual customers to determine the precise product offering that best serves the customer's needs,
- Standardized customization—a company provides a diverse set of components while maintaining control over the final product,
- Lack of customization—a company produce only standardized goods.

It is worth to mention that even earlier producing goods to individual customer order was defined as material distribution point. The definition outlined (Kisperska-Moroń and Krzyżaniak 2009):

- Engineer-to-order,
- Manufacturing-to-order,
- Assembly-to-order,
- Make-to-stock.

It is shown in Fig. 3.

Manufacturing-to-order is an effect of maintaining supplies of resources or materials, from which a lot of various products-to-order can arise, in a material chapter point.

Assembly-to-order is about maintaining a supply of components from which a customer can choose elements that he/she wants to use within a purchased product. However, its construction is usually based on a universal platform.

Make-to-stock means realisation of production on the basis of forecasts e.g. where a product is available as a final product and waits for a customer.

Research was conducted in a manufacturing company, which specializes in the assembly of very complex products made for individual orders of customers as well as their suppliers. A high level of customization was identified which means that a distribution point moved towards suppliers of materials or parts of assembly.

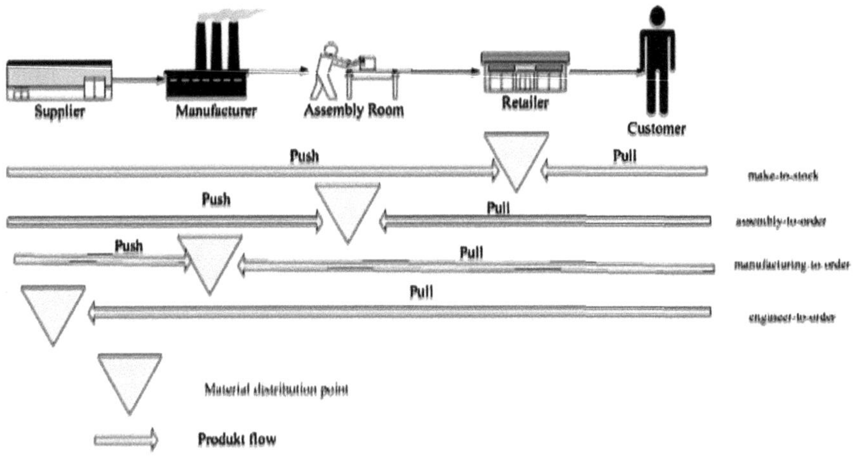

Fig. 3 Possible locations of material distribution point

Thus the demand from being dependant changes its status into independent (Klimarczyk and Masadyński 2009; Motała et al. 2008; Krzyżaniak 2005).

Both, standardized customization as well as assembly to order are a part of a selective production, in this process components are measured and grouped into several classes as they are manufactured. This approach leads to high quality assembly using relatively inexpensive components (Organizacja w produkcji 1997).

Nowadays, the industry is approaching a breakthrough which is the effect of shorter cycles of appearing innovations, growing diversifications and higher customization of goods. Obviously, it is easy to notice a process of evolution of a value chains. The Internet, which is now a base of almost every product or service, has become the key of the New Industrial Revolution called "*Industry 4.0*". The idea of the industry of the 4th generation is based on a philosophical view determining intelligent manufacturing, extended cooperation between devices working in various standards and easy flow of information (Gospodarcze 2013).

In assumptions the "*Industry 4.0*" should generate the growth of automatic production using manufacturing devices able to cooperate. The effect of this should be a product fitted to the preference of individual users. These changes are easy to notice, however the forecast points out that the nearest decade will be the one when they will become global (Stasiuk-Piekarska and Wyrwicka 2013).

5 Managing Risk Groups in the Context of Organizing Processes

Contemporary requirements associated with dynamics of the economic environment, time pressure, the scantiness of resources and the growing level of complexity of problems (Wyrwicka 2003) caused a situation when decision-makers had adopted an external strategy focused on reactions of stakeholders and building up an image of a company (Stasiuk-Piekarska and Wyrwicka 2013). Attention, directed towards approval of surroundings, is controversial, because it shows the necessity of corrections. Therefore, one can witness a serious omission in creating new posts, processes or systems which purpose should be to execute the effectiveness of an enterprise (Rummler and Brache 2000).

Referring to risk management it is possible to distinguish an internal context associated with the environment of a company and an external context linked to (PKN-ISO Guide 73:2012 2012):

- Macro- and microenvironment,
- Crucial trends and factors affecting the goals of a company,
- Activities and relations with external stakeholders.

One of model processes, which organizing is burdened with the high level of risk on the operational level (especially, in a case of an enterprise with a high level of customization) is controlling the supply chain.

The examined company is one of large enterprises functioning in the automotive industry, having its customers all over the world. The analysis of data shows, that in spite of pretty long and stable functioning on the market, the company still is developing its products and looking for new employees. The feature of the enterprise is a high level of customization, which was proved by Hadaś and Klimarczyk (2010a, b).

The operational risk is defined as a risk of loss in a situation when dealing with a customer does not meet the assumptions. Its source can be various: a company, customers or the third party involved in the transaction (Kaczmarek 2006; Stasiuk and Werner 2012). One should notice that particular stages in the supply chain are complementary—for example a production process carried out on time, when the risk of incomplete and overdue supplies was eliminated, may keep down production costs and enable to generate estimated profit.

There are two basic groups of suppliers in the examined company:

- External,
- Internal.

External suppliers form the largest group amongst analysed stakeholders. It is also possible to divide them into providers of services (who deliver the service of covering particular elements of a finished product) and providers of products (essential in finishing and producing the final item). As far as the quantity is considered, external suppliers constitute about 85–90 % of all deliveries. This affects not only identified during the process risk but also its size.

Internal suppliers are a group formed of employees from supported positions of the company and those from subsidiaries of the company (located not further than 50 km from the parent company) (Table 2).

Analysing three chosen contracts that were being fulfilled in the enterprise it was pointed out that the vast majority of components (around 96.6 %) were delivered in a standard way. Less than 3 % (2.6 %) parts were chosen by a

Table 2 Suppliers division per their background

	External suppliers (%)	Internal suppliers (%)
Production order A	96.01[a]	3.99[a]
Production order B	96.01[a]	3.99[a]
Production order C	96.14[a]	3.86[a]
On average	96.05[a]	3.95[a]

[a] All suppliers (external and internal) per indexes

Table 3 Suppliers division per their cooperation with the company

	Standard suppliers[a] (%)	Suppliers of the bevy concerned[a] (%)	Suppliers chosen pointed out by a customers (%)
Production order A	96.20	2.88	0.92
Production order B	97.42	2.38	0.19
Production order C	96.14	1.73	2.26
On average	96.6	2.6	1.1

[a] All the suppliers (internal and external) per indexes

customer from elements provided by standard suppliers. On average parts selected individually, only for the given contract, constitute the 1.10 % of indices. These elements were installed for a special order, according to strict requirements of the tender specification (Table 3).

A customer can give some details in the specification of ordered goods, such as appointing a particular supplier of a chosen component apart from standard suppliers.

In the process of analysing the risk associated with functioning of the supply chain, the following organizational risks were noticed (on the operational level) according to the division into types of suppliers. They were verified and assessed, in terms of a possibility of occurring by employees of 3 departments of the enterprise. Assessment "1" meant the least probable, and "5" very probable. In a situation, when risk does not occur during the process, one should write down "*no applicable*". The results are presented in the Table 4.

A greater risk (4.7—which means very likely) is associated with underestimating the date of delivery by a supplier who was chosen by a customer of the company. It can often cause more delays in the production. Workers of the enterprise are supposed to facilitate and enable the process of fulfilling the demand. If despite all efforts the product wasn't delivered on time, the company is not burdened with penalties for delays in the production of the final product. Most often a customer who decides to choose the untested contracting party, takes the responsibility for problems associated with fulfilling the signed agreement. So to a considerable degree, the risk is transferred to the customer.

Table 4 Analysis and estimation of probability of delays in deliveries per supplier

Reason of a delay in a supply chain	External suppliers			Internal suppliers
	Standard suppliers*	Suppliers from a bank of offered suppliers*	Suppliers imposed by customers	
Inappropriate deliveries being a fault of employees of the company				
Wrong estimation of delivery time	3	2.7	4.7	1.3
Not making an order	2.7	2.3	2.3	1.7
Change of delivery time due to change of production agenda	3	3.3	3.7	2
Wrong quantity of order products	3.3	3.3	2.7	1.7
Appropriate deliveries but				
Delivered goods damaged due to improper storage	2.7	2.7	3	2
Damage of an element during production (during process of assembly or at further stage of production)	2.3	2.7	2.7	2.7
Wrong categorization of an item	2	2	2	1.7
Deliveries delayed due to problems with the security	1.7	1.7	1.7	0.3
Supplies inconsistent with the principle 7W (7R)—supplier fault				
A supply in wrong time	2	2	2.7	1.3
Wrong quality/state	2.3	2	2.3	1
Wrong cost/price	1.7	1.3	1.7	0.7
Wrong quantity	3	2.3	2	1
Wrong place	1.7	1.7	1.7	1
Wrong product	2	1.7	1	1
Wrong customer	1.3	1.3	1.3	0.7
No delivery (causes different than mentioned above)	1	1	1	1

Analysis of given replies shows, that in a case of an external and standardized customer, there is a big likelihood of occurring of disturbances due to the fault of employees of the described enterprise. Most often (3.3), as a cause of such situation was mentioned the wrong quantity of ordered products. What is interesting, less probable (2.7) is occurring of the same situation when external suppliers of the bevy are concerned. Taking this all into consideration one may conclude that strategic vendors realize that the company depends on them, in a sense, thus it is not entirely partnership or cooperation. This situation occurs even more often when suppliers are chosen out of other contracting parties.

Damaging goods during the production turned out to be the most significant plausibility associated with internal suppliers (even though goods are delivered according to all guidelines). It is possible that internal suppliers (employees of the company) do not regard themselves as important contributors to the whole process. It is very likely that they can assume that since items are being delivered from the inside of the company, every mistake will be corrected and mended quickly and free of charge. They don't notice the risk of delays in carrying the contract out and the need to involve additional (unscheduled) resources to produce the same product once again.

Practically, all these problems have no importance if the plausibility is a below 1. After further and closer examination, they will be assessed as posing only slight threat for the continuity of functioning of the company.

Regardless the evaluation of the probability of the occurence of the given incident, employees were encourage to write down own remarks concerning the examined problem. They outlined, among others, mistakes of the operational system, changing unit specification (e.g. from the mb on the itm) which resulted in ordering wrong quantities of goods.

6 Conclusion

In the article authors reviewed chosen methodologies of risk management. They also checked norms from series 31,000 in order to distinguish characteristics of risk. The diagnosis included the evaluation of the category of risk in the enterprises.

Performed analyses for the enterprise, which was defined as one typical for production focused on individual customer order, allowed categorization of risk connected with deliveries. In all groups of suppliers, higher level of delivery disturbances were linked to the fault of employees of the examined companies. It is not possible to build the efficient supply chain, when there are still some processes and procedures in a company needed tidying up.

The aim of building up a system of risk management in a company with high level of customization is creating a tool allowing effective identification and taking counteractions in a situation of risk occurring. Risk which is very diverse and characterizes by relatively low repetitiveness. This type of risk definitely requires creating proactive solutions. The risk management is supposed to be the base which enable a company to update information concerning possible risk and how to deal with it. The effect of it should be such level of risk which will not exceed those determined in risk criteria.

References

Daliga M (2014) Przegląd międzynarodowych standardów i metodyk zarządzania ryzykiem w przedsiębiorstwie, http://www.4pm.pl/upload/artykuly/InLab.pdf. Accessed 04 Jan 2014

Hadaś Ł, Klimarczyk G (2010a) Wyniki analizy ABC/XYZ w przedsiębiorstwie montażowym o wysokim stopniu kastomizacji—studium przypadku. pp 57–58, In: Logistyka 1/2010

Hadaś Ł, Klimarczyk G (2010b) Koncepcja budowy metodyki systemu zarządzania zaopatrzeniem w przedsiębiorstwie montażowym o wysokim stopniu kastomizacji. pp 50–51, In: Logistyka 2/2010

Kaczmarek TT (2006) Ryzyko i zarządzanie ryzykiem. Ujęcie interdyscyplinarne, wyd. Diffin, Warszawa
Kisperska-Moroń D, Krzyżaniak S (ed) (2009) Logistyka, seria Biblioteka Logistyka, Poznań
Klimarczyk G, Masadyński M (2009) Wyrwicka M. Zastosowanie analizy myślenia systemowego do kierowania zmianą w firmie montażowej. In: Logistyka 2/2009
Krzyżaniak S (2005) Podstawy zarządzania zapasami w przykładach, Instytut Logistyki i Magazynowania, Poznań
Lubiński P (2005) Istota planowania w przedsiębiorstwie. In: Fertsch M, Trzcieliński S (ed) Koncepcje zarządzania systemami wytwórczymi, Wyd. Politechniki Poznańskiej, Poznań
Motała D, Pawłowski E, Pawłowski K, Trzcieliński S (2008) Designing effective management system of enterprises: concept and its verification, human factors and ergonomics in manufacturing, vol 18. Wiley Periodicals Inc, p 5
Organizacja w produkcji (1997) Moduł 3110361, REFA- Verband für Arbeitsgestaltung, Betriebsorganisation Und Unternehmensentwicklung e.V., Darmstadt
PKN-ISO 31000:2012 (2012) Polski Komitet Normalizacyjny, Zarządzanie ryzykiem. Zasady i wytyczne, Warszawa
PKN-ISO Guide 73:2012 (2012) Polski Komitet Normalizacyjny, Zarządzanie ryzykiem. Terminologia, Warszawa
Raport Specjalny (2014) Zarządzanie ryzykiem, Polrisk, http://risklogic.pl/files/CFO_Polrisk-raport.pdf Accessed 04 Jan 2014)
Rudnicki J (2012) Indywidualizacja produktu, pp 12–13, In: Logistyka Produkcji 2/2012
Rummler GA, Brache AP (2000) Podnoszenie efektywności organizacji. PWE, Warszawa
Stasiuk A, Werner K (2012) Production flow control as a key element in production management. In: Production management. Contemporary approaches. Selected aspects, Hadaś Ł (ed), wyd. Politechniki Poznańskiej, Poznań
Stasiuk-Piekarska AK, Wyrwicka MK (2013) Organizowanie- wciąż istotna funkcja zarządzania produkcją, In: Gospodarka Magazynowa i Logistyka 11-2013
WI Wiadomości Gospodarcze (2013) Wirtschaftnachrichten 02-03/2013 AHK; http://www.profibus.org.pl/ Accessed 10 Aug 2013
Wirkus M, Maciągowski D (2012) Wysoka zmienność produkcji, pp 8–11, In: Logistyka Produkcji 2/2012
Wiśniewska J, Janasz W (2011) Gospodarowanie zasobami materialnymi. In: Marek S, Białasiewicz M (ed), Podstawy nauki o organizacji. Przedsiębiorstwo jako organizacja gospodarcza, Polskie Wydawnictwo Ekonomiczne, Warszawa
Wyrwicka M (2003) Endogenne przesłanki organizacyjne rozwoju przedsiębiorstwa. Wyd. Politechniki Poznańskiej, seria rozprawy Nr 374, Poznań
Zsidisin G, Panelli A, Upton R (1999) Supply risk assessment analisys. In: practix, best practices purchasing and supply chain management, vol 2. pp 9–12

Literature Study Overseas on SCM Strategy with a State of Art SCM Strategy Model

Angela Y. Y. Chen, Yutaka Karasawa, Nobunori Aiura, Kuninori Suzuki and Keizo Wakabayashi

Abstract There are a lot of research works as well as books written about SCM strategy, but few books are based on strategic theory of management in a pure sense. As a result, theories themselves are shallow and not in depth. In other words, SCM focuses on a strategic area of supply chain activities while, however, there are few books written on a basis of the theoretical or traditional articles of management strategy in a pure sense. The aims of this article are to review the theoretical strategy formulation concepts out of 50 books published in 2004 through 1990 while we had made logistics literature studies in the middle of 1995 with almost same approach and contributed to academic journals in Japan. This type of SCM literature study is a brand new research and the first attempt of research activity in Japan. As a survey result, it has made clear that most of books paid little attention to structural aspects of SCM strategy formulation as a state of art SCM model linked with management strategy theory. Therefore, we have reviewed the contents of SCM strategy written in fifty books and finally proposed our own SCM strategy structure and formulation based on strategic theories extracted from traditional international management theory as well as strategic theory of classical management.

Keywords Management strategy · SCA strategy · Strategy formulation · Literature study · State art model

---

A. Y. Y. Chen
China University of Technology, Taipei, Taiwan

Y. Karasawa (✉)
Kanagawa University, Kanagawa, Japan
e-mail: karasawa-y@h09.itscom.net

N. Aiura
Hokkai School of Commerce, Hokkaido, Japan

K. Suzuki · K. Wakabayashi
Nihon University, Nerima, Japan

P. Golinska (ed.), *Logistics Operations, Supply Chain Management and Sustainability*, EcoProduction, DOI: 10.1007/978-3-319-07287-6_14,

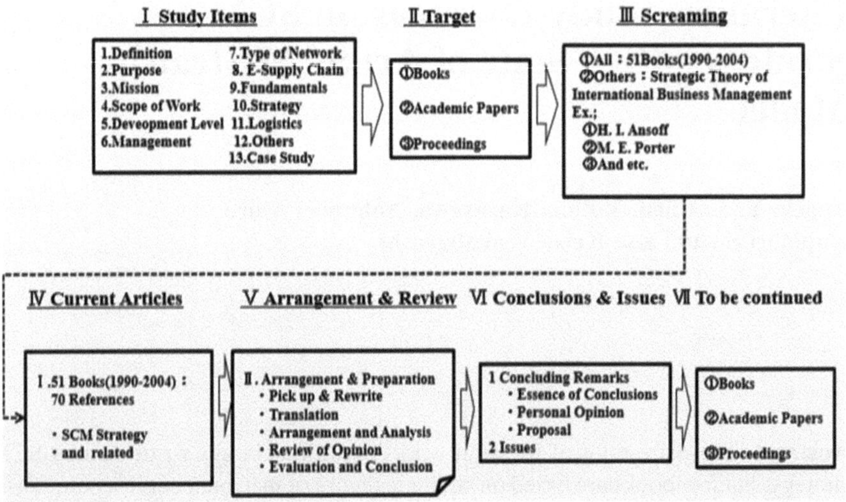

Fig. 1 Framework of literature survey

1 Background: Conclusion of Literature Survey

1.1 Framework of Literature Survey

As framework of literature survey is show in Fig. 1, 70 references were picked up from the books and articles issued from 1990 to 2004. What is studied consists of SCM Definition, Purpose, Mission, Scope of Work, Development Process, E-Supply Chain, and son. In addition relationship between Logistics and SCM was included. After making detail comparison analysis on each theme, we judged each way of thinking for SCM sub theme and finally summarized after evaluation.

1.2 Results of Literature Survey

The contents of all the articles evaluated on the basis of review criteria we set (Fig. 2).

In short we found out there were few articles of SCM strategy as well as SCM strategy based on management strategic theories. As show in conclusions, we come to the conclusions in short that there are few synthetic approach to SCM strategy only partially focused SCM strategic theory exists, and most of SCM theories are not based on traditional management strategy theory.

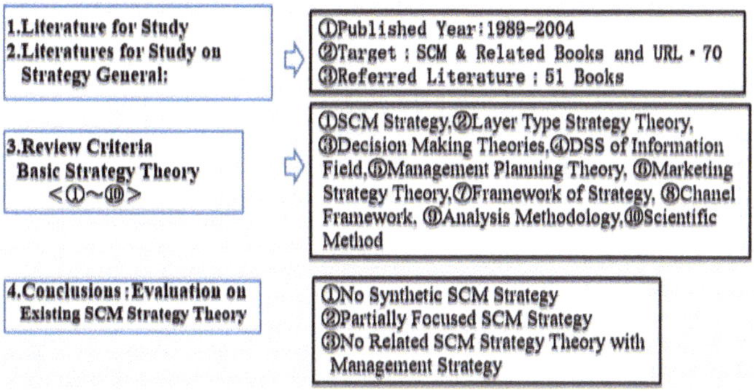

Fig. 2 Results of literature survey

1.3 General Evaluation of Study Results on SCM Strategy Theory

Briefly speaking, general comments on results are shown in Fig. 3. Current results include so much problematiques, which require a new state art model on SCM.

1.4 Evaluation Criteria for Theory of Basic Strategy <I—VIII>

Evaluation criteria are composed of following I–VIII (Fig. 4): Management control by layer, decision theory, decision support theory, long term management planning theory, marketing, management engineering, business administration and methods excluding soft ware. These elements are regarded as fundamental elements of framework production in a sense.

1.5 Results: Evaluation on SCM Strategy Theory by 12 Key Academicians

Evaluated results of main and key books are summarized in Table 1. The content on strategic theory is concerned, all other books are not qualified for review, out of question.

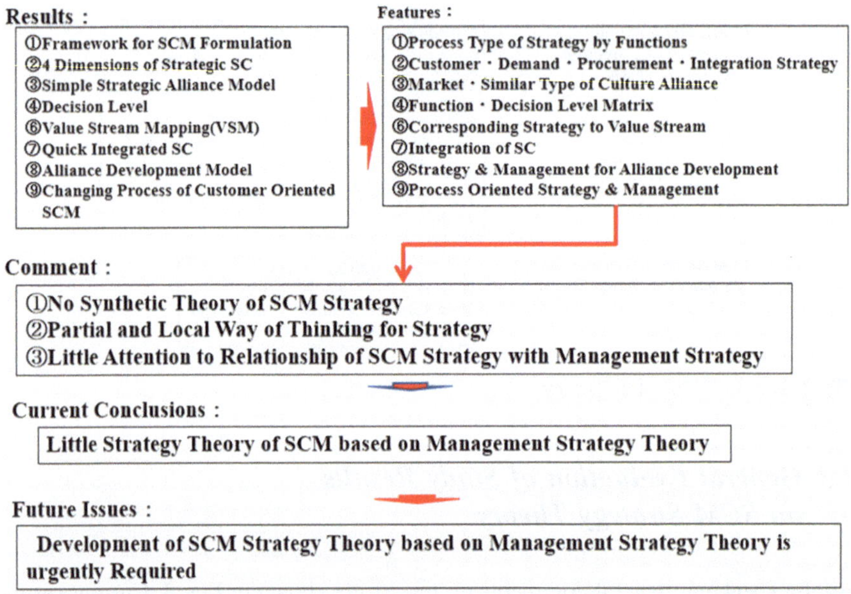

Fig. 3 General evaluation of study results on SCM strategy theory

1.6 Concluding Remarks

What is concluded by this study is summarized as follows:

1. SCM oriented theory of strategy could hardly be found out, so far as this study is concerned. In other words, the content of SCM strategic theory is so poor and insufficient although theoretical aspects of current SCM strategy and fields concerned are analyzed and made clear. In this sense research purposes are basically realized.
2. Dedicated SCM theory of strategy are investigated

   However, it is urgently required to make clear relationship between fundamental theories of strategy and SCM strategy theory. Unfortunately from SCM strategic theory view points, contents of very few books are satisfied and almost all the books concerned have insufficient explanation and framework, of which result enforces us to establish a new concept of SCM strategy theory.

2 Motivation

As already pointed out, research motivations are summarized as follows:

- There has been few literature surveys on SCM strategy based on the pure theory of business administration since SCM emerged in 1980s.

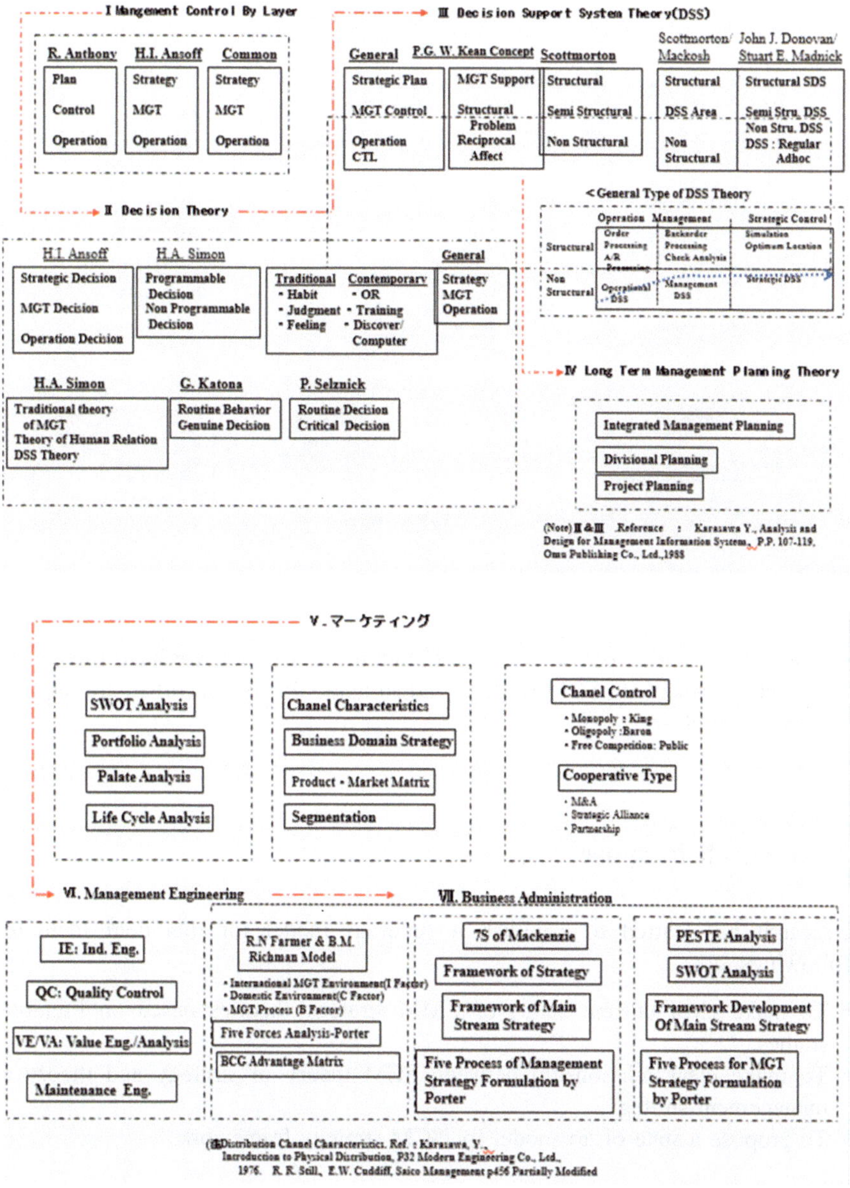

Fig. 4 Evaluation criteria for theory of basic strategy (Lysons and Gillingham 2003)

- There are few fundamental literatures discussed about strategic architecture of SCM theory based on a fundamental strategic theory of business administration.
- Establishment of common understanding for SCM strategy theory is urgently required both academic and practical viewpoints.

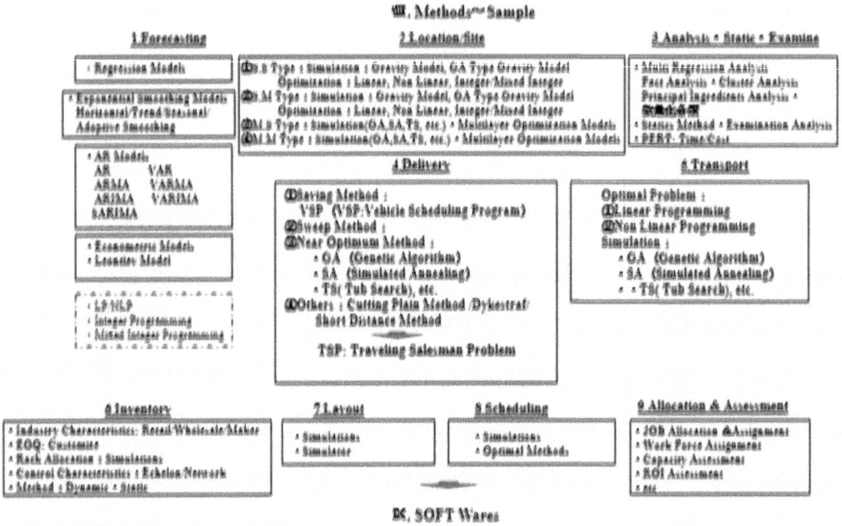

Fig. 4 continued

- Through in-depth study on oversea literatures concerned, we have recognized theoretical aspects of SCM strategy and strategic theory concerned are required to be made clear.

3 Research Purpose

Research has started to establish a state art model for this field including followings:

- To make clear current status of CMM strategy theories based on literature studies.
- To make clear relationship between SCM theory of strategy and theory of management strategy.
- To propose a state of art model for SCM strategy framework.

4 Research Preconditions and Confirmation for SCM

It should be mandatory to clear SCM definition, SCM and Logistics relationship, and SCM strategy position among management and planning of the firm before discussing about SCM strategy.

Table 1 Results: Evaluation on SCM strategy theory by 12 key academicians

No.	Author name	SCM strategy ①	Layer type strategy theory ②	DSS ③	Information DSS ④	MGT. plan theory ⑤	MKTG. theory ⑥	Strategy development framework ⑦	Chanel framework ⑧	Analysis method ⑨	Scientific method ⑩	Remarks
1	J. G.	Δ	×	×	×	×	×	×	Δ	×	×	Strategic alliance: excellent
2	R. B.	×	×	○	×	×	×	×	×	×	×	Logistics theory itself
3	P. H. et al.	×	○	×	×	×	×	×	×	×	×	Nothing but lean system
4	J. W. et al.	×	×	×	×	×	×	×	×	×	×	No strategic theory Operation guide oriented
5	R. H. et al.	×	×	×	×	×	×	×	○	×	×	Alliance theory: Refereble
6	J. A.	×	×	×	×	×	×	×	○	×	×	Design for operation: OK
7	H. S. et al.	×	×	×	×	○	×	×	×	×	×	Process oriented planning
8	T. M.	×	○	○	×	×	×	×	Δ	×	×	Production based purchasing design
9	D. B. et al.	×	○	×	×	○	×	×	×	×	×	Case oriented
10	J. G.	×	○	Δ	×	×	×	Δ	×	×	×	Not SCM, but materials supply Chain
11	K. L. et al.	Δ	○	○	×	×		○	×	Δ	×	Weak in SCM strategy, but good as a whole
12	C. P.	×	○	○	×	×	○	○	×	○	×	Materials SC oriented strategy

4.1 SCM Definition

At first definitions common to the world are referred and then we define our own definition.

1. The Council of Supply Chain Management URL 2011:

   Supply chain management encompasses the planning and management of all activities involved in sourcing and procurement, conversion, and all logistics management activities. Importantly, it also includes coordination and collaboration with channel partners, which can be suppliers, intermediaries, third party service providers, and customers. In essence, supply chain management integrates supply and demand management within and across companies.

2. APICS Dictionary (1995):

   The processes from the initial raw materials to the ultimate consumption of the finished product linking across supplier-user companies; and the functions within and outside a company that enable value chain to make products and provide services to the customer.

3. Supply Chain Council (1997):

   The supply chain encompasses every effort in producing and delivering a final product or service, from the supplier's supplier to the customer's customer. Four basic processes-plan, source, make, deliver-broadly define these efforts, which include managing supply and demand, sourcing raw materials and parts, manufacturing and assembly, warehousing and inventory tracking, order entry and order management, distribution across all channels, and delivery to the customer.

4. David Frederick Ross [] p. 9:

   The definition is as follows: Supply chain management is a continuously evolving management philosophy that seeks to unify the collective productive competencies and resources of the business functions found both within enterprise and outside in the firm's allied business partners located along intersecting supply channels into a highly competitive, customer-enriching supply system focused on developing innovative solutions and synchronizing the flow of marketplace products, services, and information to create unique, individualized sources of customer value.

   David Frederick Ross Continues:

   Based on the discussion of the definition of SCM, up to this point, it can be said that SCM consists of two dynamics. The first describes SCM as an operations management technique that enables companies to move beyond simply optimizing logistics activities only, to one where all enterprise functions-marketing, manufacturing, finance-are optimized by being closely integrated to form the foundation of a common business system David Frederick Ross, Competing Through Supply Chain Management, Kluwer Academic Publishers, P. 7, 1997.

Of particular relevance is the trend of current logistics management to assume responsibility not only for internal logistics functions but also for the challenges involved in integrating and coordinating the flow of materials and information along entire supply channel. The goal is the orchestration of an external logistics strategy that links all channels constituents together as a single operational entity. By integrating inventory, production, and distribution resources on an enterprise level as well as channel level, companies have the opportunity to realize optimal customer value while minimizing total supply chain cost. This outward-looking aspect of logistics management has been defined many experts *as supply chain management.* In fact, it is not unusual to see books and articles using the terms logistics and supply chain management as if they were synonymous, or definition where SCM is considered a subset or purely an extension of the integrated logistics concept.

David Frederick Ross, Competing Through Supply Chain Management, Kluwer Academic Publishers, P. 6, 1997. In brief, D. F. Ross explained the following comment:

Result of confusion concerning the proper definition of SCM, supply chain management, and logistics has been that instead of clarity, management teams, consulting professionals, and academics have become confused by trying to decipher what exactly is meant by SCM. Is SCM simply nothing more than the advanced application of the operations tools to be found in modern logistics management? Is SCM purely an operational tactic that seeks to extend benefits of logistics and supply chain process integration beyond the internal boundaries of the organization? Is SCM is synonymous with channel partner management? Is SCM, on the other hand, a fundamental new management paradigm, or just today's newest buzzword? Like Alice's Cheshire cat, SCM appears to mean one thing in reference to one idea, and then again something entirely different in another context, depending on who and what is being spoken about.

David Frederick Ross, Competing Through Supply Chain Management, Kluwer Academic Publishers, P. 7, 1997.

4.1.1 Brief Conclusions

- Definition: SCM is a strategic concept for supply chain
- Content: integration of supply and demand management beyond a firm or firms while supply demand management means *all the resources concerned with supply and demand management*, which includes visible and non visible goods and services, consisting of people, materials, money, information, etc.
- Span of control:
  - All the channels from upper stream to lower stream
  - Channels of forward logistics and reverse logistics

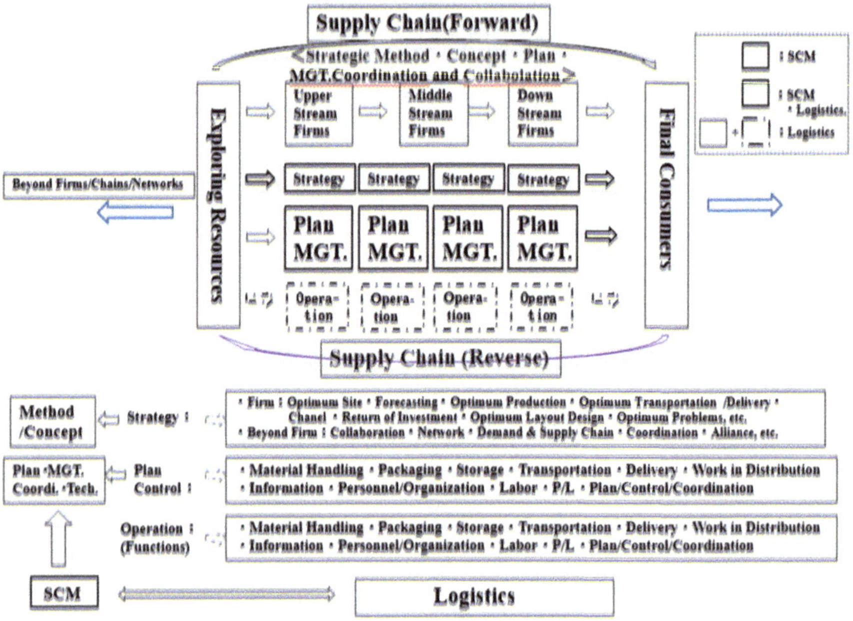

Fig. 5 SCM versus logistics

- Functional structural entity: logistics
- Activity factors and control level:
  - Basic function and support function of logistics
  - Control level corresponding to all span of operations

4.2 SCM and Logistics

4.2.1 General Comparison

General comparison of SCM versus Logistics is shown in Fig. 5, where comparison by items are analyzed:

- Functional structural entity of SCM is logistics
- Functional entity consists of materials handling, packaging, transportation, storage, information, work in process in distribution, other logistics functions
- Span: Including all the demand and supply channels with forward and reverse logistics
- Development process: Inside firm beyond firms social (e-Net) national global

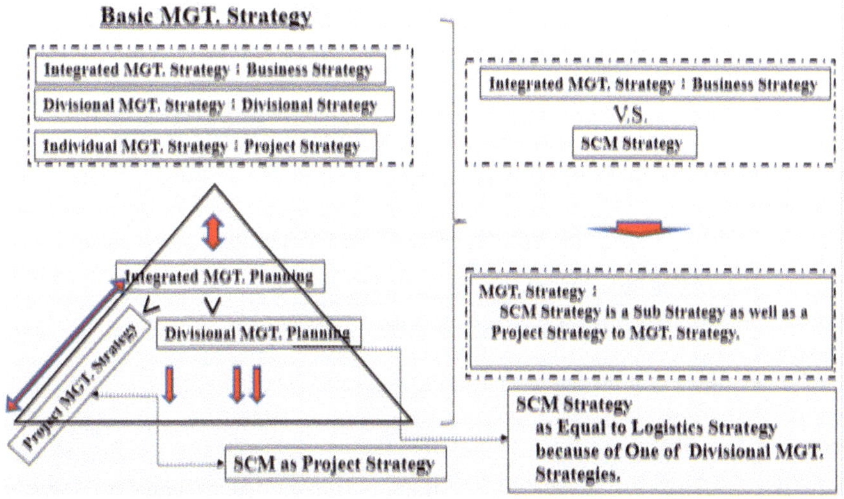

Fig. 6 SCM positioning in management strategy and planning

- Structure: Logistics strategy logistics planning logistics MGT. Logistics operation versus SCM. Strategy and management. SCM focuses on strategic aspects of logistics.

Therefore, it can be said that SCM means *a strategic way of thinking for demand and supply chains beyond firms based on logistics functions*. However, in general, as development process of logistics can be described as within firm, beyond firms, social, national, and global logistics, scope of work for SCM is equal to those of logistics.

4.3 SCM Positioning in Management Strategy and Planning

In general there are two types of management strategy: Integrated or General management strategy and project management strategy. The former has a wider structural architecture consisting of divisional management strategies and functions as a whole, while the latter is like a standalone strategy under conditions of clear target, period of termination, budget, resources, etc. Generally speaking, Logistics strategy or Logistics long range planning belongs to General management strategy and functions as a sub management strategy or divisional management strategy while SCM strategy belongs to Project wise strategy with limits of target, period of termination, budget, and resources (Fig. 6).

As shown in Fig. 6, principally SCM strategy belongs generally to project strategy while logistics strategy functions as a divisional strategy among management strategy. However, in real world, the both concepts are mixed up and so sometimes bring about confusion.

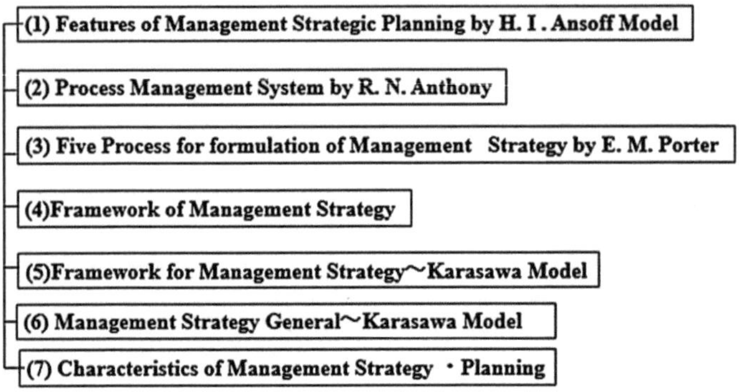

Fig. 7 Basic theory of management strategy

In this chapter, we handle SCM strategy is regarded as equal to logistics strategy so far as management strategy is concerned while SCM strategy itself is often implemented as a project strategy.

4.4 Conclusions

We come to the following conclusions. SCM strategy is: Strategic concept of demand and supply chain, Both function and structural entity are nothing but logistics, two aspects of SCM strategy: One of sub strategy of MGT. Strategy equal to logistics strategy and a project strategy for business strategy.

5 Contents

A state of art model of SCM strategy model is proposed after basic theory of management strategy is referred and discussed.

5.1 Basic Theory of Management Strategy

As to management strategy theory, Ansoff model, Anthony model, and Porter model are referred and then common framework of management strategy, Karasawa model, and other conceptual models are discussed (Fig. 7).

Table 2 Management strategic planning ~ H. I. Ansoff model

Items decision level	Problem	Characteristics of problem	Key decision item	Main feature
Strategic decision	Selection of product-market mix to realize fundamental revenue and profit in a optimum way	To allocate all the resources to the opportunities for product and market	Objectives and final objective multi business strategy expansion strategy management strategy financial strategy growth style timing for growth	What is centralized status of partial ignorance non repeatable non spontaneous generation
Management decision	Organizing company resources to bring about optimum results	Organizing, procuring, and developing resources	*Organization*	Conflicts between strategy and operation conflicts between individual target and organization target
			Organizing information	
			Authorization and job responsibility	
			Organizing resources	
			Transformation	
			Job process	
			Distribution system	Strong linkage between economic variables and social variables
			Facilities location purchasing and developing	
			Resources	Originated from strategic or operational issues
			Cash collection	
			Facility and equipment	
			Human resources	
			Law materials	
Operational decision	To activate optimum capital gain power in optimum way	Allocating budget to the main functional fields	Objectives and final objective of job	What is decentralized with risk and uncertainty repeat
			Production amount of sales price and output	
		Scheduling plan for application and utilization of resources	*Levels of performance*	Large quantity inevitability of second thinking because of complexity
			Daily production scheduling	
			Inventory amount	
		Supervising and controlling	String	
			Marketing control and strategy control	Spontaneous generation

Ref. H. I. Ansoff, management strategy, translated by T. Hirota. p. 12, Sanno College Publishing, 1981

Table 3 Process management system by R. N. Anthony

Management planning	Management control	Operation control
Selection company objective	Budgeting	Execution of budget
Organization plan	Staff allocation plan	Hireling implementation
Formulation of personnel policy	Job description production	Policy execution
Formulation of finance	Realization of capital plan	Credit term postponement control
Decision of marketing policy	Advertisement plan	Advertisement operation control
Decision of research policy	Decision of research project	Information processing services
Selection of new product	Selection of product improvement	Production planning production
Expansion, merge of company	Decision of plant reorganization	Physical distribution implementation, control
Decision of capital expenditure for regular expenditure	Regular capital expenditure decision	Inventory control
Decision of location policy	Decision rule production for operation control	Measurement, evaluation and improvement of labor efficiency
Selection of new business	Measurement, evaluation, improvement of operation	
	Performance	

Ref. Robert N. Anthony : planning control systems; a framework for analysis (Boston: Division of Research, Graduate School of Business, Harvard University, 1990), p. 19, Y. Ando et al. Kenpakusya p. 21, 1973, partially amended
Ref. Karasawa Y, Analysis and design for management information system, P. 60, Omu Publisliing Co., Ltd., 1988

1. Features of management strategic planning by H. I. Ansoff model

   The Ansoff model, well known model for hierarchical decision is shown in Table 2, where three tires decision structures are proposed. When it comes to strategic or management hierarchy three level or four level is common.

2. Process management system by R. N. Anthony

   Anthony model for hierarchical planning can be referred to Table 3, where three tiers hierarchy for planning is proposed. Only Tan Miller referred Anthony model to his own article out of 71 articles:
   Anthony (1965) offered a planning framework often quoted and employed by both theoreticians and practitioners in hierarchical production planning. A review of this framework will help to put the hierarchical characteristics of managerial decision making into perspective. A number of authors (e.g., Ackoff 1970) have termed the second category as tactical planning and the third category as operational planning and scheduling. For purpose of this book, we will

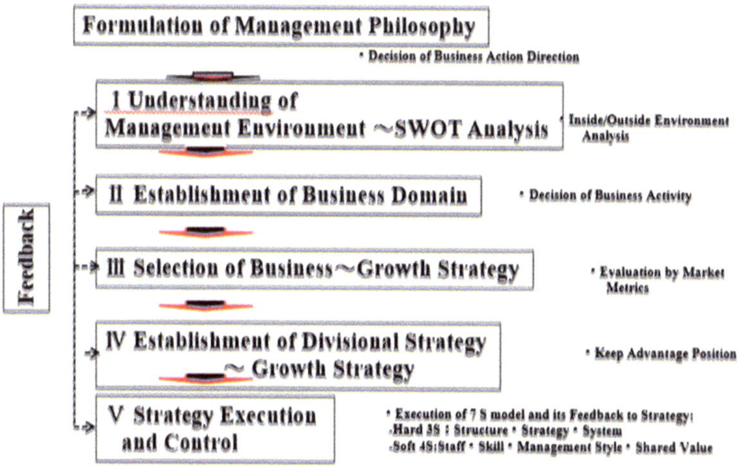

Fig. 8 Five processes of management strategy by Porter

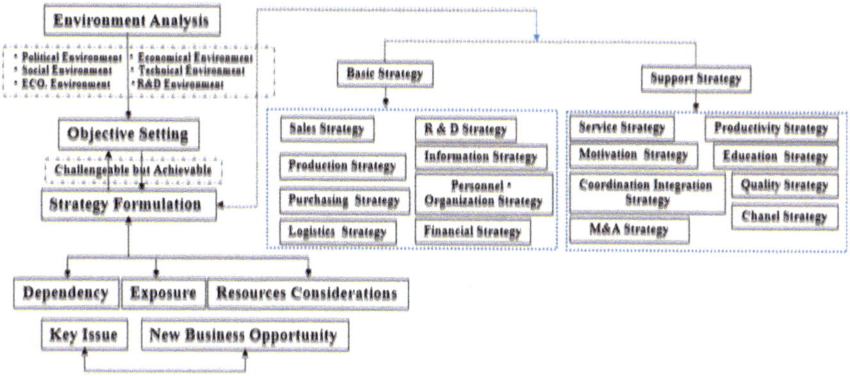

Fig. 9 Basic framework of management strategy

denote these three categories as strategic planning, tactical planning, and operational planning and scheduling.
Ref.: Tan Miller, Hierarchical operations and supply chain planning, Springer p. 1, 2003

3. Framework of management strategy

5.1.1 Five Processes of Management Strategy

Five processes of management strategy by Porter is shown in Fig. 8.

Table 4 Framework matrix for key strategy development ~Y. Karasawa

No	Structure	I general MGT. strategy	II MKTG. strategy	III Production strategy	IV Purchasing strategy	V Logistics (SCM) strategy
1	Environment analysis	① Political environment	① Market environment	① Production environment	① Materials environ.	① Logistics environment
		② Economical environment	② Product environment	② Equipment environment	② Price environment	② Eco environment
		③ Social environment	③ Chanel environment	③ Facility environment	③ Inventory environ.	③ Industry environment
		④ Cultural environment	④ Price environment	④ Design environment	④ Design environment	④ Price environment
		⑤ Technological environment	⑤ Customer environ	⑤ Maintenance environment	⑤ Technology environment	⑤ Transportation environment
		⑥ ECO environment	⑥ Technological environ.	⑥ Technology environment	⑥ Information/soft environment	⑥ Storage environment
		⑦ War/confliction environment	⑦ Industry environ.	⑦ Information/soft environment	⑦ New/alternative materials environment	⑦ Work in process environment
		⑧ Population environment	⑧ Competitive environ	⑧ Outsourcing environment		⑧ Labor environment
			⑨ Personnel environ.	⑨ Union environment		⑨ Information/soft environ.
			⑩ Import/export environ.			⑩ Facility and equipment envi.
			⑪ Trade environ.			⑪ Outsourcing environment
			⑫ Consume/ purchasing envi.			⑫ Deregulation environment
			⑬ Meeds trend environ.			
			⑭ Info- soft environment			

(continued)

Table 4 (continued)

No	Structure	I general MGT. strategy	II MKTG. strategy	III Production strategy	IV Purchasing strategy	V Logistics (SCM) strategy
2	Objective setting	① Sales-growth rate	① Sales objective by area	① Cost deduction objective	① Coat down objective	① Service up objective
		② Profit—growth rate	② Sales objective by product	② Productivity up objective	② Lead time objective	② Productivity up
		③ Market share	③ Profit objective by product	③ Quality up objective	③ No inventory objective	③ New channel development obj.
		④ Productivity-cost deduction	④ Per head revenue objective	④ Machine productivity objective	④ Productivity objective	④ Inventoryless objective
		⑤ Employees objective	⑤ Revenue objective by-customer	⑤ Loss deduction objective	⑤ Service rate objective	⑤ Transportation rationalization
		⑥ Pay roll objective	⑥ Share objective by area	⑥ Direct/indirect rate objective	⑥ Sourcing objective	⑥ Lead time speed
		⑦ Organization renovation	⑦ Share objective by area	⑦ No delay objective	⑦ Alternative material development objective	⑦ Quality & safety objective
			⑧ Service objective	⑧ No inventory objective	⑧ Joint purchasing objective	⑧ Collaborative logistics
			⑨ New customer objective	⑨ No accident objective	⑨ New supplier development objective	⑨ No accident
				⑩ Security/safety up objective		⑩ Outsouser innovation

(continued)

Table 4 (continued)

No	Structure	I general MGT. strategy	II MKTG. strategy	III Production strategy	IV Purchasing strategy	V Logistics (SCM) strategy
3	Strategy formulation	① Strategy for multi business	① Prioritized MKTG strategy	① Plant integration/ with- drawwal strategy	① Outsourcing strategy	① Center integration/ withdrawal strategy
		② Product-market mix strategy	② Core MKTG strategy	② Mechanization strategy	② Multi vending strategy	② Cost reduction strategy
		③ Market in or out strategy	③ Pinpoint MKTG. strategy	③ TPM/TQC strategy	③ Location strategy	③ Modal strategy
		④ Mechanization/ automation strategy	④ P.P.M. strategy	④ Outsourcing strategy	④ Alternative materials strategy	④ Outsourcing strategy
		⑤ Financial utilization strategy	⑤ PR strategy	⑤ Productivity strategy	⑤ Stockless strategy	⑤ Receiving strategy
		⑥ Network strategy	⑥ Chanel strategy	⑥ Stockless strategy	⑥ Service strategy	⑥ Cooperative logistics strategy
		⑦ M and A/sell out strategy	⑦ Pricing strategy	⑦ Scandinavian system strategy	⑦ Scale buying strategy	⑦ Distribution strategy
		⑧ Business alliance strategy	⑧ Customer strategy	⑧ Resources strategy	⑧ Small lot order strategy	⑧ Packaging/packing strategy
			⑨ Customer service strategy	⑨ Eco strategy	⑨ Materials exploring	⑨ Stockless strategy
			⑩ Alliance/ collaboration partnership strategy	⑩ No accident safety strategy	strategy	⑩ Collaboration strategy
			⑪ Information strategy	⑪ Information strategy		⑪ Resources utilization strategy
			⑫ Services strategy			⑫ Information/soft strategy

(continued)

Table 4 (continued)

No	Structure	I general MGT. strategy	II MKTG. strategy	III Production strategy	IV Purchasing strategy	V Logistics (SCM) strategy
4	Exposure	① Issues associated with establishment of SBU –VEU	① Priority of product development	① Joint utilization of resources	① Purchasing of oligopoly materials	① Market environment
		② Long/short term borrowing/issue of bond/capital increase	② Joint project	② Centralization of production planning	② Stability of buying international materials	② Exchange rate
		③ Outsourcing capability'	③ Import/export balance		③ Exchange rate	
5	Resources consideration	① Human resources ② Finance ③ Money ④ Information ⑤ Capability ⑥ Control level	① Human resources ② Finance ③ Money ④ Information ⑤ Capability ⑥ Control level	① Human resources ② Finance ③ Money ④ Information ⑤ Capability ⑥ Control level	① Human resources ② Finance ③ Money ④ Information ⑤ Capability ⑥ Control level	① Human resources ② Finance ③ Money ④ Information ⑤ Capability ⑥ Control level
6	Dependency	① Change of environmental condition	① Change of related fields	① Change of related fields	① Change of related fields	① Change of related fields
		② Policy change	② Changes among related fields	② Changes among related fields	② Changes among related	② Changes among related fields

Other elements:

VI New business strategy	VII Personnel/organization strategy	VIII Information strategy	IX Finance/profit strategy	0 R&D strategy

Ref Karasawa. Y. Analysis and design of MIS. p. 37, Oomu Publsher, 1988
Karasawa. Y. Introduction to physical distribution, p. 287, Yuhikaku Publishing Co. Ltd. 1989

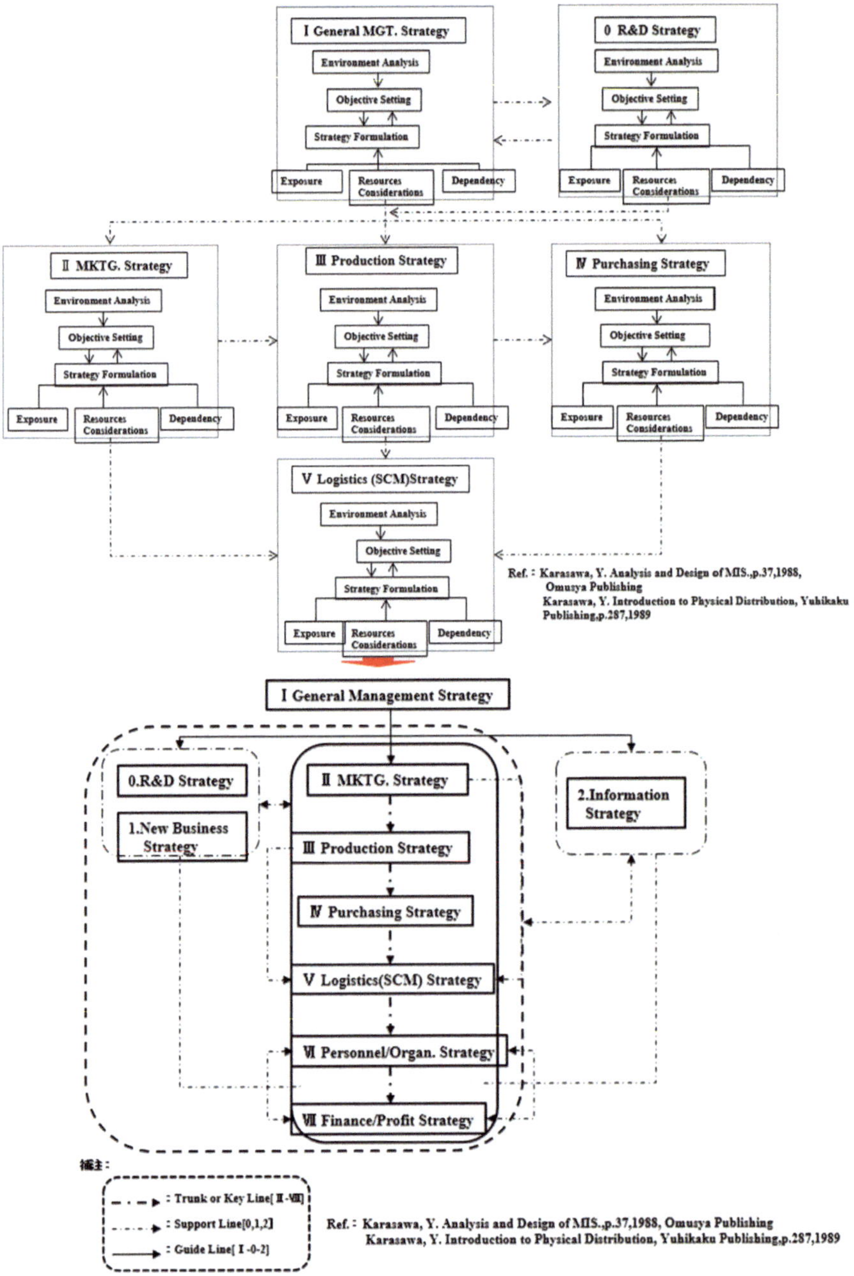

Fig. 10 Basic theory of management strategy

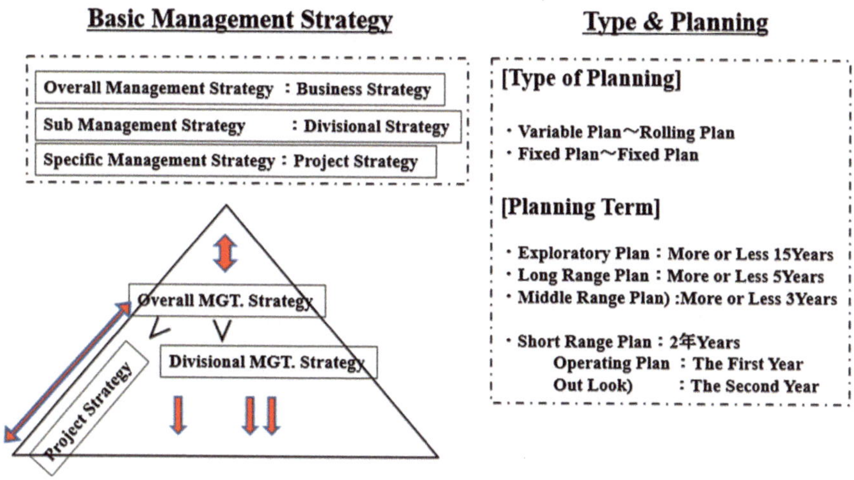

Fig. 11 Characteristics of management strategy planning

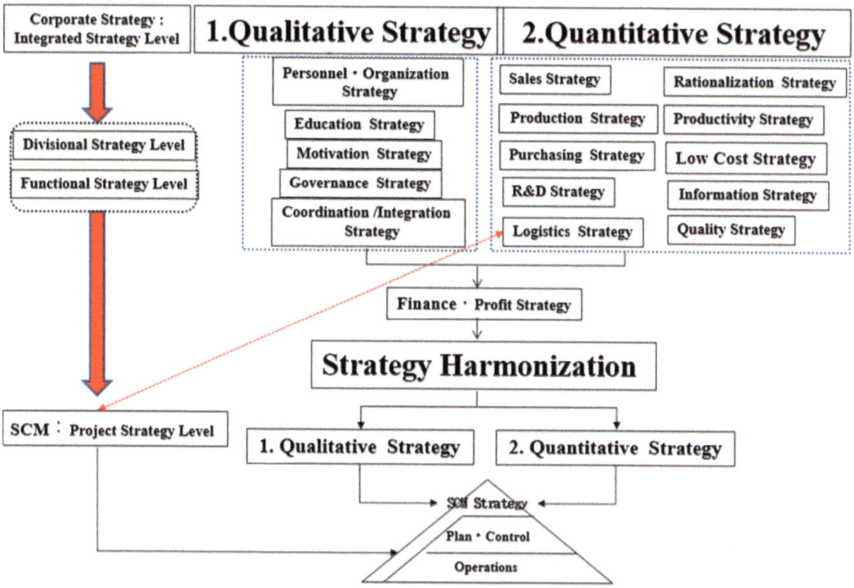

Fig. 12 Divisional strategy and project strategy

Kenneth Lysons and Michael Gillingham criticize Porter model as *A critique of Porter's five forces models*:

Porter's models such as the value chain and five forces diamond have had a lasting impact on strategic management. They are, however, subject to increasing criticism. Hines 7 alternative model of the value chain was discussed in "The Essence of Integration in Supply Chains and Reverse Supply Chains: Similarities and Differences". Other criticisms of five forces model include the following:

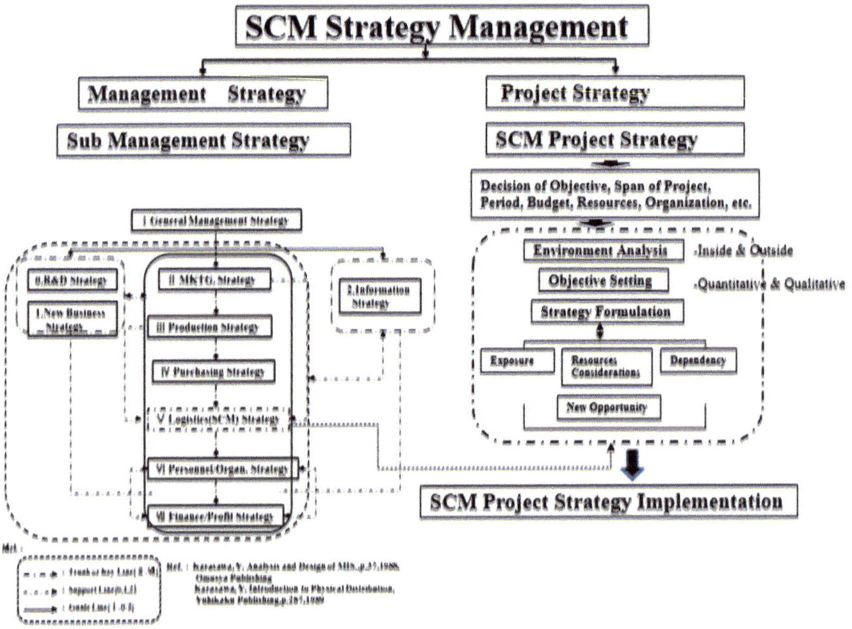

Fig. 13 SCM strategy framework

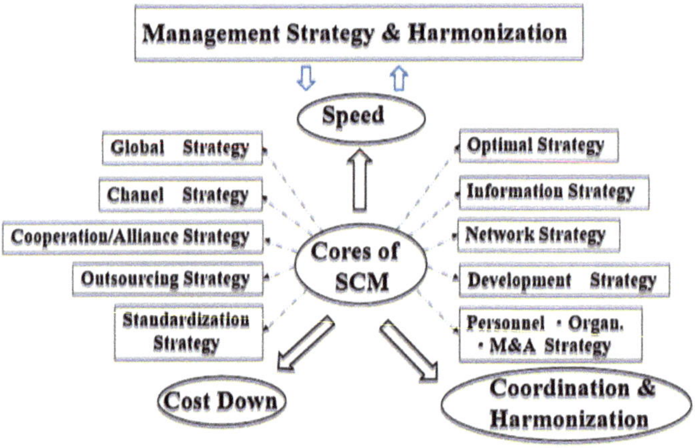

Fig. 14 Core of SCM and strategy elements

- Changed economic conditions. Porter's theories relate to the economic situation of the 1980s characterized by strong competition, inter- enterprise rivalry and relatively stable structures. They are less relevant in today's dynamic environment in which the internet and e-business applications have the power to transform entire industries.

- Identification of new forces. Downes[17] has identified digitalization, globalization and deregulation as three new forces that influence strategy.
  - Digitalization, i.e., putting data into digital form for use in a digital computer, has provided all players in a market with across to more information thus enabling even external players to change the basis of competition.
  - Globalization enables businesses to buy, sell and compare prices globally. Competitive advantage can derive from cooperation, ability to develop strategic alliance and manage extensive global networks for the mutual advantages to buyers and sellers.
  - Deregulation, i.e., A much reduced improvement of central government in the control such industry as airlines, banking, and public utilities.

Downes states that the foremost differences between what he terms the "Porter world" and "the world of new forces" is information technology. The old economy used IT as a tool for implementing change. Today technology has become the most important driver for change.

The three forces of digitalization, globalization and deregulation have effectively removed the barriers to industrial entry and enabled new competitors and new ways of competing to develop at an accelerated speed.

Note: *Beyond Porter*, in context magazine at http//www.Contextmag.com/set France Direct.asp?

Ref.: Kenneth Lysons and Michael Gillingham, purchasing and supply chain management p. 111, 6th edn, Prentice Hall, 2003.

5.1.2 General Framework

Figure 9 shows a basic framework for management strategy formulation of multinational company, of which framework can contribute to SCM strategy framework construction.

5.1.3 Framework for Key Strategy Development-Multi-National Company: Karasawa, Y

A typical framework for framework for key strategy development-multi-national company is shown in Fig. 10.

This framework covers divisional strategy framework as well as management strategy framework. Based on this framework, management strategy implementation is done while covering divisional or sub management strategy implementation like Logistics, etc. (Table 4).

Features	Monopoly	Oligopoly	Free Competition
Dominant Type	1 Company	Several Company	No Leader
Dominant Power	Extremely Strong	Strong	Weak
Target	Self Oriented	Collaborative Exclusion	Laisser Faire
Brand Power	Extremely Strong	Strong	Weak
Price Decision Power	Extremely Strong	Strong	Weak
Price Leadership	Strong ← Weak ←	Seller market V.S. Buyer Market	→ Weak → Strong

Fig. 15 Supply chanel dominant

5.1.4 Divisional Strategy and Project Strategy

General theory for basic management strategy and planning types and period is shown in Fig. 11.

SCM strategy framework concept is based on below model (Figs. 11 and 12).

Divisional strategy is always harmonized with corporate strategy since divisional strategy is a sub strategy for corporate strategy and has to keep balance and harmonization with corporate strategy.

5.2 SCM Strategy Management

A framework of SCM strategy management is shown in Fig. 13. SCM strategy management is mainly composed of divisional (sub management) strategy and project strategy (Fig. 13).

Core of SCM strategy is shown in Fig. 14. Figure 15 characterizes a supply channel dominant being composed of Monopoly, Oligopoly, and Free Competition.

Both channel and cooperative strategies are one of the most important strategies of SCM core strategy since system beyond firms requires always linkages of the system based on cooperation. Typical dominant types necessary for implementation of channel integration or linkage are shown in Fig. 15 and Table 5, which shows type of network with characteristics of dominant patterns.

Table 5 Type of network

Item	Type of network	Kingdom	Barony	Republic
1	Number of core firms	One	Several	None
2	Contracting power	High	Medium	Low
3	Network objectives	Economies of scale	Economies of scale and scope	Economies of scope
4	Network protocols	Top-down enforced	Enforced by suppliers	Mutual agreement

Source (Cooper and Slagmulder 1999, p. 116)

6 Conclusions and Issues

Rough outlines of SCM theory of strategy based on literature survey results and management strategy theory are made clear with solution of research purposes. However, paper length is so short for this type of research that pin pointed report should be required or content of research should preferably be divided.

References

Ackoff RL (1970) A concept of corporate planning. Wiley-Interscience, Chichester, New York

Anthony RN (1965) Planning and control systems: A framework for analysis. Harvard University School of Business Administration, Boston, MA

Ayers JB (2001) Handbook of the supply chain management. The St. Lucie Press/APICS series on resources management

Ballou RH (1999) Business logistics/supply chain management, 5th edn. Pearson Prentice Hall, New Jersey

Bowersox DJ, Closs DJ, Cooper MB (2002) Supply chain logistics management, McGraw Hill, Boston

Cooper R, Slagmulder R (1999) Supply chain development for the lean enterprise. IMA Publication for Applied research Inc, Portland

Gattorna JL (ed) (1998) Strategic supply chain alignment. Gower, Aldershot

Hines P, Lamming R, Jones D, Cousin P, Rich N (2000) Value stream management. Prentice Hall, London

Karasawa Y (1977) Introduction to physical distribution system. Genndai Koyukuk Sya Publishing Company

Karasawa Y (1989) Introduction to physical distribution. Yuhikaku Publishing Company, Tokyo

Karasawa Y (2000) Introduction to contemporary logistic. NTT Publishing Company

Karasawa Y, Aiura N (1999) A basic research on logistics scope of works. J Jpn Logistics Soc 7: P59–72

Karasawa Y, Aiura N (2000) a basic study on logistics definition. J Jpn Soc Logistics Syst 1(1): P15–24

Karasawa Y, Sato K, Wakabayashi K (2000) Literature study for historical development process of 3PL and its definition. J Jpn Soc Logistics Syst 9(21):11–20

Karasawa Y, Sato K, Wakabayashi K, Aiura N (2010) Literature study for logistics scope of work and functions. J Jpn Soc Logistics Syst 10(1):P37–P42

Karasawa Y, Aiura N, Suzuki K, Sato K (2012) A literature study overseas on strategy and management of SCM. J Jpn Soc Logistics Syst 10(1):P69–74
Lysons K, Gillingham M (2003) Purchasing and supply chain management, 6th edn. Prentice Hall, London
Miller T (2003) Hierarchical operations and supply chain planning. Springer, Berlin
Poirier CC (2003) The supply chain manager's problem-solver. St. Lucie Press, Boca Raton
Ross FD (1997) Competing through supply chain management. Kluwer Academic Publishers, Boston
Stadtler H, Kilger C (2002) Supply chain management and advanced planning, 2nd edn. Springer, New York
Wakabayashi K, Karasawa Y, Nakama Y, Fujita Y (2008) A basic research for the third party logistics in Japan. Proceedings of the 4th international congress on logistics and SCM systems, Bangkok, p P8–16
"Beyond Porter", in context magazine at http//www.Contextmag.com/set France Direct.asp

Part II
Reverse Logistics and Environmental Sustainability

Single Forward and Reverse Supply Chain

Ahmad E. Alozn, Moza S. Al Naimi and Omar Y. Asad

Abstract Supply chain involves the procurement structure, logistics, and inventory management from the point of origin to the end user of a product or service. In this chapter, Single Forward and Reverse Supply Chain (SFRSC) system is introduced where it is assumed that stores receive products from a single warehouse with a fixed lead times. Moreover, orders are assumed to have identical sizes and that they occur at equivalent time intervals which suggest that the natural inventory policy at warehouses and retailers is Economic Order Quantity (EOQ). Different types of costs are considered beside other social and environmental welfare costs that are outlined in this chapter. We, hence; extend the lagrangian relaxation approach that simulates the planning of supply chain to capture environmental and social factors along with economic ones.

Keywords Supply chain management · Carbon emissions · Linear programing

1 Introduction

Supply chain is related to the flow of material or information from the origin to the end customer. Whereas, adding the term management, Supply Chain Management (SCM) is basically related to approaching the long term functions improvement,

---

A. E. Alozn (✉) · M. S. Al Naimi · O. Y. Asad
Department of Engineering Systems and Management, Masdar Institute of Science and Technology, Abu Dhabi, UAE
e-mail: aalozn@masdar.ac.ae

M. S. Al Naimi
e-mail: malnaimi@masdar.ac.ae

O. Y. Asad
e-mail: oasad@masdar.ac.ae

P. Golinska (ed.), *Logistics Operations, Supply Chain Management and Sustainability*, EcoProduction, DOI: 10.1007/978-3-319-07287-6_15,

through strategic and systematic coordination of order generation, order taking and order fulfillment time; seeking the customer satisfaction at the lowest possible cost. Such functions that can be improved are manufacturing, warehousing, transporting, retailing (selling) and customers' behavior (purchasing). Moreover, SCM is dependent on all the essentials that allow for the successful product flow such as people, their activities, and available resources. Keeping this in mind, the new paradigm in SCM is the consideration of sustainability in supply chains. The aim of green supply chain is to care for the 3Ps presented in the planet corresponding to the environment, the people corresponding to social responsibilities, and the profit corresponding to the economy while selecting the operational practices aiming for the sustainable growth in a demanding environment knowing the volatile cost impact. This is important to avoid overburdening the environment and consuming the future generations' resources; that's why the environmental, social, and economical challenges are not only a single entity concern. But also it must be considered by all parties for an effective supply chain management in all sort of businesses. In this chapter, the objective is to add sustainability to a multi-echelon joint inventory-location model that simultaneously determines the location of warehouses and inventory policies at the warehouses and retailers. The contribution in this chapter would be minimizing the total cost that consists of fixed cost and holding cost for both the inventory and retailers depending on recycling, in addition to minimizing the carbon emissions while shipment (shipment mile), that is a function of the distance travelled. It is expected from the updated model that the number of resources would be reduced because of recycling and drop in shipment mile, minimizing the total cost for a friendly environment but might increase the unemployment rate (social welfare) affecting the social community negatively. To conclude, the aim of the model is to minimize the total cost considering what's best for the environment.

2 Literature Review

The competence of supply chain systems mainly depend on integrated decisions. Starting from facility location decisions plus transportation decisions, and ending with inventory strategies incorporated with all of these decisions. The literature on this area has considered each one of these decisions separately: Andersson et al. (1998), Diks et al. (1996) focuses on inventory management decisions in general, but Axsäter (2005), Cardenas-Barron (2007), went further in covering inventory management decisions especially when there are no realistic information about the market due to its emergent instability. On the other hand Karabuk (2007) focuses on transportation management decisions and specifically dealing with supply chain using simple mathematical algorithms rather than complex calculus. Some models like Tung et al. (2013) are based on the vehicle routing problem in which, routing and inventory decisions are determined simultaneously over a given time period.

Recent models have integrated these decisions like Barahona and Jensen study (Barahona 1998), which suggests that a successful criteria for locating a plant is to be of zero inventory in order to avoid the holding cost and reduce the capital expenditure. In contrast, Jones (1987) investigates the competitive advantage of having an inventory in supply only after a thorough analysis to match the demand and supply with respect to the holding cost consideration. Another study is by Epen (1979) that studies the effects of centralization, meaning the selection of the suitable location with the highest centrality measures, on reducing the expected costs. Moreover in the integrated model by Diabat and Richard (2013) as well as the model considered by Diabat et al. (2014), these models consider a multi-echelon joint inventory-location model that concurrently decides the location of warehouses and inventory policies at the warehouses and retailers. Like Diabat and Simchi-Levi (2009) model, a nonlinear mixed-integer program is used to formulate the model.

Recently some studies considered the consideration of CO_2 emissions in the design of supply chain, Abdallah et al. (2012); Diabat et al. (2013a); Diabat et al. (2014); Al Daheri et al. (2010); Diabat (2013b); Govindan et al. (2013); Abdallah et al. (2013); Le et al. (2013); Diabat et al. (in press); Diabat et al. (2009); Diabat et al. (2013c); Al Zaabi et al. (2013). Some of these studies considered developing solution algorithms to tackle the developed mathematical programs such as Diabat et al. (2013a) and Diabat et al. (2014), or considered the cost of CO_2 emissions in the oil refinery industry such as Al Daheri et al. (2010). Govindan et al. (2012) evaluated of performance measures and supply chain profit behavior under buy-back, revenue sharing, quantity flexibility and advanced purchase discount contracts versus no coordination and wholesale price systems.

Integrating all these decisions with the consideration of sustainability has been a highly demanding subject. Unfortunately, there has been relatively little work done in that perspective as described in cleaner production journal (2008), which offers good reviews on sustainable supply chain management considering 191 chapters published from 1994 to 2007. This is relatively low number, while focusing on the 3Ps towards sustainability aspects presented in Planet for the environment, People for the social community, and Profit for the economic feasibility. Linton (2007) states that sustainable development in the field of supply chain is still in its infancy and has the potential to affect future government policy, current production operations, and identify new business models. He supports that by encouraging the focus on the entire supply chain management rather including all its stages staring from the production, going through the consumption, and ending with the disposal of products as well as the customer satisfaction rather than emphasizing on the local optimization of environmental factors. Researchers' main focus when it comes to sustainability in the field of supply chain is the green houses gases and specifically the carbon emissions. Sundarakani (2010) and Al Zaabi, Al Dhaheri and Diabat (2013) found that environmental consciousness has become critical in the design and operation of globally integrated chain networks and his study looks at the practice of green supply chain management through a model that focuses on the carbon footprint across supply chains. There are several

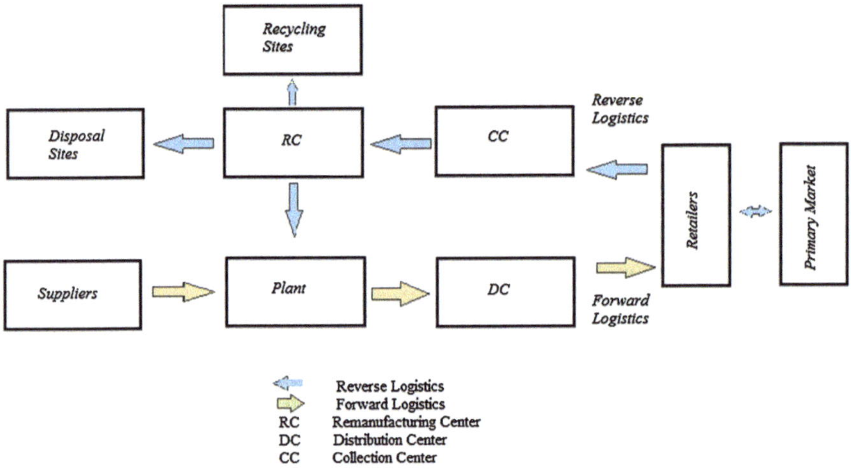

Fig. 1 Structure of CSCLSC network

thoughts to test for the severity in that context, one of them is in relation to the type of materials used in the establishment of the warehouses structure (building) that can be investigated by an environmental approach called the Life Cycle Assessment as explained by Browne (2005) . Another approach is to reduce the carbon emissions resulting from the trucks during the logistics that is a function of the distance travelled as well as the load carried as illustrated by Elhedhli (2012) or managing carbon footprints in inventory management as explained in a study by Hua (2011). The first study focuses on two aspects choosing a strategic location of the warehouse and a good planning tool for loading and distributing the products. For the second, it highlights how firms manage carbon footprints in inventory management under the carbon emission trading mechanism to derive the optimal order quantity as well as examine the impacts of carbon trade, carbon price and cap on order decisions, carbon emissions and total cost.

From literature, Diabat and Abdallah (2013) introduces carbon-sensitive closed loop supply chain (CSCLSC) network. It is a multi-echelon multi-commodity facility location problem with a trading price of carbon emissions and a cost of procurement. They included carbon emissions into location models for forward and reverse supply chains simultaneously. The CSCLSC network consists of a set of raw materials' suppliers, a set of plants each plant with different production capacity, a set of distribution centers (DCs) of different capacities, a set of retailers, a set of product types, a set of collection centers (CCs), a set of remanufacturing centers (RCs), a set of recycling centers, and a set of disposal centers.

In CSCLSC network, many raw materials are fed to the plant as shown in Fig. 1 from different suppliers and later they are transferred to different distribution centers (DCs) as new products. DCs act as a freight forwarder of products to different retailers in which they are used by primary market. Any used product will be sent to the collection centers (CCs) by the retailers in order to be delivered to the remanufacturing centers (RCs). Every single product consists of several modules and depending on the condition of each module the RCs will be responsible of remanufacturing, recycling or disposing them. There are mainly two reasons behind preventing any module from being remanufactured. Firstly, if the module has an extreme physical damage. Secondly, if it's utilization is not suitable for any remanufacturing process.

In this chapter, fuel consumption terms are not only integrated with the supply chain decisions, but also a simple recycling approach is deployed. The effect of carbon emissions on the design of a multi-echelon joint inventory-location model have been introduced, where certain modules of products are remanufactured using single supply chain system. The objective of this chapter is to develop a Single Forward and Reverse Supply Chain (SFRSC) model instead of closed loop supply chain. SFRSC capitalizes on the existing plants capacity and the transportation logistics.

3 Problem Description and Formulation

A Single Forward and Reverse Supply Chain (SFRSC) is a multi-echelon joint inventory-location problem. Unlike Al Dhaheri and Diabat (2011) model that deals with multiple products, this model deals with the distribution of a single product. The product is being distributed from a single manufacturer to a set of retailers indicated by i through a set of warehouses that can be located at various candidate locations indicated by j, where J is the set of probable sites for a warehouse. Only the manufacturer supplies each warehouse where cross supply among the warehouses is not allowed, and shortages are not permitted. Unlike Diabat et al. (2009), demand at retailers is assumed to be constant which is practical to most of the non-durable goods. Assuming single sourcing, the model consists of one warehouse, multi-retailer problem where each one has an optimal solution. The natural inventory policy at the retailers and the warehouses is an Economic Order Quantity (EOQ) policy. To simplify the model further, power-of-two inventory policy is assumed where the ratio of the time between orders at the warehouse to the time between orders at each retailer it serves is a power of two. As shown by Abdallah et al. (2012), remanufacturing and recycling is considered in this model due to its environmental and economic benefits.

In the SFRSC, each element within the supply chain network is further utilized at relatively minimal additional cost compared to CSCLSC. The product life cycle is utilized almost up to 200 % compared to CSCLSC. The normal CSCLSC approach separates between a product's delivering and remanufacturing processes. The delivering process of the product comprises raw material at the suppliers, through manufacturing at the plants, to storage at the warehouse (or distribution center) and to sale points such as retailers. The remanufacturing process of a product comprises collecting used products at collection centers, through remanufacturing at other dedicated centers, then either recycling, disposal or integrating the remanufactured parts. However, the SFRSC concept capitalizes on the

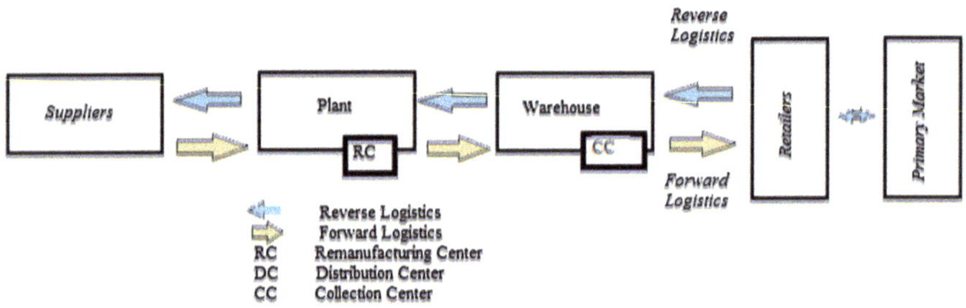

Fig. 2 Structure of the SFRSC network

suppliers, plants, warehouses (distribution centers), and transportation means between all of them.

Sustainability consists of three major parts, namely; social welfare, environmental stewardship, and economic feasibility. SFRSC concept increases the social welfare, reduces the negative effect on the environment, and enhances the economics feasibility of the supply chain system.

Economic viability in SFRSC model presents saving construction cost, fixed cost, of building new facilities like collection centers (CCs), remanufacturing centers (RCs) parallel to the plant and Distribution centers (DCs) as shown in Fig. 1. It suggests having a collection center at each warehouse and a remanufacturing center at each plant as shown in Fig. 2. The SFRSC concept capitalizes on the existing and operational warehouses by utilizing the warehouse facilities as collection centers as well. Moreover, it utilizes the plant facilities as remanufacturing centers. With this, the fixed cost of established these facilities is limited to any alterations that might be required to the existing facilities. Furthermore, this approach reduces and overhead cost and the facility running cost. Similarly, the plant is utilized as a remanufacturing center is as well. From the economic view, this is a sustainable approach. The creation of these divisions itself contributed positively toward the social welfare by new jobs creation.

Moreover, SFRSC saves the transportation cost of moving the used products from retailer to remanufacturing centers through collection centers. Instead, when the new products are delivered to each retailer, the used ones will be carried by the same transportation to be returned to warehouse from the retailers.

The Environmental side of the new model is minimizing Carbone emits by having a joint collection centers (CCs) with retailers and warehouses and a joint remanufacturing center at the plant. Reduction in energy consumed is due to economics of scope and integrated operations.

For the purpose of formulation, the cost of integrating collections centers at the warehouses and remanufacturing center at the plant, and the cost of the remanufacturing process are off-set by number of saving and gain elements. These are savings in the material cost (due to remanufacturing), savings in the disposal cost (as parts of the waste products' modules are being either remanufactured or recycled), and gains in selling the recycled modules.

The major cost components considered in the objective function of the model are as follows:

1. Fixed order cost: the cost of placing an order regardless of the size of the order,
2. Unit inventory cost: the cost of holding one unit of product for one unit of time,
3. Unit shipping cost: the cost of shipping one unit of product between facilities that is inclusive of the fuel cost, and
4. Fixed location cost: the cost of establishing and operating a warehouse.

The purpose is to determine: (i) the number of warehouses to establish; (ii) their location; (iii) the sets of retailers that are assigned to each warehouse; and (iv) and the size and timing of orders for each facility, so as to minimize the sum of inventory, shipping, ordering, and location costs, while satisfying end-customer demand.

Units

J	a set of warehouses j;
$\Im j$	set of retailers served by warehouse j;
$\hat{T}_j$	Time between orders placed by warehouse j;
T_{ij}	Time between orders placed by retailer I to warehouse j;
f_j	The fixed cost of opening and operating warehouse j;
$\hat{o}_j$	fixed cost per order at warehouse j;
o_i	fixed cost per order at retailer *i*;
$\hat{s}_j$	per unit shipping cost from the plant to warehouse *j*
s_{ij}	per unit shipping cost from warehouse *j* to retailer *i*;
$\hat{h}_j$	unit inventory holding cost per unit time at warehouse *j*;
h_i	unit inventory holding cost per unit time at retailer *i*;
$\hat{l}_j$	distance between warehouse j and plant in miles;
l_{ij}	distance between retailer *i* and warehouse j in miles;
d_i	daily demand at retailer i is constant;
t_b	base planning period;
β_{inv}	a weighting factor for inventory and ordering costs;
β_{trn}	a weighting factor for transportation costs;
a	is the fuel price in US Dollars per gallon fuel;
r	is the fuel consumption rate in gallons per mile, for 1 unit of product
g	is carbon emission in tons per mile of shipment;
GA	is a cap of carbon emiisions over the base planning period, t_b;

Variables and Decision Variables

Some variables have been used to simplify the model and they are as following:

$$\hat{k}_j = \beta_{inv}\hat{o}_j t_b$$

$$k_i = \beta_{inv} o_i t_b$$

$$b_{ij} = \beta_{trn}(\hat{S}_j + S_{ij}) d_i t_b$$

$$\hat{S}_j = \hat{s}_j + ar\hat{l}_j$$

$$S_{ij} = s_{ij} + arl_{ij}$$

$$c_{ij} = \frac{1}{2}\beta_{inv}(h_i - \hat{h}_j) d_i t_b$$

$$e_{ij} = \frac{1}{2}\beta_{inv}\hat{h}_j d_i t_b$$

$$X_j \begin{cases} 1 & \text{if a warehouse is opened at location } j \\ 0 & \text{otherwise} \end{cases}$$

$$Y_{ij} \begin{cases} 1 & \text{if retailer } i \text{ is served by the warehouse at location } j \\ 0 & \text{otherwise} \end{cases}$$

Equations

$$\min_{X,Y,Tj,Tij} \sum_{j\in J} (f_j + \frac{\hat{k}_j}{\hat{T}_j}) X_j + \sum_{j\in J}\sum_{i\in I} (\frac{k_i}{T_{ij}} + b_{ij} + c_{ij} T_{ij} + e_{ij} \max\{\hat{T}_j, T_{ij}\}) Y_{ij} \tag{1}$$

$$s.t \sum_{j\in J} Y_{ij} = 1 \quad \forall i \in I \tag{2}$$

$$Yij \leq Xj \quad \forall i \in I, \forall j \in J \tag{3}$$

$$\sum_{\substack{j\in J \\ i \in I}} \frac{t_b}{T_{ij}} g Y_{ij} (l_{ij} + \hat{l}_j) \leq G \tag{4}$$

$$\hat{T}_j / T_{ij} = 2^{N_{ij}} \quad \forall i \in I, \forall j \in J \tag{5}$$

$$Yij \in \{0, 1\} \quad \forall i \in I, \forall j \in J \tag{6}$$

$$Xj \in \{0, 1\} \quad \forall j \in J \tag{7}$$

$$N_{ij} \in Z \quad \forall i \in I, \forall j \in J \tag{8}$$

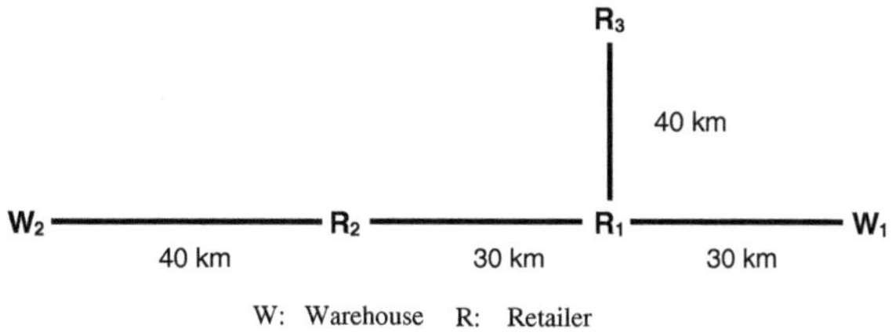

Fig. 3 Supply network example

Table 1 Fuel related parameters

Parameter	Value
Fuel cost	\$3.81 per gal
Fuel consumption	0.15625 gal/km
ECO_2 emission	0.00002433 ton/km

Table 2 Costs at warehouses

Parameter	Warehouse 1	Warehouse 2
Establishment cost	\$100,000	\$100,000
Shipment cost	\$ 4 per unit	\$ 4 per unit
Ordering cost	\$ 25,000 per order	\$ 25,000 per order
Holding cost	\$ 4 per unit	\$ 4 per unit

Table 3 Demand and costs at retailers

Parameter	Retailer 1	Retailer 2	Retailer 3
Demand	10 units	10 units	10 units
Ordering cost	\$ 15,000 per order	\$ 15,000 per order	\$ 15,000 per order
Holding cost	\$ 6 per unit	\$ 6 per unit	\$ 6 per unit

$$T_{ij} > 0 \quad \forall i \in I, \forall j \in J \tag{9}$$

$$\hat{T}_j > 0 \quad \forall j \in J \tag{10}$$

4 MJIL Problem and GAMS Code

The Multi echelon Joint Inventory Location (MJIL) problem is a nonlinear incapacitated facility location problem as shown in formula (1). Using GAMS program, the model is optimized firstly over $\hat{T}_j$ and T_{ij} for any given X and Y and secondly is optimized over X and Y for any given $\hat{T}_j$ and T_{ij} Furthermore, the model in formula (1) used to find the minimum transhipment cost and total CO2 emotion over base plan period.

To test the proposed SFRSC approach, an example of simplified network has been considered. The network includes 2 warehouses and 3 retailers as shown in Fig. 3 below. The product considered is a short-life product, milk bottles.

From Cardenas-Barron (2007), the value of base planning period t_b equal to 364 days, the weighting factor for inventory and ordering costs β_{inv} equal to 100 and the weighting factor for transportation costs β_{trn} equal to 0.1. Fuel related paramagnets are summarized below in Table 1. The rest of the constants and coefficients such as establishment cost, shipment cost and ordering costs are summarized in Tables 2 and 3 below.

5 Conclusions and Observations

In this chapter we extended Diabat and Abdalla (2013) model to accommodate Carbon emissions in the supply chain model that concurrently studies warehouses and retailers inventory policies as well as facility location decisions. We noticed that, while over much of the (β_{trn}, β_{inv}) parameter space there was no discernible benefit to integration for these particular datasets and parameter values, over specific regions of this space there was a clear benefit to integration. We further observed that the number of opened warehouses tended to increase as the transportation cost weighting factor βtrn was increased, all based on Diabat and Abdalla (2013) computational analysis.

However, our model didn't run logically over GAMS as the number of facilities opened tend always to be zero. This could be due to number of factors. First, the number of terms in the objective function are non-linear, with mixed integer variables, which complicates the model. Second, the term that requires obtaining the maximum value of two variables $\left(\max\left\{\mathrm{Tmax}\left\{\hat{T}_j, T_{ij}\right\}\right\}\right)$ is, by itself, a complication source. Trying to solve the system manually (hand calculation) was too tedious and time consuming, hence, such wasn't a feasible option.

Furthermore, Diabat and Abdalla (2013) model shows no constrains assuring that demand at all retailers is met. This might need to be re-addressed.

In conclusion, it's recommended to find alternative ways to model the supply chain system linearly.

References

Abdallah T, Diabat A, Rigter J (2013) Investigating the option of installing small scale PVs on facility rooftops in a green supply chain. Int J Prod Econ 146(2):465–477

Abdallah T, Diabat A, Simchi-Levi D (2012) Sustainable supply chain design: a closed-loop formulation and sensitivity analysis. Prod Planning Control 23(2–3):120–133

Al Dhaheri N, Diabat A (2010) A mathematical programming approach to reducing carbon dioxide emissions in the petroleum refining industry. In:Engineering Systems Management and Its Applications (ICESMA). 2nd International Conference on, IEEE

Al Zaabi S, Al Dhaheri N, Diabat A (2013) Analysis of interaction between the barriers for the implementation of sustainable supply chain management. Int J Adv Manuf Technol 68(1–4):895–905

Andersson J, Axsäter S, Marklund J (1998) Decentralized multiechelon inventory control. Prod Operations Manag 7:370–386

Axsäter S (2005) A simple decision rule for decentralized two-echelon inventory control. Int J Prod Econ 93:53–59

Barahona (1998) Plant location with minimum inventory, Math Program 83:101–111

Browne M (2005) Life cycle assessment in the supply chain: a review and case study. Transp Rev: Transnatl Transdiscipl J 25(6):761–782

Cardenas-Barron L (2007) Optimizing inventory decisions in a multi-stage multi-customer supply chain: anote. Transp Res Part E 43:647–654

Diabat A, Abdallah T, Le T A hybrid tabu search based heuristic for the periodic distribution inventory problem with perishable goods. Ann Oper Res (in press)

Diabat A, Richard JP, Codrington CW (2013a) A Lagrangian relaxation approach to simultaneous strategic and tactical planning in supply chain design. Ann Oper Res 203(1):55–80

Diabat A, Aouam T, Ozsen L (2009) An evolutionary programming approach for solving the capacitated facility location problem with risk pooling. Intern J Appl Decis Sci 2(4):389–405

Diabat A, Abdallah T, Al-Refaie A, Svetinovic D, Govindan K (2013b) Strategic closed-loop facility location problem with carbon market trading. IEEE Trans Eng Manage 60(2):398–408

Diabat A, Abdallah T, Henschel A (2013c) A closed-loop location-inventory problem with spare parts consideration. Comput Oper Res

Diabat A, Battaïa O, Nazzal D (2014) An improved lagrangian relaxation-based heuristic for a joint location-inventory problem. Comput Oper Res

Diks EB, De Kok AG, Lagodimos AG (1996) Multi-echelon systems: a service measure perspective. Eur J Oper Res 95:241–263

Elhedhli S (2012) Green supply chain network design to reduce carbon emissions. Transp Environ—Elsevier 17(5):370–379

Epen (1979) Effects of centralization on expected costs in a multi-location newsboy problem. Manag Sci 6(4):475–490

Govindan K, Diabat A, Popiuc MN (2012) Contract analysis: A performance measures and profit evaluation within two-echelon supply chains. Comput Ind Eng 63(1):58–74

Govindan K, Popiuc MN, Diabat A (2013) Overview of coordination contracts within forward and reverse supply chains. J Clean Prod 47:319–334

Hua G (2011) Managing carbon footprints in inventory managment. Int J Prod Econ—Elsevier 132(2):178–185

Jones (1987) Using inventory for competitive advantage through supply chain management. Int J Phys Distrib Logist Manag 17:94–104

Journal of Cleaner Production (2008) From a literatur review to a conceptual framework for sustainable chain management. Elsevier 16(15):1699–1710

Karabuk S (2007) A Modeling and optimizing transportation decisions in a manufacturing supply chain. Transp Res Part E 43:3221–3337

Linton J (2007) Sustainable supply chains: an introduction. J Oper Manage—Elsevier 25(6):1075–1082

Sundarakani B (2010) Modeling carbon footprints across the supply chain. Intern J Prod Econ—Elsevier 128(1):43–50

Tung L, Diabat A, Richard J-P, Yih Y (2013) A column generation-based heuristic algorithm for an inventory routing problem with perishable goods. Optimization Lett 7(7):1481–1502

A Case Study of H&M's Strategy and Practices of Corporate Environmental Sustainability

Danny C. K. Ho

Abstract This study aims to examine the degree to which a large international fashion company—H&M—has improved corporate environmental sustainability using the principle of eco-efficiency and eco-effectiveness. Case study method is employed and data are collected from its corporate annual reports and websites. The strategy and practices relating to product design, purchasing, manufacturing, transportation, retail operation, and product usage and recycling are examined. Based on the data, the study examines the extent to which cradle to cradle approach has been applied to design and manage H&M's operations. This study also explores the possibility of integrating eco-efficiency and eco-effectiveness in improving corporate environmental sustainability.

Keywords Corporate environmental sustainability · Fashion supply chain · Eco-efficiency · Eco-effectiveness

1 Introduction

The pursuit of corporate environmental sustainability is vital to the apparel industry in general and fast fashion retailers in particular for their future development. According to Sull and Turconi:

> Fast fashion describes the retail strategy of adapting merchandise assortments to current and emerging trends as quickly and effectively as possible. Fast fashion retailers have replaced the traditional designer-push model—in which a designer dictates what is "in"—with an opportunity pull approach, in which retailers respond to shifts in the market within just a few weeks, versus an industry average of six months (Sull and Turconi 2008 p. 5).

D. C. K. Ho (✉)
Department of Supply Chain Management, Hang Seng Management College,
Hang Shin Link, Siu Lek Yuen, Shatin, NT, Hong Kong
e-mail: dannyho@hsmc.edu.hk

P. Golinska (ed.), *Logistics Operations, Supply Chain Management and Sustainability*, EcoProduction, DOI: 10.1007/978-3-319-07287-6_16,

The business of fast fashion has been competitive yet profitable and growing rapidly. For example, sales of the Inditex Group (parent company of Zara) increased by 53.2 % from €10,407 million in 2008 to €15,946 million in 2012, while its net profit rose by 87.6 % from €1,262 million in 2008 to €2,367 million in 2012. During the same period, H&M achieved 35.5 % increase in sales, up from SEK1,04,041 to SEK1,40,948 million, while its net profit increased by 10.3 % from SEK15,294 to SEK16,867 million.

The fast fashion trend promoted by retailers such as Zara, TopShop and H&M has brought the trend of disposable fashion that encourages more frequent impulse purchase of cheap, in-season and non-durable garments and the tendency to keep these products for a shorter time than their real useable life (Birtwistle and Moore 2007). In addition, these garments are expected to be used less than ten times (McAfee et al. 2004), and are likely to end up as harmful, non-biodegradable wastes at landfills when they become outdated.

This study seeks to examine the strategy and practices applied by H&M to improve its corporate environmental sustainability in the areas of product design, purchasing, manufacturing, transportation, retail operation, and product usage and recycling. These strategy and practices will be analyzed against the principles of eco-efficiency and eco-effectiveness. The possibilities for integrating eco-efficiency and eco-effectiveness in improving corporate environmental sustainability will also be discussed.

2 Literature Review

According to the United States Environmental Protection Agency (2014), an estimated 13.1 million tons of textile wastes were generated in 2011, or 5.2 % of total municipal solid waste generation in the US. In the United Kingdom, an estimated £238 million-worth of textiles was threw out for waste collection and sent to landfill in 2010, according to The Waste and Resources Action Programme (2012). In 2008 around 14 million tonnes of textile waste were generated in Europe of which only 5 million tonnes were recovered (European Commission 2011). The dumping of textile waste is increasingly becoming a huge urban waste problem in developed countries. This serious environmental threat may stem partly from the unsustainable patterns of consumption and production.

2.1 Unsustainable Patterns of Clothing Consumption and Production

Past studies have shown that some young fashion-hungry consumers, who are the main targets of fast fashion retailers, have inadequate awareness and knowledge of sustainable production and consumption of clothing. For example, young female consumers in the UK agreed that "*there is a general lack of knowledge of how and where clothing is disposed of, or even how it is made, such as the environmental consequences of artificial fibres and intensive cotton production*" (Morgan and Birtwistle 2009 p. 196). Hill and Lee (2012) reported the lack of knowledge of the holistic principle of sustainability and specific adverse effects of the apparel industry in their study of US college students. In their study, about one-third of the respondents listed "*garments made of organic materials*" as a least important sustainable practice. Besides, 30 % of those surveyed considered "*garments made of biodegradable materials*" as a most important sustainable practice, whereas 28.9 % regarded the same practice as a least important sustainable practice in the apparel industry. Such conflicting perceptions were also reported for practices like "*fiber growth without pesticides*" (Hill and Lee 2012). These findings highlight the barrier of lacking awareness and knowledge (Bonini and Oppenheim 2008) that companies must remove before consumers will consider eco-friendly clothing.

Regarding the post-sale clothing disposal, despite the recent growing engagement of consumers in reuse and recycling clothes, such remedial actions should not be readily considered as an effective means to address the environmental threat of textile waste. Bianchi and Birtwistle (2010 p. 366) in a study of Scotland and Australia female consumers' disposal behavior of used clothes reported that "*the environmental consequences of production and disposal of fashion textiles were poorly understood concerns among respondents*". Nevertheless, respondents considered that donating used clothing to charities and giving used clothing to family members or friends made them ''*feel good*'' about helping other people in need. Worth noting is that Ha-Brookshire and Hodges (2009) in a study of used clothing donation found that,

> social consciousness had little, if any, impact on used clothing donation decision-making for the participants in this study. Instead, used clothing donation was just one part of the entire clothing consumption process, one that created space for future clothing purchases. Indeed, without disposal of used clothing items, new clothing items could not be purchased, and, therefore, the consumption cycle could not continue.

Ironically engagement in recycling of used clothing makes consumers feel good to sustain their unsustainable consumption patterns—increasing impulse purchasing of low-quality trendy fashion and donation of outdated clothing to someone in need before the end of the intended product lifetime—because this recycling practice involves a small change from consumers and manufacturers (Fletcher 2008). Unfortunately such unsustainable consumption patterns will drive further growth of unsustainable production patterns. To break this vicious cycle, a major change of consumers' and producers' mindset is needed from the pursuit of eco-efficiency to eco-effectiveness and from cradle-to-grave to cradle-to-cradle approach to improve sustainability in consumption and production.

2.2 *Eco-efficiency Versus Eco-effectiveness*

Eco-efficiency has become an important concept guiding companies to develop their businesses in a sustainable manner. This concept was introduced by the World Business Council for Sustainable Development (WBCSD) in its 1992 publication, Changing Course. According to WBCSD (1992):

> Eco-efficiency is achieved by the delivery of competitively priced goods and services that satisfy human needs and bring quality of life, while progressively reducing ecological impacts and resource intensity throughout the life-cycle to a level at least in line with the Earth's estimated carrying capacity.

Simply put, eco-efficiency means creating more value with less impact. Customer requirements should be satisfied with less resource depletion and pollution on a product or value basis. Companies are considered to have achieved sustainable development if they have used materials and natural resources in production more efficiently and reduced outputs of toxic substances to the environment.

Reduction of negative environmental impacts, which is a key principle of eco-efficiency, represents an initial response of most companies to address sustainability. However, eco-efficiency has been criticized as an inadequate approach to sustainability in the long run because eco-efficiency is about being 'less bad' and 'sustainable' companies can still keep depleting scarce and valuable resources and polluting the environment albeit in smaller increments (McDonough and Braungart 2002). Given that eco-efficiency can only bring relative improvements (i.e. reduced energy or resource usage per value added), it should be treated as one part of the corporate sustainability criteria, instead of as the whole (Dyllick and Hockerts 2002). Young and Tilley (2006 p. 404) summarize the flaws in the thinking behind eco-efficiency as:

> a linear, one-way, cradle-to-grave manufacturing system in which products are made and eventually discarded into a hole in the ground or a furnace is not only wasteful; it can be poisonous. Neither waste nor poisons are particularly efficient, productive or good for the environment. Making a destructive system less destructive only serves to let industry continue to destroy ecosystems and to contaminate and deplete nature more slowly. Under the influence of eco-efficiency a dystopian future lies ahead; destruction is the end game; the only choice remaining is the rate of destruction.

McDonough and Braungart (2002) argued that companies need to pursue eco-effectiveness and change the current system that caused the problem in the first place. They proposed that,

> Eco-effectiveness means working on the right things—on the right products and services and systems—instead of making the wrong things less bad. Reduction, re-use and recycling slow down the rates of contamination and depletion but do not stop these processes. The key is not to make human industries and systems smaller, as efficiency advocates propound, but to design them to get bigger and better in a way that replenishes, restores and nourishes the rest of the world (McDonough and Braungart 2002 p. 76).

2.3 Cradle-to-Cradle Approach to Corporate Sustainability

As a viable alternative to the traditional cradle-to-grave manufacturing model, McDonough and Braungart's (2002) cradle-to-cradle approach to improve eco-effectiveness and sustainability rests on three tenets including:

1. waste equals food,
2. usage of current solar income,
3. celebrating diversity.

The first tenet challenges the taken-for-granted conception of waste that there is no such a thing called '*waste*' in nature, as "*one organism's waste is food for another and nutrients flow indefinitely in cycles of birth, decay, and rebirth*" (McDonough et al. 2003 p. 436A). As such, materials should be designed as nutrients that flow through biological metabolism and/or technical metabolism in a closed-loop system in which man-made and natural resources circulate in cycles of production, use, recovery and remanufacture. The second tenet emphasizes that both energy (e.g. solar energy and wind power) and material inputs should be renewable rather than depleting. The third tenet celebrates nature's diversity and stresses that design of products, processes and systems should be integrated and interconnected with available energy and material flows in the local natural systems (McDonough et al. 2003).

In sharp contrast to the eco-efficient cradle-to-grave goal, which stresses continuous reduction in the human footprint of a product and finally achieves '*zero*' negative impacts, the eco-effective cradle-to-cradle goal combines the progressive reduction of '*bad*' with the increase in '*good*', enhancing positive footprint (McDonough and Braungart 2002).

3 Methodology

Data were collected mainly from H&M's recent sustainability reports during 2010 and 2012, as well as from its corporate website. H&M's vision and strategy of sustainability were examined first, followed by practices in five main aspects of environmental sustainability in the fashion supply chain (Fulton and Lee 2013) covering product design, process design and supply chain design (Ellram et al. 2008). The practices include:

- Fibers
  - No pesticides used in fiber growth,
  - Garments made of organic, recycled, biodegradable or recyclable materials.
- Manufacturing
  - Water usage,
  - Environmentally friendly dyes,
  - Fabric waste.

- Distribution and logistics
  - Environmental-friendly shipping containers,
  - Alternative fuels.
- Store/warehouse building efficiency
  - Building energy/efficiency,
  - Product packaging.
- Post-consumer and beyond
  - Customer sustainability program,
  - Laundering and care.

The social and economic aspects such as company donations and philanthropies as well as fair trade and human rights issues were not studied.

4 Findings

4.1 H&M's Mindset of Sustainability

> H&M's business operations aim to be run in a way that is economically, socially and environmentally sustainable. By sustainable, we mean that the needs of both present and future generations must be fulfilled. H&M's business concept is to offer fashion and quality at the best price. Quality includes ensuring that products are manufactured in a way that is environmentally and socially sustainable. We apply the precautionary principle in our environmental work and have adopted a preventative approach with the substitution of hazardous chemicals. We strive to use resources as efficiently as possible and to minimize waste. By adopting new technologies and methods, we can work preventatively to minimize our environmental footprint through improved production processes and our choice of materials. We must continuously review the company's goals and strategies to reduce the company's climate impact.

H&M has adopted the popularized approach to sustainability developed by The World Commission on Environment and Development (1987 p. 43) that refers sustainable development as "*development that meets the needs of the present without compromising the ability of future generations to meet their own needs*". On paper, H&M recognizes the interdependent nature of economy, society and environment and stresses that the pursuit of sustainability is not a balancing act or a playing of one issue against the other. H&M's approach to sustainability also reflects the precautionary principle of the Rio Declaration on Environment and Development, i.e., "*where there are threats of serious or irreversible damage, lack of full scientific certainty shall not be used as a reason for postponing cost-effective measures to prevent environmental degradation*" (United Nations General Assembly 1992).

Most importantly, central to the mindset of H&M's approach to sustainability is the eco-efficient cradle-to-grave philosophy that permeates its vision and policy, as revealed by its repeated emphasis on improving resource utilization, minimizing waste and reducing ecological footprint. To grow its business in a sustainable manner, in 2011 H&M together with several leading brands developed a roadmap to continuously eliminate the use of all hazardous chemicals and hence achieve zero discharge in all production procedures associated with the making and using of H&M products, at the latest by 2020.

4.2 H&M's Practices of Sustainability

4.2.1 Fibers

H&M has started to reduce the use of traditionally-grown cotton which involves intense use of water and pesticides, and at the same time to increase the use of organic cotton, the Better Cotton (with significantly reduction in the use of chemical fertilizer or pesticides through the program provided by the Better Cotton Initiative) and recycled cotton, aiming to reduce their environmental impact. Importantly, H&M has set a target to use only cotton coming from these sources by 2020 at the latest.

It appears that H&M is on the right track in using more sustainable cotton, as organic cotton accounted for 7.6 and 7.8 % of its total cotton use in 2011 and 2012, respectively. Such a practice has made H&M to be the number one user of organic cotton in the world, despite the slight increase in its annual use. Besides, stated in 2012, Better Cotton was used and accounted for 3.6 % of its total cotton use. The use of cotton from these two sources led to a reduced use of over 140,000 kg pesticides. H&M also used 1,450 tonnes of recycled wool and recycled polyester (equivalent to 7.9 million PET bottles) in 2012.

However, there is no data showing that recycled cotton is used currently, nor is there any specific target set for its use in the future. It is obvious that H&M's current policy on cotton use rests largely on reducing environmental footprint, instead of on closing the textile loop. The same applies to other natural and man-made fibers.

4.2.2 Manufacturing

Water Usage

H&M has developed new production processes that can save about 30 % of the water used to produce denim, and introduced a water-efficient-denim program to its suppliers to cut water use in washing processes in 2009. These new practices saved about 50, 300 and 450 million liters of water compared to traditional production processes in 2010, 2011 and 2012, respectively. Currently, about 50 % of H&M's denim is made with these techniques.

Environmentally Friendly Dyes and Fabric Waste

Given that H&M has no direct business relations with fabric mills who are only second-tier suppliers, it lacks direct control of water and chemicals use in the production of fabrics. Nevertheless, H&M encourages fabric mills to engage in its Mill Development Program voluntarily to follow good practices to improve their environmental performance. In 2012, H&M conducted a total of 58 audits on some selected mills. Currently, H&M requires all its first-tier suppliers to submit information on the fabric mill for each order. Since 2009 H&M demanded all suppliers to prove that the dyes and other chemicals they use do not contain any APEs (Alkylphenol ethoxylates) which are harmful to the environment. However, through regular tests of its products, H&M still found that some dyes and similar chemicals did actually contain APEs, even though the dyes and chemicals were certified by their producers to be APE-free. Regarding the fabric waste, H&M has not reported any action or target of improvement.

4.2.3 Distribution and Logistics

H&M uses reusable transport boxes instead of cartons to ship garments from its distribution centers to stores, making savings of more than 400,000 trees each year compared with using traditionally made cardboard boxes. By 2011, H&M managed to minimize the use of single garment packaging to almost zero when transporting products from its suppliers to distribution centers.

Since H&M does not own any transport facilities, it has to rely on third-party transport companies to minimize its transport-related climate impact. In 2012, around 90 % of H&M goods were transported by sea or rail from the production country to its distribution centers. H&M reported that emissions of carbon dioxide equivalent resulting from transport grew from 195,948 tonnes in 2010 to 203,294 tonnes in 2012, or by 3.7 %, while sales increased by 11 % during that period. There is no specific plan or target H&M developed concerning the use of alternative fuels for transport of its products.

4.2.4 Store/Warehouse Building Efficiency

Building Energy/Efficiency

H&M has set a target reducing carbon emissions by 5 % relative to sales each year covering 2010 to 2012, and uplifted the goal to absolute reductions in its operations' total emissions by 2015. Past records show that H&M by and large achieved this 5 % drop in emissions annually from 3.33 tonnes/million SEK sales in 2010 to 3 tonnes/million SEK sales in 2012. Despite the achievement of this annual target, its absolute total emissions increased from 497,264 tonnes in 2010 to 574,611 tonnes in 2012, or by 15.6 %.

Given that electricity use in H&M's stores accounted for the biggest share of its absolute emissions (e.g. 50 % in 2012), H&M has established a target to improve energy efficiency by reducing electricity use per square meter in its stores by 20 %, as compared to a 2007 baseline, by 2020. To achieve this new target, H&M has been implementing energy saving practices and installing energy monitoring devices in all stores. In 2012, H&M generated 784,200 kWh of solar energy through its own solar photovoltaic panels. In the long run, H&M seeks to source all electricity from renewable sources, although no specific target has been set.

H&M has also applied water-saving techniques such as low-flow taps to reduce water use in its stores, offices and distribution centers, and harvested about 3 million liters of rainwater for reuse through rainwater-harvesting facilities installed in its distribution centers in Europe and a store in the UK.

Product Packaging

In 2010 H&M developed Environmental Guidelines for Packaging that aims to use fewer resources and cause less waste by using recycled materials, materials from certified sources, such as Forest Stewardship Council (FSC)-certified paper and board, and single materials, which avoid mixing materials such as stickers or laminates to improve recyclability, as well as designing packaging to optimize space-use making it more efficient to transport and making packaging easy to separate for higher recyclability.

In 2011, H&M reported that 90 % of the paper used for making its mail order packages was recycled cardboard. All of H&M's standard plastic consumer bags are made of recycled material (50 % post-consumer and 50 % pre-consumer recycled polyethylene) and consumer paper bags are made of paper originated from well-managed forests certified by FSC. Thus, all of H&M's bags are recyclable. Almost all hangers are reused in stores. When they become unusable, they are sent for recycling. H&M achieved 85 % recycling rate in 2010, up from 79 % in 2009.

H&M reported that in 2012, 92 % of the wastes handled in its distribution centers were recycled, while the target of recycling rate was uplifted to be 95 % in 2013.

4.2.5 Post-consumer and Beyond

H&M claims itself as the first fashion company in the world to provide customers with the chance to bring unwanted clothes to its stores in all 48 markets for reuse as second hand clothes or recycling as cleaning cloth or insulation material, for example. However, there is no figure reported by H&M about the amount of used clothes collected and recycled.

In 2011, H&M started discussions with Ginetex, which is the owner of the current global standard care labeling system, aiming to develop a globally-applicable care label that promotes conscious garment wash and care instructions, reducing water and energy use in the 'user phase' of the product lifecycle. H&M planned to launch the new care label in summer 2013.

5 Discussion

The above findings show that the central principle guiding H&M's approach to sustainability is the eco-efficient cradle-to-grave philosophy. Based on this mindset, H&M's sustainability vision is to send zero waste from its operations to landfill. The long-term goal set by H&M toward sustainability is the pursuit of zero discharge of hazardous chemicals by 2020. In fact, this goal applies not only to H&M but also to other leading brands in the apparel industry that have joined the initiative of Roadmap to Zero.

It is clear that eco-effective cradle-to-cradle philosophy does not form the core of H&M's current sustainability policy, despite its commitment to becoming climate smart, aiming for zero waste to landfill, and using natural resources responsibly. Most of H&M's operations do not reflect the three tenets of the cradle-to-cradle philosophy.

5.1 Waste Equals Food

To turn sustainability vision into reality, H&M has started to replace traditionally-grown cotton largely by organic cotton and to a lesser extent by the Better Cotton gradually. Although this choice of materials has resulted in significantly less use of chemical fertilizer or pesticides, there is little sign that the principle 'waste equals food' has been applied. As McDonough et al. (2003 p.437A) put, "*a material should not only be nonhazardous but also provide nourishment for something after its useful life—either "food" for biological systems or high-quality materials for subsequent generations of high-tech products*". For H&M, it is about closing the textile loop by using recycled materials, be they natural or man-made fibers, as main inputs of its production system. However, no measurable target or specific plan has been established by H&M in this regard.

H&M has recently started to collect used clothes through its retail stores and then reuse and recycle them. Although this practice may help to create awareness of recycling among its customers, the recycling process H&M has engaged is down-cycling in nature because the used clothes are transformed into products of lesser value, such as cleaning cloth, which will likely be dumped to landfill finally. This recycling system is the cradle-to-grave approach to sustainability and will only prolong the life of used clothes before they become waste. H&M has not engaged in up-cycling process aggressively. Despite its efforts in using recycled polyester to produce clothes, it has not transformed the used clothes into higher quality materials, serving as inputs of the production of higher value products.

Due to the lack of information disclosure about H&M's program of recycling used clothes, consumers may have created a misconception that all used clothes are either donated to someone in need or used as materials for production of new garments. This may make consumers feel good and justify their unsustainable consumption patterns because consumers have now found an easy and proper way to deal with their unwanted clothes. However, the recycled clothes will likely become waste in landfills and not food for producing higher quality products.

5.2 Use Current Solar Income and Celebrate Diversity

H&M is on the right track in sourcing 100 % of electricity from renewable sources such as solar energy so as to reduce its scope 2 carbon emissions. The increasing use of solar energy is feasible and expected when H&M will invest in the necessary installations in new building projects or refurbishment projects of its current facilities such as distribution centers. However, H&M has not set specific deadline for achieving the target of using only renewable energy.

Regarding the tenet 'celebrating diversity', H&M has taken an initial step to harvest rainwater through its facilities, addressing the need to design its facilities that integrate with the local natural systems, using available energy and material flows in a closed loop. Despite these efforts, there is lack of evidence showing this principle has been applied in fabric and garment production by H&M and its suppliers.

The case findings also show that H&M faces major challenges in improving the sustainability performance of its second-tier suppliers. Harmful materials can still be found through H&M's regular testing of its products, despite the suppliers have provided H&M with evidence certified by an independent third party that the materials are not hazardous. Second-tier suppliers are encouraged to participate in sustainability improvement program by H&M mostly on a voluntary basis. It is unknown to what extent good practices will be adopted by these suppliers under the current arrangement.

6 Conclusion

Based on the information disclosed by H&M in its recent sustainability reports and corporate website, this study found that H&M has framed corporate environmental sustainability as reduction of negative impact through improving resource utilization, minimizing waste and reducing ecological footprint. Thus, it is plausible for H&M to develop the target of achieving zero discharge of hazardous substances in its supply chain at the latest by 2020.

To improve its environmental performance, H&M has started to:

1. use environmental-friendly materials that have less soil pollution and water consumption,
2. introduce innovative wet production processes that save water and energy use to its first-tier suppliers,
3. demand its first-tier suppliers to submit information on the fabric mill for each order,
4. use reusable transport boxes,
5. implement energy saving practices and installing energy monitoring devices in all stores,
6. source electricity from renewable sources,
7. apply water-saving techniques to reduce water use in its stores, offices and distribution centers, and harvest rainwater for reuse,
8. reuse packaging materials,
9. collect used clothes from customers for reuse or recycling,
10. develop care label that promotes reduced water and energy use in the 'user phase' of the product lifecycle.

Despite these efforts, H&M remains silent on:

1. setting a target for usage of recycled cotton as well as other natural and man-made fibers,
2. reducing fabric waste at its suppliers' factories,
3. using alternative fuels for transport of its products,
4. setting a target for sourcing all electricity from renewable sources,
5. disclosing the amount of used clothes it collected and recycled.

It is likely that H&M will reduce its negative environmental impact in the areas of reusable containers for transport, building energy efficiency and product packaging, if it further strengthens its internal environmental management practices. However, the real challenges H&M facing come mainly from its immediate suppliers and multi-tier suppliers. How to ensure that its suppliers, in particular those connected loosely to H&M, will comply with higher environmental requirements and follow good environmental practices closely in their operations is a pressing issue to be addressed by H&M, if it has to build a sustainable fashion supply chain.

As neither unsustainable production nor consumption of clothing will end in short run, H&M and other fashion companies need to rethink the viability of current cradle-to-grave, fast fashion model as a truly sustainable business model. The pursuit of eco-efficiency is inadequate to build a sustainable business. Instead, fast fashion companies should base their management philosophy upon eco-effectiveness, and influence and collaborate with supply chain partners and other stakeholders to build cradle-to-cradle, closed loop supply chain systems, leaving positive impacts to nature. As a full-scale change from eco-efficient to eco-effective operations throughout a supply chain will not happen overnight, fashion companies can start conducting pilot projects testing the applicability of cradle-to-cradle concept and practices on new product lines. The successful cases can then serve as role models for others to follow.

Further study is expected to examine how other leading fashion companies have framed corporate environmental sustainability and to what degree they have implemented environmental management practices in their supply chains to improve sustainability from the eco-effective perspective.

Acknowledgments The author thanks Hang Seng Management College for the funding supporting the attendance of The 9th International Congress on Logistics and SCM Systems ICLS 2014.

References

Bianchi C, Birtwistle G (2010) Sell, give away, or donate: an exploratory study of fashion clothing disposal behaviour in two countries. Int Rev Retail, Distrib Consum Res 20(3):353–368

Birtwistle G, Moore CM (2007) Fashion clothing–where does it all end up? Int J Retail & Distrib Manage 35(3):210–216

Bonini SM, Oppenheim JM (2008) Helping 'green' products grow, The McKinsey quarterly. 1–8 Oct

Dyllick T, Hockerts K (2002) Beyond the business case for corporate sustainability. Bus Strategy Environ 11(2):130–141

Ellram LM, Tate WL, Carter CR (2008) Applying 3DCE to environmentally responsible manufacturing practices. J Clean Prod 16(15):1620–1631

European Commission (2011) Green public procurement textiles technical background report. http://ec.europa.eu/environment/gpp/pdf/tbr/textiles_tbr.pdf. Accessed 28 Jan 2014

Fletcher K (2008) Sustainable fashion and textiles: design journeys. Earthscan, London

Fulton K, Lee SE (2013) Assessing sustainable initiatives of apparel retailers on the Internet. JFMM 17(3):353–366

Ha-Brookshire J, Hodges Nancy N (2009) Socially responsible consumer behavior? Exploring used clothing donation behavior. Cloth Text Res J 27(3):179–196

Hill J, Lee H-H (2012) Young generation Y consumers' perceptions of sustainability in the apparel industry. JFMM 16(4):477–491

McAfee A, Dessain V, Sjoeman A (2004) Zara: IT for fast fashion. Harvard Business School Publishing, Cambridge

McDonough W, Braungart M (2002) From cradle to cradle. North Point Press, New York

McDonough W, Braungart M, Anastas PT, Zimmerman JB (2003) Applying the principles of green engineering to cradle-to-cradle design. Environ Sci Technol 37(23):434A–441A

Morgan LR, Birtwistle G (2009) An investigation of young fashion consumers' disposal habits. Int J Consum Stud 33:190–198

Sull D, Turconi S (2008) Fast fashion lessons. Bus Strategy Rev 19(2):4–11

United Nations General Assembly (1992) Report of the United Nations conference on environment and development. http://www.un.org/documents/ga/conf151/aconf15126-1annex1.htm. Assessed 28 Jan 2014

United States Environmental Protection Agency (2014), Wastes—resource conservation—common wastes & materials. http://www.epa.gov/osw/conserve/materials/textiles.htm. Accessed 28 Jan 2014

Waste and Resources Action Programme (2012) Time to unlock the value of household textile waste, says WRAP. http://www.wrap.org.uk/content/time-unlock-value-household-textile-waste-says-wrap. Accessed 28 Jan 2014

World Business Council for Sustainable Development (1992) Eco-efficiency learning module. http://www.wbcsd.org/pages/EDocument/EDocumentDetails.aspx?ID=13593&NoSearchContextKey=true. Accessed 28 Jan 2014

World Commission on Environment and Development (1987) Our common future. Oxford University Press, Oxford

Young W, Tilley F (2006) Can businesses move beyond efficiency? The shift toward effectiveness and equity in the corporate sustainability debate. Bus Strategy Environ 15(6):402–415

Efficient Chemical Management in Global Paint Industry: A Case Study in Sri Lanka

T. Sunil Somasiri Gomes

Abstract Traditionally, chemicals are sold to customers, who become owners of the substances and they will be responsible for its remaining whole life cycle. Here the chemical suppliers have only an economic interest at increasing the volume of chemicals sold, which is finally related to negative releases to environment. Compared to this approach of constant increase in sales volume of chemicals, the concept of Chemical Leasing (ChL)-originated by United Nations Industrial Development Organization (UNIDO)—aims at a shift towards a much more service and value-added approach. In this new ChL business model customer pays only for the benefits obtained from chemicals, but not for the quantity of substance, hence economic success of supplier is not linked with product turnover anymore. So chemical consumption becomes a cost rather than a revenue factor for chemicals supplier, where supplier always compelled to reduce the amount consumed, which intern reduces environmental pollution. During last few years, ChL business models have been successfully applied by UNIDO in several industrial sectors worldwide such as cleaning, coating, greasing and coloring. Results obtained have clearly shown that implementation of ChL has significantly reduced the harmful emissions to environment where chemical manufacture works closely with user to meet all three sustainability dimensions of Economical, Social and Environmental. In year 2012 a study was conducted to apply the chemical leasing concept in building painting sector in Sri Lanka where the paint manufactures produced about 56,876 kl of paints annually. Here the responsibility of paint producer ends with selling of these chemicals to user, where user is totally responsible for its application and remaining whole life cycle. In this existing system observed various types of chemical wastes and complications at user's end, such as—spills during transfers to mixing containers, remains in containers, escapes to surrounding during application processes, inconsistent quality of mixed paints, high absorption of paints into walls, insufficient technical know-how on painting surface

T. Sunil Somasiri Gomes (✉)
National Cleaner Production Centre, Narahenpita, Sri Lanka
e-mail: tsunilgomes@gmail.com

P. Golinska (ed.), *Logistics Operations, Supply Chain Management and Sustainability*, EcoProduction, DOI: 10.1007/978-3-319-07287-6_17,

preparation, insufficient technical know-how on paint handling equipment, coagulation of excess paints in containers, need of more water, electricity and labor on equipment cleaning, unsatisfied labor attitudes on difficult working conditions they faced, release of many empty containers to surrounding by creating severe environmental and health hazard—are some of them. Therefore to break this vicious circle I designed and consulted the world first "*Building Painting Project on Chemical Leasing Concept*"—between Chemical Supplier—Madushika paints Industries Private Ltd and Chemical User—Sri Lanka Broadcasting. Corporation. In this new ChL methodology I arranged site visits for paint supplier to paint users sites—in order to collect technical details of the surfaces to be painted and to decide on the chemicals to be supplied. Discussions were held between paint user and paint supplier on preparation of walls, labor utilization and suitable equipment to optimize the paint usage. A new basis of service oriented payment calculated on area painted is developed to pay by chemical user to chemical supplier which was clear to check at any time. Further a new Spherical Container was designed to handle mixed paint in-between supplier and user—in order to prevent environment pollution by thrown out empty containers to surrounding. The results of this study shows that about 12 % of paint consumption, 20 % of waste generation, 15 % of water consumption, 20 % of labor involvement and 5 % of energy consumption has reduced by application of ChL concept. Finally it has brought economic advantages for both parties involved thus providing concrete solutions in sustainability of efficient chemicals management in global paint industry through Chemical Leasing Concept.

Keywords Sri Lanka building paint sector · Chemical leasing business models · Selling versus leasing of chemicals · Three sustainability dimensions · Ecodesigned spherical container

1 Introduction

1.1 What is Chemical Leasing?

In existing chemical marketing concept, the producer/supplier sells bulk quantities of chemicals to end-users, and tries to sell as much as possible in the interest of their profit, where the inefficient use of chemicals is being *"rewarded"*. Therefore the end-users becomes owners of these substances and totally responsible for rest of its life cycle such as—use, balance stock maintenance and final disposal.

Chemical Leasing is an innovative business model—originated by United Nations Industrial Development Organization (UNIDO) in year 2013—based on the resource efficient and cleaner production concept which provides practical solutions for industries to become more efficient, as well as protect human health and environment. It supports sustainable chemicals management by responding to latest changes in the international chemical policies and encourages progressive and environmentally sound industrial development while reducing unnecessary consumption of hazardous chemicals.

1.2 Difference Between Traditional Buyer–Seller Model and Service Based Chemical Leasing Model

Fundamental to the Chemical Leasing concept is that the user benefits from the services provide by chemical supplier which leads into the sustainable management of chemicals. This concept shifts the focus from increasing the sales volume of chemicals towards a more value added approach through extended producer responsibility. Here the chemical producer supplies a substance for a specific service, but retains ownership over the chemical, and also advises the user on its best use. As a result, it is in the interest of all parties to use the substances as effectively and efficiently as possible (Fig. 1).

Therefore the concept of Chemical Leasing (ChL) redefines the sustainable management of chemicals and provides a more sustainable solution, where the users only pay for services rendered by chemicals but not for volume of chemicals consumed. Here supplier mainly sells the functions performed by chemicals and functional units such as number of pieces/area painted, volume of water treated, lengths of pipes cleaned etc. are main basis for payment. Hence decoupling the payment to supplier from the consumption of chemicals, it becomes a cost rather than a revenue factor to chemical supplier and he always compels to reduce chemical consumption. The result can be seen in environmental advantages, as well as in consequential economic benefits for both chemical supplier and user.

Chemical leasing is thus in line with UNIDO Green Industry approach, which strives to decouple resource use and pollution from industrial development which promote sustainable growth of productive sectors.

1.3 UNIDO Definition of Chemical Leasing

Definition of ChL of United Nations Industrial Organization is as follows:

- Chemical leasing is a service-oriented business model that shifts the focus from increasing sales volume of chemicals, toward a value-added approach.
- The producer mainly sells the functions performed by the chemical, and functional units are the main basis for payment.
- (Functions performed by a chemical might include: number of pieces cleaned; amount of area coated, etc.).

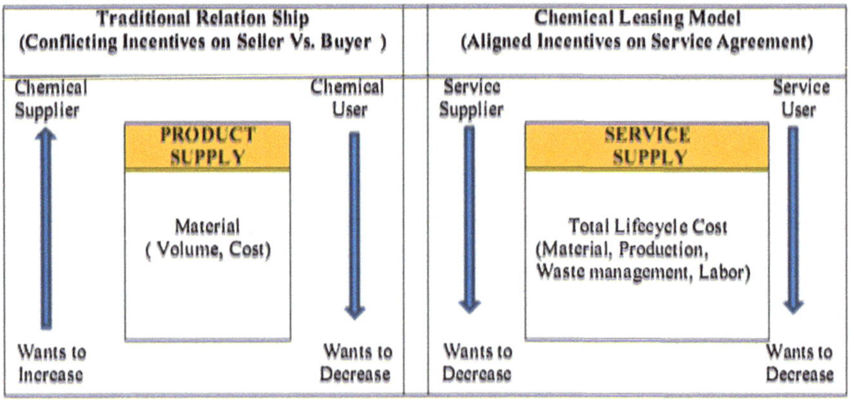

Fig. 1 Difference between product supply and service supply

- Within chemical leasing business models, the responsibility of the producer and service provider is extended and may include the management of the entire life cycle.
- Chemical leasing strives for a win–win situation. It aims to increase the efficient use of chemicals while reducing the risks of chemicals, and protecting human health. It improves economic and environmental performance of participating companies and enhances their access to new markets.
- Key elements of successful chemical leasing business models are proper benefit sharing, high-quality standards and mutual trust between participating companies.

Accordingly this innovative approach of Chemical Leasing Concept has been implemented in a number of different sectors and processes by UNIDO in close cooperation with respective National Cleaner Production Centers (NCPC) of their countries. These centers are partners of the UNIDO/UNEP global Resource Efficient Cleaner Production (RECP) network, which presently includes nearly 50 national Cleaner Production Centers globally. The UNIDO experience has shown that ChL is best applied to processes that are not the core know-how of the chemical user, such as cleaning, degreasing and painting, etc.

2 The Paint Industry of Sri Lanka

According to the Paint Manufactures Association of Sri Lanka, the local paint industry comprises of about 35 large and small manufacturers where they produced about 56,876 kl of paints annually. Among them 08 paint manufacturers are well known with brands that have a strong presence in the market. These larger companies alone have employed about 15,000 unskilled, semi-skilled, skilled and professional personnel in their organizations. The allied industries like printing, packaging, transport and marketing are the livelihood of further about 250,000 persons.

Paints in general are differentiated into decorative or architectural paints and industrial paints. The decorative or architectural paints serve the housing sector and industrial paints are included with powder coatings, high performance coating, automotive paints and marine paints. Decorative paints are primarily used on interior or exterior of homes and buildings and include coatings such as emulsions, enamels, varnishes, wood finishes, and distempers. Based on the solvents used, decorative paints are further classified into water-based and oil based where plastic, latex or emulsions are water-based while enamel are oil-based.

The local paint industry depends 99 % on imported chemicals such as binders, pigments, fillers, solvents, dispersion agents, wetting agents, additives, stabilizers and hardeners. Lead compounds are added to paints for speed drying, increase durability, retain a fresh appearance, and resist moisture that causes corrosion. According to World Health Organization the Lead is a metal with no known biological benefit to humans and these lead base chemicals in paints has been recognized as a prime toxic. The Lead in paints is responsible particularly when used on walls and items that children would lick or chew. According to researches lead-contaminated dust and soil has found in and around of old—deteriorating—painted buildings polluting environment.

The following Sri Lanka Standards (SLS) are currently available for Emulsion and Enamel Paints.

1. Emulsion paints for exterior use	–SLS 557:2009
2. Emulsion paints for interior use	–SLS 553:2009
3. Enamel paints	–SLS539:11981

According to above Sri Lanka Standards—both imported and locally manufactured all paints will necessarily have to conform to the respective standards. But in year 2009—the group partnered with India-based Toxics Link and US-based International POPs Elimination Network (IPEN)—confirmed that leaded paints are still being produced and sold exposing children and communities to this toxic substance.

In the traditional building painting processes of Sri Lanka, responsibility of paint producers ends with selling of manufactured paints and solvents to users. The users becomes owners of these chemicals, generally without having any technical knowhow on them and remains responsible for whole life cycle included their application and final disposal. Therefore the existing concept leads paint producers to have only a clear economic interest in increasing the volume of paints sold to maximize their profits.

From year 2003 I have been associated with National Cleaner Production Centre of Sri Lanka and from year 2005 working as qualified consultant in application of CP methodology in many industries on behalf of NCPC Sri Lanka and further extended my involvements as a foundation member of the Chemical Leasing Working Group of Sri Lanka. Therefore I designed and conducted a new "*Chemical Leasing Building Painting Project*"—for the Paint Supplier—Madushika paints and Chemicals Industries Private Limited and the Paint User—Sri Lanka Broadcasting Corporation—to shift the paint producer's focus from increasing the sales volume of chemicals towards a more value added approach through extended responsibility and to prevent adverse effects on environment due to mishandling of hazardous chemicals.

3 Phases of the Chemical Leasing Project

The chemical leasing project I designed for "Paint Supplier—Madushika paints and Chemicals Industries Private Limited and the Paint User—Sri Lanka Broadcasting Corporation"—was divided into stages such as data collection, planning, negotiation, contract signing and implementation and monitoring, where my contributions all over the project were as follows (Fig. 2).

3.1 Data Collection of Existing Painting Process

At the beginning of the project arranged a study painting process at paint user's end (Sri Lanka Broadcasting Corporation) in accordance with the existing painting methodology. The required chemicals and utensils were purchased from open market and the skilled and semi skilled workers attached to Sri Lanka Broadcasting Corporation were used for this exercise. The process breeds various types of uncontrollable wastages and problems as follows.

1. Wastage of excess chemicals and throw-out them to surrounding by polluting environment,
2. Wastage of paints during transfers to mixing containers,
3. Inconsistent quality of mixed paints,
4. Insufficient technical know-how on preparation of painting surfaces to optimize the paint usage,
5. Wastage of mixed paints during its application processes,
6. Insufficient technical know-how on paint handling equipment,
7. Wastage of excess paint in containers due to coagulation,
8. Use of more water, electricity and labor on equipment cleaning,
9. Unsatisfied labor attitudes due to difficult working conditions they faced,
10. Release of many plastic empty containers to the surrounding by creating severe environmental hazard.

Fig. 2 Chemical preparation at paint supplier—madushika paints and chemicals industries private limited

Table 1 Economic benefits gained by applying chemical leasing (comparison of before and after)

Before chemical leasing	After chemical leasing
Total cost of paint for 929.03 m^2 (91.7 l × Rs. 500 per liter) = Rs. 45,850	Total cost of paint = 82.3 l × Rs. 500 = Rs. 41,15 Saving (Rs. 45,850 − 41,150) = Rs. 4,700
Total coverage per liter of paint = 10.126 m^2	Total coverage per liter of paint = 11.288 m^2
Energy cost per liter Rs. 75	Paint saving for 929.03 m^2 = 9.4 l Total energy saving for 929.03 m^2 = Rs. 705
Cost for plastic containers for application of paints to 929.03 m^2 area, 05 plastic containers = Rs. 450	Saving on plastic containers use to fill paint = 05 × Rs. 97 = Rs. 457 (considered as average 05 containers)
05 printed labels = Rs. 06	
Glue for 05 Labels = Rs. 01	
Labor total coverage per day = 65.032 m^2	Labor total coverage per day = 83.613 m^2 Reduction of total labor hours for 929.03 m^2 = 25.4 h

As above this study painting process helped to identify technical problems aroused, environmental issues created and existing cost per area painted. (Table 1—Economic benefits gained by applying Chemical Leasing (comparison of before chemical leasing) (Fig. 3).

3.2 Planning Stage

The planning stage started with convincing paints supplier and paints user on advantages of Chemical Leasing Concept. In this stage the leasing model, its cost implications versus quality and environmental benefits were also discussed. Then

Fig. 3 Storing of used paints and chemicals at paint user end

process started with helping the management of paints supplier (Madushika Paints and Chemicals Industries Private Limited) to nominate a Cleaner Production Team (CP Team).

By application of Cleaner Production, the utilization of resources in Madushika Paints and Chemicals Industries Private Limited fully quantified and appraised. CP looked at input resources, how it flows through processes and helped paint supplier to eliminate or reduce creation of waste and improve efficiency of resource utilization related to water, raw materials and energy as follows.

- Educated the CP team about the chemical leasing project and its process steps.
- Carried out Walk through Audit Inspections with the CP team to identify the waste streams of the factory paint production process. This help to identify the ineffective use of energy, water and materials at paint supplier's end.
- Helped CP team to List out the paint production process steps.
- Prepared the Cleaner Production Flow Charts.
- Identified the Input Resources and Waste Streams.
- Prepared the Material Balances and quantified.
- Conducted a Brainstorming Session and cause analysis to identify CP Options
- Selected the potential CP options for improvements and helped Paint Supplier to enhance his economic performances defining the key performance indicators (KPI).
- Defined the additional resources needed to fulfill expected improvements in the processes.

3.3 Negotiation and Contracts Signing Stage

During negotiation and contracts signing stage the commercial terms, and conditions were initially discussed and continued the project as follows.

Fig. 4 Signing of the contract between partners—with the presence of representatives of UNIDO-Vienna and national cleaner production centre-Sri Lanka

- Assisted partners for negotiations considering all variable factors to initiate the building painting project on Chemical Leasing concept,
- Managed the negotiations between partners to develop a "*unit of payment—per square meter painted*"—which to be paid by paint user to paint supplier after completion of painting work,
- Arranged contract signing (Fig. 4) between partners included with new basis of service oriented payment calculate on area painted, which is very easy to understand for all parties involved and clear to check at any time.

3.4 Implementation and Monitoring Stage

In the initial point of implementation stage I arranged some field visits for paint supplier to paint user's site. Here paint supplier collected technical details of the area to be painted, conditions of surfaces to be painted, colours to be used, volume of paint requirement according to separate colours and suitable equipment to be used for painting work etc. Further paint supplier provided technical acquaintance to paint user on preparation of walls to optimise paint usage.

Finally the required paints were prepared at the factory by paint supplier and supplied to paint user for application. Again required utensils for this process were purchased from open market and skilled and semi skilled workers attached to Sri Lanka Broadcasting Corporation were used for painting operation.

During this painting process under chemical leasing concept also following technical problems were again identified as global issues.

1. Wastage of excess chemicals and throw-out them to surrounding by polluting environment,
2. Wastage of paints during transfers to mixing containers,
3. Inconsistent quality of mixed paints,
4. Wastage of excess paint in containers due to coagulation,

Fig. 5 The ecodesigned spherical container

5. Use of more water, electricity and labor on equipment cleaning,
6. Unsatisfied labor attitudes due to difficult working conditions they faced,
7. Release many plastic empty containers to the surrounding by creating severe environmental hazards.

Therefore to rise above these global technical problems it was necessary to develop an efficient and effective methodology which will minimize said uncontrollable wastages, optimize paint usage and prevent adverse effects on environment. This thinking was helped me in development of a "*Spherical Container*" to use in this exercise and its contribution and advantages are as follows.

Invention of "*Ecodesigned Spherical Container*" to paint industry.
The ecodesigned spherical container for paint industry (Fig. 5) is a new product and a process where the initial model is made out of fibre glass material which the volume is around 600 l. This single transfer—multipurpose—spherical container basically incorporated with a mouth for loading the mixed paints, a PVC Pipe—Valve system fixed to bottom to take out liquid inside and a mild steel stand to support the container during handling. Inside of the container is smooth finished and epoxy coated.

In this spherical—the volume versus inside surface area is lesser when compared to existing containers use for paint handling. Its unique spherical shape itself provides a natural effective way by not having any corners to support the coagulation and stagnancy of paints. The unit could be transportable in between the paint processing factory and painting site.

Finally the paint supplier prepared the mix of the paint for each coat in a separate Ecodesigned Spherical Container through his factory process and supplied the user for application under his guidance. The empty container was sent back to the paint factory to clean it and use for next supply.

4 Discussion

1. By introducing the ecodesigned spherical container to this painting project it eliminated existing wastages as follows.
   - Eliminates use of resources to maintain excess paints at user end,
   - Eliminates release of many plastic empty containers to surrounding at user end hence enhance environmental and health conditions,
   - Reduces use of resources to manufacture those plastic containers, labels and glue,
   - Eliminates wastage of paints during transfer processes as no mixing process available at user end, T.S.S Gomes, Efficient Chemical Management in Global Paint Industry—A Case Study in Sri Lanka
   - Optimizes paint usage and minimize wastage of mixed paint within the application processes,
   - Obtained quality consistency in mixed paints as it prepared by the supplier during the factory process,
   - Reduces the use of water—electricity and labour on cleaning processes,
   - Obtained satisfied labour attitudes due to recognition gained through quality of work, improved technical knowhow and easy handling process they involved.
2. The benefits obtained by the paint supplier "*Madushika Paints and Chemicals Industries Private Limited*" through cleaner production application in there production processes were not included in this report. Therefore calculated the financial, environmental and human health benefits gained by paint supplier and paint user through chemical leasing application were tabulated as follows (Tables 1 and 2).

5 Conclusion and Recommendation

In this new "*Chemical Leasing Building Painting Project*" chemical consumption became a cost rather than a revenue factor for paint supplier. Hence he tried always use best practices for process optimizations in order to reduce chemicals consumed at user end where all said factors were gave financial benefits for both partners involved and intern collectively reducing environmental pollution. Hence it can be applied to any such similar cases in the long term. The basis of service oriented payment by calculating the area painted and the container designed to prepare total paint requirement for each coat will be a new approach to future painting businesses.

Based on the results of the above project the Chemical Leasing is a proven innovative instrument to promote resource efficient and sustainable management of chemicals in paint industries where economic benefits do not result from

Table 2 Environmental and human health benefits by applying chemical leasing (comparison of before and after)

Before chemical leasing	After chemical leasing
Total Coverage per liter = 10.126 m^2	Total coverage per liter = 11.288 m^2
	Total paint consumption for 929.03 m^2 = 82.3 l
Total paint consumption for 929.03 m^2 = 91.7 l (Containers used = 4 × 20 l + 01 × 10 l + 02 × 01 l)	Total Saving for 929.03 m^2 = 9.4 l
According to the material balance total wastages : around 12 % (Container remains, Washouts, Spillages during paint transferring, Spillages during application, Remains in painting rollers and Brushes)	According to the material balance total wastages reduced up to 2 % Mainly reduced the said wastages using invention of Ecodesigned Spherical Container
Energy cost per liter Rs. 75	Energy savings for 929.03 m^2 = 9.4 l × Rs. 75 = Rs. 705
Water to clean empty 06 containers = 72 l	Water to clean Spherical container = 53 l Reduction of Water consumption = 19 l
Exposure of workers to fumes:	1. Introduced personal protective equipment
1. No personal protective equipment used	
2. Coverage per 09 h days work = 65.032 m^2	2. Total coverage per 09 h days work = 83.613 m^2 (Introduced proper application methods hence reduce labor exposure time during painting process and obtained reduction time around 25.4 h per 929.03 m^2)
For application of 929.03 m^2 area, around 5–6 plastic empty containers released to environment (cost of plastic container is about Rs. 90)	Due to application of ecodesigned spherical container for paint industry, there was a significant reduction of plastic containers usage for filling paints
Use of resources to manufacture those plastic containers,labels and glue	Reduces use resources to manufacture those plastic containers, labels and glue

quantity of chemicals sold, but from service rendered by chemicals. Chemical leasing brings additional non quantitative benefits also to both supplier and user. The nature of the relationship between the manufacturer and user nurtures partnership development and long-term business relationship. The regularly monitored contract yields benefits too because the contract spells out rights, duties and obligations of the parties involved. When arrange the regular exchange of product and process know-how in between chemical supplier and user will make path to generate a high quality, effective and efficient manufacturing operation. Therefore this project is a definite advantage to convince enterprises and motivate them to apply Chemical Leasing Concept for sustainable chemical management to meet all three sustainability dimensions such as Economical, Social and Environmental in global paint industry.

Reference

UNIDO (2013) United Nations Industrial Development Organization, UNIDO, Involvement of UNIDO, May 2013 (Concept of chemical leasing)

A Consideration on the Functions of Logistic Parks Against Great Disasters

Keizo Wakabayashi, Kuninori Suzuki, Akihiro Watanabe, Yutaka Karasawa and Koichi Murata

Abstract The big issue in disaster logistics such as transport relief supplies in the affected or later has been pointed out. In addition, an enormous amount of waste caused by the tsunami left in the coastal areas. It is necessary to visualize the mechanism to efficiently handle disaster waste. In this chapter, from the point of view of urban regeneration with the environment in mind, introducing previous studies, anti—disaster city planning focused on logistic parks is discussed. Firstly, the existing research of disaster logistics/SCM is introduced. In this chapter, the concept of urban planning focused on logistics parks with the environment in mind is considered. To the lessons to Sendai, which became the central areas affected by the Great East Japan Earthquake, Kobe taken up as an example. These considerations above lead the conclusions.

Keywords Disaster waste · Logistic park · Warehouse · Supply chain management

1 Introduction

The Great East Japan Earthquake that occurred on March 11, 2011, extensive damage occurs mainly in Tohoku district Pacific Ocean coastal areas, scars still remain today. Upon the occurrence of the Great East Japan Earthquake, relief supplies cannot reach far enough to the disaster area, victims of many troubled to lack daily necessities, the necessities. The big issue in disaster logistics such as

K. Wakabayashi · K. Suzuki (✉) · A. Watanabe · K. Murata
Nihon University, Izumicho, Narashino, Chiba, Japan
e-mail: QWL02764@nifty.ne.jp

Y. Karasawa
Kanagawa University, Izumicho, Narashino, Chiba, Japan

P. Golinska (ed.), *Logistics Operations, Supply Chain Management and Sustainability*, EcoProduction, DOI: 10.1007/978-3-319-07287-6_18,

transport relief supplies in the affected or later has been pointed out. In addition, an enormous amount of waste caused by the tsunami left in the coastal areas. It is necessary to visualize the mechanism to efficiently handle disaster waste. In this chapter, from the point of view of urban regeneration with the environment in mind, introducing previous studies, anti—disaster city planning focused on logistic parks is discussed. Firstly, the existing research of disaster logistics/SCM is introduced. It is shown that the study and practice in this area should be promoted further.

2 City Planning from the View of Logistics

2.1 *Definition of Logistics Facilities and Logistics Parks*

The main functions of logistics facilities are transportation, storage, distribution processing, packaging, loading, unloading, and information transmission. Examples of logistics facilities include transportation facilities such as delivery centers and storage facilities such as warehouses. The Act on the Improvement of Urban Distribution Centers stipulates that distribution business centers, also called logistics complexes, should be located in city planning areas that have been identified for improvement as distribution business zones while considering the arrangement of traffic facilities such as trunk roads and railways in order to improve distribution functions in large cities and to smoothen road traffic.

2.2 *Location Requirements for Logistics Facilities*

With regard to industrial locations, Alfred Weber considered that industrial zones were developed at specific sites from the viewpoint of minimizing labor and transportation costs. In general, the locations of modern logistics facilities are often explained as multistage industrial location problems consisting of production bases, physical distribution bases, and markets (customers) on the basis of Weber's theory of industrial location, as shown in Fig. 1.

In other words, the location of a physical distribution base and the products to be handled in the base are determined after minimizing the total cost, which is obtained by adding the operating cost of the base to the transportation costs and labor costs for loading and unloading, which are described in Weber's theory of industrial location. Moreover, a trend in recent years has been to locate a physical distribution base so as to minimize the environmental burden. To determine the location of a physical distribution base, two main scientific viewpoints are considered.

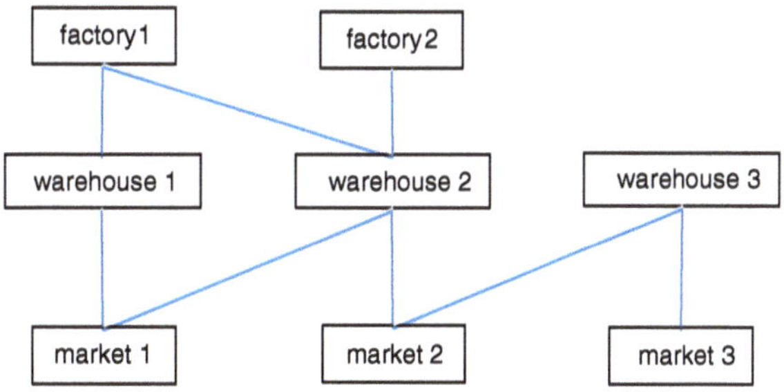

Fig. 1 Example of multi-location model

The first viewpoint is related to operations research, in which a production base, a physical distribution base, and a market are considered as nodes; links between these points are considered as transportation routes; and the network consisting of these routes is to be optimized. However, using optimization alone is insufficient for selecting the optimal location of a physical distribution base because complicated environmental measures are involved in the selection.

The second viewpoint is that after considering the overall balance of the geographical locations of a production base, a physical distribution base, and a market, the optimal location of the physical distribution base is selected. In this optimization, either the minimization of ton-kilometer (transportation volume × transportation distance) or that of the distance between the physical distribution base and the market is used as an objective function. Moreover, when the minimization of carbon dioxide emissions is used as an objective function, an environmentally friendly physical distribution base can be selected. A location where the sum of the costs required for the construction and operation (node cost) of a logistics facility and the cost required for the delivery of products (link cost) is the lowest is the optimal location of the logistics facility.

Thus, the optimization of the locations of logistics facilities can be defined as follows:

$$\text{Min } C = C^n + \sum_{j=1}^{m} C_j^l \tag{1}$$

where

m	the number of delivery points,
C	cost for logistics facility,
C^n	node cost for logistics facilities,
C_j^l	link cost with logistics facility,
$\sum_{j=1}^{m} C_j^l$	delivery cost.

Because the node cost at the center of a city is generally high, the optimal location of a logistics facility should be the suburbs of a city. Logistics facilities tend to be located close together. This is probably because logistics facilities and similar facilities such as industrial facilities share close relationships with each other. At present, it is common for freight vehicles to enter and exit logistics facilities and for forklifts to run in facilities 24 h a day. If logistics facilities are located near commercial and residential areas, many inconvenient problems such as noise may arise. Therefore, it is desirable for logistics facilities to be located close to each other in logistics complexes that are developed in industrial or exclusive industrial districts from the viewpoint of improving the urban environment. The locations of logistics complexes should have easy access to trunk roads, expressways, harbors, and airports and be far away from city centers, e.g., in the suburbs of cities.

For logistics facilities with a small number of delivery destinations, a low node cost is particularly desirable. For a logistics complex in which a facility to store daily necessities for disaster recovery is to be built, i.e., the complex is of particular importance from the viewpoint of emergency measures, it will be rational to locate the complex with a focus on the node cost rather than on the link cost. In light of disasters such as the Great East Japan earthquake, the physical distribution base should be located in a disaster-resistant environment.

The functions of logistics facilities and complexes must be secured when a natural disaster occurs. To promote disaster-resistant city planning from the viewpoint of maintaining logistics services, the following issues are crucial:

1. transportation of relief supplies (daily commodities) after the occurrence of a natural disaster (earthquake or tsunami),
2. logistics for restoration (disposal of disaster waste and supply of building materials),
3. reconstruction of logistics infrastructure (restoration of damaged logistics warehouses),
4. anti-disaster logistics management (logistics measures against aftershocks).

2.3 Core Bases for Transporting Relief Supplies After a Natural Disaster

After the Great East Japan Earthquake, cutoff of supply chains was noticed as a serious problem. When a natural disaster occurs, an important role of a logistics facility is to serve as a headquarters and supply the necessary quantity of necessary relief goods to the affected areas.

When a large-scale disaster occurs, not only warehouses and distribution centers but also land, sea, and air logistics infrastructures such as ordinary roads, trunk roads, expressways, harbors, and airports are seriously damaged and supply chains are completely paralyzed. Therefore, industrial estates must be developed in places without tsunami risks, factories and distribution centers must possess emergency support functions such as shelters for refugees, and social infrastructure systems must be secured. If these are achieved, the amount of disaster waste may be reduced and the environmental burden due to the effect of the natural disaster may be mitigated.

When the Great East Japan Earthquake occurred, logistics facilities were washed away, trucks were lost, and warehouses were flooded by tsunamis in many cases. For recovery operations, transportation routes could not be secured and necessary relief materials could not be supplied. Consequently, affected areas took a while to recover.

In addition to these catastrophic compound disasters, an accident occurred at the Fukushima Daiichi Nuclear Power Station. In its aftermath, a shipping ban was applied to prefectures if agricultural, fishery, and industrial products were found to be contaminated with radiation. The residual radioactivity needed to be inspected when exporting products from radiation-contaminated areas. Thus, the shipment of products from areas near the Fukushima nuclear accident was negatively biased. How to lower inhibitions on such shipments will be a serious problem in the future.

When a serious disaster occurs, it is necessary for logistics facilities to qualitatively and quantitatively estimate the demand for relief goods and to quickly predict the amount of daily commodities obtainable from unaffected areas and that required in disaster affected areas. Information on the supply and stock of daily necessities must be appropriately provided for and shared by refugees according to the evacuation status of residents. The transportation of materials used for temporary dwellings such as prefabricated houses must be prioritized.

Disaster waste must be rapidly disposed while restoring logistics-related infrastructures. To prepare for the occurrences of secondary and tertiary disasters, damaged supply chains must be repaired as quickly as possible. It almost seems superfluous to say that the reconstruction status of disaster-affected areas must be researched and functions to prevent problems must be surveyed from the viewpoint of logistics services.

2.4 Use of Logistics Facilities as Shelters

In previous disasters, elementary schools, junior high schools, and parks were used as emergency evaluation sites. However, in these public facilities, it is difficult to construct an environment where relief supplies can always be stored. Therefore, the idea of establishing shelters in logistics facilities located in places certain distances away from the center of disaster-affected areas and in logistics complexes where logistics facilities are located close to each other has gained ground. In other words, logistics facilities are obliged to establish facilities to store relief supplies for disasters by law, and administrative bodies manage the stock of the same for emergency use. As a matter of course, the government subsidizes the establishment of these facilities.

Based on the abovementioned issues, industrial estates must be developed in places without tsunami risks, factories and distribution centers must possess emergency support functions such as shelters for refugees, and social infrastructure systems must be secured for logistics facilities and complexes while considering environmental factors. Regarding damaged logistics facilities and complexes, it is necessary to establish a system to publicize information on the restoration status of these facilities and complexes from the websites of administrative bodies in real time.

2.5 Use of Logistics Complexes as Temporary Storage Sites for Disaster Waste

If vacant land exists in a logistics complex, it should be designed for use as a temporary storage site for disaster waste until its final disposal. Bulky refuse is often mixed in disaster waste except residential waste. A large number of temporary storage sites may be required for disaster waste until it can be transported to crushing and separation facilities. For example, when the Great Hanshin-Awaji Earthquake occurred, the number of temporary storage sites for dismantling waste from collapsed houses was insufficient, and some of the dismantled waste was left to stand on Koshien Beach for more than 2 years. After the Great East Japan Earthquake, too, some dismantled waste had been left to stand on the coasts of affected areas.

3 Case Study: Kobe City and Sendai City

To establish effective reconstruction measures in response to the Great East Japan Earthquake, reconstruction measures taken for the Great Hanshin-Awaji Earthquake can be used as an example.

Before the Second World War, many piers had already been constructed in the Port of Kobe, which was opened in 1868 as Hyogo Port. After the Second World War, the Port of Kobe has been continuously expanded and a container terminal was constructed for the first time in Japan. Until the 1970s, jetty-type wharves had been constructed. During the 1980s, linear berths such as the Port Island Berth had been constructed in extensive reclaimed lands. In 1992, the construction of Rokko Island was completed, and the Port of Kobe became a modern port with sufficient water depths for large-scale container vessels.

However, the facilities of the Port of Kobe were seriously damaged by the Great Hanshin-Awaji Earthquake that occurred in January 1995. After the earthquake, the amount of cargo handled at the Port of Kobe decreased, and it took a long time to recover from this decrease. Naturally, warehouses related to the Port of Kobe were also badly affected by the earthquake.

Warehouses related to the Port of Kobe are located in the port and coastal areas and the slightly inland areas of Higashinada-ku, Nada-ku, Chuo-ku, Hyogo-ku, Nagata-ku, and Suma-ku. For convenience, this chapter defines the East First-Fourth Reclamations and areas across their northern channel as the Tobu district, inland areas north of the Hanshin Expressway Kobe Route as the Chuo district, Shinko Pier as the Shinko district, Port Island Reclamation as the Port Island district, Hyogo Pier as the Hyogo district, and areas along both sides and south of the Hyogo Canal as the Seibu district, in addition to the Rokko Island and Suma districts. In the Tobu district, factories, houses, and warehouses are mixed. In the Rokko Island and Port Island districts, warehouses are located centering on public piers and housing estates are mixed with the warehouses.

In 2006, the Kobe Airport Island was completely reclaimed and Kobe Airport was opened. Since then, Kobe City has made an effort to develop industrial parks near the airport, which are located in reclaimed lands such as Port Island. In order to positively sell lots in the industrial parks to enterprises (i.e., attracting enterprises), Kobe City established a project organization called Kobe Enterprise Promotion Bureau.

Sendai City in Miyagi Prefecture is one of the areas that were seriously damaged by the Great East Japan Earthquake. Many retail dealers have been intensively gathered in Sendai City, where logistics facilities have also been established. Since the opening of the Sendai-Shiogama Port in 1971, three container berths and a ferry terminal have been constructed in the port. At present, the Sendai-Shiogama Port is a specially designated major port in Japan.

In Miyagi Prefecture, the area of logistics facilities in Miyagino-ku, Sendai is 382,900 m^2 and that in Makabayashi-ku, Sendai is 184,700 m^2. The area of logistics facilities in Natori City, Ishinomaki City, and Iwanuma City is approximately 30,000 m^2. In 27 municipalities, the area of logistics facilities is less than 30,000 m^2. Therefore, logistics facilities are mainly distributed in Miyagino-ku and Makabayashi-ku. In Miyagi Prefecture, areas centering on Miyagino-ku and the Arahama block in Makabayashi-ku, which face the Pacific Ocean, were seriously damaged by tsunamis that followed the Great East Japan Earthquake. Consequently, logistics services in the Tohoku district centering on Sendai City were seriously affected. When establishing an urban reconstruction plan in response to the Great East Japan Earthquake, logistics facilities must be shifted from Miyagino-ku and Makabayashi-ku to inland areas and disaster-resistant logistics complexes must be constructed. The total area of 567,600 m^2 for logistics facilities in Miyagino-ku and Makabayashi-ku can be replaced by large-size automated warehouses with a total floor area of 50,000 m^2, which have been mainly constructed in large cities. In other words, the total area of 567,600 m^2 for logistics facilities can be logically integrated into a single logistics complex.

Although the concept of disaster logistics has been studied, the study of disaster logistics is a new field in Japan. Since the sense of impending crisis regarding disaster logistics has been heightened greatly after the Great East Japan Earthquake, disaster logistics should be studied in detail in the future.

4 Conclusion

As mentioned above, the improvement of logistics infrastructure, logistics systems, and human resources is required in order to quickly respond to a disaster. When a disaster occurs, how to use logistics complexes as shelters and how to quickly transport necessary relief goods from logistics complexes to disaster-affected areas are important tasks to be examined. As a measure against disasters, chief logistics officers should be trained. In city planning, the locations of logistics complexes have been neglected. However, the restoration and reconstruction in response to the Great East Japan Earthquake were disturbed by the cutoff of supply chains, and the environmental burden increased owing to disaster waste. To cope with these disadvantages, it is necessary to evaluate a logistics complex as a key factor for promoting the construction of a disaster-resistant city in environment-considered urban planning. We think that this chapter can provide directions for future disaster logistics. To smoothly pursue a series of disaster logistics activities, a city plan focusing on environment-considered logistics complexes and facilities must be created and embodied.

A Consideration of a Reverse Logistics Network Over a Wider Area

Kuninori Suzuki, Keizo Wakabayashi, Akihiro Watanabe and Yutaka Karasawa

Abstract Regarding the lifecycle of wastes/valuable materials, the efficiency of a strategic flow, including collection, transportation, intermediate treatment, and final disposal/recycle is expected to improve from the viewpoint of environmental symbiosis. A forward logistics network is constructed mainly of physical distribution enterprises. In the case of reverse logistics, however, collection, transportation, and intermediate treatment enterprises, elements that are not involved in forward logistics play the main roles in the network's construction. Industrial wastes discharged from factories or similar facilities are collected and transported to intermediate treatment facilities for recycling. Therefore, a network that covers forward and reverse logistics is difficult to construct. Moreover, although the efficiency of forward logistics has greatly improved recently, no noteworthy progress has been seen in the efficiency of reverse logistics, even though there are several differences in the characteristics and constraint conditions between forward and reverse logistics. In this chapter, the possibility of constructing a reverse logistics network over a wide area is examined. As the example for this study, the chapter analyzes the current issues and to propose an effective reverse logistics system for used personal computers. An effective system to collect used personal computers over a wide area with a correct time window is required. The computer simulation includes a collection system algorithm with cluster-first/route-second method and local search method. This procedure consists of three factors. The first is the collection of used personal computers and their transportation to the logistics centers. The second is the improvement of the actual reverse logistics system for used personal computers. The final factor is the design of the improved reverse

K. Suzuki (✉) · K. Wakabayashi · A. Watanabe
Nihon University, Izumicho Narashino, Chiba, Japan
e-mail: QWL02764@nifty.ne.jp

Y. Karasawa
Kanagawa University, Izumicho Narashino, Chiba, Japan

P. Golinska (ed.), *Logistics Operations, Supply Chain Management and Sustainability*, EcoProduction, DOI: 10.1007/978-3-319-07287-6_19,

logistical system. This chapter clearly points out the importance of collection of used personal computers over a wide area with a correct time window with a numerical experiment.

Keywords Reverse logistics • Used personal computers • Collection and transport • Recycle • Intermediate treatment

1 Introduction

The Fundamental Principles of General Logistics Policy states that "promotion for constructing an effective reverse logistics system in which appropriate treatment and transportation are secured" is one of the principal targets of logistics administration. Regarding the lifecycle of wastes/valuable materials, the efficiency of a strategic flow, including collection, transportation, intermediate treatment, and final disposal/recycle, is expected to improve from the viewpoint of environmental symbiosis. A forward logistics network is constructed mainly of physical distribution enterprises. However, in reverse logistics, the elements that are not involved in forward logistics, such as collection, transportation, and intermediate treatment enterprises, play crucial roles in the network's construction. Therefore, a network that covers forward and reverse logistics is difficult to construct. Moreover, although the efficiency of forward logistics has recently improved in a significant way, no noteworthy progress has been seen in the efficiency of reverse logistics.

In the present study, the possibility of constructing a reverse logistics network over a wide area is examined from a viewpoint based on previous studies and practices concerning the construction of a forward logistics network.

2 Range and Problems of Reverse Logistics

The scheme of green supply chain and reverse logistics is shown in Fig. 1. Industrial wastes discharged from factories or similar facilities are collected and transported to intermediate treatment facilities for recycling. The temporary storage of industrial wastes that are moved from factories to some other place is called "*transshipment storage.*" Transshipment storage takes place immediately after industrial wastes are released from waste-discharging facilities, such as factories.

At these intermediate treatment facilities, appropriate treatments, including selection, separation, crushing, incineration, melting, and dehydration, are performed in consideration of final disposal or recycling, based on the characteristics of the waste products and the discarded materials. Any residue left behind by the

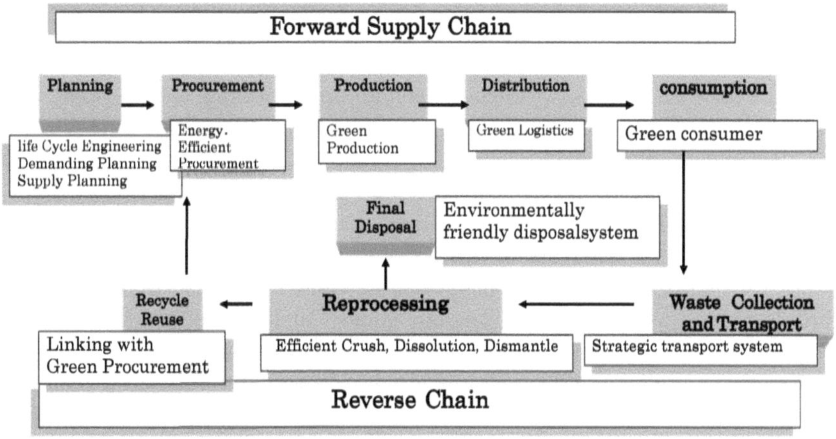

Fig. 1 The scheme of green supply chain and reverse logistics

intermediate treatment that cannot be reused is collected and transported to a final disposal site.

Consignment transportation and the companies' own transportation efforts are two methods of transporting industrial wastes. When transporting industrial wastes across prefectural borders, some prefectures and government ordinance-designated major cities have established prior consultation arrangements. The reason for the establishment of these arrangements is that, when the collection and transportation of industrial wastes extends over a wide area, the risk of illegal dumping, or some other unorthodox or illegal disposal method, increases. Consequently, the agreement and cooperation of the residents in the related regions is often difficult to acquire.

When industrial wastes are collected and transported within a small area, oftentimes transshipment storage and intermediate treatment facilities are not located in that same area and effective waste treatment cannot be achieved. Moreover, it is difficult to successfully introduce modal-shift transportation into reverse logistics and cooperative collection over a wide area.

3 Characteristics of Reverse Logistics

Several differences between the characteristics and constraint conditions of forward logistics and reverse logistics systems can be found. For modern forward logistics, short lead-times, delivery time appointments, accurate inventory strategies, thermal management, and the consolidation of physical distribution complexes are indispensable conditions needed to improve logistical efficiency. At the same time, these considerations present serious constraint conditions for the deployment of high-grade logistics. However, these constraint conditions are not necessarily applicable to reverse logistics.

In the case of reverse logistics, since commercial goods are not delivered, the lead-time for each process of collection, demolition, crushing, and recycling is not very strict. Since the inventory value of the wastes in the transshipment storage facilities is largely affected by market conditions, the accuracy of inventory strategy is also lower in reverse logistics than it is in forward logistics.

The more delicate aspects of handling operations are also less pressing in reverse logistics than they are in forward logistics; similarly, the constraint conditions for thermal management in reverse logistics are not as strict. Unlike forward logistics, which handles valuable goods, such as commercial products, competition typically plays a less important role in reverse logistics, since it handles waste articles and discarded materials. Compared to forward logistics, the need for business confidentiality in reverse logistics is also lower; unlike companies transporting valuable goods, the number of reverse logistics clients who require operators to keep waste information confidential so as to protect their business from their competitors is small.

When the construction of a reverse logistics network is considered for a single region or for multiple regions, it is often first optimized in a single region. In other words, although a reverse logistics network for each industrial accumulation area has been constructed, connecting multiple industrial accumulation areas is not often pursued.

4 Improvement of Reverse Logistics

4.1 Introduction of a Modal Shift

A modal shift, shown in Fig. 2, can be introduced into forward logistics as well as into collection logistics and reverse logistics. As mentioned above, in reverse logistics the lead-time is comparatively long and the necessity for frequent shipments of small lots and the scheduling of delivery appointments is comparatively small. Therefore, transportation costs can be markedly reduced when railways and ships are used instead of trucks.

4.1.1 Example of Introducing Reverse Logistics

The Council for Promoting Reverse Logistics in the Seto Inland Sea has been promoting reverse logistics services in the Seto Inland Sea and pursuing the construction of a green logistics system for recyclable resources. The reverse logistics services include cooperative transportation of wastes from Kansai to Kitakyushu using ships in the Seto Inland Sea. By introducing cooperative distribution and

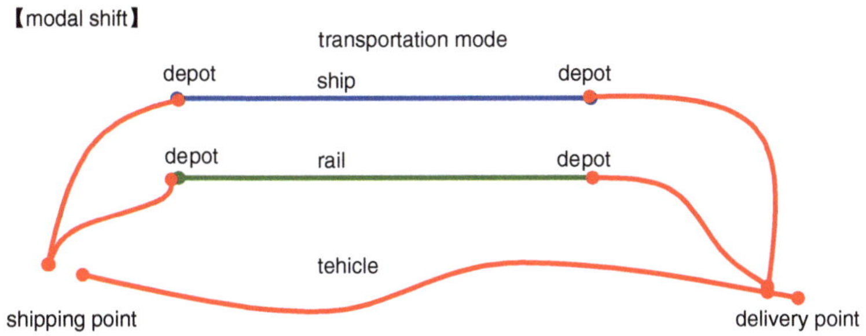

Fig. 2 Image of a modal shift

modal-shift transportation into reverse logistics, CO_2 emissions can be reduced by 75 % compared to conventional individual land transportation methods. The plan to construct a reverse logistics system has been pursued since the fiscal year 2004 due to the grant-aided project funded by the Ministry of Economy, Trade and Industry and the Ministry of Land, Infrastructure, Transport and Tourism in cooperation with the environment and port and harbor departments of Okayama, Hiroshima, Yamaguchi, Kagawa, and Ehime Prefectures. Transportation costs and CO_2 emissions can be reduced by employing cooperative transportation of large amounts of wastes using ships.[1] Although the average land transportation cost of wastes in the area of the Seto Inland Sea was calculated to be 2,900 yen/tonne (except for the loading and unloading cost), the cost of using ferries was calculated to be 2,100 yen/tonne; i.e., the transportation cost was confirmed to be reduced by using ferries.[2]

Since the wastes that are to be transported are recyclable, they can be transported using containers. In addition, while there are no restrictions on the types of waste products that can be transported using containers, depending on the shape or the amount of the waste, it might not be possible to transport some of it.

The main wastes to be transported are raw materials for cement (cinders, incineration ashes, and sewage sludge), carbonized materials for fuels (waste plastics, discarded tires, and wood chips), and scrap metals, waste oil, waste acid, waste water, and sludge for recycling. For environmental consideration, these wastes are transported in roof-open types of hermetically sealed containers, so that the wastes can be directly dumped into the containers to prevent scattering and to insure that the odor from the wastes is not released.

In order to cooperate with the requirements of the Japan Freight Railway Company, originally the standard length of the container was thought to be 12 feet

[1] The transportation division of the Council for Promoting Reverse Logistics in the Seto Inland Sea (fiscal year 2007) consisted of Tsukiboshi Logistics (representative company), Ecosystem-Japan, and Daiei Cleaner; the recycling division consisted of Ube Industries, Eco-System Sanyo, Kyoei Steel, Tokuyama Corporation, Nisshin Steel, Recycling Management Japan, and Mitsui Mining and Smelting; and the secretariat consisted of Chuden Engineering Consultants.

[2] The master conception was examined in the fiscal year 2004.

(5 deadweight tonnes). In the fiscal year 2009, the standard length of the container was changed to 20 feet (20 deadweight tonnes). Ferries were initially selected as transport ships because of the ease in loading and unloading and because of the low total cost. However, after 20 foot containers were introduced, the carrying capacity of the ferries decreased. Therefore, barges were used instead of ferries. Since companies were allowed to use ports equipped with public cranes, barges became the most cost-effective means of transportation.

The following method is used to transport the wastes:

1. the wastes are loaded into roof-open types of hermetically sealed containers at several factories;
2. collection and transportation enterprises transport the containers from several factories to a port equipped with public cranes;
3. after being retained at the port, the containers are loaded onto barges using cranes and transported;
4. after arriving at another port equipped with public cranes, the containers are loaded onto trucks using cranes and transported;
5. after arriving at a recycling factory, the containers are unloaded.[3]

4.2 Cooperative Reverse Logistics

The term "*cooperative forward logistics*" is defined as the collaboration between multiple companies for the cooperative construction of a transportation network for logistics functions, such as transportation, storage, and information systems. Moreover, participating companies also occasionally share transportation and delivery bases and logistics support information systems.

Cooperation is considered an effective option for both forward logistics and reverse logistics. If manufacturers or waste-discharging enterprises work cooperatively to collect used goods, the cost of reverse logistics can be reduced and efficiency can be improved.

Moreover, if a cooperative collection system is realized for reverse logistics, the illegal dumping of wastes may be prevented. In reverse logistics, when the distance between the collection and transportation sites is excessive, the risk of illegal dumping is said to increase. Therefore, by pursuing reverse logistics cooperatively, a system can be developed to check illegal dumping. For example, to effectively advance the reuse of wastes discharged from construction sites, the Kanto Regional Development Bureau, the Ministry of Land, Infrastructure, and Transport of Japan, has examined the construction of a cooperative collection system.

[3] See the website of the Council for Promoting Reverse Logistics in the Seto Inland Sea (www.setouchi-green-butsuryu.jp) and the brochure, "Reverse Logistics Services in the Seto Inland Sea".

5 Conclusion

Implementing effective methods for forward logistics, such as modal-shift transportation and cooperative logistics, can reduce a significant amount of carbon dioxide emissions. However, in order to introduce these methods into reverse logistics, a reverse logistics network between areas must first be constructed.

A Consideration on an Effective Reverse Logistics System for Discarded Tires

Kuninori Suzuki, Nobunori Aiura and Yutaka Karasawa

Abstract In this chapter, which focuses on used motor vehicle tires, the possibility of constructing a reverse logistics network over a wide area and the integration or aggregation of logistical bases is examined. This simulation includes a collection system algorithm and cluster-first/route-second method, and we use local research. This procedure consists of three factors. The first is the collection of used motor tires and their transportation to factories as thermal fuels. The second is the improvement of the actual reverse logistics system for used motor vehicles. The final factor is the design of the improved reverse logistical system. Here, the possibility of integrating the reverse logistical bases through a computer simulation with a collection system algorithm and cluster-first/route-second method is discussed using local research.

Keywords Reverse logistics · Used tires · Forward logistics · Collection and transport

1 Introduction

1.1 The Purpose of this Study

This study aims to analyze the present situation for collection of used tires in a series of flows in a reverse logistics network and to simulate cooperative reverse logistics. Many used automobile tires, i.e., used tires, are collected from gas

K. Suzuki (✉)
Nihon University, Izumicho, Narashino, Chiba Prefecture, Japan
e-mail: QWL02764@nifty.ne.jp

N. Aiura
Hokkai School of Commerce, Izumicho, Narashino, Chiba Prefecture, Japan

Y. Karasawa
Kanagawa University, Izumicho, Narashino, Chiba Prefecture, Japan

P. Golinska (ed.), *Logistics Operations, Supply Chain Management and Sustainability*, EcoProduction, DOI: 10.1007/978-3-319-07287-6_20,

stations, tire dealers, and wreckers. Company A, which has provided various data for use in this study for the simulation, performs intermediate treatment of collected used tires and recycles them as fuel chips in its own factories, and then distributes the fuel chips to paper and steel mills. This study reproduces this reverse logistics network for used tires and investigates the present situation and problems, and simulates future possibilities.

2 Definition and Range of Reverse Logistics

2.1 Reverse Logistics in Previous Studies

To simulate the reverse chain, firstly to clarify the definition and range of reverse logistics is necessary. Schultz (2002, pp. 3–4), in his description of reverse logistics, indicated that *although the collection process was recently added to a reference model for supply chain operations, it will be more important for supply chain management (SCM) in the future.* De Brito et al. (2003) described that *reverse logistics has spread across the world, and it will be involved in a hierarchy of various supply chain divisions.* The range of reverse chain or reverse logistics is defined as a series of processes, including collection, recycling, and final disposal of products after leaving the site of waste generation. In other words, the range indicates a series of flows, including collection, transportation, intermediate treatment, and final disposal of wastes, and the ultimate distribution of recycled products to recycling markets.

Fleischmann et al. (2003) pointed out that linking reverse logistics to a recycling system creates difficulty in defining the range of reverse logistics. In particular, when also taking into account the reprocessing, recycling, and reuse after collection and transportation of wastes, the range of the forward supply chain of products is considered to have a broad overlap with that of the reverse chain of wastes.

2.2 Range of Reverse Logistics of Used Tires

The reuse of automobile tires (i.e., for used tires to be returned to tire markets as retreaded tires after a recycling process) is currently very low. However, after collection from gas stations and wreckers, many used tires are given an intermediate treatment for conversion to fuel chips, and the fuel chips are used in boilers in cement factories and paper and steel mills as a form of thermal recycling.

In 1994, Japan Automobile Tire Manufacturers Association (JATMA) unified the names of used tires into *discarded tires.* Since fuel chips are valuable materials, their transportation and distribution belong to forward logistics instead of the reverse logistics performed by collection and transportation enterprises. However, as the range of the forward supply chain of products is considered broadly overlap that of the reverse chain of wastes, as mentioned above, this study adopts the idea that fuel chips belong to reverse logistics.

Actually, company A, which has provided various data to this study, mainly collects and transports used tire, or discarded tires. As forward logistics, the company transports and distributes only products from discarded tires, which have undergone intermediate treatment in its own factory for thermal recycling.

Figure 1 shows a basic flow for discarded tires. When enterprises such as gas stations, car shops are sources of discarded tires, the discarded tires are treated as industrial wastes. When collecting discarded tires, a person in charge of discarded tires in the source enterprise is obliged to be present at the collection point and the collection is performed using a manifesto. Therefore, the time required for collecting discarded tires in reverse logistics tends to be longer than that required for distribution in forward logistics, in which tires are only delivered. Enterprises that collect and transport discarded tires must establish a collection plan that considers these matters.

In the case where the same enterprise performs collection, transportation, and intermediate treatment of discarded tires, as well as distribution of fuel chips derived from discarded tires to paper and steel mills as thermal recycling, the enterprise must be engaged in the area of reverse logistics as well as of forward logistics, including storage and distribution of the recycled products.

3 Outlines of Analysis and Simulation of the Actual State

Reproducing the actual state of the company A's collection system for discarded tires and performing a scenario analysis of the future possibility of the collection system allows investigation of the present situation. It also allows exploration of further possibilities for the collection system for discarded tires and the measures needed to realize an efficient reverse logistics network.

Company A, which recycles discarded tires at Samukawa, Chigasaki, Kanagawa Prefecture, has licenses for collection, transportation, and intermediate treatment of discarded tires, and for transportation in forward logistics. The company recycles discarded tires into fuel chips, which it then sells to major paper and steel mills. Discarded tires are collected from tire dealers, auto wreckers, gas stations, and auto accessory stores in the Kanto district and then converted into fuel chips in the company's recycling factory.

The fuel chips are then directly transported to paper mills using 10-ton trucks or to steel mills using ships from the Port of Kawasaki. The Kanto district includes parts of Tokyo Metropolis, Chiba Prefecture, and Saitama Prefecture, centering on Kanagawa Prefecture.

A 4-ton truck collects 1,000 discarded tires from tire dealers, gas stations, car shops, and auto accessory stores every day. Since these tires cannot be collected all at once along the truck's collection route, they are collected two or three times.

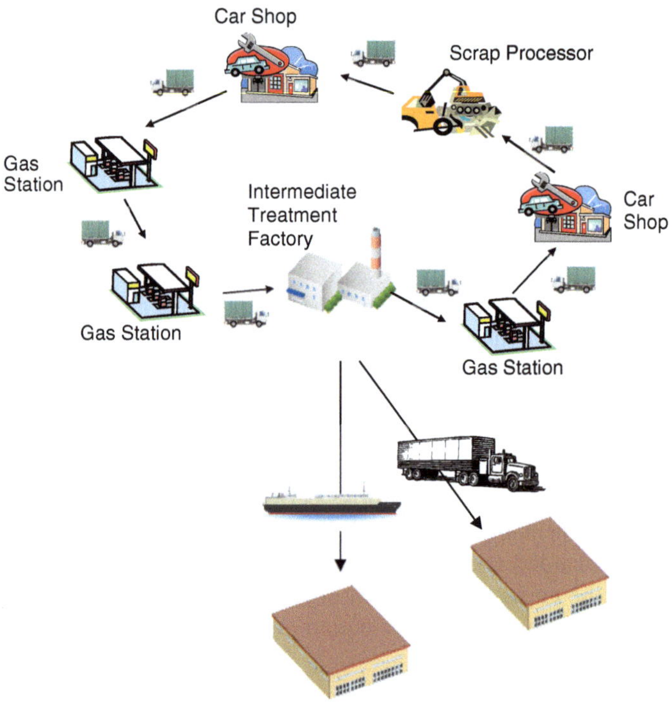

Fig. 1 Material flow of used tires

The weight of a tire is approximately 8 kg, and the tires are collected using boxes (i.e., boxes filled with tires are exchanged for empty boxes). Although a maximum of 200 tires can be put into each box, in many cases only about 150 tires are put into the box because the tires are not stacked neatly.

The business scheme of Company A can be divided into two systems (i.e., a collection system for discarded tires and a transportation and distribution system for fuel chips after collection and intermediate treatment of discarded tires for thermal recycling).

In this chapter, "Supply Chain Integration in View of Secondary Raw Materials" and "Integration Level Measurement System in Modeling Forward and Backward Supply Chains" describe the collection system for discarded tires and the transportation and distribution system for the fuel chips, and outline the simulation models for the two systems.

As Fig. 2 shows, the simulation model used in this study is composed of three echelons: the collection complex, the transshipment and storage complex, and the intermediate treatment factory. Based on the management situation of company A's collection system for discarded tires and conditions set for the scenario analysis, the following three desirable situations are investigated from the viewpoints of the environment, load, and cost of the entire collection system, using a simulation model:

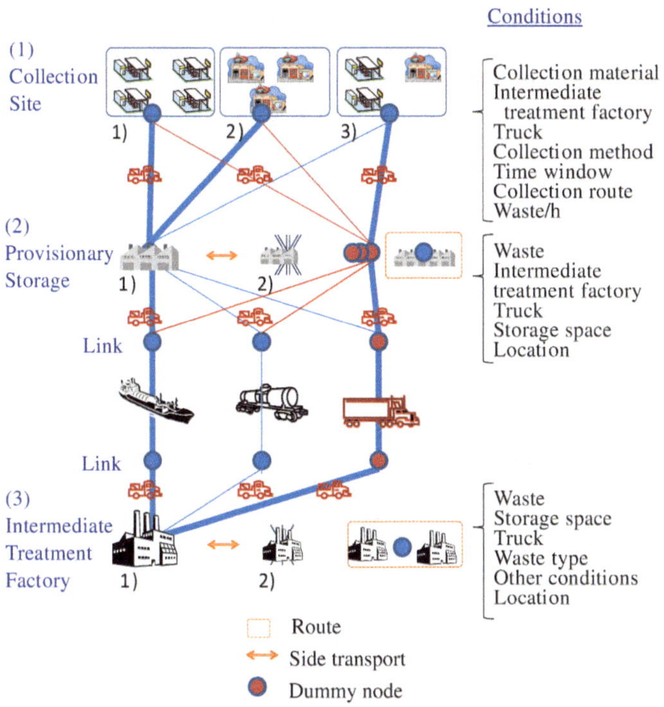

Fig. 2 Setting of echelons for used tires recycle system

(a) Desirable situations of complexes (nodes) belonging to each echelon.
(b) Desirable situation of movement (links) between complexes belonging to the same echelon.
(c) Desirable situation of movement (links) between complexes belonging to different echelons.

When the entire area of Chiba Prefecture is assumed to be the future collection area, establishing a transshipment and storage complex (temporary storage complex) in Chiba Prefecture and performing batch transportation between the temporary storage complex and an intermediate treatment factory in Kanagawa Prefecture may be more efficient than consolidating collection complexes into one or two in Kanagawa Prefecture and performing real-time collection.

4 Outlines of Data on Collection Points

At present, company A has 150 collection points for discarded tires in the Kanto district, centering on Tokyo Metropolis and Kanagawa Prefecture. In the future, the company will introduce a cooperative collection system and extend the collection area for discarded tires, concentrating on Chiba Prefecture. Assuming an increase in the number of collection points to 650, data on potential collection points are prepared in addition to data on existing collection points. A simulation is performed based on consolidation of complexes and information on collection points in the company A's collection area, to be incorporated into the cooperative collection system, and the effect of the increase in the number of collecting points on the operational efficiency is examined.

4.1 Numerical Results

To grasp the actual situation, using 150 site data, the program was run. In the mileage result, 9–11 of 4-ton trucks and 5 of 10-ton trucks are used.

In Table 1, the numerical results of truck types, driving mileage, driving time, working time (at collection sites), the number of discarded tires, and the number of trucks are shown. *DATE* means the day of the week, i.e., DATE: 0 means Monday, and DATE: 5 means Saturday. Concerning cost results by the week, the personnel expenses, the variable expenses, the fixed expenses, and the total cost are also calculated.

The scale merit and cost merit, however, cannot be found with the increase of the number of collection sites. The mileage distance increases, compared to the number of the trucks, and collection lots of discarded tires. The number of collection points that a truck could visit was restricted by the time required for the travel and the number of discarded tires. In the case where a certain number of collections were fixed, the number of vehicles was not affected by the number of collection points.

4.2 Cost-Reduction Measures

As the numerical results in Table 2, the cost merit of the increase in the number of collection points cannot be seen. That is to say, other cost-reduction measures should be considered. It is possible to reduce the working time (at collection sites).

The collection of discarded tires requires lots of time at the sites. However, the effective measures, such as collection infrastructure, can reduce its time and costs. The possible measures are as follows:

- Introduction of RFID system for the inspection of discarded tires at the collection sites,
- Introduction of Electronic Manifesto and Advanced Shipment/Collection Notice System for efficient procedures of collecting discarded tires at the collection sites,
- Introduction of removable containers for reducing the loading time and waiting time at the collection sites.

Table 1 Mileage result of the actual network

Date	Truk type	Driving mileage (km)	Driving time (h)	Working time (at collection sites) (h)	The number of discarded tires	The number of trucks
DATE: 0	4-ton	1,499	50	45	1,999	11
DATE: 1	4-ton	1,212	40	44	1,624	10
DATE: 2	4-ton	1,203	40	45	2,048	9
DATE: 3	4-ton	1,499	50	45	1,999	11
DATE: 4	4-ton	1,212	40	44	1,624	10
DATE: 5	4-ton	1,203	40	45	2,048	9
DATE: 0	10 ton	1,057	35	8	1,638	5
DATE: 1	10 ton	824	27	10	1,220	5
DATE: 2	10 ton	802	27	8	1,298	5
DATE: 3	10 ton	1,057	36	8	1,638	5
DATE: 4	10 ton	824	27	10	1,220	5
DATE: 5	10 ton	802	27	8	1,298	5
	Total	13,193	440	321	19,654	90

Table 2 Cost result of the potential network

Date	Truk type	Variable expense (driving) (yen)	Variable expense (at collection site) (yen)	Personnel expense (yen)	Fixed expense (yen)	Total (yen)
DATE: 0	4-ton	127,270	26,182	401,400	168,000	722,852
DATE: 1	4-ton	145,803	25,389	435,800	186,000	792,992
DATE: 2	4-ton	152,394	23,535	441,200	189,000	806,129
DATE: 3	4-ton	127,270	26,182	401,400	168,000	722,852
DATE: 4	4-ton	145,803	25,389	435,800	186,000	792,992
DATE: 5	4-ton	152,394	23,535	441,200	189,000	806,129
DATE: 0	10 ton	12,913	1,325	41,250	25,000	80,488
DATE: 1	10 ton	25,127	1,757	69,000	40,000	135,884
DATE: 2	10 ton	28,914	1,586	74,500	45,000	150,000
DATE: 3	10 ton	12,913	1,325	41,250	25,000	80,488
DATE: 4	10 ton	25,127	1,757	69,000	40,000	135,884
DATE: 5	10 ton	28,914	1,586	74,500	45,000	150,000
	Total	984,842	159,548	2,926,300	1,306,000	5,376,690

Based on these cost-reduction measures, the scenarios for the simulations are made. Compared to the actual network, the costs for each scenario can be reduced. *Full complement* includes RFID system, electronic manifesto and ASN, and removable containers (Table 3).

Table 3 Cost result of the measures

Cost reduction measure	Total cost (yen)
Actual state	1,306,250
RFID system	1,221,332
Electronic Manifesto and ASN	1,128,018
RFID system, electronic manifesto, and ASN	1,087,960
Removable containers	1,021,432
Full complement	928,032

5 Conclusion

This study used actual data for the collection system of discarded tires to perform a simulation that clarified the actual state of the collection system, and that pointed out problems that reduced the efficiency of reverse logistics. When a model for the consolidation of distribution complexes, which has been an effective tool to improve the distribution efficiency in forward logistics, was directly applied to reverse logistics, the expected results were not realized, because the time required for tasks at the complexes is longer in reverse logistics than in forward logistics. A cooperative reverse logistics network will require the establishment of a new scheme while utilizing accumulated knowledge in forward logistics.

In the future, the transportation and distribution system for fuel chips should be analyzed for cost reduction, while considering a reduction in environmental load. A simulation for the transportation and distribution system of fuel chips should include a number of factors including collection complexes in overseas regions such as China and South Korea in addition to Japan, intermediate treatment complexes, and consumption areas for the recycled goods.

References

De Brito, Marisa P, Dekker R (2003) A Framework for reverse logistics, Reverse logistics, Springer, Heidelberg, pp 3–27

Fleischmann et al (2003) Reverse logistics network design, Reverse logistics, Springer, Berlin, pp 68–69

Schultz G (2002) The recurring side of supply chain management, APICS—The performance advantage, APICS, Oklahoma

Analysis of Effective Recycle System for Used Personal Computers in Japan

Akihiro Watanabe, Kuninori Suzuki, Keizo Wakabayashi and Yutaka Karasawa

Abstract Regarding the lifecycle of wastes/valuable materials, the efficiency of a strategic flow, including collection, transportation, intermediate treatment, and final disposal/recycle is expected to improve from the viewpoint of environmental symbiosis. A forward logistics network is constructed mainly of physical distribution enterprises. In the case of reverse logistics, however, collection, transportation, and intermediate treatment enterprises, elements that are not involved in forward logistics play the main roles in the network's construction. Industrial wastes discharged from factories or similar facilities are collected and transported to intermediate treatment facilities for recycling. Therefore, a network that covers forward and reverse logistics is difficult to construct. Moreover, although the efficiency of forward logistics has greatly improved recently, no noteworthy progress has been seen in the efficiency of reverse logistics, even though there are several differences in the characteristics and constraint conditions between forward and reverse logistics. In this chapter, the possibility of constructing a reverse logistics network over a wide area is examined. As the example for this study, the chapter analyzes the current issues and to propose an effective reverse logistics system for used personal computers. An effective system to collect used personal computers over a wide area with a correct time window is required. The computer simulation includes a collection system algorithm with cluster-first/route-second method and local search method. This procedure consists of three factors. The first is the collection of used personal computers and their transportation to the logistics centers. The second is the improvement of the actual reverse logistics system for used personal computers. The final factor is the design of the improved reverse logistical system. This chapter clearly points out the importance of collection of used personal computers over a wide area with a correct time window with a numerical experiment.

A. Watanabe (✉) · K. Suzuki · K. Wakabayashi
Nihon University, Izumicho Narashino City, Chiba Prefecture, Japan
e-mail: QWL02764@nifty.ne.jp

Y. Karasawa
Kanagawa University, Yokohama, Kanagawa Prefecture, Japan

P. Golinska (ed.), *Logistics Operations, Supply Chain Management and Sustainability*, EcoProduction, DOI: 10.1007/978-3-319-07287-6_21,

Keywords Reverse logistics · Used personal computers · Collection and transport · Recycle · Intermediate treatment

1 Introduction

Recent years have seen a movement away from a society based on mass-production, mass-consumption, and large-scale disposal of products toward a society with environmentally sound material recycling options and a smaller environmental burden. In many industries, however, the introduction of an efficient reverse logistics system is still urgently required.

This chapter clearly identifies and analyzes the current need for an effective collection and transport system for used personal computers and proposes a reverse logistics system that can be implemented over a wide area. Toward that end, this chapter presents a computer simulation that relies on a numerical experiment to recommend collection site arrangements and time window adjustments. The study's results reveal that the suggested reverse logistics improvement measures can function effectively.

2 Recycle Market for Used Personal Computers

In this chapter, the actual situation and the issues of used personal computers in a reverse logistics system are analyzed and considered. Figure 1 shows the collection route for used personal computers. As can be seen, the used computers discharged from private homes and business offices have been increasingly collected for recycling via electronic retail stores. However, used personal computers can be disposed of in a variety of way and the computers can then be distributed as valuable items. For example, used personal computers found in private homes can be directly returned to major PC manufacturers. Used personal computers are also collected through the 3R promotional program, which is offered by the General Incorporated Association of Personal Computers and other companies that collect waste materials. Nevertheless, if personal computers are not recycled, many end up being stored in private homes, in landfills, or in dead storage facilities; the total accumulated number of personal computers that meet that fate could be as high as 20 million in 2015. There are several reasons why so many personal computers may end up in dead storage instead of being recycled; first, personal computers are compact enough to store in a closet at home and second, personal computers contain a great deal of personal information that cannot be easily eliminated or erased from the hard drive. Additionally, the period from dead stock to discharge tends to be rather long; even though the typical product cycle of a personal computer is not long; on average, the product cycle is 10.82 years from purchase to discharge (Fleischmann et al. 2003).

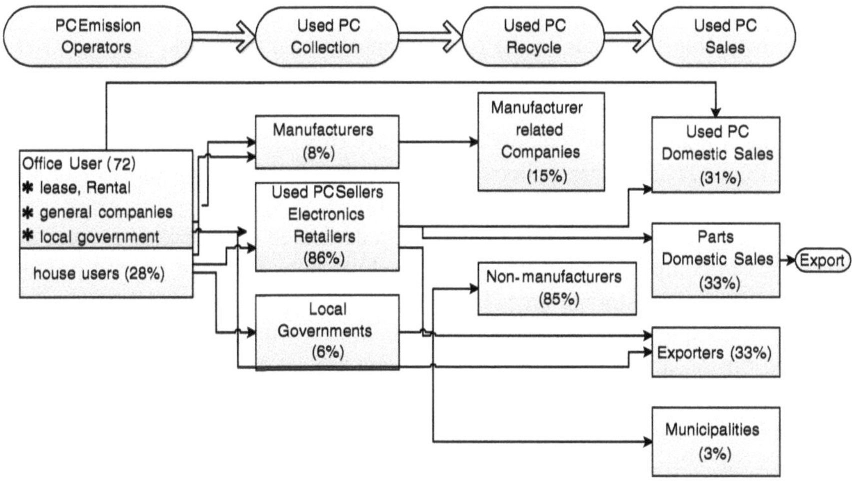

Fig. 1 Used computers collection route

On the other hand, the demand for used personal computers in the domestic market is continuously strong; in fact, it has grown more than 10 % a year; and, the proportion of note book personal computers is especially high (Suzuki 2010a, b). In such circumstances, in order to prepare for the intensification of price competition in the domestic market where deflation has progressed, major electronic retail stores have started to improve the efficiency of trading in and recycling used personal computers, either directly through their stores or through a collaborative effort with partner companies. As specifications and performance of personal computers have improved every year, the replacement of a personal computer seems to be a significant concern for consumers when considering how to discharge their old products. That is, if more used personal computers are traded in when a new computer is purchased, the number of used personal computers in dead storage would certainly decrease. On the basis of this background information, this chapter analyzes the recycling system of used personal computers that are traded in at electronic retail stores.

3 Time Window for Collection and Transport

Used computers are primarily collected directly from offices and private households. Recently, however, there have been a growing number of cases in which electronics retail stores have launched a trade-in system for used computers that customers can use when they purchase a new computer. Used computers brought to the electronics retail stores under the trade-in system are collected by waste collection and transportation service providers. Personal data are erased from the computers at a reuse factory, parts are replaced or repaired, if necessary, and they become available for resale in the used computers sales network.

In Japan, used computers are normally collected from electronic retail stores in the mornings, according to the waste collection custom mentioned above. However, if the collection time window doesn't have to be fixed in the mornings, and if it is possible to schedule the collection time for the afternoon, or from 9 to 17, then it becomes possible to achieve an efficient collection system.

To realize the 9–17 collection timeframe, a dedicated or temporary storage area for the waste computers needs to be prepared separate from the general products pickup/delivery and storage area. Furthermore, a diagram collection system that indicates the expected arrival time of a vehicle needs to be introduced.

Based on the above considerations, this current study uses a computer simulation to show how the efficiency of the waste collection system can be improved if waste collection is scheduled all day instead of just during the mornings.

4 Outlines of the Simulation

Company A, a recycling company, collects trade-in used computers at major electronics retail stores. Distributive processing, including parts replacement and data erasing, is performed at the company's three distribution centers in Tokyo, Saitama, and Tochigi prefectures. Company A then places the computers into the used computer sales network and sells them as a wholesaler and as a retailer. Here, used computers are collected as valuables and although no license is required to collect and transport them, Company A has secured a waste collection and transportation license so it can deal with the potential risk that these computers might not be defined as valuables in the future due to changing market conditions. For this chapter, an efficient collection system used to collect used computers at 244 electronics retail stores in Tokyo, Saitama, Chiba, and Tochigi prefectures is examined, as shown in Fig. 2. The electronics retail stores are all clients of Company A. The changes in collection efficiency are simulated by changing the depot arrangement and the collection time windows at a collection site.

The purpose of the simulation is (1) to streamline the number of collection sites and (2) to improve the collection time of waste collection logistics.

Specifically, the study shows whether collection efficiency can be improved by aggregating three collection depots into one depot, with the prospect of a wider collection area. The study also shows how much improvement can be made in collection efficiency when the collection time is changed from the current morning hours (9–12) to all day (9–17).

Company A's 244 collection sites are divided into the morning collection (9–12) group and the all-day collection (9–17) group. The collection efficiency between the two groups is then compared. For this study, it is assumed that collection is made every day except Mondays. It is also assumed that the

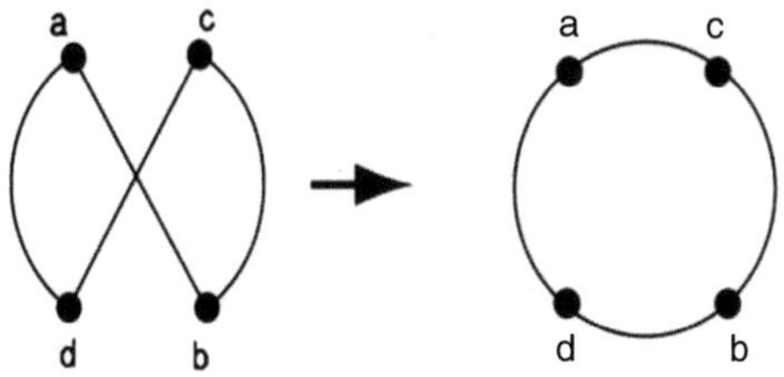

Fig. 2 The 2-opt method reference drawing

collections are made from 244 points with three distribution centers in Tokyo, Saitama, and Tochigi serving as the collection depots.

It is assumed that the collection is done once a week (Tuesday through Friday) at a specified time (9–17), and that the collection time windows are not set by the collection site. With regard to the conditions at each collection site, it is assumed that there are no restrictions as to the vehicle size, and the loading time is set to 20 min, based on a survey of the personnel involved.

Although there are no restrictions on the departure and arrival times at any of the depots, some restrictions do occur, depending on the time constraints at the collection sites. Each collection site is covered by the closest distribution center (collection depot) and it is assumed that there is no interoperation between the centers according to the load. A collection with three centers and a collection with one center are simulated.

As for the routing plan, random numbers are generated using Excel's Rand function, and these are used to create a routing group among the adjacent wards. A Hamilton path is used to chart the collection sites and the depots for each of the groups; moreover, each group is assigned a driver who can operate a four-ton truck.

5 Adoption of the 2-opt Method

As a simulation model, the 2-opt method is used. The collection sites are divided into several groups and the total amount to be collected from the collection sites within the group is arranged so as not to exceed the loading capacity of the trucks.

Here, the 2-opt method is an algorithm, as follows:

1. Create an arbitrary tour.
2. Remove two edges (*ab*, *cd*) from the tour.
3. Reverse one of the edges that is created.
4. If $ac + bd < ab + cd$, then connect *ac* and *bd*.
5. Remove edges until there are no edges that satisfy (4).
6. Repeat (2) through (4) until the end condition is satisfied.

6 Simulation Results

6.1 Simulation Concerning Depot Arrangement

The collection conditions, based on the number and location of the depots, are compared. The collection sites for one week are known in advance and the vehicles are allocated on a weekly basis. Only the number of depots and their locations are changed.

In general, if the distance from a depot to a collection site is close, the distance required for collection becomes shorter and individual the traveling distance and traveling time are both reduced. On the other hand, if there are more depots, the number of collection sites per a single depot will be small, reducing the economies of scale.

The simulation results are shown in Table 1. In terms of vehicle operation, three depots that are located close to one another show less efficiency than one depot due to the reason mentioned above. When depots are located at an appropriate distance from one another, delivery efficiency is improved. Therefore, from the view point of collection efficiency, covering the collection area with three depots rather than one depot reduces the total travel distance and the vehicle operation requirements; consequently, it also reduces the collection cost.

6.2 Comparison of the Methods Used to Determine the Collection Routes

In reality, the collection sites are not known prior to the collection day. If the collection sites are to be determined in advance, it is possible to prepare a more efficient collection route. Therefore, we have simulated the following two cases and compared the results.

Case 1 The details of the collection for the week are determined in advance and vehicle allocation can be made. Vehicle allocation simulation for a week is done at one time and for each day; a route is assigned later based on available manpower.

Case 2 The collection sites are not known until the day before the collection date. The collection sites are divided into four groups using random numbers, and each group is simulated as the collection site for the given day of the week.

Table 2 shows the simulation results. In comparison to Case 1, the number of collection sites for one simulation is 1/4 in Case 2, and the economies of scale are not easily realized and vehicle allocation efficiency is reduced.

However, in Case 2 the collection efficiency may be higher with one depot for some situations. Furthermore, in terms of operation, the result is expected to

Table 1 Simulation result concerning depot arrangement

Condition	Plan	Number of routes	Operation minutes	Operation average time (minutes)	Traveling distance (km)	Traveling distance (average/km)
1 Depot: 9–17	Fixed plan	18	11,073	615	4,051	225
3 Depots: located to close to one another: 9–17	Fixed plan	20	11,230	562	3,914	196
3 Depots: 9–17	Fixed plan	19	10,521	554	3,156	166
Uniform distribution	Fixed plan	18	10,413	579	2,993	166

1 depot collection at one depot; *3 depots located close to one another* collection when three depots are arranged close to one another; *3 depots* collection when three depots are arranged at an appropriate distance from one another; uniform distribution: collection when three depots are distributed uniformly

Table 2 Simulation of comparison of methods for determining collection route (morning)

Condition	Number of routes	Total traveling distance (km)	Total traveling time (minute)
3 Depots (9–12): fixed plan	45	4,893	7,750
3 Depots (9–12): flexible plan	61	7,492	11,219

Table 3 Simulation of comparison of methods for determining collection route (all day)

Condition	Number of routes	Total traveling distance (km)	Total traveling time (minute)
3 Depots (9–17): fixed plan	19	3,156	5,641
3 Depots (9–17): flexible plan	30	6,187	9,844
1 Depot (9–12): fixed plan	43	6,660	8,873
1 Depot (9–12): flexible plan	56	9,680	12,801

change depending on the number of collection sites and whether the collection sites are flexible about the collection date.

From the simulation results, in the case in which collections are made at three depots (9–17), the travel distance in Case 1 is 3,156 km, as shown in Tables 2 and 3, whereas in Case 2, the travel distance is 6,187 km, resulting in a 49 % reduction in the distance traveled. In the case in which morning collection is possible, the travel distance in Case 1 is 4,893 km, whereas in Case 2 it is 7,492 km, resulting in a 35 % reduction in the distance traveled.

6.3 Collection Time

A simulation is performed to examine how the vehicle allocation efficiency changes depending on whether the collection is performed in the morning only (9–12) or all day (9–17).

Generally, if the collection time window is short, the number of collection sites that can be visited in one route will be small. The number of routes will then increase and the collection efficiency will decrease.

Since the collection includes the travel time to and from the collection sites, the longer the distance between a depot and a collection site, the lower the efficiency will be. Therefore, the effect of the reduction in the collection time window is larger when the number of depots is smaller.

The simulation of the comparison of the collection times revealed that the total travel distance is 9,680 km when the collection route is determined every day under the current three depots, whereas the total travel distance is 4,039 km when the collection route is determined in advance and the collection is made between 9 and 17. This indicates that the total travel distance can be reduced by 58 %. Similarly, in terms of vehicle operation, the time is reduced from 17,681 min to 11,117 min, enabling a 37 % reduction in the travel time. The number of routes is also reduced from 56 to 18 routes, enabling a 68 % reduction for that variable.

7 Conclusion

In this study, the efficiency of collection logistics of used computers is investigated from the perspectives of depot arrangement and improvement in collection time, using a computer simulation. The results revealed that instead of merely aggregating collection depots, it is better to have several depots and to plan efficient collection routes avoiding a route that is longer than necessary.

As for the collection time window, the travel distance for an all day collection is nearly half of the travel distance for a morning collection and significant improvement can be made in efficiency if an all day collection system is used.

Furthermore, when preparing a collection route, if a route can be fixed in advance on a per week basis instead of preparing a route daily, the number of routes, routing time, travel distance, and service time, as well as the number of collection sites, can be reduced significantly.

To implement a collection system that uses a fixed route and all day collection, it is necessary to prepare a dedicated or temporary storage area for used computers separate from the general products pickup/delivery and storage area. Furthermore, to minimize the waiting time at each collection site, a diagram collection system that shows the expected arrival time of a vehicle is needed. The formulation of a detailed introductory plan is an issue that needs to be addressed.

References

Fleischmann et al (2003) Reverse logistics network design, reverse logistics, pp 68–69
Suzuki K (2010a) A Consideration of reverse logistics network between regional industrial hubs. In: Proceedings of the 13th J.S.L.S. national conference, The Japan society of logistics systems, pp 15–18
Suzuki K (2010b) Green supply chain no Sekkei to Kochiku, Hakuto Shobo

Optimal Reutilization of the Leased Products in a Closed Loop Supply Chain

Hsiao-Fan Wang and Chang-Fu Hsu

Abstract The life cycle of products has become much shorter because of the rapid technological development. When the products are out of market, it can be attributed to not only the physical deterioration but also the functional obsolescence. Therefore, product recovery remains valuable. Nowadays, take-back legislation based on extended producer responsibility is critical to a company. Moreover, the business models have gradually changed from traditional product selling to functional sales. Leasing is an example of functional sales that provides an alternative business strategy to cope with the uncertainty in quantity, quality, and time for the returned products. In this chapter, we propose an analytical model, namely Periodical Leasing Model (PLM), to solve the reutilization planning problem under a Closed-loop Supply Chain (CLSC) system which includes multiple products with common components in multiple-periods. The PLM is divided into two phases. The first phase is a pricing model in the form of Integer Nonlinear Program (INLP), and the second phase is a production model in the form of Integer Linear Program (ILP). According to the different characteristics of the components, the model can be used to support the manager's decision regarding the optimal combinations of components from different products for reuse and upgrade possibilities. Finally, numerical results are presented to illustrate the procedure and the validity of the proposed model with the sensitivity analysis.

Keywords Remanufacturing/reprocessing · Upgrading · Leasing · Pricing · Closed-loop supply chain · PLM

H.-F. Wang (✉) · C.-F. Hsu
Department of Industrial Engineering and Engineering Management, National Tsing Hua University, No. 101, Section 2, Kuang-Fu Road, Hsinchu 30013, Taiwan Republic of China
e-mail: hsiaofanwang@gmail.com

P. Golinska (ed.), *Logistics Operations, Supply Chain Management and Sustainability*, EcoProduction, DOI: 10.1007/978-3-319-07287-6_22,

1 Introduction

The process of remanufacturing has been in existence since the Second World War. After the remanufacturing, the products are expected to be as good as new. At that time, most of the industries in the developed countries focused on military production. With the increasing burden on the environment, many companies began to invest in remanufacturing for two reasons: First, the people with the environmental consciousness have increased. In addition, the government likewise sets regulations to mitigate the impact on the environment. Second, the recovery management is profitable, as the technology development that the costs of reusing, remanufacturing, repairing, and recycling are lower than the market prices of the products (Östlin et al. 2008). At present, demand for green products has significantly increased. Enterprises not only earn profits, but also obtain a good reputation through producing green products, which enables enterprises to gain a competitive edge in the globalized market.

The life cycle of products, especially electronic products, becomes shorter due to rapid technological development (Ramani 2010). Such short life cycle results in plenty of electrical and electronic wastes as well as the consumption of an enormous amount of natural resources. In 2003, the European Union issued the Directive on Waste Electrical and Electronic Equipment (WEEE) to cope with the increasing electrical and electronic wastes in order to reduce the burden on landfills or incinerators, and prevent hazardous substances from contaminating the environment (Directive 2002). The directive requires that manufacturers actively act on the collection, recycling, and disposal of electrical and electronic products waste as so called the Extended Producer Responsibility. Numerous countries have been striving to minimize wastes (Wang and Huang 2013).

When products are in the market, their short life cycle can be attributed to physical deterioration, as well as functional obsolescence. Therefore, product recovery is still valuable. A large number of common component designs have been developed in recent years due to the diversity of products. Recycled products will pass through the process of disassembly and cleaning. Several components can be reprocessed for reuse, which can then be used for other types of products and effectively reduce wastes and the negative impact on the environment. Decision makers have to consider both the ease of disassembly at the design stage; and also the change in business models to execute easier reprocessing. It can be observed that current business models have gradually changed from traditional product-selling to functional product-leasing. Leasing strategy is suitable for the durable products with the properties of high-price, recovery saving, and high added-value services (Aras et al. 2011).

If the components are still functional, they could be remanufactured or upgraded; otherwise, they will be disposed. Leasing helps companies manage the quantity and the quality of used products effectively for reprocessing; and exercise the Customer Relation Management (CRM).

Although many studies have focused on CLSC problems, it still lacks of a systematic planning on the Product Recovery Management (PRM) for functional sales. Due to short life cycle of products, value of product recovery, and take-back legislation, providing an alternative business model for decision makers is necessary. This is our aim of this study, and the proposed model is called PLM.

2 Literature Review

Due to the environmental awareness, many literatures have been focused on the recycle problem in a CLSC system which is integrated with a forward and reverse flow (Guide and Van Wassenhove 2009). The roles of CLSC include manufacturer, retailer, distribution center, and recovery facility with logistics. The main aspects in closed-loop management are to acquire the economical and environmental benefit. A stable closed-loop logistics is important for companies. Hence, Wang and Hsu (2010) proposed a closed-loop logistic model with a cyclic logistics network problem. While Savaskan et al. (2004) proposed appropriate CLSC models with remanufacturing for Original Equipment Manufacturer (OEM) to achieve the sustainable development; Wang and Huang (2013) presented a disassembly programming for PRM, of which the recovery options include reuse, repair, and remanufacturing.

The functional sale has become more prevalent nowadays, and its emergence is mainly market-driven in current business patterns. Functional sales mean service-oriented product sales. Östlin et al. (2008) proposed seven types of closed-loop relationships for product remanufacturing. Ownership-based type is the one emphasized the relationship between manufacturer and customer. The company product is more in line with customer requirements, which means that mass production becomes mass customization. The key factor for this relationship to success is making the customers feel that the products and services are reliable (Östlin et al. 2008).

Geyer et al. (2007) presented the impact for the cost in the production system under different durability of components and different life cycles of products. Results show that a more durable component is more valuable for recycling. The main difference between durable and non-durable is uncertainty of the timing in the disposal, recovery, and processing (Linton et al. 2005).

When the new products turn into the end of use, it is often caused by the functional obsolescence or physical deterioration. The measure of the recycled components is the degree of Fitness for Extended Utilization (FEU). Based on fuzzy logic, the concept of "*being acceptable for reutilization*" of recycled components can be evaluated by using the values ranging between zero and one (Xing and Lee 2009). We can likewise define the remanufacturing cost function of the remanufacturing cost ratio to calculate the expected cost at each quality level (Guo and Liang 2011). Returned products can be graded and classified into different quality levels through a list of nominal quality metrics (Van Wassenhove and Zikopoulos 2010).

In summary, from the related literatures in a CLSC system with functional sales, remanufacture process, and upgrading process, we have discovered certain issues needed to be investigated, which will be our core of current study and stated in the next section.

3 A Leasing Model in a Closed-Loop Supply Chain System

3.1 Problem Description

Consider a multi-period and multi-product assembly line in a CLSC system. At the beginning of the planning horizon, the potential market demands are estimated by OEM to determine the optimal prices for the leased products on issue a contract. Based on the contracts, OEM then determines the production planning with the reprocessing and upgrading processes in order to satisfy the demand of products.

As shown of the problem framework in Fig. 1. If the system has no end-of-leasing products during the early periods, OEM needs to obtain new components through the production or the procurement in $S1^a$ only. The components are then assembled into the products in *S*2 to satisfy the demands of the next period. When the lease expires after the leasing periods, the products would be returned to *S*3 by the collection center. The products are then disassembled into individual components in *S*4. The indicator of fitness to extended utilization is applied for evaluating the components, which are classified into 'fit', for recycling in *S*5 and 'unfit', for disposal.

We classify the original design of the products as the first generation and the upgraded products as the second generation. OEM can assemble the new products (1st generation) by using newly processed and procured components to satisfy the demands, and assemble new upgraded products (2nd generation) by using newly processed components, along with the procured upgraded components. If OEM has the reprocessed and upgraded components in $S1^b$, then it can obtain the new components through the processing or procurement in $S1^b$ to make up for the reassembling insufficiency. The returned products from the primary market can likewise be upgraded to become products of a new version (2nd generation).

In this study, we develop an analytical model, namely PLM, to analyze such complicated system and to support the manager's decisions on the optimal prices and quantity for the leased products, as well as the optimal combinations of components from procurement, processing, reprocessing, and upgrading in a leasing system.

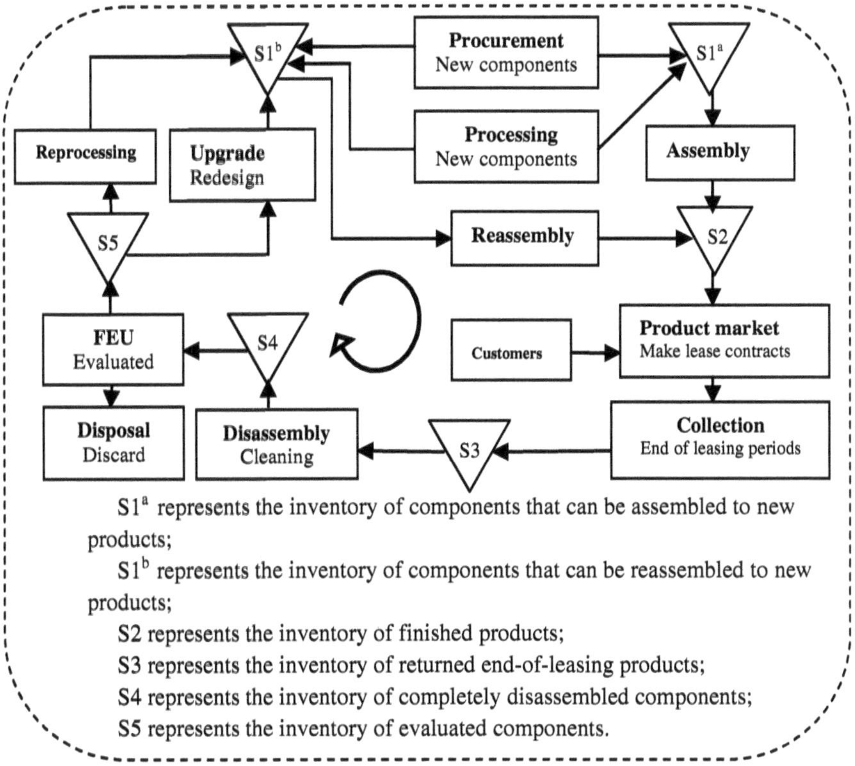

Fig. 1 Illustration of framework and system boundary

3.2 Assumptions

In this section, an analytical model is proposed base on the following assumptions:

1. The demands which are satisfied by new and reprocessed products are known.
2. The cost parameters are all known constants.
3. The length of usage time in market is fixed by the contract.

3.3 Pricing for the Leased Products

For customers, the biggest difference between leasing and purchasing is that they do not have to buy the product, and so, not own the product. Products or services can be used by the customers at the lower threshold of leasing charges. Because the apportioned purchasing price is relatively cheaper than that of leasing, therefore, the price difference between these two would affect the leasing demand. We consider a linear pricing function of the demand of leased product by the leasing price and the difference between the apportioned purchasing price and the leasing price, which shows that the lower apportioned price of the purchased products triggers demand leakage from the leased products to the purchased products. Therefore, the leasing price and its difference from the purchasing price depend on the demand function, as shown in Eq. 1.

$$x_{price}^{t,h} = a - b1^h \cdot x_{price}^{t,h} - b2 \cdot (x_{price}^{t,h} - \frac{h}{DL} \cdot PA) \tag{1}$$

where:

- a is the amount of potential demand, $a \in \mathbb{N}$
- $b1^h$ is the sensitivity for the leased price of product in h periods of leasing time, $b1^h \in \mathbb{Q}$,
- $b2$ is the sensitivity for the difference price between lease and purchase of product, $b2 \in \mathbb{Q}$,
- $x_{price}^{t,h}$ is a decision variable of unit price of the leased product at period t in total h periods of leasing time,
- DL and PA are the designed life span and the purchased price respectively for the product, $DL \in \mathbb{N}$, $PA \in \mathbb{N}$.

3.4 Measures of Fitness to Extended Utilization

Indicator of fitness to extended utilization is composed of Physical Reusability (PRe) and Functional Reusability (FRe), and measured as degrees of both physical and functional suitability of a component, i, as shown in Eq. 2.

$$f_{FEU}(i) = \min[PRe_i, FRe_i]^{1-\max(\mathrm{PRe_i},\mathrm{FRe_i})/2}, \quad i \in \{1,\ldots,N\} \tag{2}$$

Based on the fuzzy concept of '*being acceptable for reuse*', the PRe is defined as (3) [19]

$$PRe_i = \frac{\max[0, e^{-FA^i(WH^i)\cdot r} - RE^i]}{1 - RE^i}, \quad i \in \{1,\ldots,N\}, \quad r \in \{0,1,\ldots,Z\} \tag{3}$$

where

- r is the number of periods in market,
- FA^i is a constant failure rate of component i, $FA^i \in \mathbb{Q}$,
- RE^i is the minimum expected reliability of component i, $RE^i \in \mathbb{Q}$,
- WH^i is the working hours per period for component i, $WH^i \in \mathbb{N}$.

The indicator of FRe is defined as (4) (Xing and Lee 2009) and originates in the fuzzy concept of '*being desirable*'. The Effective Technology Life (ETL) is defined as (5).

Table 1 Score assignment of component functionality level (Xing and Lee 2009)

Linguistic value	Score (FL^i)
Very high	0.90–1.00
High	0.70–0.89
Medium	0.50–0.69
Low	0.30–0.49
Very low	0.00–0.29

$$FRe_i = \frac{1}{1+\exp[-\tau^i \cdot (ETL_i - TC^i/24)]}, \quad i \in \{1,\ldots,N\} \tag{4}$$

$$ETL_i = \max\left[0, \frac{TC_i}{12} \cdot e^{-(1-FL^i)} - \frac{t}{12}\right], \quad i \in \{1,\ldots,N\}, \quad t \in \{0,1,\ldots,Z\} \tag{5}$$

where

- t is the time period,
- TC^i is the technology cycle of component i, $TC^i \in \mathbb{N}$,
- FL^i is the functionality level of component i, $FL^i \in \{FL^i | FL^i \in \mathbb{Q}, 0 \leqq FL^i \leqq 1\}$,
- τ^i represents the slope or steepness of curve at functional membership function for component i,$\tau^i \in \{\tau^i | \tau^i \in \mathbb{Q}, 0 \leqq \tau^i\}$.

The functionality level in Table 1 is the quality to function compared with the enterprise benchmarks (Xing and Lee 2009). If the component is assessed as the best design in the market, then the linguistic value of the component would be given higher, and its score is closer to one. This case represents that ETL is closer to the technology cycle. On the contrary, ETL is actually smaller than the technology cycle for the component with the low scores.

3.5 Classification of the Quality

To obtain the minimum threshold value of each quality level for the recycled components, we are based on the reprocessing cost of nominal quality metrics (Van Wassenhove and Zikopoulos 2010) to provide the appropriate range of FEU value for each classification. The degree of FEU is based on the concept of '*being acceptable for reutilization*', and the reprocessing process aims to prolong the use-life for the returned products. Therefore, we use the cost ratio to obtain the interval of quality level. The FEU value is divided into several intervals. Quality level k for component i in u status is defined as Eqs. 6–8, where $C^i_{repr,k}$ is the cost of reprocessing component i in u status at quality k level.

- The interval of quality level 1 in FEU:

$$\left(1-\frac{\sum_{k=1}^{k} C^{i}_{repr,k,u}}{\sum_{k=1}^{G} C^{i}_{repr,k,u}},1\right], \quad \begin{array}{c} i\in\{1,\ldots,N\},\\ k\in\{1,\ldots,G\},\\ u\in\{0,1\}\end{array} \tag{6}$$

- The interval of quality level 2 $\sim$ G-1 in FEU:

$$\left(1-\frac{\sum_{k=1}^{k} C^{i}_{repr,k,u}}{\sum_{k=1}^{G} C^{i}_{repr,k,u}},1-\frac{\sum_{k=1}^{k-1} C^{i}_{repr,k,u}}{\sum_{k=1}^{G} C^{i}_{repr,k,u}}\right], \quad \begin{array}{c} i\in\{1,\ldots,N\},\\ k\in\{1,\ldots,G\},\\ u\in\{0,1\}\end{array} \tag{7}$$

- The interval of quality level G in FEU:

$$\left(0,1-\frac{\sum_{k=1}^{k-1} C^{i}_{repr,k,u}}{\sum_{k=1}^{G} C^{i}_{repr,k,u}}\right], \quad \begin{array}{c} i\in\{1,\ldots,N\},\\ k\in\{1,\ldots,G\},\\ u\in\{0,1\}\end{array} \tag{8}$$

3.6 The Proposed Leasing Model

Based on the problem we described in the previous sections, we develop a mathematical model with the notations, solution procedure, two phases of PLM, and model properties defined in Tables 2 and 3:

To facilitate solution, a two-phased modeling structure is proposed. At the first phase, the input data are potential demands, and its output data are demands and prices for the leased products which are the input data of the second phase. The structure is shown as Fig. 2.

The first phase of PLM is to decide the leasing price to issue the contracts. The pricing model in the form of integer non-linear program (INLP) is developed as follows (Tables 4 and 5):

3.6.1 1st Phase: Pricing model

Objective function

Max LRV (Leasing Revenue)

$$\sum_{t}\left[\sum_{h}\left[\sum_{u}\left[\sum_{l}\left[(x^{l,t,h}_{dema,u})\cdot(x^{l,t,h}_{pric,u}+\pi^{l,h}_{u})\right]\right]\right]\right] \tag{9}$$

Table 2 Parameters

Parameter	Structure of product
$A^l_{i,u}$	The number of component i in u status of product l, $I = \{1, 2, \ldots, i, \ldots, N\}$ $L = \{1, 2, \ldots, l, \ldots J\}$, $U = \{u \mid u = 0, 1\}$
Upgrade time	
t^l	The period that the product l can be upgraded
Demand function and related revenue	
$a^{l,t}_u$	The amount of potential demand for product l in u status at period t
$a^{l,t,h}_u$	The amount of potential demand for product l that the length of leasing time are h periods in u status at period t
$b1^l_u$	The price sensitivity of the customer for product l in u status
$b2^l_u$	The price sensitivity of the customer for product l in u status
$p^l(h)$	The probability function of the length of leasing time h for product l
$p^{l,h}_{low,u}$	The lower ('low') bounds of leasing price of the product l that the length of leasing time are h periods in u status
$p^{l,h}_{upp,u}$	The upper ('upp') bounds of leasing price of the product l that the length of leasing time are h periods in u status
$\pi^{l,h}_u$	Unit profit for services of the product l that the length of leasing time are h periods in u status
PA^l_u	Purchasing price for product l in u status
DL^l_u	Designed life time for product l in u status
Related fitness to valuated utilization	
FA^i_u	The constant failure rate for component i in u status
RE^i_u	The minimum expected reliability of reprocessing for component i in u status
WH^i_u	The average working hours per period for component i in u status
FL^i_u	The functionality level for component i in u status
TC^i_u	The technology cycle for component i in u status
τ^i_u	The value represents the slope or steepness of curve at functional membership function for component i in u status
$F^i_{k,u}$	The minimum threshold value at quality k for component i in u status, $K = \{1, 2, \ldots, k \ldots, G\}$
Cost	
$C^l_{reas,u}$	The cost of reassembling ('reas') for product l in u status
$C^l_{disa,u}$	The cost of disassembly ('disa') for product l in u status
$C^l_{coll,u}$	The cost of collection ('coll') for product l in u status
$C^l_{s2,u}$	The cost of inventory in S2 for product l in u status
$C^l_{s3,u}$	The cost of inventory in S3 for the returned product l in u status
Cp^l	The cost of the non-durable part of product l
C_{pena}	The penalty ('pena') cost of unsatisfied demand
$C^i_{process,u}$	The cost of processing ('process') the new component i in u status
$C^i_{clea,u}$	The cost of cleaning ('clea') for component i in u status
$C^i_{repr,k,u}$	The cost of reprocessing ('repr') component i in u status at quality k level
$C^i_{up,k}$	The cost of upgrading ('up') component i in u status at quality k level

(continued)

Table 2 (continued)

Parameter	Structure of product
$C^i_{disp,u}$	The cost of disposal ('disp') for component i in u status
$C^i_{s2,u}$	The cost of inventory in S2 for component i in u status
$C^i_{s4,u}$	The cost of inventory in S4 for the returned component i in u status
$C^i_{s5,k,u}$	The cost of inventory in S5 for component i in u status at quality k level
$C^i_{s1,u}$	The cost of inventory in S1 for component i in u status
$C^i_{proc,u}$	The cost of procurement ('proc') for component i in u status
Capacities	
U^t	The total capacity for processing/reprocessing at period t
U^i_u	Processing/Reprocessing capacity per unit of component i in u status
U^t_u	The total capacity for upgrading process at period t
U^i_{up}	Upgrade ('up') capacity per unit of component i

Table 3 Decision variables

Pricing for product	
$x^{l,t,h}_{pric,u}$	Unit price ('pric') of the product l that the length of leasing time are h periods in u status at period t, $H = \{1, 2, \ldots, h, \ldots, Z\}$
Demand for product	
$x^{l,t,h}_{dema,u}$	Demand ('dema') of the product l that the length of leasing time are h periods in u status at period t
$\delta^{l,t,h}_u$	The amount of unsatisfied demand of product l that the length of leasing time are h periods in u status from product market at period t
Processing/procurement for assembly processes	
$x^{l,t,h}_{reas,u}$	The amount of the product l that the length of leasing time are h periods for reassembly ('reas') in u status at period t
$x^{i,t}_{process,u}$	The processing ('process') volume of new component i in u status at period t
$x^{i,t}_{proc,u}$	The amount of the component i in u status by procuring ('proc') at period t
Collection/disassembly for end of leasing products	
$x^{l,t,h}_{coll,u}$	The amount of the product l that the length of leasing time are h periods in u status for collection ('coll') at period t
Evaluated the degree of utilization for recycled components	
$x^{i,t}_{disp,u}$	The amount of the component i in u status for dispose ('disp') at period t
$x^{i,t}_{disp,r,v,u}$	The amount of the component i in u status that are processed or procured at period v and its' lifetime in the market is r periods for disposal ('disp') at period t
The recovery options for recycled components	
$x^{i,t}_{repr,k,u}$	The amount of component i in u status for reprocessing ('repr') at quality k level at period t
$x^{i,t}_{repr,r,k,v,u}$	The amount of component i in u status that are processed or procured at period v and its' lifetime in the market is r periods for reprocessing ('repr') at quality k level at period t
$x^{i,t}_{up,k}$	The amount of component i for upgrading ('up') at quality k level at period t
$x^{i,t}_{up,r,k,v}$	The amount of component i that are processed or procured at period v and its' lifetime in the market is r periods for upgrading ('up') at quality k level at period t

(continued)

Table 3 (continued)

Inventory	
$x^{i,t,l,h}_{s2,u}$	The amount of component i in u status are designated as materials for product l that the length of leasing time are h periods in inventory S2 at period t
$x^{i,t}_{s2,r,v,u}$	The amount of component i in u status that are processed or procured at period v and its' lifetime in the market is r periods in inventory S2 at period t
$x^{i,t,l,h}_{s2,r,v,u}$	The amount of component i that are processed or procured at period *v* and its' lifetime in the market is r periods in u status are designated as materials for product l that the length of leasing time are h periods in inventory S2 at period t
$x^{i,t,l,h}_{s3,r,v,u}$	The amount of component i that are processed or procured at period *v* and its' lifetime in the market is r periods in u status are designated as materials for product l that the length of leasing time are h periods in inventory S3 at period t
$x^{i,t}_{s3,r,v,u}$	The amount of component i in u status that are processed or procured at period v and its' lifetime in the market is r periods in inventory S3 at period t
$x^{i,t}_{s4,u}$	The amount of component i in u status in inventory S4 at period t
$x^{i,t}_{s4,r,v,u}$	The amount of component i in u status that are processed or procured at period v and its' lifetime in the market is r periods in inventory S4 at period t
$x^{i,t}_{s5,k,u}$	The amount of component i in u status at quality k level in inventory c at period t
$x^{i,t}_{s5,r,v,u}$	The amount of component i in u status that are processed or procured at period v and its' lifetime in the market is r periods in inventory S5 at period t
$x^{i,t}_{s5,r,k,v,u}$	The amount of component i in u status that are processed or procured at period v and its' lifetime in the market is r periods at quality k level in inventory S5 at period t
$x^{i,t,l,h}_{s1,u}$	The amount of component i in u status are designated as materials for product l that the length of leasing time are h periods in inventory S1 at period t
$x^{i,t}_{s1,r,v,u}$	The amount of component i in u status that are processed or procured at period v and its' lifetime in the market is r periods in inventory S1 at period t
$x^{i,t,l,h}_{s1,r,v,u}$	The amount of component i that are processed or procured at period v and its' lifetime in the market is r periods in u status are designated as materials for product l that the length of leasing time are h periods in inventory S1 at period t
Binary variables	
$y^{i,t}_{r,k,v,u}$	=1, when the FEU value of component i in u status that are processed or procured at period v and its' lifetime in the market is r periods is greater than or equals the minimum threshold value of k level at period t = 0, Otherwise
$y^{i,t}_{r,v,u}$	=1, when the FEU value of component i in u status that are processed or procured at period v and its' lifetime in the market is r periods is equal to zero at period t = 0, Otherwise

Constraints

Pricing constraints

$$x^{l,t,h}_{dema,u} = \max\{0, a^{l,t,h}_u - b1^{l,h}_u \cdot x^{l,t,h}_{pric,u} - b2^l_u \cdot (x^{l,t,h}_{pric,u} - \frac{h}{DL^l_u} \cdot PA^l_u)\}, \quad \forall l \in L, u \in U, h \in H, t \in T \tag{10}$$

$$a^{l,t,h}_u \le a^{l,t}_u \cdot p^l(h), \quad \forall l \in L, u \in U, h \in H, t \in T \tag{11}$$

$$a^{l,t,h}_u + 1 > a^{l,t}_u \cdot p^l(h), \quad \forall l \in L, u \in U, h \in H, t \in T \tag{12}$$

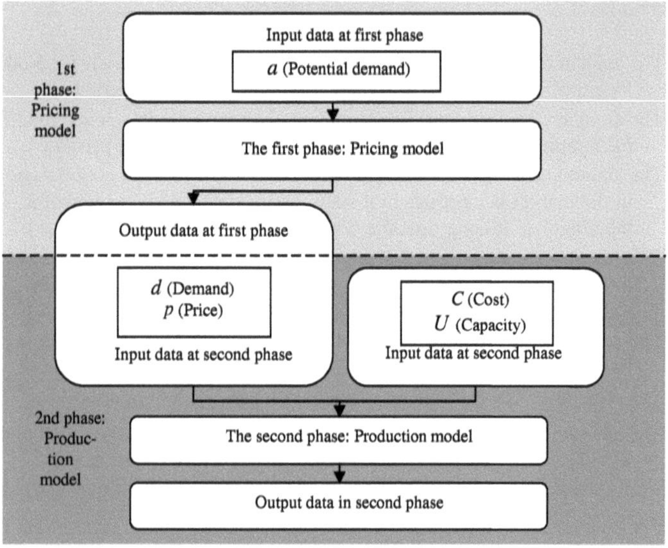

Fig. 2 Structure of the proposed PLM

Table 4 Periodic potential demands for the illustrative example

(Units)	Period t					
	1	2	3	4	5	6
Product I	1250	1595	1885	2007	1876	1509
Up. Product I	0	0	0	1120	1461	1794
Product II	1250	1595	1885	2007	1876	1509
Up. Product II	0	0	0	1120	1461	1794

Table 5 The scale of the illustrative example

Description	Value	Description	Value
Type of products (l)	2	Times periods (t)	6
Type of components (i)	5	Type of upgrading status of components or products (u)	2
Type of quality levels (k)	6		

$$x_{pric,u}^{l,t,h} \geq p_{low,u}^{l,t}, \quad \forall l \in L, u \in U, h \in H, t \in T \tag{13}$$

$$x_{pric,u}^{l,t,h} \leq p_{upp,u}^{l,t}, \quad \forall l \in L, u \in U, h \in H, t \in T \tag{14}$$

Non-negativity constraints

$$x_{pric,u}^{l,t,h}, x_{dema,u}^{l,t,h} \geq 0, \text{ and integer}, \quad \forall l \in L, u \in U, h \in H, t \in T \tag{15}$$

The output data of first phase which are the prices and quantity for the leased products are the input of the second phase. Then, the second phase of PLM is to decide the optimal production planning with the most profitable volumes of processing, reprocessing, procurement, and upgrading of each period in order to satisfy the demand. The production model in the form of Integer Linear Program (ILP) is formulated as follows:

3.6.2 2nd Phase: Production model

Objective function

$$\begin{aligned} \text{Max } & LRV\,(x^{l,t,h}_{dema,u}, x^{l,t,h}_{pric,u}) - RAC\,(x^{i,t}_{process,u}, x^{i,t}_{proc,u}, x^{l,t,h}_{reas,u}) - CDC\,(x^{l,t,h}_{coll,u}) \\ & - CNC\,(x^{l,t,h}_{coll,u}, x^{i,t}_{disp,u}) - RPC\,(x^{i,t}_{repr,k,u}) - UGC\,(x^{i,t}_{up,k}) - DPC\,(x^{i,t}_{disp,u}) \\ & - INC\,(x^{l,t,h}_{dema,u}, x^{l,t,h}_{coll,u}, x^{i,t.l,h}_{s2,u}, x^{i,t}_{s4,u}, x^{i,t}_{s5,k,u}, x^{i,t}_{s1,s,v,u}) - PNC\,(\delta^{l,t,h}_{u}) \end{aligned} \tag{16}$$

where

LRV (Leasing Revenue)

$$LRV = \sum_t \left[\sum_h \left[\sum_u \left[\sum_l \left[(x^{l,t,h}_{dema,u} - \delta^{l,t,h}_u) \cdot (x^{l,t,h}_{pric,u} + \pi^{l,h}_u) \right] \right] \right] \right] \tag{16.1}$$

RAC (Reassembly Cost)

$$RAC = \sum_t \left[\sum_u \left[\begin{array}{l} \sum_i \left[C^i_{process,u} \cdot x^{i,t}_{process,u} \right] + \sum_i \left[C^i_{proc,u} \cdot x^{i,t}_{proc,u} \right] \\ + \sum_l \left[(C^l_{reas,u} + Cp^l) \cdot \sum_h x^{l,t,h}_{reas,u} \right] \end{array} \right] \right] \tag{16.2}$$

CDC (Collection and Disassembly Cost)

$$CDC = \sum_t \left[\sum_u \left[\sum_l \left[(C^l_{coll,u} + C^l_{disa,u}) \cdot \sum_h \left[x^{l,t,h}_{coll,u} \right] \right] \right] \right] \tag{16.3}$$

CNC (Cleaning Cost)

$$CNC = \sum_t \left[\sum_u \left[\sum_i \left[C^i_{clea,u} \cdot \sum_l \left[\sum_h \left[x^{l,t,h}_{coll,u} \right] \cdot A^l_{i,u} - x^{i,t}_{disp,u} \right] \right] \right] \right] \tag{16.4}$$

RPC (Reprocessing Cost)

$$RPC = \sum_t \left[\sum_u \left[\sum_i \sum_k \left[C^i_{repr,k,u} \cdot x^{i,t}_{repr,k,u} \right] \right] \right] \tag{16.5}$$

UGC (Upgrade Cost)

$$UGC = \sum_t \left[\sum_i \sum_k \left[C^i_{up,k} \cdot x^{i,t}_{up,k} \right] \right] \tag{16.6}$$

DPC (Disposal Cost)

$$DPC = \sum_t \left[\sum_u \left[\sum_i \left[C^i_{disp,u} \cdot x^{i,t}_{disp,u} \right] \right] \right] \tag{16.7}$$

INC (Inventory Cost)

$$INC = \sum_t \left[\sum_u \left[\begin{array}{l} \sum_l \left[C^l_{s2,u} \cdot \sum_h \left[x^{l,t,h}_{dema,u} \right] + C^l_{s3,u} \cdot \sum_h \left[x^{l,t,h}_{coll,u} \right] \right] \\ + \sum_i \left[\begin{array}{l} C^i_{s2,u} \cdot \sum_h \sum_l \left[x^{i,t,l,h}_{s2,u} \right] + C^i_{s4,u} \cdot x^{i,t}_{s4,u} \\ + \sum_k \left[C^i_{s5,k,u} \cdot x^{i,t}_{s5,k,u} \right] + \sum_s \sum_v C^i_{s1,u} \cdot x^{i,t}_{s1,s,v,u} \end{array} \right] \end{array} \right] \right] \tag{16.8}$$

PNC (Penalty Cost)

$$PNC = \sum_t \left[\sum_u \left[\sum_l \left[\sum_h C_{pena} \cdot \delta^{l,t,h}_u \right] \right] \right] \tag{16.9}$$

Constraints

Collection constraints

$$\begin{array}{c} x^{l,t+h-1,h}_{coll,u} = x^{l,t,h}_{dema,u} - \delta^{l,t,h}_u \\ \forall l \in L, u \in U, h \in H, t \in T \end{array} \tag{17}$$

Degree of reutilization of the returned components constraints

$$\begin{array}{l} [f_{FEU}(FA^i_u, RE^i_u, WH^i_u, TC^i_u, FL^i_u, \tau^i_u) - F^i_{k,u} - S_{\min}] \cdot y^{i,t}_{r,k,v,u} \geq 0 \\ \forall i \in I, u \in U, k \in K, r > 0 \in R, v \in V, t \in T \end{array} \tag{18}$$

$$x^{i,t}_{s5,r,k,v,u} \leq x^{i,t}_{s4,r,v,u} \quad \forall i \in I, u \in U, k \in K, r > 0 \in R, v \in V,\ t \in T \tag{19}$$

$$x^{i,t}_{s5,r,k,v,u} \leq x^{i,t}_{r,k,v,u} \cdot M_{\max} \quad \forall i \in I, u \in U, k \in K, r > 0 \in R, v \in V,\ t \in T \tag{20}$$

$$\begin{array}{l} f_{FEU}(FA^i_u, RE^i_u, WH^i_u, TC^i_u, FL^i_u, \tau^i_u) \cdot y^{i,t}_{r,v,u} = 0 \\ \forall i \in I, u \in U, r > 0 \in R, v \in V,\ t \in T \end{array} \tag{21}$$

$$x^{i,t}_{disp,r,v,u} \leq x^{i,t}_{s4,r,v,u} \quad \forall i \in I, u \in U,\ r > 0 \in R, v \in V,\ t \in T \tag{22}$$

$$x^{i,t}_{disp,r,v,u} \leq y^{i,t}_{r,v,u} \cdot M_{\max} \quad \forall i \in I, u \in U,\ r > 0 \in R, v \in V,\ t \in T \tag{23}$$

$$\sum_k [y^{i,t}_{r,k,v,u}] + y^{i,t}_{r,v,u} = 1 \quad \forall i \in I, u \in U,\ r \in R, v \in V,\ t \in T \tag{24}$$

$$x^{i,t}_{disp,u} = \sum_{r \leq t} \sum_{v \leq t-r} x^{i,t}_{disp,r,v,u} \quad \forall i \in I, u \in U,\ t \in T \tag{25}$$

$$x^{i,t}_{s4,r,v,u} = x^{i,t}_{s5,r,v,u} + x^{i,t}_{disp,r,v,u} \quad \forall i \in I, u \in U, r \in R, v \in V,\ t \in T \tag{26}$$

$$x^{i,t+h,l,h}_{s3,r+h,v,u} = x^{i,t,l,h}_{s1,r,v,u} - x^{i,t,l,h}_{s2,r,v,u} \quad \begin{array}{l} \forall l \in L, i \in I, u \in U, h \in H, \\ r \leq R - h \in S, v \in V, t \in T \end{array} \tag{27}$$

$$x^{i,t}_{s2,r,v,u} = \sum_h \sum_l x^{i,t,l,h}_{s2,r,v,u} \quad \forall i \in I, u \in U, r \in R, v \in V,\ t \in T \tag{28}$$

$$x^{i,t}_{s3,r,v,u} = \sum_h \sum_l x^{i,t,l,h}_{s3,r,v,u} \quad \forall i \in I, u \in U, r \in R, v \in V,\ t \in T \tag{29}$$

$$x^{i,t}_{s1,r,v,u} = \sum_h \sum_l x^{i,t,l,h}_{s1,r,v,u} \quad \forall i \in I, u \in U, r \in R, v \in V,\ t \in T \tag{30}$$

$$\begin{array}{l} x^{i,t+1}_{s3,r,v,u} + \sum_k [x^{i,t}_{s5,r,k,v,u} - x^{i,t}_{repr,r,k,v,u} - x^{i,t}_{up,r,k,v}] = x^{i,t+1}_{s4,r,v,u}, \\ \forall i \in I, r \in R, v \in V, t \in T, u \in \{0\} \end{array} \tag{31}$$

$$\begin{array}{l} x^{i,t+1}_{s3,r,v,u} + \sum_k [x^{i,t}_{s5,r,k,v,u} - x^{i,t}_{repr,r,k,v,u}] = x^{i,t+1}_{s4,r,v,u} \\ \forall i \in I, r \in R, v \in V, t \in T, u \in \{1\} \end{array} \tag{32}$$

$$x^{i,t}_{s5,r,v,u} = \sum_k x^{i,t}_{s5,r,k,v,u} \quad \forall i \in I, u \in U, r \in R, v \in V,\ t \in T \tag{33}$$

$$x^{i,t}_{s5,k,u} = \sum_{r \leq t} \sum_{v \leq t-r} x^{i,t}_{s5,r,k,v,u} \quad \forall i \in I, u \in U, k \in K,\ t \in T \tag{34}$$

$$x^{i,t}_{s4,u} = \sum_{r \leq t} \sum_{v \leq t-r} x^{i,t}_{s4,r,v,u} \quad \forall i \in I, u \in U, t \in T \tag{35}$$

Reprocessing/upgrade constraints

$$x^{i,t}_{repr,r,k,v,u} \leq x^{i,t}_{s5,r,k,v,u} \quad \forall i \in I, u \in U, r \in R, k \in K, v \in V, t \in T \tag{36}$$

$$x^{i,t}_{repr,r,k,v,u=0} + x^{i,t}_{up,r,k,v} \leq x^{i,t}_{s5,r,k,v,u=0} \quad \forall i \in I, r \in R, k \in K, v \in V, t \geq t^l \in T \tag{37}$$

$$x^{i,t}_{repr,k,u} = \sum_{r \le t} \sum_{v \le t-r} x^{i,t}_{repr,r,k,v,u} \quad \forall i \in I, u \in U, k \in K, t \in T \tag{38}$$

$$x^{i,t}_{up,k} = \sum_{r \le t} \sum_{v \le t-r} x^{i,t}_{up,r,k,v} \quad \forall i \in I, k \in K, t \ge t^l \in T \tag{39}$$

$$\sum_{k} x^{i,t+1}_{repr,r,k,v,u} + x^{i,t}_{s2,r,v,u} = x^{i,t+1}_{s1,r,v,u} \quad \begin{array}{c}\forall i \in I, u \in U, r \in \{0\},\\ v \in V, t \in T\end{array} \tag{40}$$

$$\sum_{k} x^{i,t+1}_{repr,r,k,v,u} + \sum_{k} x^{i,t+1}_{up,r,k,v} + x^{i,t}_{s2,r,v,u} = x^{i,t+1}_{s1,r,v,u} \quad \begin{array}{c}\forall i \in I, u \in \{1\}, r \in R,\\ v \in V, t \in T\end{array} \tag{41}$$

$$x^{i,t}_{s1,r=0,v=t,u} = x^{i,t}_{process,u} + x^{i,t}_{proc,u} \quad \forall i \in I, u \in U, t \in T \tag{42}$$

Reassembly constraints

$$A^{l}_{i,u} \cdot x^{l,t,h}_{reas,u} \le x^{i,t,l,h}_{s1,u} \quad \forall i \in I, l \in L, u \in U, h \in H, t \in T \tag{43}$$

$$x^{i,t}_{s2,r,v,u} = x^{i,t+1}_{s1,r,v,u} \forall i \in I, u \in U, r \in \{0\}, v \in V, t \in T \tag{44}$$

$$x^{i,t,l,h}_{s1,u} = \sum_{r \le t} \sum_{v \le t-r} x^{i,t,l,h}_{s1,r,v,u} \quad \forall i \in I, l \in L, u \in U, h \in H, t \in T \tag{45}$$

$$x^{i,t,l,h}_{s2,u} = \sum_{r \le t} \sum_{v \le t-r} x^{i,t,l,h}_{s2,r,v,u} \quad \forall i \in I, l \in L, u \in U, h \in H, t \in T \tag{46}$$

$$A^{l}_{i,u} \cdot x^{l,t,h}_{reas,u} = x^{i,t,l,h}_{s1,u} - x^{i,t,l,h}_{s2,u} \quad \forall i \in I, l \in L, u \in U, h \in H, t \in T \tag{47}$$

$$x^{i,t,l,h}_{s1,r,v,u} \ge x^{i,t,l,h}_{s2,r,v,u} \quad \forall l \in L, i \in I, u \in U, r \in R, v \in V, t \in T \tag{48}$$

$$x^{i,t}_{s1,r,v,u} \ge x^{i,t,}_{s2,r,v,u} \quad \forall i \in I, u \in U, r \in R, v \in V, t \in T \tag{49}$$

$$x^{i,t}_{reas,u} = x^{i,t+1,h}_{dema,u} - \delta^{l,t+1,h}_{u} \quad \forall l \in L, u \in U, h \in H, t \in T \tag{50}$$

Capacity constraints

$$\sum_{i} \sum_{u} \left[U^{i}_{u} \cdot \left(x^{i,t}_{process,u} + \sum_{k} x^{i,t}_{repr,k,u} \right) \right] \le U^{t} \quad \forall t \in T \tag{51}$$

$$\sum_{i} \sum_{k} [U^{i}_{up} \cdot x^{i,t}_{up}] \le U^{t}_{up} \quad \forall t \in T \tag{52}$$

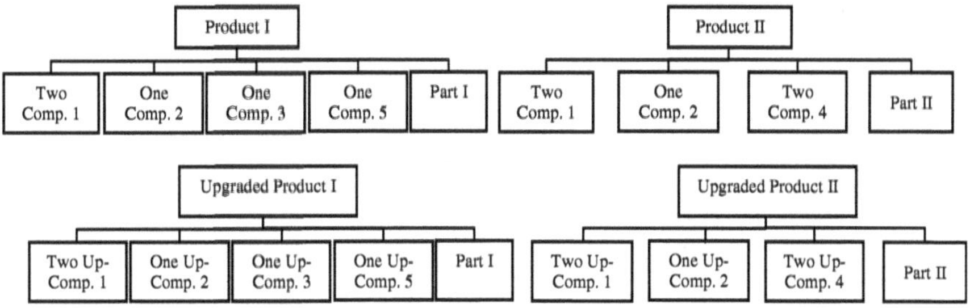

Fig. 3 The description of the products

Binary and non-negativity constraints

$x_{pric,u}^{l,t,h}, x_{dema,u}^{l,t,h}, x_{reas,u}^{l,t,h}, x_{process,u}^{i,t}, x_{proc,u}^{i,t}, x_{coll,u}^{l,t,h}, x_{disp,u}^{i,t}, x_{disp,r,v,u}^{i,t}, x_{repr,k,u}^{i,t}, x_{repr,r,k,v,u}^{i,t}, x_{up,k}^{i,t},$
$x_{up,r,k,v}^{i,t}, x_{s2,u}^{i,t,l,h}, x_{s2,r,v,u}^{i,t}, x_{s2,r,v,u}^{i,t,l,h}, x_{s3,r,v,u}^{i,t,l,h}, x_{s3,r,v,u}^{i,t}, x_{s4,u}^{i,t}, x_{s4,r,v,u}^{i,t}, x_{s5,k,u}^{i,t}, x_{s5,r,v,u}^{i,t}, x_{s5,r,k,v,u}^{i,t},$
$x_{s1,u}^{i,t,l,h}, x_{s1,r,v,u}^{i,t}, x_{s1,r,v,u}^{i,t,l,h}, \delta_u^{l,t,h} \geq 0$ and integer, $y_{r,k,v,u}^{i,t}, y_{r,v,u}^{i,t} \in \{0, 1\}$, $M_{\max}$ is large positive number, $S_{\min}$ is small positive number. $\forall i \in I, l \in L, u \in U, k \in K, r \in R, v \in V, t \in T$

4 An Illustrative Example

In this section, a numerical example is adopted to demonstrate the proposed model, which is solved by the ILOG/CPLEX 7.1.0 optimizing software.

4.1 Data Description

Assume that there are two similar products, five types of components in the system, and the first and second generations of each product will appear in the market, as shown in Fig. 3. The potential demands of 6 periods are shown in Table 2. The scale of the example is shown in Tables 3.

4.2 Numerical Result

We illustrate the validity of our model with this example. The output is summarized in Table 6. The optimal solution emerges, with the objective being \$59,037,540. Table 7 and Fig. 4 show the output data of the first phase, optimal demand volume and the optimal leasing price of each product respectively. Figure 5 indicates the processing volumes and procurement volumes for components, for which all

Table 6 Output of the detailed revenue and cost for the illustrative example

Variable	Value	Variable	Value
Objective value (profit)	$59,037,540	Upgrading cost	$1,421,515
Leasing and service revenue	$111,758,580	Reassembly cost	$2,191,760
Processing and procurement cost	$39,269,000	Disposal cost	$1,165,550
Reprocessing cost	$2,080,840	Collection and disassembly cost	$1,525,580
Cleaning and inventory cost	$5,066,795		

Table 7 Periodic market demands for the illustrative example

(Units)	Period t					
	1	2	3	4	5	6
Product I	682	1008	1277	1399	1268	921
Up. product I	0	0	0	560	900	1233
Product II	695	1030	1320	1421	1311	943
Up. product II	0	0	0	567	907	1222

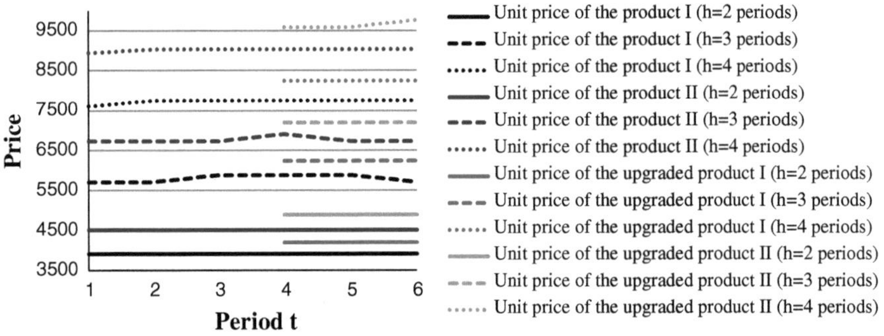

Fig. 4 Comparison between the unit prices for products

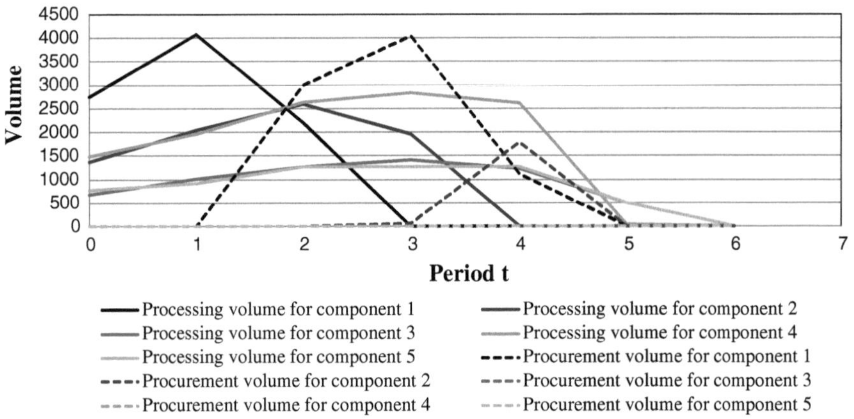

Fig. 5 Comparison between the processing volumes and procurement volumes for components

components are processed by OEM at the beginning of period. Because the given production capacities are not sufficient, the procurement of new components 1 and 2 at period 2, 3, and 4 is necessary in order to satisfy the demand.

5 Summary and Conclusions

Considering the variation of demands in a Closed-loop Supply Chain (CLSC) system, this study proposed the Periodic Leasing Model (PLM) for OEM. To facilitate the solution, the PLM is divided into two phases, the pricing model is developed in the form of INLP to determine the prices and the quantity of the leased products at the first phase; and the production model is developed in the form of ILP to determine the amount of production for the periodic demands at the second phase.

In summary, the major contributions of this study are to support a manufacturer to determine the optimal prices and required quantity of products of each period along the planning horizon in order to maximize the revenue, as well as to develop an optimal production plan which provides the best combinations of components from different products with effective multiple reuses of components in order to minimize the production cost.

Acknowledgments The authors acknowledge the financial support from National Science Council of Taiwan, ROC with the project number 100-2221-E007-062-MY3.

References

Aras N, Güllü R, Yürülmez S (2011) Optimal inventory and pricing policies for remanufacturable leased products. Int J Prod Econ 133(1):262–271

Directive 2002/96/EC of the European Parliament and of the Council of 27 January 2003 on Waste Electrical and Electronic Equipment. European Union

Geyer R, Van Wassenhove LN, Atasu A (2007) The economics of remanufacturing under limited component durability and finite product life cycles. Manage Sci 53(1):88–100

Guide VDR Jr, Van Wassenhove LN (2009) The evolution of closed-loop supply chain research. Oper Res 57(1):10–18. doi:10.1016/j.jenvman.2009.09.037

Guo J, Liang C (2011) Models for recycling and production policy for a closed-loop system. J Qual 18(4):303–314. doi:10.1111/j.1540-5915.2012.00359.x

Linton J, Scott Yeomans J, Yoogalingam R (2005) Recovery and reclamation of durable goods : a study of television CRTs. Resour Conserv Recycl 43(4):337–352. doi:10.1016/j.resconrec.2004.06.008

Östlin J, Sundin E, Björkman M (2008) Importance of closed-loop supply chain relationships for product remanufacturing. Int J Prod Econ 115(2):336–348. doi:10.1016/j.ijpe.2008.02.020

Ramani K, Ramanujan D, Bernstein WZ, Zhao F, Sutherlamd J, Handwerker C, Choi JK, Kim H, Thurston D (2010) Integrated sustainable life cycle design: a review. J Mech Des 132(9), 091004:1–15 (2010). doi: 10.1002/aic.13914

Savaskan R, Bhattacharya S, Van Wassenhove LN (2004) Closed-loop supply chain models with product remanufacturing. Manage Sci 50(2):239–252. doi:10.1287/mnsc.1030.0186

Van Wassenhove LN, Zikopoulos C (2010) On the effect of quality overestimation in remanufacturing. Int J Prod Res 48(15):5263–5280. doi:10.1080/00207540903150585

Wang HF, Hsu HW (2010) A closed-loop logistic model with a spanning-tree based genetic algorithm. Comput Oper Res 37(2):376–389. doi:10.1016/j.cor.2009.06.001

Wang HF, Huang YS (2013) A two-stage robust programming approach to demand-driven disassembly planning for a closed-loop supply chain system. Int J Prod Res 51(8):2414–2432. doi:10.1080/00207543.2012.737940

Xing K, Lee L (2009) Modelling and evaluation of product fitness for service life extension. J Eng Des 20(3):243–263. doi:10.1080/09544820701834512

Part III
Modeling and Optimization of the Manufacturing Operations

A Pareto-Archived Differential Evolution Algorithm for Multi-Objective Flexible Job Shop Scheduling Problems

Warisa Wisittipanich and Voratas Kachitvichyanukul

Abstract This chapter presents an efficient evolutionary algorithm, called multi-objective differential evolution algorithm (MODE) to find a Pareto front for multi-objective flexible job shop scheduling problems. The objective is to simultaneously minimize makespan and total tardiness of jobs. The MODE framework adopts the idea of the Elite group to store solutions and utilizes those solutions as the guidance of the vectors. Five mutation strategies with different search behaviors are proposed in MODE algorithms in order to search for the good quality Pareto front. The performances of the MODE algorithms with different mutation strategies are evaluated on a set of benchmark problems and compared with results obtained from an existing algorithm. The experimental results demonstrated that the MODE algorithms are highly competitive approaches which are capable of providing a set of diverse and high-quality non-dominated solutions.

Keywords Pareto front · Differential evolution · Multi-objective · Flexible job shop scheduling

1 Introduction

Flexible job shop scheduling problem (FJSP) is an extension of the classical job shop scheduling problem (JSP), which allows an operation of a job to be processed on one machine selected from a set of available machines. In FJSP, some machine

W. Wisittipanich (✉)
Chiang Mai University, 239 Huay Keaw Road, Muang District, Chiang Mai 50200, Thailand
e-mail: warisa@eng.cmu.ac.th

V. Kachitvichyanukul
Asian Institute of Technology, P.O. Box 4, Klong Luang, Pathumthani 12120, Thailand
e-mail: voratas@ait.ac.th

P. Golinska (ed.), *Logistics Operations, Supply Chain Management and Sustainability*, EcoProduction, DOI: 10.1007/978-3-319-07287-6_23,

stations are allowed to be visited more than once or not at all. This additional requirement to determine the appropriate assignment of operations on the machines makes FJSP a more complex problem than JSP. Over the past two decades, many researchers have studied and proposed several algorithms based on metaheuristics to solve FJSP. The early research on FJSP was focused on neighborhood-based metaheuristics like Tabu Search (TS) and Simulated Annealing (SA). However, the performance of these algorithms highly depend on the initial solution and are more vulnerable to be trapped into local optima. Consequently, most recent research efforts have been devoted to develop population-based metaheuristics or Evolutionary Algorithms (EAs) such as Genetic Algorithm (GA) and Particle Swarm Optimization (PSO). Zhang and Gen (2005) presents two algorithms based on Genetic Algorithm (GA) to minimize total tardiness in FJSP. Wisittipanich and Kachitvichyanukul (2011, 2013) demonstrated a potential of Differential Evolution (DE) approach to minimize makespan and total tardiness of jobs in FJSP.

Recently, multi-objective (MO) optimization toward Pareto-based approach have gradually captured interests from both practitioners and researchers due to its reflection to most real-world problems containing multiple conflicting objectives. Compared to an aggregated weighted approach, Pareto-based method offer advantage for decision makers to simultaneously find the trade-offs solutions on a Pareto front in a single run without prejudice. Even though population-based metaheuristics has shown its efficiency in solving FJSP, only few research works on Pareto-based approaches have been proposed for solving multi-objective FJSP (MOFJSP). Kacem et al. (2002a, b) presented a hierarchical approach for multi-objective FJSP using either of the weighted summation of objectives or Pareto approaches by applying GA controlled by the assignment model which is generated by the approach of localization. Zhang and Gen (2005) proposed a multistage operation-based GA (moGA) to solve multi-objective FJSP. Their algorithm is based on a hierarchical approach where PSO is used for operation assignment and then SA is implemented for operation sequencing. Gao et al. (2007) developed the hybrid algorithm of GA and bottleneck shifting for multi-objective FJSP to fully exploit the global search ability of GA and the local search ability of bottleneck shifting heuristic. Moslehi and Mahnam (2011) proposed a hybridization approach of PSO and LS algorithm to find a set of Pareto solutions in MOFJSP.

Differential Evolution (DE), proposed by Storn and Price in 1995 (Storn and Price 1995), is one of the latest EAs which has been widely applied and shown its efficiency in many application areas. Nevertheless, the research works of DE algorithm for solving MO problems are still limited. Abbass et al. (2001) presented a Pareto-frontier DE approach (PDE) where only non-dominated solutions are allowed to participate in generating new solutions. However, if only few non-dominated solutions are found in the beginning, the chances of discovering new better solutions may be limited and the solutions may get trapped at local optima. Madavan (2002) introduced the combination of the newly generated population and the existing parent population which results in double population size.

The non-dominated solutions are then selected from this combination to allow the global check among both parents and offspring solutions. The drawback of this approach is the requirement of additional computing time in the sorting procedure of the combined population. Qian et al. (2008) proposed a memetic algorithm based on DE algorithm (MODEMA) to maintain a set of non-dominated solutions and simultaneously balance exploration and exploitation ability in solving MO-FJSP. Wisittipanich and Kachitvichyanukul (2014) presented a novel multi-objective DE algorithm, called MODE, to find a Pareto front in MO problems. The emphasis of MODE algorithm is on developing several mutation strategies as the guidance of DE population to search for high quality non-dominated solutions. MODE has been successfully applied for solving many continuous and combinatorial optimization problems.

This chapter extends the successful implementation of MODE in solving MOJSP (Wisittipanich and Kachitvichyanukul 2013) to find a Pareto front for MOFJSP. The remainder of this chapter is organized as follows. The description and formulation of MOFJSP are provided in Sect. 2. Section 3 presents the MODE algorithm and its application to multi-objective FJSP. Experimental results are reported in Sect. 4. Finally, conclusion and further research are provided in Sect. 5.

2 Multi-Objective Flexible Job Shop Scheduling

Similar to the classic JSP, FJSP schedule a set on n jobs on a set of m machines. However, the problem in FJSP is more difficult because it requires proper assignment of each job operation to a machine from a set of capable machines. In FJSP, the set of machines is denoted as M, $M = \{M_1, M_2, \ldots, M_3\}$. Each job i consists of a sequence of n_i operations. Each operation $O_{i,j}$ ($i = 1, 2, \ldots, n$: $j = 1, 2, \ldots, n_i$) of job i is required to be processed on one machine M_k out of a set of given compatible machines $M_{i,j}$ (for $M_k \in M_{i,j}$, $M_{i,j} \subseteq M$). Each machine is independent from one another. Machine set up time and transfer time between operations are negligible. Machine breakdown is not considered and the pre-emption is not allowed in this problem. Each machine can process at most one operation at a time, and there are no precedence constraints among operations of different jobs. The goal is to determine both assignment and sequence of operations on machines in order to optimize certain objectives subjected to the constraints. In this chapter, the objective is to simultaneously minimize the makespan and total tardiness of jobs in FJSP.

- Makespan: makespan $= \max\ \{C_{i,j}\}$ where $C_{i,j}$ is the completion time of operation $O_{i,j}$,
- Total tardinesss of jobs: Total tardiness $= \sum \max\ \{0,\ (s_{i,j,k} + p_{i,j,k}) - D_i\}$ where

- $p_{i,j,k}$ is processing time of operation $O_{i,j}$ on machine k,
- $s_{i,j,k}$ is start time of operation $O_{i,j}$ on machine k,
- D_i is due date of job i.

3 MODE Algorithm

3.1 MODE Framework

As previously mentioned, this study employs an MODE algorithm, proposed by Wisittipanich and Kachitvichyanukul (2014), and its framework is illustrated in Fig. 1. Similar to the classic DE algorithm, MODE starts with randomly generated initial population of size N of D-dimensional vectors. The population continues to evolve through repeated cycles of three main operators; mutation, crossover, and selection until stopping criterion are met. In MODE, the population experience is stored in an external archive, called Elite group, as a set of non-dominated solutions. In this framework, instead of applying the sorting procedure to every single move of a vector, the sorting is only performed on the set of newly generated trial vectors after all vectors completed one move to identify the group of new non-dominated solutions. The reason is to reduce computational time. This sorting procedure applies to the group of new solutions and current solutions in the external archive and store only non-dominated solutions into an archive for the Elite group. Then, Elite group screens its solutions to eliminate inferior solutions. As a result, the Elite group in the archive contains only the best non-dominated solutions found so far in the searching process of the MODE population.

Another critical decision in MO problems is how to select the candidates among the Elite group as guidance toward the Pareto frontier since the existence of multiple candidates in the archive offers a large number of options on the movement of vectors and the quality of the final solutions will be strongly influenced by the movement behavior adopted by the population. Five mutation strategies are proposed in MODE framework as the movement guidance in order to obtain the high quality Pareto front. Each mutation strategy possesses distinct search behavior which directs a vector in the DE population in different ways with the purpose of reaching the Pareto optimal front.

- MODE-ms1: search around solutions located in the less crowded areas
- MODE-ms2: pull the current front close to the true front
- MODE-ms3: fill the gaps of non-dominated front
- MODE-ms4: search toward the border of the non-dominated front
- MODE-ms5: explore solution space with multiple groups of vectors

Mutation strategies MODE-Ms1 and MODE-Ms3 aim to improve the distribution of solutions on the front. Mutation strategy MODE-Ms2 intends to pull to the current front toward the true Pareto front, and mutation strategy MODE-Ms4

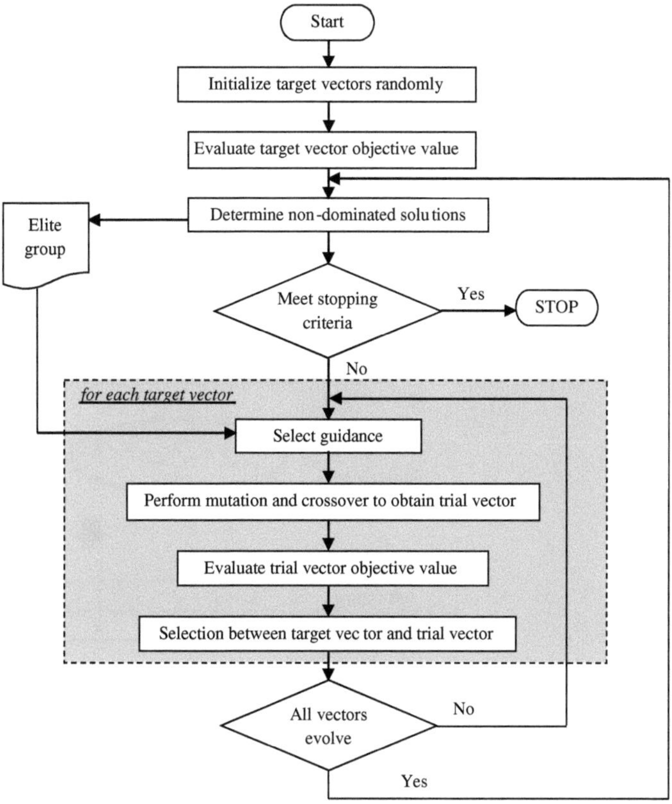

Fig. 1 Multi-objective differential evolution (MODE) framework

try to explore more solutions around the border to increase the spread of the non-dominated front. Each mutation strategy performs different search behavior with its own strengths and weaknesses. Thus, mutation strategy MODE-Ms5 aims to extract the strengths of various DE mutation strategies and to compensate for the weaknesses of each individual strategy in order to enhance the overall performance by combining four groups of vectors with multiple mutation strategies in one population. For more details, see Wisittipanich and Kachitvichyanukul (2013).

3.2 Solution Representation

A solution of FJSP can be represented using a vector with dimensions equal to the total number of operations required for all jobs.

Consider an example of FJSP in Table 1, job 1 and job 2 consists of three and two operations respectively. Therefore, the total number of dimension is set to be equal to the total number of operations of all jobs which is 5. This study employs

Table 1 FJSP with two jobs and three machine

Job	$O_{i,j}$	Processing time		
		M1	M2	M3
1	$O_{1,1}$	3	4	5
	$O_{1,2}$	–	1	2
	$O_{1,3}$	3	6	–
2	$O_{2,1}$	8	7	9
	$O_{2,2}$	–	2	3

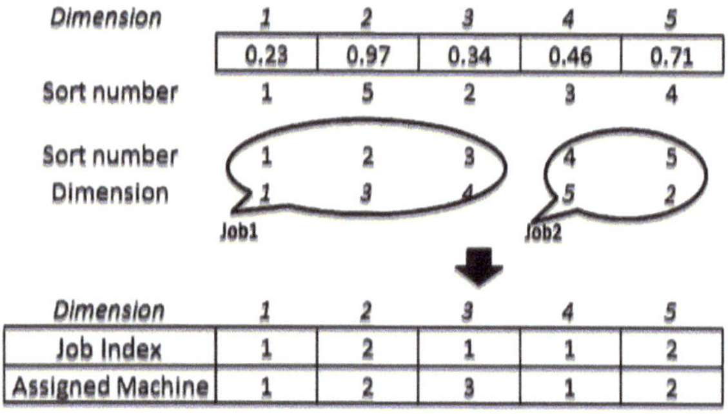

Fig. 2 Solution mapping procedures for FJSP

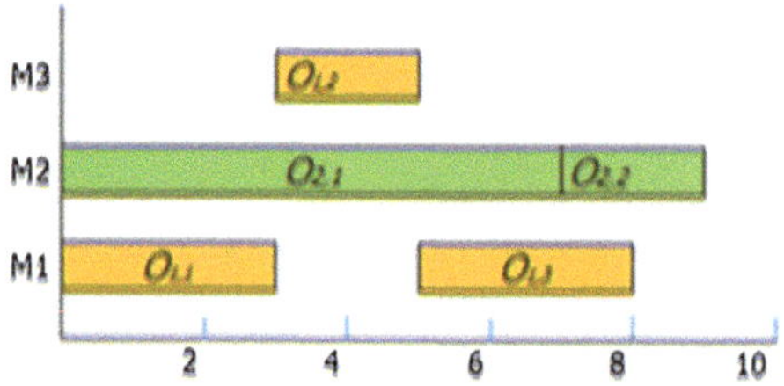

Fig. 3 An active schedule after decoding process

random key representation encoding scheme (Bean 1994) where each value in a vector dimension is initially generated with a uniform random number in range [0, 1]. Next, the permutation of n-repetition of *n* jobs (Bierwirth 1995) with a sorting list rule is adopted to determine a sequence of operations. The advantage of this approach is that any permutation of this representation always provides a feasible schedule. The procedures of solution mapping for FJSP used in this study are illustrated in Fig. 2.

Like mention earlier, in FJSP, an operation is allowed to be processed by one machine selected from a set of capable machines. After a sequence of operations is determined, an operating machine for each operation must also be decided. In this study, the machine assigned for each operation is selected based on the earliest completion time of the machine required to process that operation. If there are two or more machines having the same completion time, one machine is randomly selected as an assigned machine. Finally, the operation-based approach is employed in order to generate an active schedule as shown in Fig. 3.

4 Computational Experiment

4.1 Parameter Setting

In this study, the population size and number of iterations are set as 100 and 1,000 respectively to provide an adequate number of function evaluations in the search process. The value of scale factor F is set to random value to allow variation in the scaled difference and thus retains population diversity throughout the search process. After some preliminary experiments, the value of F set to be uniformly randomized between 1.5 and 2 provides generally good solution quality in all mutation strategies, and the use of exponential crossover yields better results than binomial crossover. Crossover rate (C_r) is linearly increased from 0.1 to 0.5 to maintain the characteristic of generated trial vectors at the beginning of the search. As the search progress, increasing value of C_r yields more deviations for the generated trial vectors and helps the solution to escape from being trapped at local optima. In the MODE, fixed members in Elite archive are set as maximum as 100. If the number of new non-dominated solutions found by the population exceeds the limit of this archive, the solutions with lower crowding distance will be removed. In MODE-ms5, the ratio of numbers of vectors in each group is set as 1:1:1:1 respectively.

4.2 Experimental Results

The performance of MODE is evaluated using two benchmark FJSP data sets. Each instance is characterized by problem size: number of jobs (n) x number of machines (m) x total number of operations (O). Since these benchmark instances are originally designed for determining the makespan, in order to evaluate the total tardiness of a schedule, a due date of each job must be specified. Generally, the due date of jobs in FJSP can be determined by the following equation.

$$D_i = r_i + (t \, * \, \Sigma_i \bar{p}_{i,j,k}) \tag{1}$$

The due date of job i depends on the release date of job i, r_i, the tardiness factor, t, and the average processing time of operation j of job i on its alternative machines $\bar{p}_{i,j,k}$. In this study, the tardiness factor t used to generate due date data is calculated as the following formula (He et al. 1993).

$$t = \frac{N * M}{1000} + 0.5 \tag{2}$$

where N is the number of jobs, M is the number of machines, and r_i of all jobs are set to be 0.

The performances of the MODE are particularly compared to an effective multi-objective particle swarm optimization-based (MOPSO) approach, proposed by Nguyen and Kachitvichyanukul (2010), which has demonstrated its efficiency for solving many continuous optimization problems by generating superior results over several optimization approaches. It is noted that, for a fair comparison, the performance evaluation is based on the use of same number of function evaluations in both MODE and MOPSO. In addition, the encoding and decoding procedures used in both algorithms are identical. The experimental results are shown in Tables 2 and 3. The notation (x, y) represents (makespan, total tardiness). For each instance, the non-dominated solutions are determined from ten independent runs.

According to the results in Tables 2 and 3, MOPSO, MODE-ms1, MODE-ms2, MODE-ms3, MODE-ms4, and MODE-ms5 perform well in generating a large set of non-dominated solutions. The makespan values are very close to the optimal results; while minimum value of the total tardiness found are also close to the minimum value of total tardiness as reported in Wisittipanichkul and Kachitvichyanukul (2013). This means that these algorithms are able to yield a good spread of the non-dominated front. In addition, MODE-ms1, MODE-ms-2, MODE-ms3, MODE-ms4, and MODE-ms5 clearly outperform MOPSO since the majority of the solutions obtained by MOPSO are dominated. Even though solutions obtained by MODE-ms4 are all dominated by MOPSO in some cases, solutions generated from MOPSO are mostly dominated by those from MODE-ms4 in most instances.

Figure 4 illustrates the comparison of non-dominated solutions generated from MOPSO and all MODE strategies. It is important to note that the "Pareto front" curve indicates the best non-dominated solutions found by all algorithms. In most cases, the non-dominated solutions obtained from MOPSO are mostly dominated by those from all MODE strategies. Moreover, it is observed that the solutions on the Pareto front are mainly obtained from those by MODE-ms1, MODE-ms2, MODE-ms3, and MODE-ms5.

4.3 Performance Measurement

The performance evaluation of MO problem is different from the single objective optimization problem since, in MO problems, a set of solutions is generated which make it difficult for comparison. This study uses $\tilde{C}$ metric to compare the set of non-dominated solutions obtained from the MODE strategies with those obtained from MOPSO. The $\tilde{C}(A,B)$ metric measures the fractions of members of B that are dominated by members of A.

$$\tilde{C}(A,B) = \frac{|\{b \in B : \exists a \in A, a \succ b\}|}{|B|} \tag{3}$$

where $|B|$ is the number of solutions in B. Therefore $\tilde{C}(A,B) = 1$ means that each solutions in B is dominated by some solutions in A. On the other hand, $\tilde{C}(A,B) = 0$

Table 2 Comparison results on Brandimarte's data set (Brandimarte 1993)

Instance	Problem size	MOPSO	MODE-ms1	MODE-ms2	MODE-ms3	MODE-ms4	MODE-ms5
MK04	15 × 8 × 90	(66, 508.46)	(64, 445.46)	(64, 470.46)	(65, 494.46)	(65, 449.46)	(66, 455.46)
		(67, 459.46)	(65, 439.46)	(66, 416.46)	(66, 443.46)	(68, 432.46)	(67, 399.46)
		(68, 450.46)	(66, 401.46)	(67, 408.46)	(67, 418.46)		(68, 396.46)
		(69, 447.46)	(67, 390.46)	(68, 398.46)	(68, 417.46)		(69, 388.46)
		(70, 426.46)	(74, 386.46)	(69, 392.46)	(69, 410.46)		
		(72, 416.46)			(71, 406.46)		
					(72, 403.46)		
MK07	20 × 50 × 100	(147, 1821.29)	(144, 1492.49)	(143, 1789.49)	(146, 1718.49)	(146, 1717.57)	(145, 1760.49)
		(148, 1722.49)	(146, 1476.49)	(144, 1542.49)	(147, 1702.49)	(148, 1591.49)	(146, 1573.49)
		(149, 1764.49)	(147, 1465.49)	(146, 1522.49)	(148.1539.49)	(150, 1585.49)	(148, 1540.49)
		(150, 1675.49)	(148, 1463.49)	(147, 1495.49)	(150, 1501.49)	(152, 1526.49)	(149, 1489.49)
		(151, 1665.49)	(150, 1444.49)	(149, 1477.49)	(152, 1492.49)	(153, 1516.49)	(150, 1457.49)
		(152, 1553.49)	(152, 1419.49)	(150, 1424.49)	(156, 1489.14)	(157, 1492.49)	(151, 1452.49)
		(153, 1523.49)		(152, 1413.49)	(157, 1488.49)		(153, 1438.49)
		(155, 1516.49)		(154, 1341.18)	(158, 1479.49)		(170, 1424.49)
		(156, 1486.49)			(160, 1445.49)		
		(157, 1474.49)					
MK09	20 × 10 × 203	(314, 3727.10)	(307, 3234.10)	(307, 3504.10)	(309, 3368.10)	(311, 3539.10)	(307, 3216.10)
		(315, 3715.10)	(309, 3206.10)	(311, 3249.10)	(312, 3359.10)	(314, 3491.10)	(315, 3190.10)
		(318, 3671.10)	(311, 3191.10)		(315, 3272.10)	(316, 3447.10)	(403, 3166.10)
		(319, 3646.10)			(318, 3268.10)	(317, 3440.10)	(409, 3084.10)
		(320, 3527.10)			(323, 3260.10)	(327, 3423.10)	(418, 2993.10)
		(323, 3516.10)			(327, 3236, 10)	(329, 3386.10)	
		(325, 3496.10)			(339, 3229.10)	(330, 3380.10)	

Table 3 Comparison results on Dauzère-Pérès and Paulli's data set (Dauzère-Pérès and Paulli 1997)

Instance	Problem Size	MOPSO	MODE-ms1	MODE-ms2	MODE-ms3	MODE-ms4	MODE-ms5
02a	10 × 5 × 196	(2473, 17761.65)	(2426, 16125.65)	(2415, 16641.65)	(2448, 16603.65)	(2456, 17287.65)	(2412, 16073.65)
		(2501, 17149.65)	(2445, 16031.65)	(2419, 16428.65)	(2451, 16337.65)	(2465, 16969.65)	(2413, 15994.65)
		(2526, 16634.65)	(2446, 15858.65)	(2424, 15884.65)	(2466, 15666.65)	(2493, 16584.65)	(2452, 15905.65)
		(2579, 16632.65)	(2452, 15684.65)	(2433, 15857.65)	(2502, 15630.65)	(2505, 16489.65)	(2463, 15663.65)
		(2616, 16614.65)	(2456, 15679.65)	(2435, 15754.65)	(2526, 15581.65)	(2509, 16288.65)	(2549, 15662.65)
		(2653, 16578.65)	(2521, 15637.65)	(2471, 15602.65)	(2547, 15552.65)	(2524, 16101.65)	(3051, 15599.65)
		(2661, 16491.65)		(2481, 15478.65)		(2528, 16063.65)	(3195, 15403.65)
		(2668, 16209.65)		(2513, 15452.65)		(2554, 15909.65)	(3352, 15339.65)
		(2716, 16153.65)					
09a	15 × 8 × 293	(2199, 21611.3)	(2193, 21572.3)	(2168, 21022.3)	(2184, 20993.3)	(2184, 20856.3)	(2170, 20909.3)
		(2225, 21334.3)	(2195, 21007.3)	(2174, 20228.3)	(2190, 20979.3)	(2187, 20840.3)	(2172, 20734.3)
		(2236, 21264.3)	(2198, 20931.3)	(2191, 20101.3)	(2191, 20710.3)	(2194, 20776.3)	(2173, 20691.3)
		(2245, 21157.3)	(2201, 20886.3)	(2196, 20098.3)	(2194, 20479.3)	(2206, 20298.3)	(2178, 20669.3)
		(2247, 20952.3)	(2207, 20748.3)	(2198, 20088.3)	(2204, 20257.3)	(2232, 20227.3)	(2180, 20638.3)
		(2255, 20895.3)	(2210, 20678.3)	(2200, 20068.3)	(2208, 20106.3)	(2234, 20221.3)	(2181, 20605.3)
		(2269, 20798.3)	(2212, 20512.3)	(2201, 20015.3)	(2212, 20088.3)	(2247, 20093.3)	(2182, 20529.3)
		(2275, 20757.3)	(2215, 20457.3)	(2204, 19650.3)	(2213, 20062.3)		(2183, 20246.3)
		(2281, 20685.3)	(2226, 20350.3)	(2210, 19578.3)	(2224, 20013.3)		(2193, 20099.3)
		(2290, 20644.3)	(2234, 20197.3)	(2253, 19464.3)	(2229, 19927.3)		(2206, 20074.3)
		(2299, 20459.3)	(2238, 19911.3)		(2248, 19753.3)		(2207, 19600.3)
							(2220, 19430.3)
11a	15 × 8 × 293	(2319, 22980.61)	(2274, 21720.61)	(2247, 21397.61)	(2281, 21425.61)	(2289, 22764.61)	(2234, 21190.61)
		(2344, 22627.61)	(2276, 21624.61)	(2272, 21134.61)	(2325, 21248.61)	(2305, 22404.61)	(2238, 21109.61)
		(2350, 22617.61)	(2311, 21425.61)	(2273, 21073.61)	(2349, 21147.61)	(2307, 22178.61)	(2312, 21015.61)
		(2354, 22604.61)	(2314, 21385.61)	(2282, 20537.61)		(2325, 21943.61)	(2325, 20767.61)
		(2356, 22240.61)	(2315, 21251.61)	(2307, 20496.61)		(2345, 21850.61)	(2328, 20763.61)

(continued)

Table 3 (continued)

Instance	Problem Size	MOPSO	MODE-ms1	MODE-ms2	MODE-ms3	MODE-ms4	MODE-ms5
		(2374, 21858.61)	(2345, 21250.61)	(2330, 20385.61)		(2349, 21652.61)	(2866, 20301.61)
		(2431, 21602.61)	(2363, 21142.61)			(2350, 21411.61)	(2951, 20296.61)
		(2472, 21456.61)	(2369, 21002.61)			(2375, 21381.61)	(3152, 20226.61)
16a	20 × 10 × 387	(2718, 34813.83)	(2582, 33004.83)	(2599, 33336.83)	(2661, 33825.83)	(2719, 34957.83)	(2605, 33128.83)
		(2748, 34757.83)	(2584, 32930.83)	(2602, 33062.83)	(2680, 33217.83)	(2733, 34493.83)	(2610, 32426.83)
		(2791, 34526.83)	(2607, 32909.83)	(2611, 32964.83)	(2707, 33176.83)	(2740, 34217.83)	(2636, 32159.83)
		(2795, 34111.83)	(2635, 32450.83)	(2626, 32839.83)	(2740, 33105.83)	(2782, 34177.83)	(2644, 32057.83)
		(2814, 34008.83)	(2642, 32009.83)	(2631, 32809.83)	(2751, 32841.83)	(2783, 33897.83)	(2658, 31897.83)
		(2822, 33726.83)	(2643, 31674.83)	(2649, 32645.83)	(2779, 31948.83)		(3380, 31709.83)
		(2985, 33594.83)	(2644, 31577.83)	(2716, 32630.83)			(3432, 31657.83)
		(3102, 33063.83)	(2652, 31294.83)	(2777, 32594.83)			(3530, 31626.83)
		(3267, 32773.83)	(2657, 31187.83)				(3552, 31130.83)
18a	20 × 10 × 387	(2286, 28386. 92)	(2247, 26721.92)	(2238, 27556.92)	(2266, 26519.92)	(2256, 27918.92)	(2227, 26698.92)
		(2288, 28023. 92)	(2271, 26684.92)	(2248, 27479.92)	(2277, 26324.92)	(2261, 27841.92)	(2229, 26431.92)
		(2294, 27542. 92)	(2280, 26630.92)	(2251, 26097.92)	(2281, 26225.92)	(2263, 27753.92)	(2233, 26428.92)
		(2320, 27100. 92)	(2284, 26567.92)	(2253, 25754.92)	(2284, 26139.92)	(2264, 27729.92)	(2234, 26363.92)
		(2341, 26894. 92)	(2287, 26418.92)	(2254, 25651.92)	(2286, 25977.92)	(2265, 27660.92)	(2237, 26354.92)
		(2421, 26814. 92)	(2297, 26392.92)	(2256, 25447.92)	(2287, 25964.92)	(2268, 27539.92)	(2238, 26135.92)
		(2429, 26600. 92)	(2308, 25794.92)	(2276, 25307.92)	(2295, 25970.92)	(2272, 26983.92)	(2258, 25972.92)
		(2431, 26567. 92)	(2402, 25653.92)	(2277, 25190.92)	(2299, 25752.92)	(2274, 26851.92)	(2273, 25447.92)
		(2459, 26395. 92)	(2556, 25622.92)	(2278, 25123.92)	(2315, 25350.92)	(2289, 26733.92)	(2312, 25281.92)
		(2461, 26355. 92)	(2725, 25314.92)	(2280, 24991.92)	(2319, 25188.92)	(2292, 26416.92)	(2324, 25266.92)
		(2482, 26331. 92)	(2789, 24078.92)	(2282, 24789.92)		(2302, 26081.92)	(2327, 25183.92)

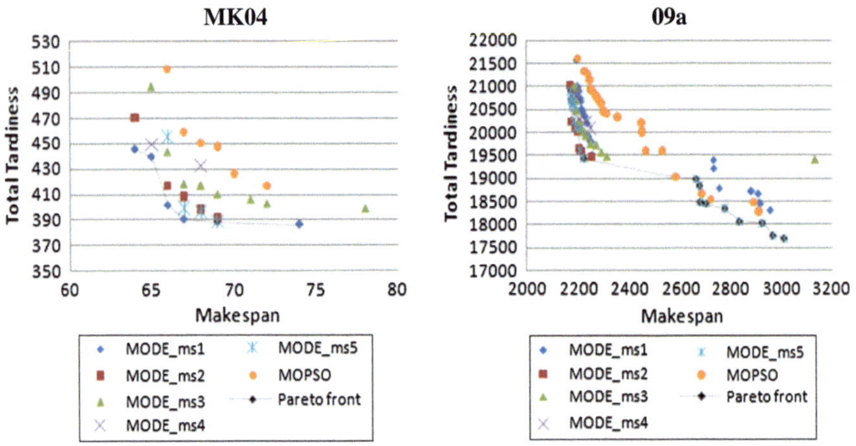

Fig. 4 Comparison of non-dominated solutions

Table 4 The *C* metric comparison between MODE strategies and MOPSO

Instance	C(P, ms1)	C(ms1, P)	C(P, ms2)	C(ms2, P)	C(P, ms3)
MK04	0	1	0	1	0
MK07	0	1	0	1	0.333
MK09	0	0.8	0	0.8	0
2a	0	0.769	0	0.923	0
9a	0.391	0.783	0	0.783	0.067
11a	0	0.818	0	1	0
16a	0	0.947	0	0.474	0
18a	0.176	0.773	0	0.864	0
Instance	**C(ms3, P)**	**C(P, ms4)**	**C(ms4, P)**	**C(P, ms5)**	**C(ms5, P)**
MK04	0.667	0	0.667	0	1
MK07	0.727	0.167	0.727	0	1
MK09	0.8	0.5	0.7	0.4	0.8
2a	0.923	0	0.692	0.25	0.692
9a	0.783	0	0.609	0	0.956
11a	0.727	0	0.818	0	0.909
16a	0.632	0.2	0.263	0.273	0.737
18a	0.818	0	0.545	0	0.818

represents that all solutions in B are non-dominated by any solution in A. The lower the ratio $\tilde{C}(A,B)$ is, the better the solution set in B is.

The comparison between MOPSO and MODE strategies and the comparison among MODE strategies are shown in Tables 4 and 5. It is noted that P, ms1, ms2, ms3, ms4, and ms5 are represented for MOPSO, MODE-ms1, MODE-ms2, MODE-ms3, MODE-ms4, and MODE-ms5, respectively.

It is shown from Table 4 that MOPSO is inferior to the MODE strategies. The MOPSO mostly obtains the C metric values [$C(ms1, P)$, $C(ms2, P)$, $C(ms3, P)$,

Table 5 The *C* metric comparison among MODE strategies

Instance	C(ms1, ms2)	C(ms2, ms1)	C(ms1, ms3)	C(ms3, ms1)	C(ms1, ms4)	C(ms4, ms1)	C(ms1, ms5)	C(ms5, ms1)	C(ms2, ms3)	C(ms3, ms2)
MK04	1	0	1	0	1	0	0.75	0	1	0
MK07	0.5	0.333	1	0	1	0	1	0	1	0
MK09	1	0	1	0	1	0	0	0.333	0.571	0
2a	0	0.667	0.333	0.167	1	0	0.25	0.333	0.833	0
9a	0	0.609	0.2	0.478	0.286	0.391	0	1	0.933	0
11a	0	0.889	0	0.222	0	0	0	1	1	0
16a	0.75	0	1	0	1	0	0.636	0.083	0.833	0
18a	0.083	0.529	0.091	0.529	0.818	0	0	0.941	0.909	0
Instance	C(ms2, ms4)	C(ms4, ms2)	C(ms2, ms5)	C(ms5, ms2)	C(ms3, ms4)	C(ms4, ms3)	C(ms3, ms5)	C(ms5, ms3)	C(ms4, ms5)	C(ms5, ms4)
MK04	0.5	0	0.25	0.6	0.5	0.125	0	0.75	0.25	0.5
MK07	1	0	1	0	0.833	0.111	0.125	0.889	0	1
MK09	1	0	0	1	1	0	0	1	0	1
2a	1	0	0.375	0.25	1	0	0.25	0.5	0	1
9a	1	0	0.286	0.2	0.714	0.133	0	1	0	1
11a	1	0	0.3	0.333	0.9	0	0	1	0	1
16a	1	0	0.091	0.75	1	0	0	1	0	1
18a	1	0	0.364	0.167	0.545	0	0.045	0.909	0	1

C(*ms*4, *P*), and *C*(*ms*5, *P*), respectively] of 1 or close to 1 whereas MODE-ms1, MODE-ms2, MODE-ms3, MODE-ms4, and MODE-ms5 generally obtain the *C* metric values [*C*(*P*, *ms*1), *C*(*P*, *ms*2), *C*(*P*, *ms*3), *C*(*P*, *ms*4), and *C*(*P*, *ms*5), respectively] of 0 or close to 0. This demonstrates that the majority of solutions generated from MOPSO are completely dominated by those from all MODE strategies, and there are no or only few solutions, found by the MODE algorithms are dominated by those from MOPSO.

According to results from Table 5, it can be concluded that MODE-ms1, MODE-ms2, and MODE-ms5 outperforms MODE-ms3 since most values of *C*(*ms*3, *ms*1), *C*(*ms*3, *ms*2), and *C*(*ms*5, *ms*2) are equal to zero and lower than the value of *C*(*ms*1, *ms*3), *C*(*ms*2, *ms*3), and *C*(*ms*2, *ms*5), respectively. MODE-ms5 demonstrates outstanding results over MODE-ms1 and MODE-ms2 since the MODE-ms5 is able to generate a better quality of non-dominated solutions. The performance of MODE-ms1 and MODE-ms2 are relatively equivalent because the difference between the *C* metric indicating that the MODE-ms1 is dominated by MODE-ms2 [*C*(*ms*2, *ms*1)] and MODE-ms2 is dominated by MODE-ms1 [*C*(*ms*1, *ms*2)] is not significant. Obviously, the MODE-ms4 is the worse strategy among the proposed MODE strategies. Its *C* metric values are mostly equal to one compared to others indicating that the quality of solutions generated from the MODE-ms4 is inferior.

5 Conclusion

This study presents an implementation of MODE for solving multi-objective flexible job shop scheduling problems (MOFJSP) with the objective to simultaneously minimize makespan and total tardiness of jobs. The MODE framework uses Elite group to store solutions and utilizes those solutions as the guidance of the vectors. Five mutation strategies with different search behaviors are proposed in order to search for the Pareto front. The performances of MODE are evaluated on a set of benchmark FJSP problems and compared with results from an effective MOPSO algorithm. The experimental results demonstrate that MODE algorithms provide outstanding results compared to MOPSO. All MODE mutation strategies are capable of finding a set of diverse and high quality non-dominated solutions. The future research aims to extend implementations of MODE not only in scheduling problems but other optimization domains.

References

Abbass HA, Sarker R, Newton C (2001) PDE: a Pareto-frontier differential evolution approach for multi-objective optimization problems. In Proceedings of the 2001 congress on evolutionary computation, vol 2, pp 971–978

Bean JC (1994) Genetic algorithms and random keys for sequencing and optimization. ORSA J Comput 6(2):154–160

Bierwirth C (1995) A generalized permutation approach to job shop scheduling with genetic algorithms. OR Spectr Spec Issue Appl Local Search 17(2–3):87–92

Brandimarte P (1993) Routing and scheduling in a flexible job-shop by tabu search. Ann Oper Res 41(3):157–183

Dauzère-Pérès S, Paulli J (1997) An integrated approach for modeling and solving the general multiprocessor job-shop scheduling problem using tabu search. Ann Oper Res 70:281–306

Gao J, Gen M, Sun L, Zhao X (2007). A hybrid of genetic algorithm and bottleneck shifting for multiobjective flexible job shop scheduling problems. Comput Ind Eng 53(1):149–162

He Z, Yang T, Deal DE (1993) A multiple-pass heuristic rule for job-shop scheduling with due dates. Int J Prod Res 31(11):187–199

Kacem I, Hammadi S, Borne P (2002a) Approach by localization and multi-objective evolutionary optimization for flexible job shop scheduling problems. IEEE Trans Syst Man Cybertics Part C 32(1):1–13

Kacem I, Hammadi S, Borne P (2002b) Pareto-optimality approach for flexible job-shop scheduling problems: hybridization of evolutionary algorithms and fuzzy logic. Math Comput Simul 60(3–5):245–276

Madavan NK (2002) Multiobjective optimization using a Pareto differential evolution approach. In: Proceedings of the evolutionary computation, pp 1145-1150, 12-17 May 2002

Moslehi G, Mahnam M (2011) A Pareto approach to multi-objective flexible job-shop scheduling problem using particle swarm optimization and local search. Int J Prod Econ 129(1):14–22

Nguyen S, Kachitvichyanukul V (2010) Movement strategies for multi-objective particle swarm optimization. Int J Appl Metaheuristic Comput 1(3):59–79

Qian B, Wang L, Huang DX, Wang X (2008) Scheduling multi-objective job shops using a memetic algorithm based on differential evolution. Int J Adv Manuf Technol 35(9–10):1014–1027

Storn R, Price K (1995) Differential evolution—a simple and efficient adaptive scheme for global optimization over continuous spaces. Technical Report TR-95-012. International Computer Science, Berkeley, CA

Wisittipanich W, Kachitvichyanukul V (2011) Differential evolution algorithm for makespan minimization in flexible job shop scheduling problem. In: Proceeding of the 12th Asia Pacific industrial engineering and management systems conference, Beijing, China

Wisittipanich W, Kachitvichyanukul V (2013) Total Tardiness minimization in flexible job shop scheduling problem by differential evolution algorithm. In: Proceeding of operations research network conference, Thailand

Wisittipanich W, Kachitvichyanukul V (2014) Mutation Strategies toward Pareto front for multi-objective differential evolution algorithm. Int J Oper Res 19(3):317–337

Wisittipanich W, Kachitvichyanukul V (2013) A Pareto-based differential evolution algorithm for multi-objective job shop scheduling problems. In: Proceeding of the institute of industrial engineers Asian conference, pp 1117–1125

Zhang H, Gen M (2005) Multistage-based genetic algorithm for flexible job-shop scheduling problem. J Complex Int 11:223–232

Sugarcane Harvest Scheduling to Maximize Total Sugar Yield with Consideration of Equity in Quality Among the Growers

Kanchana Sethanan, Somnuk Theerakulpisut and Woraya Neungmatcha

Abstract This chapter proposes a farm harvest scheduling to provide cane equity among growers while maximizing the total sugar yield over the entire mill area. A heuristic algorithm, called two-phase cane harvest scheduling (CANEHAS), was developed to solve a practical large-size problem. In Phase I, the CANEHAS algorithm was developed to maximize the total sugar yield. In Phase II, the growers' equity, in terms of cane quality, was proposed. The solutions of both phases were improved using Tabu search. The mill could use the proposed algorithms, either Phase I or Phase II, to make decisions when providing the queue cards to the contracted grower groups in order to increase profitability.

Keywords Sugarcane harvesting · Tabu search · Heuristic · Supply chain management · Mathematical model

1 Introduction

Sugar production requires the integration of supply chain management, incorporating inbound logistics activities (i.e. growing, harvesting, and transporting to the sugar mill) and internal logistics activities which involve milling to produce raw sugar. Hence, raw material is considered to be of the utmost importance for the

K. Sethanan (✉) · W. Neungmatcha
Research Unit on Advanced Productivity Improvement and Logistics Management, Industrial Engineering Department, Faculty of Engineering, Khon Kaen University, Khon Kaen 40002, Thailand
e-mail: ksethanan@gmail.com

S. Theerakulpisut
Mechanical Engineering Department, Faculty of Engineering, Khon Kaen University, Khon Kaen 40002, Thailand
e-mail: somthe@kku.ac.th

P. Golinska (ed.), *Logistics Operations, Supply Chain Management and Sustainability*, EcoProduction, DOI: 10.1007/978-3-319-07287-6_24,

production of sugar from sugarcane. Generally, farmers make a crop selection based on various criteria, including prices, experience, availability of inputs such as labor (both household and hire), availability of seeds, government policy, and a host of environment factors, such as soil and climatic conditions. Under certain unfavorable conditions, some farmers may not choose to grow sugar cane. Such a decision in crop production may gradually lead to a decline in sugarcane production. For a given quantity of cane contracted, the mill must seek opportunities to increase the sugar volume of production by increasing the total sugar yield. Within the mill area, the sugar yields obtained for a given harvest date and crop age depend on crop variety, crop class and geographical area. Also, sugar yields vary with seasonal climate conditions from year to year. Harvesting is one area considered to have potential in enhancing net revenue gain in various mill areas. Therefore, the harvesting date is considered a key factor for sugar yield and net profits (Higgins et al. 1998; Higgins 1999). The field schedules can exploit differences in CCS (the percentage of extractable sugar from cane) at harvesting dates and require an efficient method be developed to capture this opportunity. In Thailand, the sugarcane payment scheme is called a 70:30 revenue sharing system. Under the 70:30 revenue sharing system, the net industry proceed is allocated to growers and millers at the ratio of 70 to 30. This system will mainly provide gains in net revenue to the mill from increased sugar yield and to the growers from both increased sugar yield and increased CCS. To maximize the profitability at the mill area level by increasing the total sugarcane yield, strategies are hence focused on cane harvesting and improvement in breeding. Technical and managerial strategies can be practiced. Technical improvement approaches currently considered are (1) design of efficient tools and equipment for better harvesting activities in order to reduce the length of the cane stubble, which is the cane part containing the highest content of sucrose, and (2) breeding new cane varieties to suit the conditions of the growing areas (i.e. soil, climate, and irrigation), whereas the managerial approach normally employs techniques aiming for low costs but requiring more time to implement. The managerial technique in harvesting considered in this study is how to make the harvesting schedule of the sugarcane fields in a mill area efficient.

Recent studies have been made in order to enhance the sugarcane harvesting system. Unfortunately, those models cannot be totally implemented in Thailand because they are somewhat complicated and were developed under different environments regarding sugarcane production system, transportation modes, harvesting modes, strategic management of supply, and potential of growers considering education and skill. In Australia, a research team has developed a model that defines optimal harvest dates and crop cycle length in order to maximize the total sugar yield and the gains in terms of net revenue at the mill region (Higgins et al. 1998; Higgins 1999). Higgins et al. (1998) develop an optimization model applied with block productivity data to explore options for better scheduling the sugarcane harvest in order to maximize net revenue for the entire mill region. In 1999, Higgins (Hildebrand 2002) developed a large-scale integer model to optimize the decisions of harvest date, length of crop cycle, and whether to fallow for all the fields within a mill region in order to maximize net revenue over a planning horizon of several years. However, the large model is developed as a random based heuristic to solve for an industrial size and to improve the solution using Tabu search. The harvesting and transport system in the sugarcane value chain has been regarded as the highest priority in an independent evaluation of the sugar industry in Australia (Hildebrand 2002). Higgins et al. (2004) developed a modeling framework identifying the key drivers and links so that a suite of models representing the components can be used to explore the potential benefits of system-wide opportunities along the value chain. In 2006, Higgins (2006) improved harvesting and transport within a sugar value chain. A modeling framework was developed which considered the logistical and economic complexity of the harvesting and transport systems. This model framework was used to address some key industry issues in rationalizing rail track infrastructures and re-organizing harvesting. The models in the literature mostly considered only how to increase the total sugar yield over the entire mill area. Although maximizing the total sugar yield of the entire mill region will vary the gains among growers, it does not guarantee the average revenue per ton of cane among growers resulting in less sustainability of the sugar production system in the near future. Therefore, this research focuses on the development of a sugarcane harvest scheduling with the objective to balance and increase profitability among all stakeholders in the sugarcane supply chain, especially the mill and growers, to ensure that growers obtain both a fair revenue and an approximately equal proportion of cane throughout the harvest season, while the total sugar yield over the entire mill areas is maximized. In the next section, the case-study approach which was adopted in this study is discussed. The first model which was an optimizing algorithm developed prior to this study is described in Sect. 3. This is followed by the development of a heuristic for solving the problem in Sect. 4. Section 5 presents the numerical illustration of the solution to the problem by the heuristic approach. Finally, a summary of the main findings and discussion is given in Sect. 6.

2 Case Study Approach

The research was motivated by a case-study of a sugar growing area in north-eastern Thailand. In this region there are 14 sugar mills out of the total 46 sugar mills in the entire country. In the production of sugar, there are various stakeholders involved in the sugarcane supply chain which regulates the flow of material from growers to millers in both quality and quantity (Lejars et al. 2008; Le Gal et al. 2008). To make the production system more efficient and sustainable, a reduction of excessive supplies during the peak of the harvest season is needed. Also, managing the sugarcane supplied to the mill according to its maximum crushing capacity, by providing the harvest schedules to growers so that they harvest sugar cane at different times while fulfilling the required equity among the growers, will ensure a smooth flow of sugarcane the mills. As part of the sugarcane payment system, the sugar industry in Thailand will pay for sugarcane based on the CCS and the ton of cane delivered to the mill. A premium is paid for CCS, which is a measure of sugarcane quality. Since Thai growers are paid for their cane crop after it has been harvested and processed in the mills, the equity system also provides a consistent payment throughout the harvest season. On the other hand, equity regulation is a formal agreement among growers that aims to prevent any farms from late payment for their crops and from having any unfair disadvantage of harvest time. Hence, during the peak of a harvest season, the mill currently practices an equity policy to manage sugarcane supplied to the mill called "Queue Locked System (QLS)". This equity regulation permits each grower to have an approximately equal proportion of his sugarcane cut during each round throughout the harvest season. Besides providing the equity among the growers and managing sugarcane supply corresponding to the mill's maximum crushing capacity, the QLS was developed to ensure the quality of the cane supplied to the mill. The system will guard against sugarcane being dry or deteriorated in cases of burnt cane. The QLS system permits growers to supply their sugarcane based on the queue card from the mill which provides the harvest schedules for each grower. In the queue card, the quantity, the date, and the time for sugarcane supplied to the mill are specified. To practice the equity regulation, the mill determines the numbers of queue cards depending on the total amount of sugarcane contracted with the growers, the number of crushing days, and the maximum crushing capacity of the mill. Unfortunately, using the QLS, the mill can only control the amount of harvested sugarcane supplied to the mill by each grower group, but it cannot entirely control the sugarcane quality in terms of CCS of the harvested sugarcane because the sugarcane will be harvested in rotation from each field where each grower will decide on his own preference depending on farmer equity, field preparation, and amount of sugarcane in the fields.

3 Optimization Algorithm

The mathematical model for doing just this will be presented in this section. Parameters and variables used in formulating the mathematical model are defined as follows:

Indices

p field indices; 1, 2, ..., P

q grower group indices; 1, 2, ..., Q

t harvest date indices; 1, 2, ..., T

Parameters

MC_t maximum crushing capacity of the mill at time period t

$CANE_q$ amount of cane contracted to the mill of grower group q

$CCS_{p,q,t}$ CCS value of the qth grower group of the pth field at time period t

$B_{p,q}$	amount of cane available of the pth field of grower group q
$CANE_{p,q,t}$	amount of cane available of the pth field of grower group q at time period t
G_AvgCCS	average CCS per ton of cane obtained from the entire mill area

Decision variables

$QTY_{p,q,t}$	amount of cane cut at the pth field of grower group q at time period t
$TotalYield$	total sugar yield over the entire mill area
$AvgCCS_q$	actual average CCS per ton of cane of grower group q
$DIFF_q$	the difference between the average of actual CCS per ton of cane delivered to the mill of grower group q and the average CCS per ton of cane over the entire mill area throughout the harvest season
$TotalDIFF$	total difference between the average of actual CCS per ton of cane delivered to the mill of grower group q and the average CCS per ton of cane over the entire mill area throughout the harvest season

Binary variable

$$X_{p,q,t} = \begin{cases} 1, & \textit{if cane cut at the } p\text{th} \textit{ field of grower group } q \\ & \textit{at time period } t \\ 0, & \textit{otherwise} \end{cases}$$

Objective Function

$$\textit{Minimize } Z = TotalDIFF \tag{1}$$

Subject to

$$\sum_{p=1}^{P}\sum_{q=1}^{Q} X_{p,q,t} \cdot QTY_{p,q,t} \le MC_t \quad ;\forall t \tag{2}$$

$$\sum_{p=1}^{P}\sum_{t=1}^{T} X_{p,q,t} \cdot QTY_{p,q,t} \le CANE_q \quad ;\forall q \tag{3}$$

$$\sum_{p=1}^{P} X_{p,q,t} \cdot B_{p,q} \le CANE_q \quad ;\forall q \tag{4}$$

$$\sum_{t=1}^{T} X_{p,q,t} \cdot CANE_{p,q,t} \le B_{p,q} \quad ;\forall p, \forall q \tag{5}$$

$$TotalYield \le \sum_{p=1}^{P}\sum_{q=1}^{Q}\sum_{t=1}^{T} (CCS_{p,q,t} \cdot QTY_{p,q,t} \cdot X_{p,q,t}) \quad ;\forall p, \forall q, \forall t \tag{6}$$

$$G\_AvgCCS \leq TotalYield \Big/ \sum_{p=1}^{P} \sum_{q=1}^{Q} \sum_{t=1}^{T} QTY_{p,q,t} \quad ;\forall p, \forall q, \forall t \tag{7}$$

$$AvgCCS_q \leq \frac{\sum_{p=1}^{P} \sum_{t=1}^{T} (QTY_{p,q,t} \cdot CCS_{p,q,t} \cdot X_{p,q,t})}{\sum_{p=1}^{P} \sum_{t=1}^{T} QTY_{p,q,t}} \quad ;\forall q \tag{8}$$

$$DIFFq \leq \sum_{q=1}^{Q} (G\_AvgCCS - AvgCCS_q) \quad ;\forall q \tag{9}$$

$$TotalDIFF \leq \sum_{q=1}^{Q} DIFFq \tag{10}$$

$$QTY_{p,q,t} \geq 0 \quad ;\forall p, \forall q, \forall t \tag{11}$$

The objective function minimizes the total difference between the average sugar yield of each grower group and the average of the total sugar yield obtained from the mill region. Constraints (2) ensure that the total amount of cut cane supplied to the mill must not exceed the maximum crushing capacity at each period. Constraints (3) ensure that the total amount supplied to the mill by each grower group must not exceed its amount of cane contracted to the mill. Constraints (4) indicate that the total available amount of cane from all fields of each grower group must not exceed the total amount of cane contracted to the mill. Constraints (5) ensure that the total amount of cane cut from each field during all periods must not exceed the total available amount of cane at the beginning. Constraints (6) determine the total sugar yield obtained from the entire mill region throughout the harvest season, while Constraints (7) determine the average of the total sugar yield value obtained from the entire mill region. Constraints (8) determine the average of the total sugar yield of each grower group. Constraints (9) determine the difference between the average of actual CCS per ton of cane delivered to the mill of grower group q and the average CCS per ton of cane over the entire mill area throughout the harvest season. Constraints (10) determine the total difference between the average of actual CCS per ton of cane delivered to the mill of grower group q and the average CCS per ton of cane over the entire mill area throughout the harvest season. The mathematical model was implemented on a standard PC with a Pentium CPU 1.90 GHz and 1.37 GB RAM using ILOG's OPL Studio 3.6.1 as the modeling environment and CPLEX 8.1 as the solver. Although an optimal solution was obtained for the problem, the computational time was excessive. Attempts to solve larger problems were unsuccessful as they required too much CPU time. Hence, a heuristic algorithm was developed to obtain a near-optimal solution for realistic sized problems.

4 Heuristic Development

In this section, a two-phase algorithm was developed. This heuristic was termed "*CANE HArvest Scheduling*": CANEHAS. The CANEHAS is a two-phase algorithm. The first phase (i.e., Phase I) consists of a constructive heuristic developed to obtain an initial solution to maximize the total sugar yield over the entire mill region. The second phase, referred to as Phase II, sugarcane harvesting scheduling attempts to reduce the $DIFF_q$ values obtained from the first phase in order to create equity in terms of average CCS per ton of cane among the growers. The key idea of the Phase II CANEHAS algorithm starts with determining the expected total sugar yield of each grower group ($YIELD_q$) by multiplying the G_AvgCCS by the contracted amount of cane of grower group q ($CANE_q$). At the beginning of the scheduling, for each grower group obtaining the queue cards, the heuristic schedules the field with the lowest positive difference value between the CCS of that field ($CCS_{p,q,t}$) and the G_AvgCCS if the cane of that field is in the early-mature periods. This means the cane in this field will be higher in CCS when it will be in mid-mature stage. On the other hand, the algorithm will schedule the field with the highest positive difference value between the CCS of that field ($CCS_{p,q,t}$) and the G_AvgCCS if the cane of that field is in the late-mature period, because the CCS of the cane in this field will be less as time goes by. Every time the cane is harvested, the algorithm will update the remaining total sugar yield of each grower group and attempt to find the field with the $CCS_{p,q,t}$ closed to the average CCS per ton of cane that the grower group will obtain ($CCS_{q,t}$) instead of G_AvgCCS. The process is repeated until all the fields of the grower groups are harvested at the time according to the schedules from the mill. Hence, the algorithm attempts to evenly share the average CCS per ton of cane supplied to the mill among all grower groups over the entire harvest season. The problem parameters and notations used in the development of the heuristics are defined below.

Parameters

$CCS_{p,q,t}$ CCS value of the qth grower group of the field at time period t

$DCCS_{p,q,t}$ the differences of the CCS value of field pth belonging to grower group qth between time period t and time period $t + 1 = CCS_{p,q,t} - CCS_{p,q,t+1}$

$QUE_{q,t}$ maximum amount of sugarcane of the qth grower group can be supplied to the plant at time period t (tons)

$CANE_{p,q,t}$ amount of available sugarcane of the qth grower group of the pth field at time period t (tons)

$CANE_{q,t}$ amount of available sugarcane of the qth grower group at time period t (tons) $= \sum_{p=1}^{P} CANE_{p,q,t}$

Decision Variables

$QTY_{p,q,t}$ the actual amount of sugarcane of the qth grower group of pth field supplied to the mill at time period t

$YIELD_{p,q,t}$ Actual sugar yield of the qth grower group determined from multiplying the CCS value by the amount of sugarcane of the pth field supplied to the mill at time period $t = CCS_{p,q,t} \cdot QTY_{p,q,t}$

$YIELD_q$ Actual sugar yield of the qth grower group throughout the harvest season $= \sum_{p=1}^{P} \sum_{t=1}^{T} YIELD_{p,q,t}$

$TotalYield$ Total sugar yield obtained over the entire mill areas throughout the harvest season $= \sum_{q=1}^{Q} YIELD_q$

$Expct\_Yld_{q,t}$ the expected total sugar yield of the qth grower group at time period $t = G\_AvgCCS \cdot CANE_{q,t}$

$CCS_{q,t}$ the average CCS per ton of cane of the qth growergroup of at time period $t = \frac{Expected\_Yield_{q,t}}{CANE_{q,t}}; \forall q$

$DIFF_q$ the difference between the average of CCS per ton of cane of grower group q and the average CCS per ton of cane over the entire mill area throughout season $= \frac{YIELDq}{CANEq} - G\_AvgCCS$

$TotalDIFF$ Total difference between the average CCS per ton of grower group q and the average CCS per ton of cane over the entire mill area throughout the season $= DIFF_q; \forall q$

$AvgCCS_q$ Actual average CCS per ton of grower group $q = \frac{YIELDq}{CANEq}; \forall q$

Scheduling of CANEHAS algorithm

Phase I: Find the Total sugar Yield (TotalYield) over the entire mill area

WHILE QUEq,t $\neq$ 0;$\forall$q, t or CANEp,q,t $\neq$ 0;$\forall$p,q,t

FOR each q, find the highest CCSp,q,t value.

IF there is more than one CCSp,q,t , THEN find DCCSp,q,t with the highest positive DCCSp,q,t

IF there is more than one positive DCCSp,q,t value, THEN select the highest CANEp,q,t value

IF there is more than one highest of the CANEp,q,t , THEN select any of them

END

IF $QUE_{q,t}$ - $CANE_{p,q,t}$= 0 THEN $QTY_{p,q,t} = CANE_{p,q,t}$; $QUE_{q,t} = CANE_{p,q,t} = 0$

ELSE IF $QUE_{q,t}$ - $CANE_{p,q,t} > 0$ THEN

$QTY_{p,q,t} = CANE_{p,q,t}$; $QUE_{q,t} = QUE_{q,t} - CANE_{p,q,t}$; $CANE_{p,q,t} = 0$.

ELSE IF $QUE_{q,t}$ - $CANE_{p,q,t} < 0$ THEN

$QTY_{p,q,t} = QUE_{p,q,t}$;$CANE_{p,q,t} = CANE_{p,q} - QUE_{q,t}$; $QUE_{q,t} = 0$.

Calculate $YIELD_{p,q,t} = CCS_{p,q,t} \cdot QTY_{p,q,t}$

END

Determine the $QTY_{p,q,t}$, $YIELD_q$, $AvgCCS_q$, $TotalYield$, G_AvgCCS, $DIFF_q$, and $TotalDIFF$

Phase II: Smoothen out the $DIFF_q$ values

FROM the procedure of *Phase II* THEN
Calculate $Expected\_Yield_{q,t}$
WHILE $QUE_{q,t} \neq 0; \forall q, t$ or $CANE_{p,q,t} \neq 0; \forall p,q,t$
Find p' for each grower group q
IF p' is the first field to be harvested, THEN go to Label1,
ELSE go to Label2.
Label1:Find $p' \leftarrow D\_CCS_{p',q,t} = \max_{\forall p}\{ CCS_{p,q,t} - G\_AvgCCS \}$,
IF there is more than one $D\_CCS_{p,q,t}$, THEN
go to Label3; ELSE go to Label4.
Label2: FOR each q at time period t, calculate $\overline{CCS}_{q,t}; \forall q$
Find $p' \leftarrow D\_CCS_{p',q,t} = \max_{\forall p}\{ CCS_{p,q,t} - \overline{CCS}_{q,t}\}$
IF there is more than one $D\_CCS_{p,q,t}$. THEN go to Label3,
ELSE go to Label4.
END

Labe2: Select $p' \leftarrow CANE_{p'',q,t} = \max_{\forall p'} CANE_{p',q,t}$, then set $p' = p''$
Labe3: FOR grower group q, harvest the cane at field p' at time t.

Update the values of $CANE_{p,q,t}$ and $QUE_{q,t}$
IF$QUE_{q,t}$ - $CANE_{p,q,t} = 0$, THEN
$QTY_{p,q,t} = CANE_{p,q,t}$; $QUE_{q,t} = CANE_{p,q,t} = 0$ 10.2
ELSE IF $QUE_{q,t}$ - $CANE_{p,q,t} > 0$, THEN
$QTY_{p,q,t} = CANE_{p,q,t}$; $QUE_{q,t} = QUE_{q,t} - CANE_{p,q,t}$
$CANE_{p,q,t} = 0$.
ELSE IF $QUE_{q,t}$ - $CANE_{p,q,t} < 0$, THEN
$QTY_{p,q,t} = CANE_{p,q,t}$;
$CANE_{p,q,t} = CANE_{p,q}$ - $QUE_{q,t}$ and $QUE_{q,t} = 0$
Update $CANE_{q,t} = CANE_{q,t}$ - $QTY_{p',q,t}$
$YIELD_{p,q,t} = CCS_{p,q,t} \cdot QTY_{p,q,t}$;
$Expct\_Yld_q = Expct\_Yld_q - Yld_{p',q,t}; \forall q$
END
Determine $QTY_{p,q,t}$, $YIELD_q$, $AvgCCS_q$, *TotalYield*, *G_AvgCCS*, $DIFF_q$, *Total-DIFF*

Once the CANEHAS algorithm was developed in order to obtain an initial solution for the problem, the Tabu Search Heuristic (TSH) was then developed to improve the quality of the solutions for both Phases I and II. In the algorithm, a move is insertion-type and is defined by a vector V = Ti(D, p1, q1, p2, q2, ..., pn, qn); i = 1, 2, ..., t, where Ti = actual harvest schedule of day i, D = day of planned harvest schedule obtained from the CANEHAS algorithm, pn = field number n to be harvested on day D, and qn = amount of cane to be harvested at field number n. For the tabu list, it consists of two dimensional arrays. A pair of actual harvest dates (i, j); i, j = 1, 2,...,t, and i $\neq$ j, will be stored. The tabu list size is fixed and set to be 5. The neighborhood size (*d*) depends on the number of actual harvest dates defined as

$$d = \lfloor t/2 \rfloor - 1 \; for \; t \leq 30 \; and \; d = (\lfloor t/2 \rfloor / 2) \times (c/4)$$

where

c is the value between 1 and 4. After the experiments, $c = 2$ was used in this study. For the aspiration criterion, if a move is in the tabu list and it gives a better solution than the best objective function value obtained so far, then this move is applicable. In the terminating process, an iteration number is used as termination rule and set to be 100 [as suggestion in Kinney et al. (2007)].

5 Experimental Results

Table 1 shows the essential on-farm data for the 2011/2012 crop year of the case study. Five out of eleven growing regions were selected, and twenty grower groups were selected from each growing region.

In the experiments, the algorithm was tested to see if there was an improvement in sugar yield and equal profitability among all stakeholders in the sugarcane supply chain to ensure that growers obtain both fair revenue and approximately equal proportion of cane. To accomplish this, the two-phase algorithm was compared with the actual total sugar yield and the average CCS per ton of cane obtained by the current scheduling practice. The heuristic algorithm was coded in JAVA and run on a 1.90 GHz PC, with 1.37 GBytes of RAM. The equity in terms of average CCS obtained among growers. The total sugar yield for the *TSH* algorithms of Phases I and II versus that of the current practice is shown in Table 1. It is evident from the table that the *TSH* algorithms of the two phases give better values of total sugar yield and average CCS per ton of cane than the mill's current practice for all growing regions. The improvements on the total sugar yield of Phase I CANEHAS algorithm are in the range of 19.72–33.20 %, with an average of 24.74 %, while those of Phase II CANEHAS algorithm are in the range of 4.77–23.36 %, with an average of 15.12 %, and the improvements on an average CCS per ton of cane among the growers of Phase I and Phase II CANEHAS algorithm were 9.42–18.60 and 4.77–13.14 %, respectively. Considering the grower equity in terms of the average CCS among the growers, the variance were shown that the average CCS per ton of cane among the growers for all growing regions from the Phase II CANEHAS algorithm was lowest.

6 Conclusions and Discussion

This research determines a better sugarcane harvest scheduling with the objective being to create equity among growers in terms of both cane quantity and quality while maintaining the maximum total sugar yield of the entire mill area. Firstly, a

Table 1 The total sugar yield and average CCS per ton of cane between growers for the heuristic algorithms versus current practice sugarcane harvesting

Growing region	Current practice		Phase I CANEHAS		Phase II CANEHAS		*%GAP* on total yield		*%GAP* on avg. CCS	
	Total yield	Avg. CCS	Total yield	Avg. CCS	Total yield	Avg.CCS	Currentpractice versus Phase I CANEHAS	Current practice versus Phase II CANEHAS	Current practice versus Phase I CANEHAS	Current practice versus Phase II CANEHS
1/1	83,443	10.10	111,143	11.05	102,935	10.60	33.2	23.36	9.42	4.77
1/2	89,353	9.82	115,618	11.37	105,076	11.01	29.39	17.6	15.8	10.79
2/1	92,739	9.57	111,273	10.91	103,106	10.49	19.99	11.18	14	8.72
3/2	63,452	9.74	77,026	11.55	69,662	11.22	21.39	9.79	18.6	13.14
6	94,632	9.20	113,291	10.89	107,564	10.48	19.72	13.67	18.36	12.14
Min	63,452	9.20	77,026	10.89	69,662.00	10.48	19.72	9.79	9.42	4.77
Max	94,632	10.10	115,618	11.55	107,564.00	11.22	33.20	23.36	18.60	13.14
Avg.	84,724	9.69	105,670	11.15	97,668.60	10.76	24.74	15.12	15.24	9.91
Var.		0.1105		0.1128		0.0859				

mixed integer programming model was developed to obtain an optimal solution. Unfortunately, it can only deal with small-size problems. As a result, the two-phased heuristic algorithm was developed to solve the practical larger size problem. In this chapter, the average CCS per ton of cane of each grower group was examined to ensure that all grower groups obtain the average CCS per ton of cane equally. The variance show that the average CCS per ton from the Phase II CANEHAS algorithm was lowest, which mean the average CCS per ton of cane among the growers was not different for all growing regions. Under the 70:30 revenue sharing system implemented in Thailand, the present approach helps by providing net revenue to the mill from increased sugar yield and by providing net revenue to the growers by increasing both sugar yield and CCS. The mathematical and heuristic models developed in this chapter are easily adaptable and should prove to be beneficial to other sugar industries in Thailand and around the world.

References

Higgins AJ (1999) Optimising cane supply decisions within a sugar mill region. J Sched 2:229–244

Higgins A (2006) Scheduling of road vehicles in sugarcane transport: a case study at an Australian sugar mill. Eur J Res 170:987–1000

Higgins AJ, Muchow RC, Rudd AV, Ford AW (1998) Optimizing harvest date in sugar production: a case study for the Mossman mill region in Australia. Field Crops Res 57:153–162

Higgins A, Antony G, Sandell G, Davies I, Prestwidge D, Andrew B (2004) A framework for integrating a complex harvesting and transport system for sugar production. Agric Syst 82:99–115

Hildebrand C (2002) An independent assessment of the sugar industry. Report to the Hon Truss MP, Minister for Agriculture, Fisheries and Forestry, Australia

Kinney G, Barnes JW, Colletti B (2007) A reactive tabu search algorithm with variable clustering for the unicost set covering problem. Int J Oper Res 2(2):156–172

Le Gal P-Y, Lyne PWL, Meyer E, Soler L-G (2008) Impact of sugarcane supply scheduling on mill sugar production: a South African case study. Agric Syst 96:64–74

Lejars C, Gal P-YL, Auzoux S (2008) A decision support approach for cane supply management within a sugar mill area. Comput Electron Agricu 60(2):239–249

Production Scheduling in Food Freezing Process Under the Effect of Freezer-Door Opening

Pachara Chatavithee and Supachai Pathumnakul

Abstract In this chapter, a scheduling problem of a freezing process in frozen food industry has been addressed. Freezing times of products are not uniform. Every time that the freezer is interrupted due to freezer-door open either for loading or unloading product trolleys, the freezer temperature will increase caused by the heat transfer. The freezing time of jobs being processed in the freezer need to be extended to cover the temperature loss. Therefore, the loading and unloading schedule of product trolleys in the freezing station affects to the production time and cost. A mathematical model and a heuristic algorithm were developed to determine the suitable production scheduling. The objective function was the minimization of the total cost, including cost of jobs waiting before freezing process, cost of extended processing time due to freezer-door opening, and cost of jobs that need to be delay in the freezer after freezing process has been done. The performance of the developed heuristic was assessed under various scenarios. The results were optimal for all small-size problems and outperformed the traditional First-In First-Out (FIFO) method in practical size problems.

Keywords Freezer · Ready times · Single machine · Scheduling

1 Introduction

This chapter considers a scheduling problem of a freezing process in frozen food industry. During the food processing, the products need to be frozen to maintain the quality and extend its shelf life before being exported to the customers. The freezer could process multiple product trolleys (jobs) simultaneously as shown in

P. Chatavithee · S. Pathumnakul (✉)
Department of Industrial Engineering, Faculty of Engineering, Supply Chain and Logistic Systems Research Unit, Khon Kaen University, Khon Kaen, Thailand
e-mail: supa_pat@kku.ac.th

P. Golinska (ed.), *Logistics Operations, Supply Chain Management and Sustainability*, EcoProduction, DOI: 10.1007/978-3-319-07287-6_25,

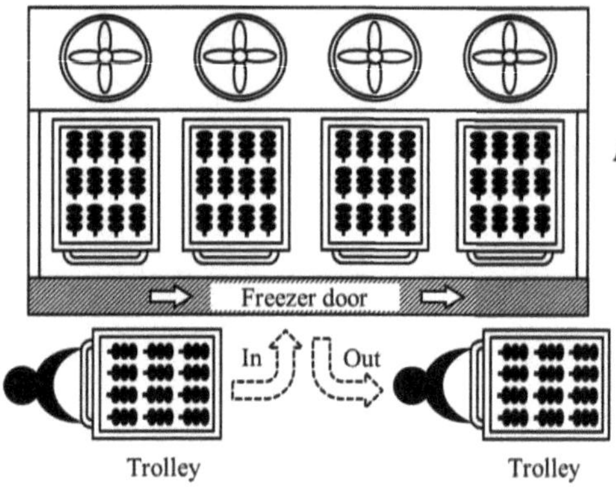

Fig. 1 Air blast freezing machine

Fig. 1. Jobs are not available at the same time. Job arrives to the freezer after be completed from the previous workstation. Processing times or the freezing times of jobs are not uniform depended on the product type. Every time that the freezer is interrupted due to freezer-door open either for loading or unloading jobs, the freezer temperature will increase caused by the heat transfer. The freezing time of jobs being processed in the freezer needs to be extended to cover the temperature loss. It has been found that the freezer's temperature is variant and approximately 10 % of energy used in freezing station due to the freezer-door open (Liu et al. 2004). Therefore, the loading and unloading schedule of jobs at the freezing station affects to the production time and cost. An example of production scheduling Gantt chart is as in Fig. 2. In practice, to reduce unnecessary freezer-door opens, job may not be loaded once it arrives to the freezing station even though the freezer capacity is available. It can be delayed to wait for the next job(s) arrives in order to load several jobs at the same time or waiting for processing job(s) to be completed and unloaded from the machine. Similarly, jobs may not be necessary to be unloaded from the freezer once it has been completed its freezing process. It may be waited in freezer until there are other jobs completed and unloaded at the same time or there are the other jobs ready to be loaded. The suitable loading and unloading schedule should be considered under various factors, especially the costs of jobs waiting to be loaded and unloaded, and also the cost related to the extended freezing time when the freezer is interrupted due to the freezer-door opening for load and unload other jobs.

This problem can be considered as the single machine scheduling problem where the machine can process multiple jobs concurrently. The problem is also called single machine scheduling with multiple job processing ability (Wang and Tang 2011) which has been found in various industries such as in the steel industry (Tang and Wang 2008, 2009; Tang and Zhao 2008; Wang and Tang 2009). For such problems, machine could process more than one job simultaneously. All jobs

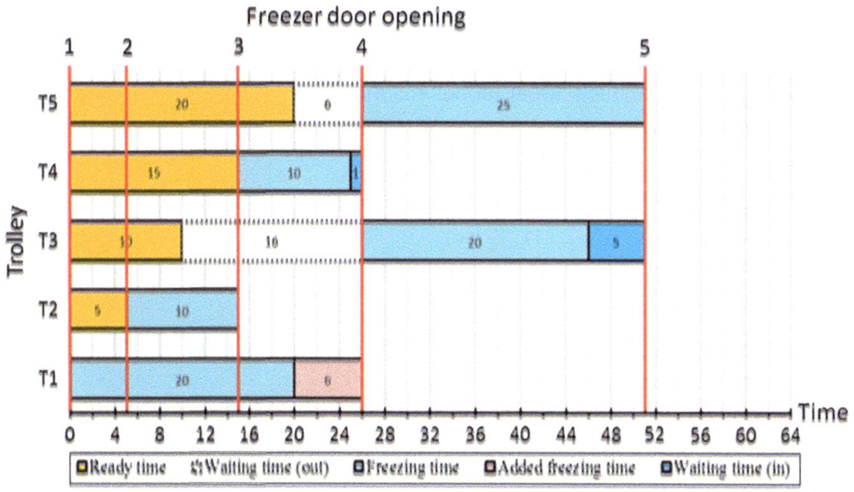

Fig. 2 An example of the production scheduling Gantt chart of the addressed problem

enter and leave the machine in batch form. When each batch starts, it could not be pre-empted and other batches could not be loaded into the machine until the processing of this batch has been finished. All jobs in the same batch have the same start time and the finish time. In (Wang and Tang 2011), the authors studied single machine scheduling with multiple jobs processing ability which jobs do not necessary enter and leave the machine in batch form. It could enter and leave machine one by one. This problem is quite similar to our proposed problem, but we consider the ready time of jobs and the jobs do not necessary to enter and leave the machine continuously one by one.

Most of the problems in the literature were solved by the mathematical model and the heuristic algorithm approaches. The mathematical model was used to determine the optimal solution for the small size problems while the heuristic algorithm was employed for the practical size problems. The problem in this chapter is different from the problem in the literature. In this problem, the jobs could be loaded into and out of the machine independently. There is no batch during processing. Each job has its individual start time and finish time, though the machine could process multiple jobs at the same time.

In this chapter, a mathematical model and a heuristic algorithm were developed to determine the suitable loading and unloading schedule of jobs at the freezing machine. The objective function was the minimization of the total costs, including cost of jobs waiting before freezing process, cost of extended processing time due to freezer-door open, and cost of job that need to be delay in the freezer after freezing process has been done.

2 Problem Formulation

In this section, the mathematical model for this problem is formulated.

2.1 Notations

Indices

J	number of jobs (product trolleys)
T	number of freezer opens
J	index of job, $j = 1, 2, \ldots, J$
t	index of freezer open, $t = 1, 2, \ldots, T$

Parameters

F_C	extended time cost (cost per unit time)
Wc^{in}	cost of job waiting to be unloaded from the freezer after the freezing process has been completed (cost per unit time)
Wc^{out}	cost of job waiting to be loaded into the freezer after arriving to freezing station (cost per unit time)
Tr_j	arrival time or ready time of job j at the freezing station
Ta_j	extended time of job j due to the freezer-door open (unit time per open)
Tf_j	freezing time or processing time of job j if there is not interrupted by freezer-door open (unit time)
Cap	maximum capacity of a freezer (jobs)
δ	minimum time duration allowed between freezer-door opens
M	large positive number
$Y_j^{Zero} \in \{0, 1\}$	$Y_j^{Zero} = 1$, if job j has been loaded into a freezer before planning starts, otherwise $Y_j^{Zero} = 0$

Decision variables

X_j^{in}	time to load job j into freezer
X_j^{out}	time to unload job j from freezer
$Y_{j,t}^{in} \in \{0, 1\}$	$Y_{j,t}^{in} = 1$, if job j is loaded into a freezer at the tth freezer-door opening, otherwise $Y_{j,t}^{in} = 0$
$Y_{j,t}^{out} \in \{0, 1\}$	$Y_{j,t}^{out} = 1$, if job j is unloaded out from a freezer at the tth freezer-door opening, otherwise $Y_{j,t}^{out} = 0$

Intermediate variable

O_t	opening time at the *t*th freezer-door opening
$Z_t \in \{0, 1\}$	$Z_t = 1$, if there exists loaded or unloaded job at the *t*th freezer-door opening, 0, otherwise
Z^{Max}	total number of freezer-door opens
Rm_t	number of processing jobs in freezer after the *t*th opening
F_j	extended freezing time of job *j* due to the freezer-door open
W_j^{in}	waiting time of job *j* to be unloaded after the freezing process is completed
W_j^{out}	waiting time of job *j* to be loaded after it arrives to freezing station

The proposed mathematical model is written as follows:
Minimize:

$$\sum_j (Wc^{out} W_j^{out} + FcF_j + Wc^{in} W_j^{in}) \tag{1}$$

Subject to

$$W_j^{out} = X_j^{in} - Tr_j \quad \forall j \tag{2}$$

$$F_j = Ta_j \left(\left(\sum_t tY_{j,t}^{out} - \sum_t tY_{j,t}^{in} \right) - 1 \right) \quad \forall j \tag{3}$$

$$W_j^{in} = X_j^{out} - (X_j^{in} + Tf_j + F_j) \quad \forall j \tag{4}$$

$$Y_j^{zero} + \sum_t Y_{j,t}^{in} = 1 \quad \forall j \tag{5}$$

$$\sum_t Y_{j,t}^{out} = 1 \quad \forall j \tag{6}$$

$$MZ_t \geq \sum_j (Y_{j,t}^{in} + Y_{j,t}^{out}) \quad \forall t \tag{7}$$

$$Z_t \leq M \sum_j (Y_{j,t}^{in} + Y_{j,t}^{out}) \quad \forall t \tag{8}$$

$$Z^{Max} \geq tZ_t \quad \forall t \tag{9}$$

$$Z^{Max} = \sum_t Z_t \tag{10}$$

$$Rm_t = Rm_{t-1} - \sum_j Y_{j,t}^{out} + \sum_j Y_{j,t}^{in} \quad \text{for } t > 1 \tag{11}$$

$$Rm_t = \sum_j Y_j^{zero} - \sum_j Y_{j,t}^{out} + \sum_j Y_{j,t}^{in} \quad \text{for } t = 1 \tag{12}$$

$$Rm_t \leq Cap \quad \forall t \tag{13}$$

$$O_t \geq X_j^{in} + M(Y_{j,t}^{in} - 1) \quad \forall j, t \tag{14}$$

$$O_t \leq X_j^{in} + M(1 - Y_{j,t}^{in}) \quad \forall j, t \tag{15}$$

$$O_t \geq X_j^{out} + M(Y_{j,t}^{out} - 1) \; \forall j, t \tag{16}$$

$$O_t \leq X_j^{out} + M(1 - Y_{j,t}^{out}) \quad \forall j, t \tag{17}$$

$$O_{t-1} + \delta \leq O_t \quad \text{for } t > 1 \tag{18}$$

The objective function (1) is to minimize total costs consisting of cost of jobs waiting before freezing process, cost of extended processing time due to freezer-door open, and cost of job that needs to be delay in the freezer after freezing process has been done. Constraint (2) estimates that the waiting time to be loaded equals to the time that a job is loaded into a freezer minus the time that the job arrives to freezer. The extended freezing time is calculated by constraint (3). It equals to the product of extended time per freezer-door open and the total number of freezer-door opens. Constraint (4) estimates the waiting time of job to be unloaded after the freezing process is completed. Constraints (5) and (6) ensure that each job must have one loaded and one unloaded operations at the freezing station. Constraints (7) and (8) define the relationship of variables $Y_{j,t}^{in}$, $Y_{j,t}^{out}$, and Z_t. Constraints (9) and (10) ensure that the index t is continuously sequenced. The number of jobs processing in freezer is balanced by constraints (11) and (12). Constraint (13) ensures that the number of jobs in the freezer must not exceed the freezer's capacity. Constraints (14), (15), (16), and (17) define the relationships of variables $X_{j,t}^{in}, X_{j,t}^{out}, Y_{j,t}^{out}, Y_{j,t}^{out}, Z_t$ and O_t. Constraint (18) ensures that time between two consecutive freezer-door opens must be larger than the minimum time duration allowed between two opens.

3 Heuristic Algorithm

The problem addressed in this chapter is classified as single-machine scheduling problem with ready time constraint. It is known as an NP-hard problem (Lenstra et al. 1977). Optimal solutions for NP-hard problems are difficult to obtain by exact algorithms, especially for large problem sizes. Our problem involves with many jobs loading and unloading at freezing station; thus, an efficient approximation algorithm that generates optimal or near optimal solutions to the problem in a reasonable computational time should be developed. In this study, we develop a heuristic, which is based on the relaxation of the proposed mathematical model. The relaxation approach and relax-and-fix heuristic have been applied to solve various types of the lot sizing and scheduling problems such as in (Ferreira et al. 2009, 2010).

The heuristic starts with scheduling all jobs based on the First-In, First-Out (FIFO) policy. The $Y_{j,t}^{in}$ and $Y_{j,t}^{out}$ are obtained. Then the initial FIFO solution is decomposed into N small sub-problems based on the number of freezer-door opens. For example, if there are ten opens based on FIFO and let $N = 5$. It means that the problem will be decomposed into five sub-problems. The first sub-problem will consist of the 1st and 2nd opens, while the 9th and 10th opens will be included in the fifth sub-problem, respectively. The solution will be improved by employing the proposed mathematical model to solve a pair of closed sub-problems simultaneously while the rest sub-problems are kept unchanged. Let P is the number of closed sub-problems to be considered for improvement. M is the set of sub-problems will be concurrently solved by the mathematical model and F is the set of sub-problems whose schedules are fixed as the previous solution.

In the case where we set $N = 5$ and $P = 4$, the problem is decomposed into five sub-problems, called A_1, A_2, A_3, A_4, A_5, and there will only two sets of sub-problems that closed to each other that we would consider to improve solution together are $\{A_1, A_2, A_3, A_4\}$, $\{A_2, A_3, A_4, A_5\}$. The set such as $\{A_1, A_2, A_3, A_5\}$ will not be considered, since the A_3 and A_5 are not closed to each other. Based on sets $\{A_1, A_2, A_3, A_4\}$, $\{A_2, A_3, A_4, A_5\}$, we may obtain 9 possible solutions by solving a pair of sub-problems as follows:

- Schedule 1: $M = \{A_1, A_2\}$, $F = \{A_3, A_4, A_5\}$
- Schedule 2: $M = \{A_1, A_3\}$, $F = \{A_2, A_4, A_5\}$
- Schedule 3: $M = \{A_1, A_4\}$, $F = \{A_2, A_3, A_5\}$
- Schedule 4: $M = \{A_2, A_3\}$, $F = \{A_1, A_4, A_5\}$
- Schedule 5: $M = \{A_2, A_4\}$, $F = \{A_1, A_3, A_5\}$
- Schedule 6: $M = \{A_2, A_5\}$, $F = \{A_1, A_3, A_4\}$
- Schedule 7: $M = \{A_3, A_4\}$, $F = \{A_1, A_2, A_5\}$
- Schedule 8: $M = \{A_3, A_5\}$, $F = \{A_1, A_2, A_4\}$
- Schedule 9: $M = \{A_4, A_5\}$, $F = \{A_1, A_2, A_3\}$

In the Schedule 1, the sub-problems A_1, A_2 will be resolved by the mathematical model while the sub-problems A_3, A_4, A_3 are kept unchanged as the previous solution. The other schedules can be described in the same manner. The best solution of the set of 9 solutions is selected to be the solution of this iteration. If the best solution obtained is better than the previous solution, then the heuristic is iterated by the same approach until no better solution is obtained. The steps of the heuristics can be described as in Fig. 3.

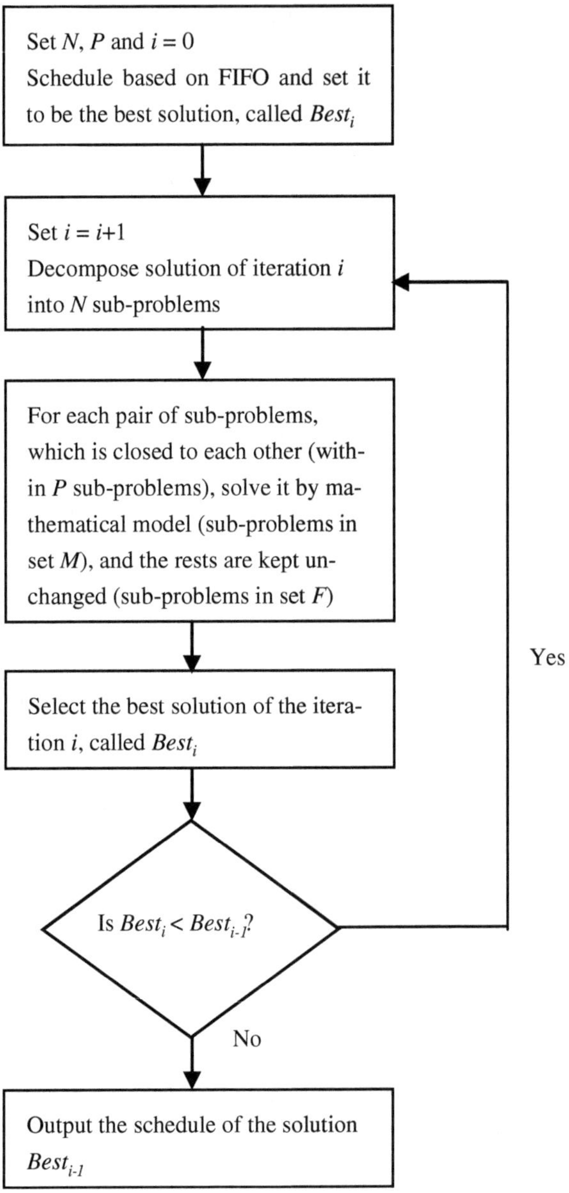

Fig. 3 The flow chart of the proposed heuristics

4 Results and Discussion

The performance of the proposed heuristic was evaluated by comparing its solutions to the solution from the mathematical model for small-size problems using four problem sizes with variations in the number of jobs (5–8 jobs). For each

Table 1 Comparative results

Number of jobs	Average total cost (unit cost)		Average computational time (s)		Percentage of cost deviation
	Optimal solutions	Heuristic solutions	Optimal solutions	Heuristic solutions	
5	216.05	216.05	414.40	69.30	0
6	193.35	193.35	2134.50	115.80	0
7	246.10	246.10	2523.70	241.30	0

Table 2 Computational result of heuristic algorithm

Number of jobs	Percentage of solutions from heuristic better than FIFO			Computational time (s)		
	Average	S.D.	Max	Average	S.D.	Max
10	11.35	8.60	25.43	308.90	154.39	547.00
20	15.93	6.21	23.28	1006.90	385.19	1604.00
30	22.17	5.24	30.38	1828.00	530.04	2916.00
40	24.80	8.54	40.02	2556.30	855.83	4156.00
50	18.42	4.03	23.89	2895.40	961.92	4125.00

problem size, ten sample problems were generated and tested; therefore, there were a total of 40 datasets solved by the mathematical model. In an acceptable computational time (2 h), 30 optimal solutions were obtained for the problems which the number of jobs is not greater than 7. The optimal solutions and their computation time were compared to the solutions obtained by the heuristic. From the computational results, the proposed heuristic could produce optimal solutions for all 30 problems as shown in Table 1. For large-problem size, the solutions from the heuristic were compared to the traditional method (First-In First-Out or FIFO) which is commonly used in the industry. The results showed that the solutions from the heuristic were at least 10 % better than solutions from FIFO (see Table 2). The sample problems were solved using LINGO 13.0 optimization software linked with Microsoft Excel on a PC with Core(TM) 2 Quad CPU Q 8200 2.33 GHz and 4 GB of RAM.

5 Conclusions

In this chapter, a mathematical model and a heuristic algorithm were developed to solve a production scheduling in food freezing process under the effect of freezer-door opening. The objective is to determine suitable loading and unloading schedule of product trolleys (jobs) in the freezing process. The proposed mathematical model could optimally solve only small-size problems. The heuristic is developed based on the relaxation approach to solve industrial-size problems. The results indicate that the proposed heuristic could produce optimal solutions for all small-size instances. The heuristic also outperformed the First-In First-Out method which is the traditional scheduling approach employed by the industry in the practical size problems.

Although, the proposed heuristic could practically solve the industrial-size problem but their computational result is still quite long. Therefore, modifying the heuristic in order to reduce the computation time should be concerned in future research.

Acknowledgments This research was funded by The National Research University of Khon Kaen University Fund under project FC 6.1.8 Ph.D. and Betagro Science Center Company Limited. The authors also acknowledge the company's help in collecting data, providing invaluable guidance and making this research possible.

References

Ferreira D, Morabito R, Rangel S (2009) Solution approaches for the soft drink integrated production lot sizing and scheduling problem. Eur J Oper Res 196:697–706

Ferreira D, Morabito R, Rangel S (2010) Relax and fix heuristics to solve one-stage one-machine lot-scheduling models for small-scale soft drink plants. Comput Oper Res 37:684–691

Lenstra JK, Rinnooy Kan AHG, Brucker P (1977) Complexity of machine scheduling problems. Ann Discret Math 1:343–362

Liu DY, Chang WR, Lin JY (2004) Performance comparison with effect of door opening on variable and fixed frequency refrigerators/freezers. Appl Therm Eng 24:2281–2292

Tang L, Wang G (2008) Decision support system for the batching problems of steelmaking and continuous-casting production. Omega 36:976–991

Tang L, Wang X (2009) Simultaneously scheduling multiple turns for steel color-coating production. Eur J Oper Res 198:715–725

Tang L, Zhao Y (2008) Scheduling a single semi-continuous batching machine. Omega 36:992–1004

Wang X, Tang L (2009) A simplified scatter search for a special single machine scheduling problem to minimize total weighted tardiness. In: IEEE conference on decision and control, pp 6250–6255

Wang X, Tang L (2011) Scheduling a single machine with multiple job processing ability to minimize makespan. J Oper Res Soc 62(8):1555–1565

Redefinition of Tasks to Increase the Process Capacity of Bottlenecks: Adjustment to a Real Case of Cutting Process of Structural Profiles of Carbon Steel

Clemente Lobato Carral and Carlos Andrés Romano

Abstract Once identified the bottleneck process, that one that limits the total capacity of the system, the following logical step is tied to the decision of how to optimize it. It will be necessary to know all the tasks that influence the speed of the process. The limitations can appear as materials, information, persons, machinery or any another aspect. It is very usual to connect the limitation of the process to aspects that, though indirectly if they determine the same one, they are not really those on which it is necessary to work hard. Therefore, the efficiency in the redesign of the process does not have the desired result. The detailed analysis of the procedure of each task that takes part in the process at some point of its accomplishment, allows knowing with accuracy the proper restrictions as well as the critical points on which it is necessary to work. This article provides a guide to the implementation of process improvements (regardless of the process in question) by identifying the constraints and solving them. The solutions are intended to increase the process speed, which is converted to a capacity increase of the whole organization. In the business of transformation of structural steel we meet the process of cutting as the common bottleneck of different organizations located in this sector. The standard product supplied by the manufacturers of steel is characterized by standard lengths. For the construction of structures, where the lengths are variable, it needs additional cutting to obtain specific lengths. This transformation of the base product is habitually done by appropriate machinery, and the capacities of cutting are based with the own process time of cutting. The optimization of this one task, measure as the increase of the number of hours of effective cutting, is obtained reducing to the maximum all the restrictions that limit the speed of the process. In our practical case, the solution to the bottleneck of cutting to allow the increase of sales could be solved by subcontracting, buying

C. L. Carral (✉) · C. A. Romano
Polytechnic University of Valencia, Camino de Vera s/n E-46022, Valencia, Spain
e-mail: clelocar@doctor.upv.es

C. A. Romano
e-mail: candres@omp.upv.es

P. Golinska (ed.), *Logistics Operations, Supply Chain Management and Sustainability*, EcoProduction, DOI: 10.1007/978-3-319-07287-6_26,

new tangible fixed assets or optimizing the process. The latter aspect is the developed one in this case of study. The cutting program is subordinated to the optimization of the bottleneck. We took measurements before the redesign of the process as well as after the changes. The result is cutting capacities much better than the initial values. This case of study is used in different degrees and master class as an example of the steps to identify, to analyze and to re-design a process. In addition, we emphasize the importance of measuring to understand the successfulness of the changes. For the real organizations, the redesign is fundamentally based on the learning and growth perspective. People who take part in the process are instructed in order to obtain a coherent adjustment with the needs of the new way of processing.

Keywords Process Capacity · Bottlenecks · Process improvement

1 The Industrial Process of Profiles Cutting of Carbon Steel

The objective of the particular business of profiles cutting process is geared towards producing specific pieces to sell since it works to order (on request). The standard product supplied by manufacturers of steel is characterized by lengths (6, 10, 12, 15, 16 and 18 m) so that in the actual construction of structures, where lengths are different, is necessary the subsequent cutting process (Fig. 1) to achieve specific lengths. This transformation of the base product is usually carried out by tape or disk saw machines and cutting capabilities are associated to the effective cutting time process.

The objective is always easier to reach if the productivity of the process is higher, understood as “the ability to cut as much time as possible at the lowest cost”. Such productivity is achieved by removing the constrictions that prevent the machine running time is as close to the total time of use of the machine.

2 Elements of the Process of Profiles Cutting of Carbon Steel

The components used in the process of cutting of profiles (Fig. 2), as in any other process (Teixidó and Parreño 1997) are:

- The Inputs
  Information coming from the customer directly or through the sales department (drawings, delivery time required), materials coming from long Standard structural profiles (price, quality), energy, etc.

Fig. 1 Tap cutting machine for cutting of profiles

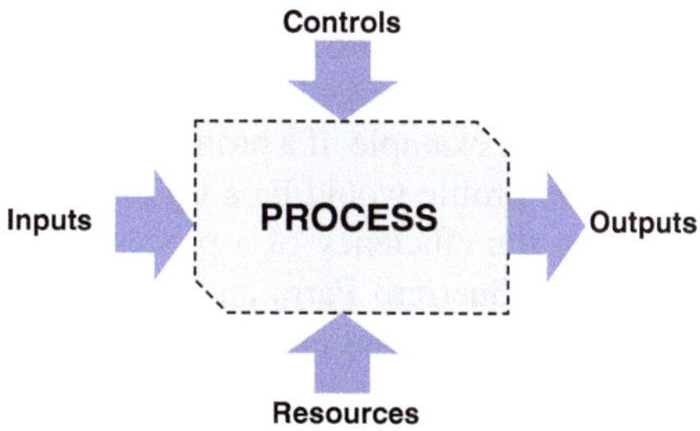

Fig. 2 The elements of a process

- The Controls
  Measurement of resource used (quality of process), level of customer satisfaction (indicators), etc.
- The Resources
  People, cutting machinery (tap cutting in our case of study), computers (adapted to the specific process of cutting), etc.
- The Outputs
  Product (profile cut to size), information that customers need (certifications, delivery notes, invoice), etc.

3 Measurement Parameters and Initials Values

It is a well-known fact the importance of measuring a process to manage (Robert and David 1996). Therefore any assessment in this chapter is quantitatively developed. Only in this way it is possible to know the real evolution of the implemented improvements.

Another important aspect is the fact that this study is not based on the identification of the bottleneck. The identification work was further developed in a previous study. The aim of this article is to describe the analysis of the restriction that generates the bottleneck as well as actions to reduce them.

Data collection was conducted during the months of February to July 2004, and study the following starting values (Fig. 3) were extracted.

Where the values that mean optimizing the process are associated with the following items:

- *Percentage machine cutting* is the percentage of time that the machine is actually performing a cut in a structural profile as compared to the total time the machine is available for this. In the particular case of our data collection there are 16 h per day of time available. The 27.77 % indicates that the machine is actually cutting 16 min and 40 s of every hour available. The rest of the activities are performed while the machine is stopped.
- *Percentage of tips* is the volume of the ends of beams which are available for use in workshop related to the total volume of cut profiles. A tip is a part of beam greater than or equal to 1 m of the base profile used leftover after being cut for sale. For example, if a profile over 15 m is used to cut two bars of 6.5 m, the rest of the profile would be a tip of 2 m. It is necessary to measure this parameter since the efficiency of a process is also based on achieving a stock of higher value (Guerrero Parra and de Stocks 2008), and the tip is a product of lower value to the basic profile. A higher % of tips represent the worst process optimization. In our case of study for the measurement time, there was an average of 10.72 % of tons that mean that there was 107.2 kg of tips for each 1000 kg cut.
- *Percentage scrap* is the volume of structural profiles sold as scrap since the possibilities of reuse in future orders is very low over the total volume of structural profiles sold. It is understood that scrap is any piece of 1 m lower profile leftover basis used after being cut for sale. For example, if a beam over 15 m is used to cut one bar of 6.5 m and other of 8 m, the rest of the profile that is left would be 0.5 m and thus being sold as scrap. The reason to measure this value is based in the same parameters of the tips. The scrap is a product of lower value to the basic profile. A higher % of scrap represents the worst process optimization. In our case study the measured scrap has an average of 2.96 % over the total volume sold. It means that there is 29.6 kg of scrap for each 1000 kg cut.
- *Percentage of quality service* is the % of lines of delivery notes that have been delivered according to the delivery date commitments agreed with customers. A lower % of quality service would reduce the assets of the company losing

	2004			2005			
	Oct	Nov	Dec	Jan	Feb	Mar	Total
Days	21	21	19	20	19	23	123,0
Working hours	336,0	336,0	304,0	320,0	304,0	368,0	1968,0
Effective hours	98,3	92,4	84,9	91,5	83,8	95,6	546,5
% machine cutting	29,26%	27,50%	27,93%	28,59%	27,57%	25,98%	27,77%
Tips (ton)	87,0	93,0	90,0	90,0	87,0	93,0	540,0
% of tips	12,43%	9,11%	10,47%	11,71%	10,05%	11,33%	10,72%
Total Sales (ton)	699,9	1.021,20	859,5	768,8	866,1	820,6	5036,1
Sales per hour (ton/h)	2,1	3,0	2,8	2,4	2,8	2,2	2,6
Scrap (ton)	22,6	24,9	28,3	19,25	32,54	21,24	148,83
Scrap (%)	3,23%	2,44%	3,29%	2,50%	3,76%	2,59%	2,96%
Nº delivery notes	370	463	340	345	428	407	2353
Lines in delivery notes	1182	1544	1303	1211	1331	1274	7845
Lines in time	1.040	1.312	1.212	993	1.211	1.147	6.915
Quality service	88,00%	85,00%	93,00%	82,00%	91,00%	90,00%	88,15%

Fig. 3 Measurement parameters and initials values

customers and, therefore, account customers. In our case of study the measured quality service has an average of 88 % over the total volume of lines of delivery notes.

- A *sale per hour is* the parameter that shows the number of tons that are cut to order for each hour of time available. This value is not a real indicator of the optimization process since tons of cut vary depending on the length required by customers, not the time it takes to cut it. For example, a cutting process based on a profile of 15 m to get 2 bars 7 m requires the same time that if the cutting order was 2 bars of 6 m, but in the first case the weight would be higher. However, we give importance to this parameter for two reasons: on the one hand because it is the reference value with which people work on this kind of business and moreover that although the example described are possible, the mixture of profile orders is the same so more cutting time would become greater tonnages cut. In our real case of study initially are cut 2.6 ton/h of cutting machine available.

4 Analysis of the Cutting Process

To perform process analysis we focus on the aspects that cause the machine to be stopped since the reduction of the time of such restrictions (Goldratt 2004) is converted directly into machine running time and thus optimizes the process. For the explanation of the reasons listed below stop machine perform a simple sketch in Fig. 4.

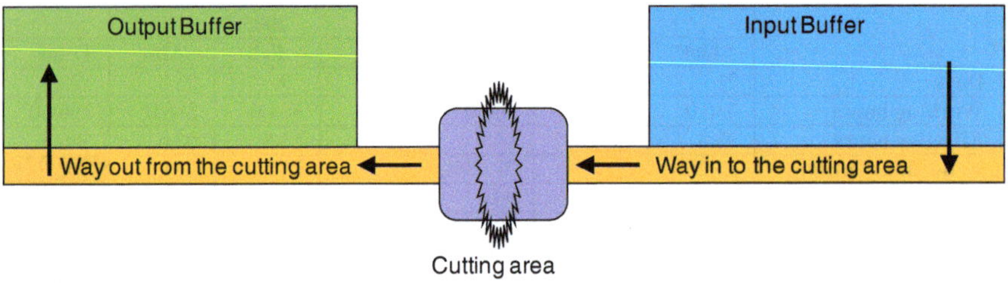

Fig. 4 Sketch of the flow of materials and installation for cutting structural profiles

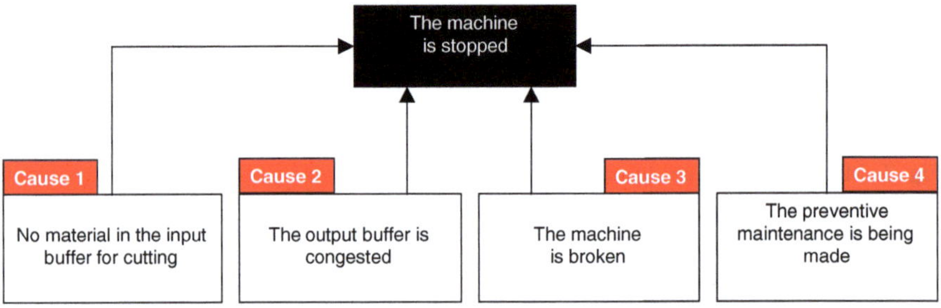

Fig. 5 Reasons why the machine is stopped (cause-effect diagram)

The machine is stopped for the following reasons:

1. No material in the input buffer for cutting.
2. Although there is material in the input buffer of the machine, cutting area is busy with the final cut profile, which cannot be placed in the output buffer due to that area is filled with material waiting to be ejected.
3. The machine is broken.
4. Preventive maintenance.

Figure 5 shows a cause-effect diagram (Goldratt 1994) of these reasons.

4.1 Analysis of the Cause 1: No Material in the Input Buffer for Cutting

When no material prepared in the input buffer there are several aspects that favor this problem, all due to the high difficulty and time required to locate the tips to be used as base material as shown in Fig. 6.

As detailed in Sect. 3, the tips are items that have less value for the company as sales opportunities and use are lower than standard long profile. Therefore, it is essential to search tips for cutting instead of standard lengths. In the case of not having the desired profile tips it would be necessary to use standard long profiles.

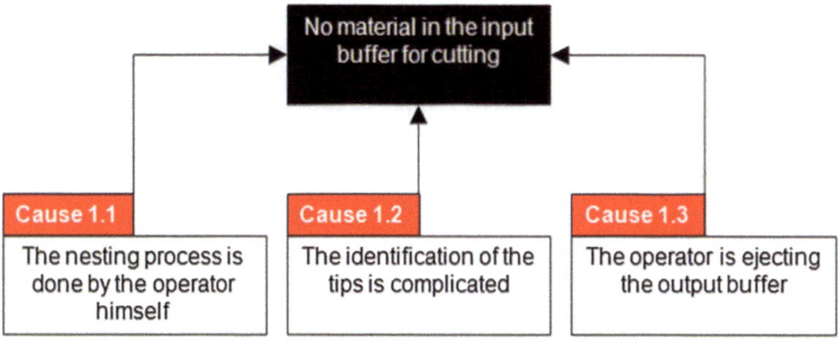

Fig. 6 Reasons why the machine is stopped (cause-effect diagram)

4.1.1 Cause 1.1: Nesting Process Is Done by the Operator Himself

A determining factor in many measurement parameters on the efficiency of the process is based on the fact that it is the operator himself who performs the nesting of the parts to be cut on the available beams that there are in the workshop. The operator receives a list of the different orders so that is the same person who makes the most appropriate link from his/her point of view. The shop conditions and the simultaneous operations run by the operator go to a wrong result.

To improve this situation nesting operation moved to another time and place. The operator went on to perform operations directly related to effective cutting. All the tasks it was possible to make them at another time or place were changed, this especially because of the great time that was used by the operator.

Parallel some aspects were modified to optimize the development time of the operation although it was made at office. For example, a task that required a lot of time was sorting the materials by type of profile later to make the pieces fitting standard profile.

Each order has different lengths for different types of profiles. For example, a request A may need a 6.5 m bar of IPE-300 profile and a 6.5 m bar of UPN-120 profile. Alternatively, another request B may need a 5.5 m bar of IPE-300 profile and a 5.5 m bar of UPN-120 profile. It seems logical that it is more effective grouping based on the type of profile cutting that depending on the order because in the same 12 m bar of IPE-300 profile both orders would be cut for this kind of profile. The same action would be made with the UPN-120 profile. Therefore, order list ordered by profile type and date of delivery is incorporated. The date of delivery helps us to always be aware of the quality service.

Information was computerized so we could work with it on the measure of our needs. The new order established is a *new software development* based on a bigger efficiency.

Although the cost may rise slightly in office revenue generated by the increase in cutting capacity are much higher. Optimization bottleneck will increase sales due to the increased capacity. The workshop operators would have this task already done and only have to locate the profile and place it in the input buffer.

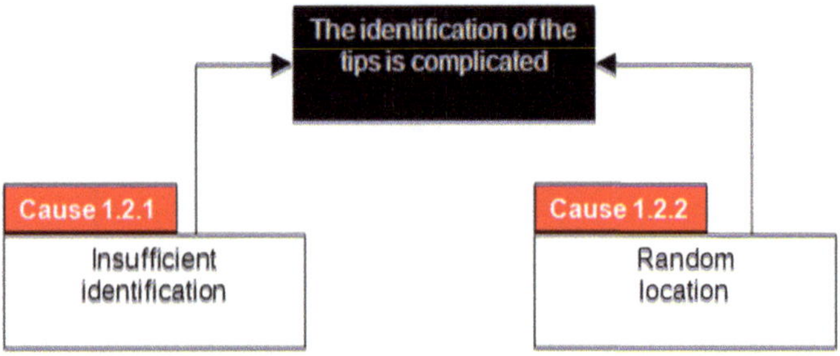

Fig. 7 Reasons why the identification of the tips is complicated (cause-effect diagram)

In addition to increasing the chances of effective machine cutting time (effective hours), performing the nesting process in the office allows more time and the condition to improve the uses of base materials which would reduce the % of tips and the % of scrap generated. At the same time, we analyzed different software on the market for this type of operation with the introduction of one of them to improve the measurement parameters while considerably reducing programming time.

4.1.2 Cause 1.2: Identification of the Tips Is Complicated

This case is the most meaningful of all studied. Its effect is very negative for the speed of the process, but the same solution can be very simple. Order and organization are the basis on which work to reduce the restriction as far as possible.

Why is it difficult locating tips in the workshop? Why require such a long time? Two causes were identified (Fig. 7): insufficient identification and random location.

Cause 1.2.1: Insufficient Identification

The tips are not identified with the length and the type of profile. The operator has to measure long to see if the length is optimal for customer requested and width to confirm the type of profile. This operation, in other conditions can be simple but in a warehouse with pile of tips becomes complicated (long time) and dangerous.

As can be seen in the photos in Fig. 8, initially the tips were not identified. Therefore, the identification of the tips in the output buffer is determined, when it is safe and easy for the operator. This identification is based on length and type of profile (for example 2240-IPE300). The improvement reduces the time of identification considerably being required only a visual inspection.

Initial appearance Improved appearance

Fig. 8 Improvement in the identification of profiles

Initial appearance

Fig. 9 Disordered warehouse of tips

Cause 1.2.2: Random Location

The warehouse of tips is cluttered and the tips are not located in any particular place, as shown in Fig. 9, but there is a big area where they could be. Besides increasing the time and location of the material, this aspect has several negative effects for the company (Alonso García 1997), and more specifically for the workshop, among which include: underutilization of available space, inventory is very complicated and requires too much time and increases the probability of accidents.

Conversely, at the beginning the improvements have the following advantages: a logical order increase space optimization, inventories will be simple, fast and with less error and the likelihood for accidents is reduced.

The order that was implemented has various details of great impact on the search time of the material such as: hips by type of profile and identified zones by location and lengths.

Initial appearance Improved appearance

Fig. 10 Piles by type of profile

Fig. 11 Specific areas for each type of profile

- Piles by type of profile
  Piles by type of profile (Fig. 10) allow quick identification and the ability to carry the entire pile of the same profile to the input buffer by machining to reduce the movement time.
- Identified zones by location and lengths
  The improvement, based on the order in the warehouse identifies specific areas for each type of profile (Fig. 11) and also in several hips of specific lengths (Fig. 12) according to analysis of different order groups.

The new layout will aim to establish ordered workspaces (defined locations) and the equipment that is most economical for work and the most safe and satisfying for employees. Some of the objectives achieved are: fewer accidents, more satisfied employees, less movement of materials, increased time of cutting, easier monitoring, space saving, more flexibility, etc.

Fig. 12 Several hips of specific lengths

Fig. 13 Appearance of open spaces

Following the start of the improved order, some early qualitative symptoms are the appearance of open spaces (Fig. 13) due to a better optimization of it, and decongesting the buffers (Fig. 14).

4.1.3 Cause 1.3: The Operator Is Evicting the Output Buffer

It is not possible for an operator to ensure the machine cutting, while supplying the input buffer, output buffer dislodges and makes nesting in the most appropriate for the efficient achievement of the value structure of the company.

To improve this situation a support operator in some task is required. Thus, the machine will reduce its stoppages due to the possibility of no action by the main operator to cut.

As we've seen improvements made in this cause have been based on the concept of 5S techniques: The visual Workplace (Hirano 1996) which will not be

Fig. 14 Decongesting buffers

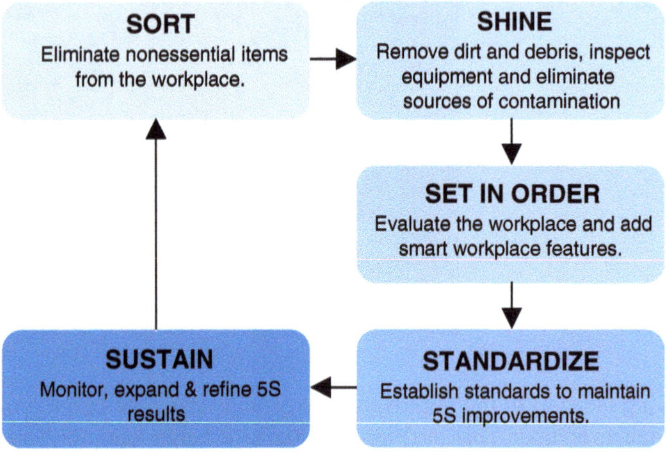

Fig. 15 The 5S of a workplace organization

discussed because it is a support tool and not the main topic of this article. Nevertheless then we summarize the general features of the tool.

The basic elements of a 5S workplace organization, as shown in Fig. 15, are sort (seiri), set in order (seiton), shine (seiso), standardize (seiketsu) and sustain (shitsuke). Successful improvement activities depends on these and represent the foundation for achieving zero defects, cost reductions, improvements in safety and zero accidents. The five pillars seem a simple concept that often people lose sight of its importance.

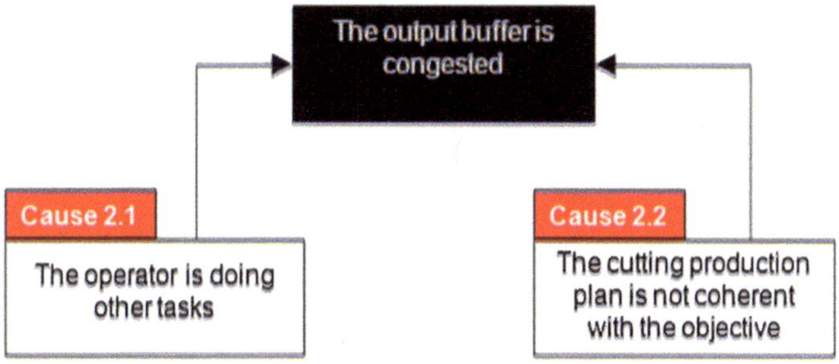

Fig. 16 Reasons why the output buffer is congested (cause-effect diagram)

4.2 Analysis of the Cause 2: The Output Buffer Is Congested

The material does not flow from the input buffer because the output buffer is congested. The fact that the output buffer is congested was reasons in two totally different ways (Fig. 16). First, as has been detailed in cause 1, the main operator cannot be cutting while ejecting material from the output buffer neither doing other tasks. Furthermore, it is essential to prepare coherent cutting production plans in order to increase the % of effective time of cutting.

Regarding the first aspect, it has already been defined in case 1 the need for another operator to perform support tasks such as ejection of material from the output buffer and feeding the input buffer.

Regarding the cutting production plan it is necessary to know that there are cutting operations on short cycle and long cycle. Short cycle appears in profiles of small web, whereas long cycle appears in profiles of big web. Also, some types of profiles require more cycle time than others. If the cutting production plan is only made with short-cycle programs there will probably not be enough time to leave the output buffer without machine stops because the cutting times are shorter than the time of extraction of the material. Therefore, it is necessary to combine short-cycle programs with long-cycle programs that allow sufficient time for the ejection of the output buffer. This would also be valid to have time to feed the input buffer while the machine is cutting.

In the following Figs. 17, 18, 19 and 20 there are some features of dimensions (Unión de Almacenistas de Hierros de España 2001), mass and area of the more characteristic structural profiles.

As can be seen, the cutting area determines the cutting speed. Therefore, in order to make a cutting production plan that can make feeding times and extraction, it is necessary to combine short-cycle programs (small area) with long-cycle programs (large area). The objective is to have a higher % cutting machine time (effective time) combined with the need to meet the commitments of delivery date of the customers.

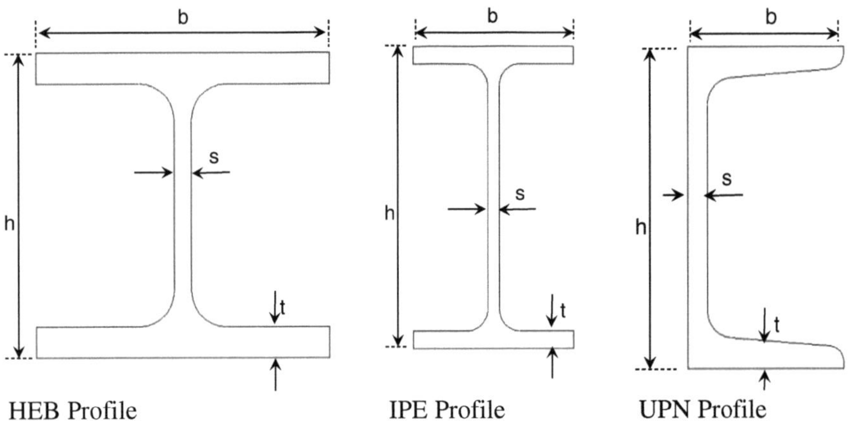

Fig. 17 More characteristic structural profiles

Profile	Web		Flange		Area	Weight	Cutting time
	h	s	b	t	kg/m	kg/m	Seconds
HEB 100	100	6	100	10	26,00	20,40	45,00
HEB 120	120	6,5	120	11	34,00	26,70	60,00
HEB 140	140	7	140	12	43,00	33,70	75,00
HEB 160	160	8	160	13	54,30	42,60	90,00
HEB 180	180	8,5	180	14	65,30	51,20	112,50
HEB 200	200	9	200	15	78,10	61,30	135,00
HEB 220	220	9,5	220	16	91,00	71,50	142,50
HEB 240	240	10	240	17	106,00	83,20	157,50
HEB 260	260	10	260	17,5	118,40	93,00	90,00
HEB 280	280	10,5	280	18	131,40	103,00	112,50
HEB 300	300	11	300	19	149,10	117,00	135,00

Fig. 18 HEB profiles—measures and section data

Profile	Web		Flange		Area	Weight	Cutting time
	h	s	b	t	kg/m	kg/m	Seconds
IPE 80	80	4	46	5,2	7,64	6,00	22,50
IPE 100	100	4	55	5,7	10,30	8,10	22,50
IPE 120	120	4	64	6,3	13,20	10,40	27,00
IPE 140	140	5	73	6,9	16,40	12,90	27,00
IPE 160	160	5	82	7,4	20,10	15,80	30,00
IPE 180	180	5	91	8	23,90	18,80	30,00
IPE 200	200	6	100	8,5	28,50	22,40	37,50
IPE 220	220	6	110	9,2	33,40	26,20	45,00
IPE 240	240	6	120	9,8	39,10	30,70	60,00
IPE 270	270	7	135	10	45,90	36,10	75,00

Fig. 19 IPE profiles—measures and section data

Profile	Web		Flange		Area	Weight	Cutting time
	h	s	b	t	kg/m	kg/m	Seconds
UPN 80	80	6	45	8	11,00	8,65	30,00
UPN 100	100	6	50	8,5	13,50	10,60	35,00
UPN 120	120	7	55	9	17,00	13,40	37,50
UPN 140	140	7	60	10	20,40	16,00	45,00
UPN 160	160	7,5	65	11	24,00	18,80	60,00
UPN 180	180	8	70	11	28,00	22,00	75,00
UPN 200	200	8,5	75	12	32,20	25,30	90,00
UPN 220	220	9	80	13	37,40	29,40	112,50
UPN 240	240	9,5	85	13	42,30	33,20	135,00
UPN 260	260	10	90	14	48,30	37,90	135,00
UPN 280	280	10	95	15	53,30	41,80	142,50
UPN 300	300	10	100	16	58,80	46,20	157,50

Fig. 20 UPN profiles—measures and section data

4.3 Analysis of the Cause 3: The Machine Is Broken

It is essential to do a detailed analysis of the repairs to the machinery used in the process, in our case the tap cutting machine, for two main reasons. First of all, usually chronic or systematic failures can be identified. If so, the company can carry on with modifications, adaptations or simply changes of any part of the machine that causes the damage, minimizing their appearance. In addition, the identification of some systematic failures makes preventive maintenance more efficient.

4.4 Analysis of the Cause 4: Preventive Maintenance Is Being Made

Analyzing the tasks that are performed by different operators we observe that there is not a checklist of maintenance tasks to perform (Rizzo 2008). Not all operators perform the same operations and there is no temporal sequence of it.

A series of discussions with operators, heads of workshop equipment and the supplier of the machinery was carried out, and we concluded that the machine was stopped sometimes due to tasks that were not needed. Also, sometimes the machine was stopped because operators did *internal activities* (those that cannot be done with the machine running) some of the activities that were really *external* (those that can be done with the machine running). To minimize machine stoppage a checklist of tasks were defined highlighting which activities were internal and which were external (Shingo and Productivity Press Development Team 2011).

It was established in writing and publicly on the panels next to the machine the preventive maintenance tasks and their sequence. Also, a defined list of tools and other resources (oil, grease machine, etc.) that were required to do all required

	2005			
	May	June	July	Total
Days	22	22	21	65,0
Working hours	352,0	352,0	336,0	1040,0
Effective hours	124,9	131,5	121,7	378,1
% machine cutting	35,48%	37,36%	36,22%	36,36%
Tips (ton)	107,8	104,9	118,6	331,3
% of tips	8,87%	8,97%	9,10%	8,98%
Total Sales (ton)	1215,4	1169,8	1303	3688,2
Sales per hour (ton/h)	3,5	3,3	3,9	3,5
Scrap (ton)	31,3	29,1	34,8	95,2
Scrap (%)	2,58%	2,49%	2,67%	2,58%
Nº delivery notes	517	496	588	1601,0
Lines in delivery notes	1684	1596	1910	5190,0
Lines in time	1.576	1.473	1.758	4807,0
Quality service	93,59%	92,29%	92,04%	92,62%

Fig. 21 Values after implementation of improvements

task. In this way, the operator can prepare all the necessary equipment for the maintenance and not use a longer time than necessary, overall for internal operations.

5 Values of Measurement Parameters After Implementation of Improvements

As mentioned, a data collection and analysis process was conducted from October 2004 to March 2005 (see point 3 for initial values). During the month of April all points evaluated were launched and returned to measure in the months of May–June–July. In order to better know if the improvements were efficient, sales area was encouraged to attract more orders.

Figure 21 shows the measured values in these 3 months.

Comparing the new values to the initial situation we have the Fig. 22:

- The % of cutting machine time has increased by 8.59 % reaching 36.36 % of the time available.
- The % of tips has fallen 1.74 % even though the absolute value has increased from a monthly average of 90 tons to a final value of 110.4 tons. It is logical due to increased cut volume.
- The % of scrap has followed a similar evolution to % of tip because it has been reduced by 0.37 % even though the absolute value has increased from a monthly average of 24.8 tons to a final value of 31.7 tons.
- The quality service has increased by 4.47 % of the order lines.
- The sales per hour (ton/h) have increased by 0.99 ton/h, from 2.6 to 3.5 ton/h.

	Total	Improved	Variation
Days	123,0	65,0	
Working hours	1968,0	1040,0	
Effective hours	546,5	378,1	
% machine cutting	27,77%	36,36%	8,59%
Tips (ton)	540,0	331,3	
% of tips	10,72%	8,98%	-1,74%
Total Sales (ton)	5036,1	3688,2	
Sales per hour (ton/h)	2,6	3,5	0,99
Scrap (ton)	148,83	95,2	
Scrap (%)	2,96%	2,58%	-0,37%
Nº delivery notes	2353	1601,0	
Lines in delivery notes	7845	5190,0	
Lines in time	6.915	4807,0	
Quality service	88,15%	92,62%	4,47%

Fig. 22 Value comparison

One aspect to note is that has not missed any order delivery and processing capacity cut has not been saturated.

6 Conclusions and Educational Considerations

Although it may seem that a bottleneck process is fully optimized, the detailed analysis of all its restrictions can increase the capacity of the minimally whatever. To do this, all decisions shall be consistent with the ultimate goal of increasing capacity and subjecting them to achieve this.

This case study we is used in degrees and master's degrees as an example of the steps to identify, analyze and redesign a process. Furthermore, in this study highlight the need to measure as a means to determine whether the improvements made are really positive or not (Kelvin 1889). Focus on increase capacity through the continuous improvement of processes stands above the addition of new assets (machinery, transport, etc.) or process outsourcing.

For the companies where these types of studies are done, the redesign is based primarily on the training of all those involved in the process used in order to achieve a consistent fit with the needs of that process. It is not based on the steps to optimize the bottleneck but target the functions and tasks of the people towards the new approach. Process measurements and the results of improvement act positively on people.

References

Alonso García A (1997) Conceptos de Organización Industrial. Editorial Marcombo, Barcelona
ARCELORMITTAL, Product ranges-sales programme, pp 70–100 http://amsections.arcelormittal.com//products-services/products-ranges.html
Goldratt EM (1994) It's not luck. North River Press, Great Barrington
Goldratt EM (2004) The goal: a process of ongoing improvement, Revised 2nd edn. North River Press, Great Barrington
Guerrero Parra F, de Stocks G (2008) Editorial ESIC
Hirano H (1996) 5S for operators: 5 pillars of the visual workplace. Productivity Press, Portland
Kelvin WTB (1889) Popular lectures and addresses, vol 1. Electrical units of measurement. McMillan and Co, London, p 73 https://archive.org/stream/popularlecturesa01kelvuoft#page/72/mode/2up
Rizzo K (2008) Total productive maintenance, American Printer Review, September 2008 www.americanprinter.com
Robert SK, David PN (1996) The balanced scorecard: translating strategy into action. Harvard Business School Press, Harrington
Shingo S, Productivity Press Development Team (2011) Quick changeover for operators: the SMED system. Productivity Press, Portland
Teixidó E, Parreño A (1997) Reingeniería de procesos y costes ABC/ABM: (introducción al Management del Denominador), Consulting Adim
Unión de Almacenistas de Hierros de España (2001) Prontuario UAHE

Examining Effect of JITP Implementation on Performance of Jordanian Firms

Abbas Al-Refaie and Nour Bata

Abstract This research examines the impact of just-in-time purchasing (JITP) implementation on firms' performance in Jordan using structural equation modeling (SEM). A structural model is proposed that relates JITP with firm performance. JITP is then measured by six dimensions while firm performance is measured by three dimensions. The data is collected from 250 firms. SEM results showed that top management commitment, employees' training, and employees' relations positively influence the JITP implementation and firms' performance. However, supplier value-added practices, transportation practices, and quantities delivered practices have insignificant effect on both JITP and firms' performance. These results indicate that Jordanian firm have taken the first steps in JITP implementation, however they need to develop action plans to implement JITP practices to improve their performance. In conclusion, this research provides decision makers a valuable feedback about their route of JITP implementation and assesses to what extent JITP dimension influence firms' performance.

Keywords Firm performance · JITP · Structural equation modeling

1 Introduction

Confronting the challenges of global competition, firms world-wide are forced to find ways to reduce costs, improve quality, and meet the ever changing needs of their customers. One successful solution has been the adoption of just-in-time (JIT) manufacturing systems (Balci et al. 2007; Barlow 2002; Marsh and Conard 2008). JIT purchasing is integral and is typically incorporated when describing JIT

A. Al-Refaie (✉) · N. Bata
Department of Industrial Engineering, The University of Jordan, Amman 11942, Jordan
e-mail: abbas.alrefai@ju.edu.jo

P. Golinska (ed.), *Logistics Operations, Supply Chain Management and Sustainability*, EcoProduction, DOI: 10.1007/978-3-319-07287-6_27,

management practices. With JIT purchasing (JITP), suppliers become extended operations of the JIT manufacturer. The importance of JIT purchasing in the overall JIT management system is demonstrated by the magnitude of parts by a typical manufacturer (Gunasekaran 1999; Waters-Fuller 1995). The fundamental aim of JITP is to ensure that production is as close as possible to a continuous process from receipt of raw materials/components through to the shipment of finished goods. JITP is an important part of the overall JIT program and can produce benefits of reduced lead times, reduced inventories, improved quality, improved lead time reliability, reduced material costs and improved flexibility. JITP requires small but frequent deliveries of total quality parts from single sourced, local suppliers, with whom there is a close relationship rounded on mutual dependency and frequent communication.

In practice, several dimensions are used to measure the degree of JITP implementation, including top management commitment, employees training, employees' relations, supplier value added practices, transportation practices, and quantities delivered. Jordan is classified as an emerging market. Sharp competition has forced companies to continually improve their performance. Therefore, industries implement several improvement management programs to strengthen their competitive advantages. Among such programs is the JITP framework. Further, structural equation modeling is widely deployed for statistical analysis in many managerial applications (Al-Refaie and Hanayneh 2014; Al-Refaie et al. 2012a, b, c). Therefore, this research attempts to examine the effects of JITP implementation on organization performance using structural equation modeling.

2 Literature Review

Garg et al. (1996) examined critically JIT purchasing in Indian context. There was an indication that Indian industries were giving importance to JIT attributes, facing some problems in implementing JIT, and expecting an overall benefit on an average 59.8 % if JIT purchasing is fully implemented. Tests confirmed that the scope of JITP implementation in India was fair and it was independent of the type of industries, layout, and number of employees. Radovilsky et al. (1996) stated that using JIT philosophy in purchasing yields a number of positive results in manufacturing and service organizations. The primary areas of improvement included inventory reduction, increased quality, and an overall reduction in total costs. Relationships with the suppliers and their support were the major focal point in running JITP system. Garg et al. (1997) found the importance of identified JITP purchasing and supplier evaluation criteria attributes in Indian industries. In addition, he conducted a case study of an Indian automobile industry which has implemented JITP. Landry et al. (1997) conducted a case study on a medium-sized Canadian industry, manufacturing weather strips for cars. The industry has integrated material requirements planning, kanban and bar-coding systems to achieve JIT procurement. Benefits achieved by the company include drastic reduction of total inventory levels, purchase lead times, better first in first out control. Gunasekaran (1999) proposed a framework to design a JITP system using the application of conductive education and stated that the application of information technologies can play a significant role in achieving JITP. Garg and Deshmukh (1999) presented the results of a case study of an Indian tractor assembly and the significant benefits that were achieved are reductions in inventory, material movement, space, manpower, work-in-process and lead-time and an increase in productivity and quality as a result of JITP implementation. Kaynak (2002) examined the effects of JITP implementation on firm performance in 214 organizations operating in the US and encompassed various sizes of firms and large variety of industries. Kaynak and Pagan (2003) analyzed the technical productive efficiency effects of implementing JITP techniques in the US manufacturing industry through modeling technical efficiency using a stochastic frontier production function. Results showed that supplier value-added and transportation issues do not appear to be associated with increasing productive efficiency. Kaynak and Hartly (2006) used replication research to validate JIP construct development. Six JITP factors were examined including: top management commitment, employee relations, training, supplier quality management, transportation, and quantities delivered using two different data sets and testing the first- and second-order structure.

3 Conceptual Framework

Top management must make JITP a priority for the whole organization and communicate this commitment to its employees clearly. Daft (1998) indicated that top management support is pivotal to having resources for training employees in any new implementation initiative in an organization. Beer (2003) indicated that top management commitment makes JITP a priority and communicate its importance to all employees and that top management must obtain support from all functions and all management levels.

Employees' training is also a key part in JITP to improve performance and to acquire problem solving and planning skills in JITP. Training also includes the process of teaching specialized job-related skills and knowledge to employees in the organization. Dong et al. (2001) stated that JIT requires streamlining purchasing processes and working closely with suppliers and those activities require buyers to develop new skills. Therefore, buyers must be trained to manage supplier quality.

Employee relations encompass various organizational development techniques; such as, employee participation in JITP decisions, employee recognition for improvement, teamwork, the creation of awareness of the organizational goals of JITP implementation among employees through communication and the communication of quality improvement to all employees, to facilitate change. Jayaram and Vickery (1998) found that empowerment of employees is positively correlated to procurement lead-time performance. Ford and Fottler (1995) stated that employee relations represented by recognition and empowerment are significantly important in JITP implementation.

Another dimension of JITP is supplier value-added practices, which include establishing long-term cooperative relations with fewer suppliers, evaluating and selecting suppliers based on quality and delivery performance rather than price, establishing thorough supplier rating system and engaging suppliers in product/service development, in order to obtain quality supplied materials and/or services on time. Nassimbeni (1996) found that securing suppliers' commitment to timely delivery of high-quality materials and parts prior to relying on frequent deliveries in small lots and in required quantities is essential.

Transportation practices play a critical role in effective JITP implementation. Practices include establishing buyer driven delivery schedule, involvement by purchasing department in designation of inbound carrier and routing and establishing reliable company-owned or contracted-carrier transportation services, in order to secure on-time delivery. Finally, practices related to quantities delivered are of main importance in JITP implementation, which include frequent delivery of purchased parts in small lots and in required quantities and delivery of materials and parts on a just-in-time basis.

Consequently, the following hypotheses are proposed:

H_1 Top management commitment is positively related to successful JITP implementation.

H_2 Employees' training is positively related to successful JITP implementation.

H_3 Employee relations are positively related to successful JITP implementation.

Further, quantities delivered in small lots of required quantities must meet quality and delivery specifications to avoid costly downtime. Vonderembse et al. (1995) stated that JITP requires changes in transportation practices. That is, logistics systems must deliver materials on time so as not to disrupt operations. Hence, it is hypothesized that:

H_4 Supplier value added is positively related to successful JITP implementation.

H_5 TP is positively related to successful JITP implementation.

Finally, Handfield (1993) stated that JITP organizations shall commit to quantity delivered practices. That is, products are produced in small lot sizes when needed, so materials and parts are frequently delivered in small lots and these lots must be delivered in exact quantities. If materials are delivered in smaller quantities than needed, the shortage of materials disrupts production. Delivery of larger quantities than needed defeats one of the purposes of JITP, which is reducing inventory carrying costs.

Consequently, the following hypothesis is proposed:

H_6 Quantity delivered practices is positively related to successful JITP implementation.

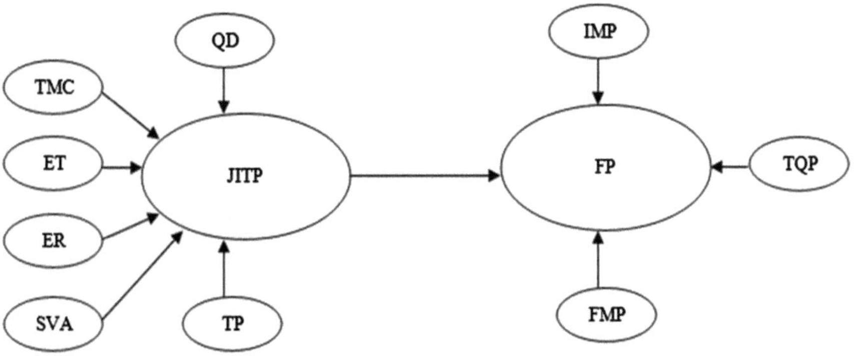

Fig. 1 The proposed model

Firm performance can be divided into three categories including: (i) inventory management performance, which is reflected in increased purchase material turnover and total inventory turnover, (ii) *time-based quality performance,* which is reflected in improved product/service quality, improved productivity, reduced delivery lead time and reduced scrap/rework costs, and (iii) *financial and market performance*, which is reflected in increased sales and competitive position. It is believed that JITP dimensions affect positively organization performance. Therefore, it is hypothesized that:

H_7 JITP dimensions are positively related with firm performance.

The proposed model is shown in Fig. 1. The item measures for JITP and firm performance are displayed in Tables 1 and 2, respectively.

4 Data Collection and Analysis

A pilot study is conducted to check the suitability of factors and measurements. The survey is reviewed by specialized persons in industry as well as in academia. Random sampling is collected through a survey distributed either by hand or email to Jordanian companies which vary in structure, size, capacity and goals, providing a rich environment for systematically studying the level of JITP implementation. Targeted respondents' positions include Chief Executive Officers, senior managers, directors, supervisors, and consultants and engineers. The total number of distributed surveys is 250, useable responses obtained are 208 surveys, the remaining are either not replied on or with incomplete information. Consequently, a response rate of 83 % is attained.

The survey distributed to collect data is composed of three main parts, including demographic and descriptive information about the respondent and the company where he/she works. The first part consists of basic information about the

Table 1 Item measures of JITP

Factor	Item measure
Top management commitment (TMC)	Top management communicates their commitment to the implementation of JITP to the employees
	Top management provides necessary resources for the implementation of JITP
	Involvement of major department heads in JITP implementation
	Top management has had training in the implementation of JITP
	Training the employees to improve performance and solve problems in JITP
Employee training (ET)	Training the employees to acquire planning skills in the JITP process
	Commitment of top management to train employees on JITP
	Availability of resources for employee JITP training in the organization
	Participation in JITP decisions by hourly/nonsupervisory employees
	Amount of feedback provided to employees on their performance quality in the JITP process
Employee relations (ER)	Employees are recognized and rewarded for superior quality performance in JITP process
	Employees are well aware of the organization's purpose of JITP implementation
	Degree of employee cooperation to attain the organization's purpose of implementing JITP
	Long-term relationships are offered to fewer, high quality suppliers
	Suppliers are evaluated based on quality and delivery performance rather than price
	Involvement of the supplier in the product/service development process
Supplier value added (SVA)	Thoroughness of the organization's supplier rating system, to evaluate and select suppliers
	Suppliers conform to the exact quality attributes required on your incoming parts
	Amount of JITP-related education and technical assistance provided to the suppliers by your organization
	Involvement of purchasing/transportation/logistics departments in designation of inbound carrier
Transportation practices (TP)	Involvement of purchasing/transportation/logistics departments in designation of inbound Routing
	Delivery dates and times are met according to buyer's schedule
	Transportation services are reliable
	Carriers' consistency of delivering parts on schedule
	Deliveries are made in small lot sizes
Quantity delivered (QD)	Deliveries are received in exact quantities ordered
	Deliveries are made on a daily (and fast) basis
	Vendors supply on a just-in-time basis

Table 2 Measures of firm performance

	Code	Item measure
Inventory management performance (IMP)	IMP1	Purchase material turnover
	IMP2	Total inventory turnover
Time-based quality performance (TQP)	TQP1	Product/service quality
	TQP2	Productivity
	TQP3	Deliveries lead time of finished product/service to customer
	TQP4	Cost of scrap and rework as percent of sales
Financial and market performance (FMP)	FMP1	Market share
	FMP2	Market share growth
	FMP3	Sales growth
	FMP4	Profit
	FMP5	Profit growth
	FMP6	Return on investment (ROI)

organization/company. The second part is related to the respondent information. The last part consists of the model's measures. The item measures are closed ended using a five-point Likert scale ranges from strongly disagree to strongly agree (1 = strongly disagree, 2 = disagree, 3 = neither agree nor disagree, 4 = agree, 5 = strongly agree).

4.1 Demographic Data Profile Analysis

Figure 2 shows analysis results for demographic parameters for company. Figure 3 shows analysis results for demographic parameters for respondent.

4.2 Validity Analyses

To check the validity of the models, test of reliability for the measurement variables and model fit were conducted. Results of the analyses for the models are presented as follows.

4.2.1 Multicollinearity Test

Samples of inter-item correlation matrix and inter-item covariance are calculated and provided in Tables 3 and 4, respectively. It is found that all the inter-item correlations and covariance do not exceed the threshold of 0.9. Hence, the multicollinearity type problems do not exist.

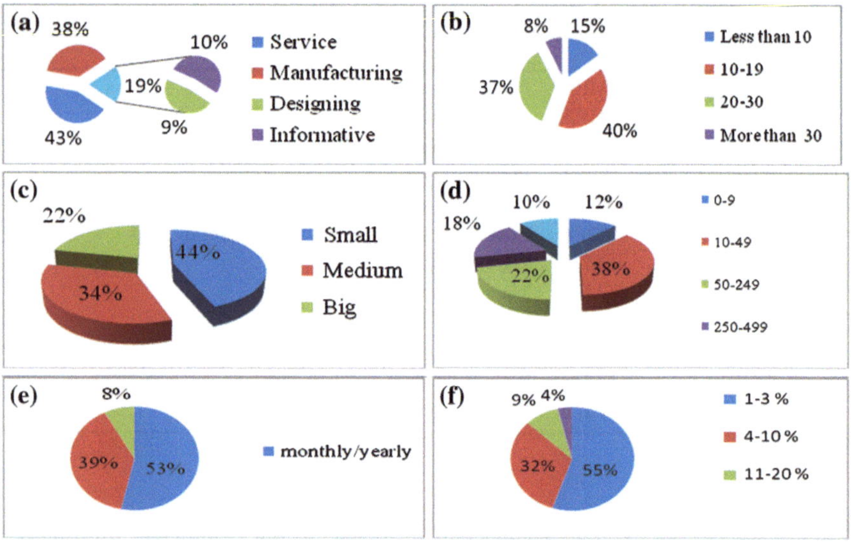

Fig. 2 Analysis of demographic parameters for firms. **a** Company type **b** Company's Age (years) **c** Capital of the company **d** Number of employees **e** How often supplies needed are ordered **f** Percentage of products Are defective at the last inspection point

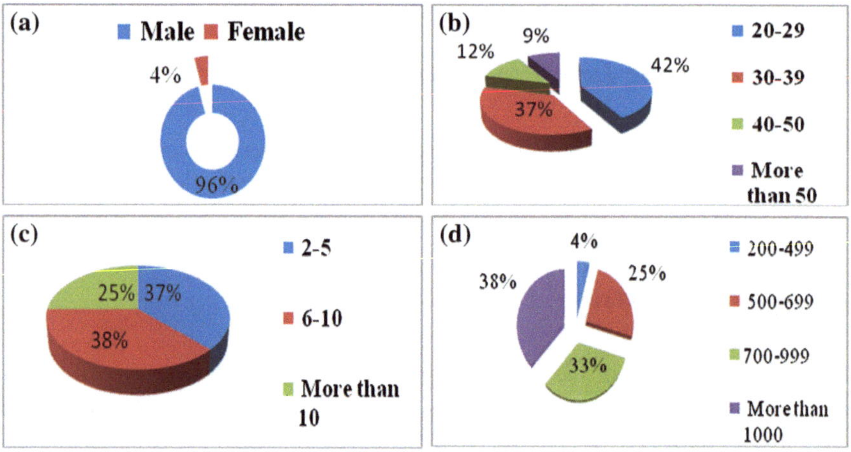

Fig. 3 Analysis of demographic parameters for respondents. **a** Gender of Employee **b** Age of employee **c** Years of experience of employees **d** Salary of employee

4.2.2 Test of Reliability

The Cronbach's α is adopted to assess internal consistency between the item measures of each model dimension as well as all items measures. Table 5 shows the α values for all dimensions. A value of α that is greater than 0.6 indicates

Table 3 Sample data of inter-correlation matrix

Measure	QD	ER	ET	TP	SVA
ER	0.648				
ET	0.414	0.149			
TP	0.699	0.371	0.741		
SVA	0.034	0.151	0.249	0.167	
TMC	0.814	0.693	0.632	0.841	0.031

Table 4 Sample data of inter-covariance matrix

Measure	QD	ER	ET	TP	SVA
ER	0.058				
ET	0.046	0.149			
TP	0.056	0.371	0.174		
SVA	0.024	0.132	0.239	0.152	
TMC	0.071	0.369	0.236	0.480	0.169

Table 5 Estimated Cronbach's α values

Model dimension	Cronbach's α value
TMC	0.783
ET	0.897
ER	0.857
SVA	0.780
TP	0.811
QD	0.622
IMP	0.784
TQP	0.846
FMP	0.955

internal consistency in item measures. In Table 5, the value of the lowest α value (=0.622) corresponds to quantities delivered, which is larger than 0.6. In addition, the overall α value is calculated and is equal to 0.876. These values indicate the reliability of these models.

4.2.3 Evaluating the Model

Structural equation modeling (SEM) is used to provide empirical support for the effectiveness of the models. Table 6 represents the values of fit measures for the measurement model.

In Table 6, the values of GFI, AGFI, and RMSEA are 0.949, 0.896, and 0.06, respectively. These values indicate the validity of the measurement model. In order to draw conclusions about models hypotheses, the structural models are

Table 6 Fit summary for the measurement model

Model	GFI	AGFI	RMSEA	χ^2	χ^2/DF
Measurement	0.949	0.896	0.060	69.4	1.83
Structural	0.979	0.940	0.055	16	1.6

analyzed. Solving the structural model, the minimum is achieved with a Chi-square divided by degrees of freedom equals 1.6 and Probability level of (*p*-value = 0.021). Table 6 also summarizes the values of fit measures for the structural model, where the values of GFI, AGFI, and RMSEA are 0.979, 0.94, and 0.055, respectively. These values indicate the validity of the structural model.

5 Results Discussion and Implications

The structural model is analyzed and the results are displayed in Table 7, where it is found that JITP implementation is positively influenced by top management commitment (TMC), employees' training (ET), and employees' relations (ER) because the corresponding *p*-values are less than 0.05. Therefore, in order to improve the degree of JITP implementation, the organization should continually develop improvement plans related to TMC, TR and ER. Moreover, supplier value-added practices (*p-value* = 0.074), transportation practices (*p-value* = 0.061) and quantities delivered practices (*p*-value = 0.059) have insignificant effects on the degree of JITP implementation. This indicates that organizations do not take serious actions towards implementing JITP practices. Also, employee training (*Estimate* = 1.516) is the most affecting factor on the degree of JITP implementation followed by employee relations and top management commitment. While supplier value-added practices, transportation practices, and quantities delivered practices are the least affecting factors. Further, firm Performance is positively influenced by top management commitment, employees' training and employees' relations, with *p*-values less than 0.05. Whereas, supplier value-added practices (*p-value* = 0.074), transportation (*p*-value = 0.057) and quantities Delivered (*p-value* = 0.063) have insignificant effect on firms' performance. Furthermore, employee training (*Estimate* = 1.875) is the most affecting factor on firm performance followed by employee relations and top management commitment. While supplier value-added, transportation, and quantities delivered practices are the least affecting factors. Finally, firm performance is positively influenced by time-based quality performance, and financial and market performance, with *p*-values less than 0.05. However, inventory management performance (*p-value* = 0.058) has insignificant effect on firm performance. It is noted that the financial and market performance (*Estimate* = 1.185) is the most affecting factor on firm performance followed by time-based quality performance.

Table 7 Results of structural model

Hypothesis	Estimate	Standard error	P-value	Result
H_1: TMC has a positive influence on JITP	1.165	0.201	0.031	Supported
H_2: TR has a positive influence on JITP	1.516	0.003	0.004	Supported
H_3: ER has a positive influence on JITP	1.421	0.487	0.015	Supported
H_4: SVA has a positive influence on JITP	0.165	0.156	0.074	Unsupported
H_5: T has a positive influence on JITP	0.962	0.040	0.061	Unsupported
H_6: QD has a positive influence on JITP	0.761	0.157	0.059	Unsupported
H_A: TMC has a positive influence on FP	1.326	0.006	0.021	Supported
H_B: ET has a positive influence on FP	1.875	0.001	0.038	Supported
H_C: ER has a positive influence on FP	1.439	0.027	0.008	Supported
H_D: SVA has a positive influence on FP	0.233	0.031	0.074	Unsupported
H_E: TP has a positive influence on FP	0.020	0.050	0.057	Unsupported
H_F: QD practices have a positive influence on FP	0.134	0.041	0.063	Unsupported
H_G: IMP has a positive influence on FP	0.031	0.038	0.058	Unsupported
H_H: TQP has a positive influence on FP	1.130	0.190	0.023	Supported
H_I: FMP has a positive influence on FP	1.185	0.009	0.041	Supported

This research contributes to the JITP theory by validating the effects of JITP implementation on firm performance. In Jordan, the importance of strong management commitments to successful JITP implementation is clear. Firms that have not benefited substantially from JITP implementation are those suffer weak commitments to JITP techniques that might be diluting their effectiveness. This research also highlights the fact that effective employees' training and relations are crucial JITP development for long-term cooperative relations with suppliers. In this research, it is found that Jordanian firms with committed management and employees training and relations have resulted in improved performance measured by time-based quality and financial/market indicators. However, they did not direct significant efforts to plan, improve, and maintain effective long-term supplier value-added, transportation, and quantities delivered practices. In return, JITP implementation does not significantly affect inventory management performance. In order to achieve sustainable competitive advantage managers must integrate JITP into their strategic planning, develop operations strategy that should support effective JITP implementation, and support, coordinate, and maintain effective supplier value-added, transportation, and quantities delivered practices to avoid delivery problems that could impede operations and production that could retard improving firms performance and weaken organizational commitment to effective JITP implementation.

6 Conclusion

This research successfully examined the effects of JITP implementation on firms performance in Jordan using structural equation modeling. JITP implementation is examined via six key dimensions, including top management commitment, employees' training, employees' relation, supplier value-added practices, transportation practices, and quantities delivered practices. Firm performance is measured by three dimensions, involving inventory management performance, time-based quality performance, financial and market performance and time-based quality performance. Results showed that Jordanian firms assured top management commitment, employees' training, and employees' relations, but they still lack implementing effective practices for successful JITP, which are related to supplier added-value, transportation, and quantities delivered. Moreover, inventory management performance has insignificant effect on firm performance. Therefore, Jordanian organizations should start developing plans to improve their inventory management performance if they wish to retain their competitive position.

References

Al-Refaie A, Hanayneh B (2014) Influences of TPM, TQM, six sigma practices on performance in Jordan. Int J Prod Qual Manag 13(2):219–234

Al-Refaie A, Jalham IS, Li M-H (2012a) Factors influencing the repurchase intention and customer satisfaction: a case of Jordanian telecom companies. Int J Prod Qual Manag 10(3):374–387

Al-Refaie A, Ko JH, Li M-H (2012b) Examining the factors that affect tourists' satisfaction, loyalty, WOM and intention to return using SEM: evidence from Jordan. Int J Leisure Tourism Mark 3(2):179–197

Al-Refaie A, Li M-H, Ko JH (2012c) Factors affecting customer linking capabilities and customer satisfaction in CRM: evidence from Jordanian hotels. Int J Customer Relationship Mark Manag 3(4):16–30

Balci D, Kesen SE, Baykoc OF (2007) Adaptability of just-in-time (JIT) philosophy to service systems: a case study. Simulation 83:631–642

Barlow GL (2002) Just-in-time: implementation within the hotel industry—a case study. Int J Prod Econ 80:155–167

Beer M (2003) Why total quality management programs do not persist: the role of management quality and implications for leading a TQM transformation. Decis Sci 34:623–643

Daft R (1998) Organization theory and design. South-Western College, Cincinnati

Dong Y, Carter CR, Dresner ME (2001) JIT purchasing and performance: an exploratory analysis of buyer and supplier perspectives. J Oper Manag 19:471–483

Ford RC, Fottler MD (1995) Empowerment: a matter of degree. Acad Manag Exec 9(3):21–29

Garg D, Deshmukh SG, Kaul ON (1996) Critical analysis in JIT purchasing in Indian context. Prod J 37:271–279

Garg D, Deshmukh SG, KAUL ON (1997) Supplier evaluation in JIT purchasing: inferences from a survey. Prod J 38:322–326

Garg D, Deshmukh SG (1999) JIT purchasing: literature review and implications for Indian industry. Prod Plann Control 10(3):276–285

Gunasekaran A (1999) Just-in-time purchasing: an investigation for research and applications. Int J Prod Econ 59:77–84

Handfield R (1993) Distinguishing features of just-in-time systems in the make-to-order/assemble-to-order environment. Decis Sci J 24:581–602

Jayaram J, Vickery SK (1998) Supply-based strategies, human resource initiatives, procurement lead time and firm performance. Int J Purch Mater Manag 34(1):12–23

Kaynak H (2002) The relationship between just-in-time purchasing techniques and firm performance. IEEE Trans Eng Manag 49(3):205–217

Kaynak H, Hartly JL (2006) Using replication research for just-in-time purchasing construct development. J Oper Manag 24(6):868–892

Kaynak H, Pagan JA (2003) Just-in-time purchasing and technical efficiency in the US manufacturing sector. Int J Prod Res 41(1):1–14

Landry S, Duguay CR, Chausse S, Themens J-L (1997) Integrating MRP, kanban and barcoding systems to achieve JIT procurement. Prod Inventory Manag J 37:8–12

Marsh F, Conard A (2008) A pull system for delegating knowledge work. Oper Manag Res 1:61–68

Nassimbeni G (1996) Factors underlying operational JIT purchasing practices: results of an empirical research. Int J Prod Econ 42(3):275–288

Radovilsky Z, Gotcher JW, Mistry R, Yip R (1996) JIT purchasing: analyzing survey results. Ind Manag 38(6):17–21

Vonderembse MA, Tracey M, Tan CL, Bardi EJ (1995) Current purchasing practices and JIT: some of the effects on inbound logistics. Int J Phys Distrib Logist Manag 25(3):33–48

Waters-Fuller N (1995) Just-in-time purchasing and supply: a review of the literature. Int J Oper Prod Manag 15:220–236

Analysis and Improvement of the Process Engineer's Levels of Competence in a Manufacturing Company

Małgorzata Spychała

Abstract Analysis of tasks in given job position is essential to develop employee competence profile. In this section of the article the main stages of the analysis and improvement of Process Engineer competency levels were presented. Technical competences can be examined in terms of quality and quantity, it is far more difficult to analyze engineer's social competences. Using a 360 degree method competency profile for the engineer was developed and then an improvement plan for competency gaps was presented.

Keywords Process Engineer's competences • Competency profile • Development

1 Introduction

Development and improvement of employee competencies is a fundamental task in innovative organizations because *High-performing people are critical for high-performing organizations* (Rodriguez et al. 2002). Modern technologies, means of production and objects of labor influence on changes of production processes (Lachiewicz and Matejun 2011) and thus on performers' new tasks. Knowledge of these tasks is necessary to prepare the employees in terms of the required competences, namely *features (properties) of unit that underlie effective action or behavior at work* (Slivinski and Miles 1996).

Identification of employees' competences is not an easy task. While technical competences can be examined in terms of quality and quantity, it is far more difficult to analyze social competences (Jurek 2008; Spychała 2013; Walkowiak 2008). *However, in order to manage training and development efficiently,*

M. Spychała (✉)
Poznan University of Technology, Strzelecka 11, 60-965 Poznan, Poland
e-mail: malgorzata.spychala@put.poznan.pl

P. Golinska (ed.), *Logistics Operations, Supply Chain Management and Sustainability*, EcoProduction, DOI: 10.1007/978-3-319-07287-6_28,

managers need to understand what constitutes human competence at work. Without such an understanding, competence development cannot be managed effectively and, therefore, effectiveness in organizations cannot be achieved (Sandberg 2000). This is why the identification and definition of the competences of employees is the base of preparation of the development plans for tasks performers.

The aim of this chapter of monograph is to develop methods for testing and improving levels of employee competences. In the first part the main steps for identifying the knowledge and skills of employees will be presented. In the empirical part of the article all stages of the research and improvement of the competences levels based on example of Process Engineer at selected manufacturing company will be presented. For analyzing employee competencies the method of 360 degrees has been used and expanded with observations and interviews. The most important part of this chapter is the design of development plan for the Process Engineer.

2 Method of 360 Degrees as a Way to Study and Evaluate of the Professional Competences of Employees in Innovative Enterprises

In the literature there is a number of methods and techniques described to study the competences of employees (Dale and Iles 1993; Dubois and Rothwell 2004; Filipowicz 2004; Jurek 2008; Sidor-Rządkowska 2006; Smółka 2008; Spencer et al. 1990; Walkowiak 2008).

Dubois and Rothwell believe that the competence of employees are possible to be identified by:

- identifying behaviors or measurable results achieved through the use of those competences taking into account the nature of the work,
- using behavioral indicators—describe action or set of actions which may be expected to occur when a person uses effectively owned by him competences to do the job (Dubois and Rothwell 2004).

One of the methods used to identify and assess the competence of workers in modern organizations is the method of 360 degrees. The procedure for assessing take into account the opinions of the persons involved in a variety of job positions, having the opportunity to observe the employee from different perspectives: senior manager, direct supervisor, different teams managers, HR staff, colleagues, team members, subordinates, internal and external customers. In Fig. 1 a group of people evaluating Process Engineer's competences is shown. Number of people participating in the assessment process is undefined—employee competences can be identified and assessed by several or even dozens of people. However it is believed that satisfactory objectivity and reliability of evaluation could be obtained when at least four sources of evaluation participate in assessment where at least

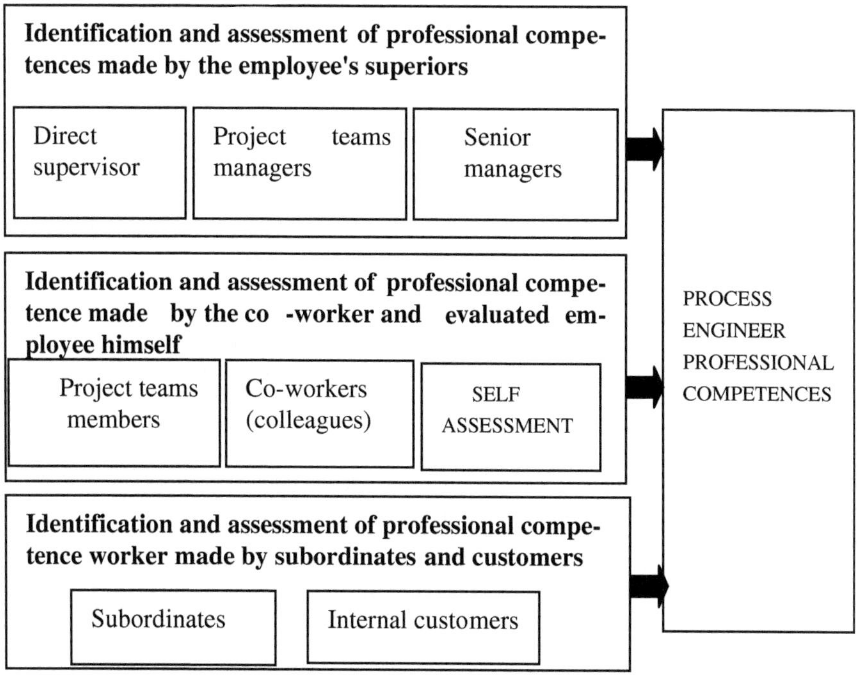

Fig. 1 Diagram of identification and assessment of Process Engineer's professional competences with use of 360 degrees method. *Source* own based on (Spychała 2011)

two of them are represented by four or more persons. The total number of people assessing a person should not be less than six (Jurek 2008). Each of those groups assess the competences of a person differently, looking at a different angle at his work. For example, assessment of Process Engineer's competences made by a senior manager will not be fiction only when the supervisor will have a direct, professional contacts with subordinate. In this case he will be able to assess a person in the perspective of a larger piece of the organization. Employee competences assessment made by the leaders of project teams will make sense when an employee will be involved in several projects—then he will be able to be assessed by managers of each of the teams.

360 degree feedback provides a complete picture about how people perceive the behavior of assessed person. In the presented method the very important element is self-assessment of competences, as it enables the employee to identify competency gaps, which sometimes are a source of problems arising during work and the factors that enable solving them. It also affects employee's sense of responsibility for the way of performing tasks (Bieniok 2006). Provision of feedback by 360 degrees method is considered by employees as one of the most accurate, comprehensive, objective and acceptable evaluations. It is based on the fundamental principle that the greater the number of sources of assessment, the more objective and reliable assessment of competence is calculated on the their basis (Jurek 2008).

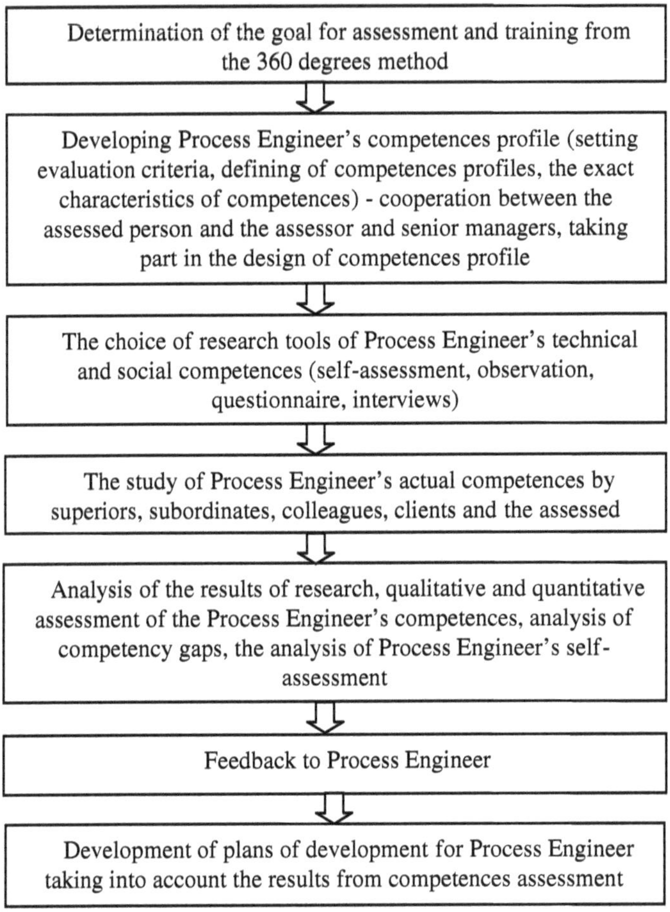

Fig. 2 Diagram of design and assessment of employees' professional competences. *Source* own study based on data from the conducted research

Figure 2 shows a scheme of examining, evaluating and improving of Process Engineer's competences. The first step is to determine the purpose of the study and train a group of employees, who will participate in the evaluation of Process Engineer of the method of 360 degrees. Then it is necessary to develop a competency profile for the assessed job position or to modify an existing one. In addition to determining the evaluation criteria, necessary action is to define all the tested competencies at the workplace (see Table 1). Each evaluator must know the detailed definitions of competences, in order to objectively determine the level of assessed employee's skills.

The next step is the selection of tools for examining the competences of Process Engineer. In this example, a questionnaire and interview was used as well as observation during performing tasks. The assessed employee also performed self-assessment of his knowledge and skills.

Table 1 Definitions of competences required for the position of Process Engineer

Process Engineer professional competences	Definitions of competences required for the position of Process Engineer
Knowledge of production processes in the organization	Knowledge and skills associated with the applicable means of production (tools, machinery) and objects of labor (raw materials, materials, semi-finished products, energy) and the technology used in the production process
Ability to solve production problems	The ability to diagnose the problems in production processes and the ability to solve breakdowns in production process. It is also the ability to carry out corrective actions, the ability to develop temporary solutions enabling continuity production and the ability to assess the effectiveness of the solution
Ability of designing of technological processes	Knowledge of the general principles of design of the process and its elements. Knowledge of methods of processes design and the ability to use them in practice. Knowledge of the structure of production lines and the ability to select equipment for the production line
The ability to manage documentation	Knowledge and skills related to the processing of documents, document sharing and collaboration during documents creation enabling effective tasks performance
The ability to communicate within the organization	Ability to effective and appropriate communication in speech and writing, while also having the ability understand easily speech of others
Ability to manage a project team	The knowledge and practical abilities in the field of organizing project teams, motivation, delegation, supervision of employees in such a way that their work in the project team generate synergy, creativity, sense of responsibility and raised efficiency in accomplishing the goals of the enterprise
Ability to learn fast	Ability to quickly assimilate new knowledge and its proper use at a given time
Analytical skills	The ability to identify, analyze and improve complex workflows. Ability to solve problems by analyzing, separation into components, interpreting data and finding the cause
Ability of creative thinking (creativity)	Knowledge of the techniques of creative thinking and the ability to apply them in work processes. Using the skills and imagination to create new work process improvement solutions
Organizational skills	Planning skills of own and employees' effective time management through appropriate distribution and evaluation of activities to ensure the execution of short time-consuming tasks with the right effect at the right time
Innovation	Ability to apply new knowledge in the production process

Source own study based on data from the conducted research

Having investigated the Process Engineer's levels of competences, the competency profiles design specialist compiled the report, which included an analysis of qualitative and quantitative Process Engineer's levels of competences, competency gap analysis and the analysis of self-assessment of his knowledge and skills. Then Engineer received detailed feedback on his competences. The final step was working out of the development plans for key competencies of evaluated Engineer and those where his competency gaps were greatest.

3 Modeling of Tools for Assessment of Process Engineer's Professional Competences

The first step in modeling of tools for assessment of Process Engineer's professional competences is a detailed analysis of the tasks performed by the Engineer Process:

- Designing of technological processes.
- Design, qualification, implementation and development of production processes.
- Development and supervision of the list of software, manufacturing equipment and process documentation.
- Installation supervision, commissioning and start-up of equipment and tooling.
- Defining technological time standards.
- Determining the areas of continuous improvement of production processes with particular emphasis on improving the quality and performance optimization.
- Solving the most serious problems associated with the production process.
- Preparing the plan, supervise the implementation and report of non-series production run.
- Participation in audits conducted by the customer, provide all the necessary technical information as part of the manufacturing process, for which he is responsible.
- Participation in the preparation of production floor layout.

These tasks are the basis for a detailed analysis of the tasks performed in a job position by the Process Engineer (see Table 2). On the basis of these tasks technical and social competences have been developed (see Table 2).

The next step is to develop a model of competences for Process Engineer. In the analyzed organization there are three job positions (levels): Process Engineer I, Process Engineer II and Process Engineer III. Each of these positions has different competency levels. These levels were established by the Engineering Manager, the employee that performs tasks in assessed job position and a specialist in the field of designing of competency profiles. Figure 3 shows the differences in levels of competences between the Engineer II and III.

Table 2 Example of specific tasks performed by the Process Engineer with assigned professional competence

Task	Detailed tasks	Required professional and social competences
Designing of technological processes	The analysis of process specifications and process documentation, requirements for the quantity of products, operations, operations time consideration, what are the special characteristics, inspections Determination of the process efficiency, determination of bottlenecks, preparation of manufacturing processes diagrams, cooperation with suppliers of machinery (search for suppliers, carry out the process of selection of suppliers and bidding, managing the contract with supplier (e.g., equipment construction) and current agreements)	*Technical competence* Knowledge of the process of cutting, filling and capping: knowledge of the process, knowledge of the means of production, knowledge of the objects of labor in the process, the ability to use required tools, the ability to operate machinery, knowledge of raw materials, semi-finished products Knowledge of process documentation The ability to determine the efficiency of the process The ability to determine the bottlenecks in the process Knowledge of process flow diagrams, ability to prepare manufacturing processes diagrams Knowledge of programs: Word, Excel, Power Point, Visio, Outlook, Web browsers, *Interpersonal and social competencies* Analytical skills Organizational skills Negotiation skills Communication skills: • Cooperation with machinery suppliers • Ability to work in a project team • Ability to work with other departments • Knowledge of the English language in speaking and writing Creativity Innovation

Source own study based on data from the conducted research

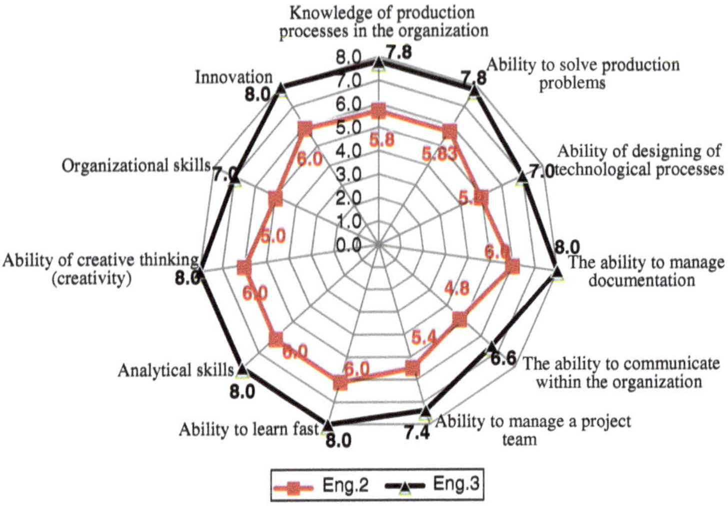

Fig. 3 A comparison of competences levels required for levels Engineer 2 and Engineer 3. *Source* own study based on the research in enterprise X

4 A Comparison of Average Ratings of Process Engineer Competences Levels Required for Engineer 2 and Engineer 3 Levels

After initial training regarding the competences evaluation criteria, detailed definitions and rules of application of the 360 degrees method, the evaluators completed survey questionnaires. The study took place in January 2014, in a large manufacturing company. There were 12 people evaluating Process Engineer (3 supervisors, 5 colleagues, 2 subordinates, 2 clients) and he made a self-assessment as well. Table 3 shows the average levels of Process Engineer competences assessed by all the evaluates (colleagues, clients, subordinates, superiors, self-assessment) and the requirements for the Engineer 2 and Engineer 3. Negative results on the evaluated Process Engineer competence gap compared with the Engineer 2 represent the excess of competences. This means that, according to all evaluators Engineer has an excess of competence in the area of: knowledge of production processes in the organization, skills of, production problems solving, design of technological processes, communication skills in organization, project team management skills, ability to learn quickly, analytical skills, creative thinking (creativity), organizational skills and innovation.

Some excesses of competences are greater than 1, which means that there is potential of utilization of certain competence in other areas of the organization. These are the following competencies: the ability to design processes, communication skills in organization, creative thinking (creativity) and organizational

Table 3 Average results of Process Engineer levels of competences

Process Engineer professional competences	Average	Engineer 2	Competences gap for Engineer 2	Engineer 3	Competences gap for Engineer 3
Knowledge of production processes in the organization	6.5	5.8	−0.7	7.8	1.3
Ability to solve production problems	6.4	5.8	−0.6	7.8	1.4
Ability of designing of technological processes	6.7	5	−1.7	7	0.3
The ability to manage documentation	5.9	6.0	*0.1*	8.0	2.1
The ability to communicate within the organization	6.0	4.8	−1.2	6.6	0.7
Ability to manage a project team	5.6	5.4	−0.3	7.4	1.7
Ability to learn fast	6.4	6.0	−0.4	8.0	1.6
Analytical skills	6.8	6.0	−0.8	8.0	1.2
Ability of creative thinking (creativity)	7.2	6.0	−1.2	8.0	0.8
Organizational skills	6.1	5.0	−1.1	7.0	0.9
Innovation	6.8	6.0	−0.8	8.0	1.2

Source own study based on data from the conducted research

skills. These are very good results, because they are necessary to perform the tasks of a Process Engineer 3.

Analyzing the average scores of 13 people assessing levels of Process Engineer competences one observes only one very small competence gap (0.1) related with records management capabilities and in particular technical knowledge and skills to develop documentation of production processes, test trials, changes, devices, equipment measurement equipment and the ability to release those documents. Co-workers and superiors evaluated this competence at required or below the required level for the Engineer 2. The other evaluators believe that this competence is at sufficient level or higher than required.

Comparing the scores of all evaluators with the demands for the position of Engineer 3 one may notice one excess competence: knowledge of the English language both written and spoken 6.92 where the required level for the Engineer 3 is 6. The smallest competency gaps relate to (from 0.7 to 0.9): the ability to communicate within the organization, creative thinking (creativity) and organizational skills. There is only one competence gap greater than 2, which relates to document management (Fig. 4).

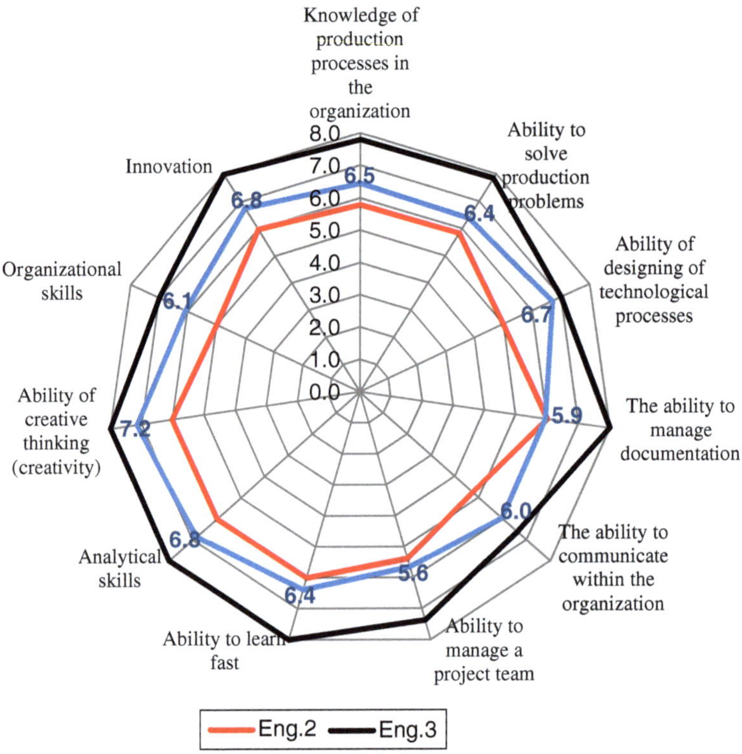

Fig. 4 A comparison of average ratings of Process Engineer competences levels required for levels Engineer 2 and Engineer 3. *Source* own study based on the research in enterprise X

5 Analysis of Ratings of Process Engineer Competences Levels Depending on the Evaluation Group

Analyzing ratings of Process Engineer competences levels depending on the evaluation group one can conclude that the lowest level of competence is the level 5 (or "competence absorbed in a good level, which allows for good fulfillment of the tasks in a given area and passing own experiences on to others") while the highest level 8 (expert level, the ability to creatively share knowledge and skills with other employees and managers) (Table 4).

Colleagues believe that the strongest side of Process Engineer is creativity (7.2) and the ability to design processes (7.0), on the other hand the weakest side is management of the project team (5.3) and document management (5.4).

Customers have different perceptions of Process Engineer competences: for them his strongest side are: the ability to communicate within the organization (7.1), knowledge of production processes (7.0), the ability to design processes (7.0) and the ability to manage documentation (7.0). None of the customers rated

Table 4 Assessment of Process Engineer competences levels from different groups of assessors

Process Engineer professional competences	Colleagues	Clients	Subordinates	Superior	Self assessment	Engineer 2	Engineer 3
Knowledge of production processes in the organization	6.1	7.0	6.2	6.7	6.6	5.8	7.8
Ability to solve production problems	5.8	6.7	6.8	6.8	6.7	5.8	7.8
Ability of designing of technological processes	7.0	7.0	6.5	6.7	6.0	5.0	7.0
The ability to manage documentation	5.4	7.0	6.6	5.6	6.3	6.0	8.0
The ability to communicate within the organization	5.6	7.1	6.7	5.0	6.5	4.8	6.6
Ability to manage a project team	5.3	6.0	6.7	5.2	5.7	5.4	7.4
Ability to learn fast	6.0	6.5	6.5	6.3	8	6.0	8.0
Analytical skills	6.8	6.5	7.0	7.0	7	6.0	8.0
Ability of creative thinking (creativity)	7.2	6.5	7.5	6.7	8	6.0	8.0
Organizational skills	5.8	6.5	7.5	5.0	7	5.0	7.0
Innovation	6.4	6.5	7.0	6.7	8	6.0	8.0

Source own study based on data from the conducted research

Process Engineer below level 6, which means that evaluated Engineer assimilated all competencies at a very good or excellent level, which eventually allows for very good or excellent fulfillment of his tasks in a given area.

Another group are the subordinates who assessed the evaluated Process Engineer competences in the range of 6.2–7.5. This is a very good and targeting notes that even catch the expert level (7.5). Subordinates recognize that the Engineer has the highest competence in the field of creativity (7.5), organizational skills (7.5), innovation (7.0) and analytical skills (7.0). It should be emphasized that, according to subordinates Process Engineer has team management skills at the level of 6.7—that is, to a degree almost perfect, where at the same time colleagues and supervisors assessed these skills at a good level (5.2–5.3). At the level of 6.2 (this is the lowest note in this group) subordinates assessed the knowledge of production process.

The superiors decided that the strongest competences are analytical skills (7.0) and the ability to solve production problems (6.8). At the same time lowest rank from of all the people involved in the survey assessed the competence of Process Engineer in communication skills in the organization (where customers—that is, Production, said that this is the strongest side of the test), organizational skills (where subordinates evaluated the competence of the expert level—7.5) and the ability to manage a team of (5.2).

A very important result is the self-test Process Engineer. He believes that his strongest skill is creativity, innovation and speed of learning. All of these skills rated at the highest level 8 These are the powers of personality. At level 7 test rating analytical skills. It is worth noting that the subordinates and superiors also assessed the competence at the same level, and co-workers and customers at much lower (6.8 and 6.5). Ability to problem solving is another competence, which was very much praised by the majority of investigating, at the level of 6.7. Studied rated competence ability to design processes at level 6 while all other assessed the competence of the above within the limits of 7.0–6.5. The test considers that the weakest of his competence is the ability to manage a project team (5.7) are of a different opinion subordinates who assessed the competence level higher (6.7).

6 Improving of Process Engineer's Competency Levels

The main objective of the analysis of professional competences in Process Engineer position is to improve levels of competency. Evaluated employee has already reached all levels of competences necessary to perform the tasks of an Engineer 2 and at the same time obtained the excess of competences in many areas. Therefore, on the basis of studies development plan for each of the key competencies required for the position of Process Engineer 3 was designed (see Table 5).

Other competencies that have not been included in Table 5 have a competence gap of less than 1, which means very little deviation from the required level for the Engineer Process 3.

Table 5 Proposed development plan for the Process Engineer in the position of Process Engineer 3

Process Engineer professional competence	Process Engineer behaviors demonstrating competence gap	Proposed plan for the development of the competencies
The ability to manage documentation Competence gap 2.1	• Too detailed documentation • Very technical language in documents, incomprehensible to the readers • Using shortcuts • Good reviewer of documents	Process Engineer should: • Write clear, understandable for documentation readers • Avoid shortcuts • Make sure that the recipient understands the technical language included in the documentation • Help technicians develop documentation
The ability to manage project team Competence gap 1.7	• Not enough involvement of production dept representatives in designing process • Failing in noticing all needs of team members • Problem with tasks delegation	Process Engineer should: • Deeper involve the production representatives in design • Motivate team members • Outsource simple tasks to team members and check if they completed tasks
Ability to solve production problems Competence gap 1.4	• Problem with the approval of corrective actions • Problem with deadlines • Problem with completing and taking over responsibility for the solution of the team • The problem of legibility of documentation related with corrective actions, difficult language	• Determine the scope of responsibility for approval of corrective actions, who should do it • Co-workers should know who is responsible for the approval of corrective actions and corrective process Process Engineer should also: • Determine the actual date of implementation of corrective actions • Use simple language during documenting corrective actions • Avoid difficult phrases • Feedback is required for the interpretation of correction documents

(continued)

Table 5 (continued)

Process Engineer professional competence	Process Engineer behaviors demonstrating competence gap	Proposed plan for the development of the competencies
Knowledge of production processes in the organization Competence gap 1.3	Requirements for Process Engineer 3 position regarding the knowledge of the use of technology, knowledge of the means of production, raw materials, semi-finished products and the ability to use given tools and machines are operating at the level of an expert. This means that evaluated Process Engineer is required to continuously improve these areas	Therefore, the Engineer should: • Attend in the fair related with modern means of production and objects of work and the latest technology which can be used in the organization • Participate in the internships in companies involved in the latest technology and production equipment • Participate in training with the latest software, such as for the design of manufacturing processes • Read industry literature • Participate in various projects where he can learn from other experienced engineers
Analytical skills Competence gap 1.2	No aware teaching others of analytical competences	Evaluated Engineer during participation in project teams should consciously "teach others" analytical skills
Innovation Competence gap 1.2	No aware teaching others of innovation	It is recommended that others participating in projects followed Process Engineer, or he should be their mentor in this regard

Source own study based on data from the conducted research

7 Summary

Competence profiles for individual job positions are the leading factor in the development of skills of workers in modern organizations. They allow to estimate the level of competences achieved and to identify competency gaps.

The base for preparation of employees' competences improvement programs is a detailed analysis of the tasks of the job and the technical and social competences. Identification of competences at the given work position is of a major importance for the development of employees, because it is the basis for an objective assessment of competences and allows to estimate the gap between the possessed and desirable competences (Witaszek 2011).

Developing competences profiles are also standard in estimating the effects of competence development. They can also be the basis for the design of plans of internal and external training of employees.

References

Bieniok H (2006) Red system zarządzania zasobami ludzkimi przedsiębiorstwa, AE, Katowice

Dale M, Iles P (1993) Assessing management skills: a guide to competencies and evaluation techniques. Kogan Page, London

Dubois D, Rothwell W (2004) Competency-based human resource management. Davies-Black Publishing, Palo Alto

Filipowicz G (2004) Zarządzanie kompetencjami zawodowymi. PWE, Warszawa

Jurek P (2008) Analiza wybranych metod oceny kompetencji zawodowych. [w:] Witkowski A, Listwan T (red.) Kompetencje a sukces zarządzania organizacją. Difin, Warszawa

Lachiewicz S, Matejun M (2011) Rola kierownictwa średniego szczebla w procesie stymulowania przedsiębiorczości technologicznej, [w:] Krzakiewicz K. (red.) Praca kierownicza w nowoczesnym zarządzaniu, Zeszyty Naukowe UE w Poznaniu, nr 189, Poznań s, pp 125–133

Rodriguez D, Patel R, Bright A, Gregory D, Gowing MK (2002) Developing competency models to promote integrated human resource practices. Hum Resour Manag 41(3):309–324

Sandberg J (2000) Understanding human competence at work: an interpretative approach. Acad Manag J 43(1):9–25

Sidor-Rządkowska M (2006) Kompetencyjne systemy ocen pracowników. Przygotowanie, wdrożenie i integrowanie z innymi systemami ZZL. Wolters Kluwer—Oficyna Ekonomiczna, Kraków

Slivinski L, Miles J (1996) The wholistic competency profile: a model. Personnel Psychology Centre, Public Service Commission, Ottawa

Smółka P (2008) Kompetencje społeczne, metody pomiaru i doskonalenia umiejętności interpersonalnych. Oficyna, Kraków

Spencer LM, McClelland DC, Spencer SM (1990) Competency assessment methods: history and state of the art. In: Paper presented at the American Psychological Association annual conference, Boston, Massachusetts

Spychała M (2011) An analysis of methods of monitoring and evaluation of workers' occupational competencies in a modern enterprise. w: Branowska A, Siemieniak P, Spychała M (eds) Workers' occupational competencies in a modern enterprise. Publishing House of Poznan University of Technology, Poznan

Spychała M (2013) Kompetencje zawodowe menedżera przyszłości w wielkopolskim sektorze komunalnym. Acta universitatis lodziensis, Folia oeconomica, Łódź

Walkowiak R (2008) Problemy metodologiczne pomiaru i raportowania kompetencji. [w:] Witkowski A, Listwan T (red.) Kompetencje a sukces zarządzania organizacją. Difin, Warszawa

Witaszek Z (2011) Rozwój kompetencji menedżerskich przesłanką sukcesu organizacji. Zeszyty Naukowe akademii Marynarki Wojennej Rok lii nr 4(187):303

Modeling and Performance Improvement: The Remedy to Treat Social and Environment Issues for Enterprises in Today's Difficult Economic Climate

Paul-Eric Dossou and Philip Mitchell

Abstract European economies have been deeply affected by different crises. The impact of the economic crisis on enterprises is now recognized by everybody. Enterprises need to reorganize in order to be better adapted to this situation and to integrate new dimensions in their development. Reduction of cost is not the only way for making enterprises more efficient. It is now clear that a mono-criterion analysis is not adapted to the actual enterprise situation. Enterprises need a multi-criteria analysis by combining quality, cost, lead time but also carbon management, social societal and environmental dimensions. If QCD criteria are already considered as necessary for obtaining the optimum enterprise system, it remains difficult to convince the enterprise management of the opportunity to integrate social, societal and environmental dimensions for improving cost. This need still needs to be clearly demonstrated. This chapter introduces concepts for showing that enterprises would be more efficient, better-organized and adapted to the new changes in society. The reduction of cost is necessary, the increase in enterprise turnover too, but it is also indispensable to change the structure of enterprises. Enterprise modeling (GRAI Methodology) and a tool will be used for illustrating the concepts presented through a detailed case study.

Keywords Carbon management · Sustainable supply chain management · Quality management · Knowledge management · Energy reduction

P.-E. Dossou (✉) · P. Mitchell
ICAM VENDEE, 28 Boulevard d'Angleterre, 85000 La Roche-Sur-Yon, France
e-mail: paul-eric.dossou@icam.fr

P. Mitchell
e-mail: philip.mitchell@icam.fr

P. Golinska (ed.), *Logistics Operations, Supply Chain Management and Sustainability*, EcoProduction, DOI: 10.1007/978-3-319-07287-6_29,

1 Introduction

No one can dispute the present plight of many European countries. However it would be a mistake to regard the entire performance in purely economic terms. In the actual global reference model (capitalism), the main enterprise performance criterion is based on economic values. It is clear that we are nowadays at the limits of this model. For instance, in Europe, debt has increased by 450 billion € despite cost reductions imposed by the International Monetary Fund (IMF) and European governments. The only prescription offered to the PIIGS and other European countries is an increase in taxation and a reduction in government expenditure. Furthermore, the countries concerned are exposed to speculation and no alternative solution to the actual situation has been offered to them. Indeed, it is quite different in enterprises because of the introduction of quality and lead time as criteria for completing cost [Quality, Cost and lead time (QCD)]. The choice of performance criteria is important for an enterprise. Indeed, this enterprise could evaluate its performance according to these criteria and prepare action plans if necessary. Since 2008 and the beginning of the crisis, the failings of the capitalism system have become more apparent. Then in parallel to European countries, enterprises are searching for the ideal solution. Nobody knows exactly how to solve the situation not only in terms of the global economy but also in terms of performance for enterprises.

Many of these enterprises have understood that for being efficient, they need to satisfy not only customers, shareholders and but also suppliers and employees. The case of an enterprise is not just limited to production for satisfying customer demand and shareholders. Nowadays an enterprise is a multi-criteria system integrating social, societal and environmental dimensions in addition to QCD for its improvement (Fig. 1). The definition of a new sustainable reference model for enterprises integrating these criteria is necessary.

Enterprise modeling is regularly used to prepare enterprises for the outcome of the crisis. Enterprises need to find the best way to resist the present crisis and then to improve in order to be more efficient. GRAI Methodology is one of the three main methodologies (with PERA, CIMOSA) of enterprise modeling. To support this methodology different tools have been developed. GRAIMOD is the latest one being developed by using JAVA technology, JADE and JESS platforms, and an open architecture and structure.

This chapter summarizes the research done at ICAM Vendee in this area. Two different objectives could be defined.

- The Industrial and societal organization research team is working on the elaboration of a much-needed new reference model to replace the capitalist one. What type of society do we want in the future, what balance, which optimum to have. What organization should be adopted by local authorities taking into consideration the present parameters and how to define tomorrow's enterprise in accordance with the society we want to live in?

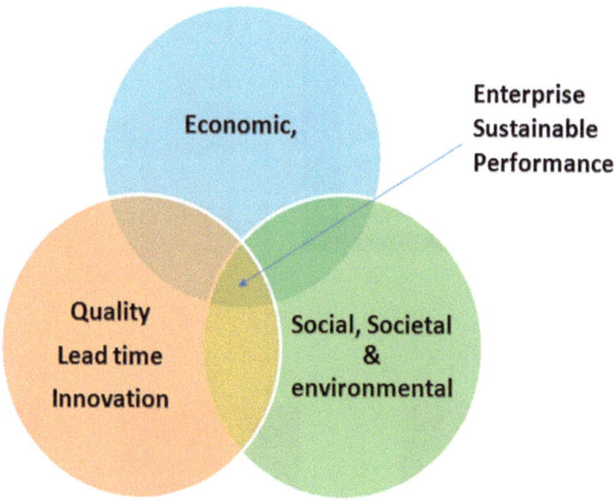

Fig. 1 Enterprise sustainable performance

- The use of actual organization for defining how to progressively improve enterprises. In this context, reference models are being elaborated according to activity domains. A supporting tool is also being developed according to the concepts of GRAI Methodology.

The elaborated concept aims to consider future manufacturing, supply chain, enterprise, local authority needs etc. The choice made is to define these concepts by introducing sustainable values in the proposed changes.

2 GRAI Methodology and GRAIMOD

The objective is to use reasoning (e.g. CBR or decomposition), enterprise typology, expert systems, Multi-agents systems, enterprise knowledge for defining a tool (GRAIMOD) destined to improve enterprise performance. This tool will support GRAI Methodology (Doumeingts and Ducq 1999).

The objective of this research is to:

- Allow enterprises to evaluate their performance and to drive the change of their economic model by integrating the social, societal and environmental aspects.
- Aid enterprises towards ecologic and energy transition,
- To improve progressively and sustainably enterprise supply chains.

GRAI Methodology is one of the three main methodologies used for analyzing and designing enterprises. The GRAI approach is composed of four phases: An initialization phase to start the study, a modeling phase where the existing system is described, an analysis phase to detect the inconsistencies of the studied system and a design phase during which the inconsistencies detected are corrected and a new system proposed. These concepts could be used to ensure the transformation

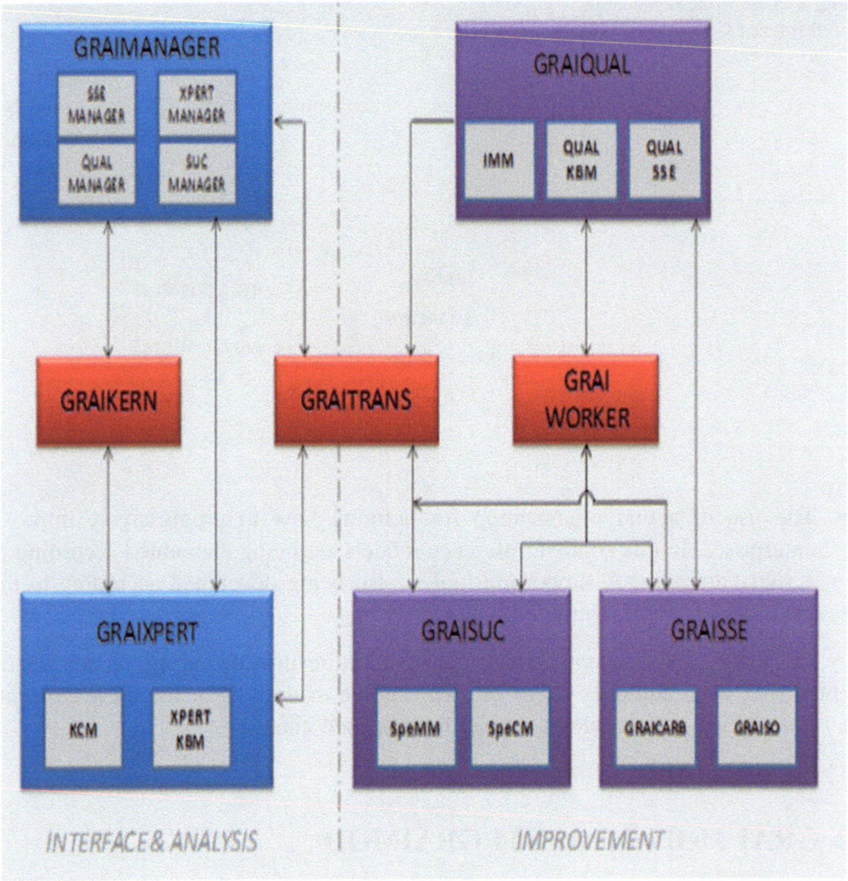

Fig. 2 Architecture of GRAIMOD

of enterprises to meet the real market needs (globalization, relocation, capacity to be proactive, cost optimization, lead time, quality, flexibility, etc.) and need to be adapted. An enterprise is completely described according to GRAI Methodology by finding five models: functional (functions of the enterprise and their links), physical (the production system), informational (the network, tools and informational flows), process (suite of sequences or tasks) and decisional (structure of orders, hierarchic organization). Then these models could be improved to enhance enterprise performance.

GRAIMOD is a new tool being developed by ICAM Engineer School for proposing concrete solutions to improve enterprises according to new market evolutions. At present, it contains five modules working around three sub modules (Fig. 2). The tool is divided into two parts: the interface containing modules for modeling enterprises and the analysis and improvement part for changing the existing system and proposing new organizations.

GRAIKERN, a graphic editor used for representing the different models associated to GRAI methodology, is an interface. GRAIWORKER is the work base elaborated for managing, modifying and capitalizing knowledge about the studied case. GRAITRANS is a Transfer Interface used for putting the new case in GRAIXPERT in order to improve its Cases Base. The reference model elaborated for each enterprise domain will be improved by the acquisition of this new model in GRAIXPERT between the different modules.

GRAIXPERT is a hybrid expert system for managing the analysis of the existing system and proposing a new system (Dossou and Mitchell 2009; Suh 1990; Yahia et al. 2000). It is composed of two sub-modules in interaction with GRAIKERN: the Knowledge Capitalization (KCM) and the Knowledge Based System (XPERTKBM). GRAIMANAGER is a management module used for organizing the different interactions between the modules of GRAIMOD. It controls and manages the system's interactions with the users. GRAISUC is a module used for managing the choice of an ERP or SCM tool for an enterprise. It is composed of two sub-modules SpeMM and SpeCM. The Specification Management Module (SpeMM) is used for choosing the appropriate ERP or SCM Tool of an enterprise. The specifications obtained are capitalized in the Specification Capitalization Module (SpeCM). GRAIQUAL is a module used for managing quality approach implementation or quality improvement in an enterprise (Rad 2006). It contains two sub-modules IMM and QUALKBM. The Improvements Management Module (IMM) is used for managing the different quality action plans of the enterprise. The Quality Knowledge Base Module (QUALKBM) is being elaborated for containing the rules related to quality certifications in order to use them for improving or elaborating quality in an enterprise.

A new module GRAICARB is being added to GRAIMOD in order to pinpoint the environmental, societal and social dimensions in enterprises (Shamah 2009). This module would integrate for example changes associated to carbon management, ISO 26000, ISO 14000 implementations, social and societal evolution impact not only on enterprises but also territorial collectivities (states, associations, local authorities, etc.).

Unlike GRAIQUAL whose goal is to improve enterprise performance by using the criteria defined, GRAISSE focuses only on the social, societal and environmental aspects (Colin 2002; Dossou and Mitchell 2009, 2010). This module is composed of two sub modules:

- GRAICARB to calculate enterprise carbon footprint and to propose environmental improvements according to ISO 26000 norm.
- GRAISO focusing on the improvement of social and societal aspects in enterprises.

Carbon management is an approach based on the setting up of a project for the evaluation and reduction of gas emissions consisting of six main stages:

1. Awareness of Greenhouse gas emissions
2. Definition of the area to be studied

3. Data acquisition
4. Exploitation of findings
5. Establishment of reduction action plan
6. Execution of reduction action plan.

In GRAICARB the ADEME (French environmental agency) method is promoted according to ISO 14000 norms. This method proposes the use of a step by step approach of calculation rules, of calculation software and of associated documentation. The database used is "Carbon base" and external data could also be used by the users.

This method is used in three stages:

- The preparation for defining the perimeter of the study.
- The accounting analysis for collection of precision data.
- And strategic analysis classifying the critical areas and those at risk. The action plans are elaborated after that for correcting the situation.

The concepts contained in the ADEME Method are combined to the Greenhouse Gas (GHG) Protocol, the regular guide for Gas management and data from the *coach carbone* tool. The GHG protocol describes principles and requirements for quantifying enterprise activities, gas emissions and defining accounting and reporting principles.

The methodological guide describes the way to follow to obtain an official Gas emission management.

The *coach carbone* tool is developed by Nature and Human Foundation (FHN) and ADEME. It allows to visualize the energy consumption in Kwh and liters of fuel and to compare it with other enterprises. The tool also allows to generate savings in Euros and in CO2.

GRAISO is dedicated to social and societal aspects according to concepts of ISO 26000 and Lucie Label, which will be presented later. With ISO 26000 there is no certificate contrary to Lucie which issues a certificate to validate the work done.

3 Enterprise Typology and Reference Models

In regards to the actual situation of enterprises and new constraints of the market, the use of a structured approach for improving enterprises is pertinent. GRAIMOD is being elaborated by using different types of reasoning and applying them to enterprises: decomposition reasoning, Case Based Reasoning (CBR), Rules Based reasoning, transformation reasoning. The objective is to define formalized processes of enterprise improvement and to be able to manage each step of these processes and defining action plans for short, middle and long terms.

Then, three modes of knowledge representation are used in GRAIMOD:

- The reference models show the standard for a given sector of activity. They allow to define an ideal for each sector of activity, which can be used as a reference in the elaboration of the future model (TO BE model).
- The cases studied are capitalized in order to enrich the knowledge capitalization module of GRAIXPERT with the objective being to improve the use of Case Based Reasoning (CBR).
- The rules are used throughout the different phases of the operation of GRAI methodology. Not only do they serve to elaborate the modules concerning the existing situation of the enterprise (AS IS) but also to detect the malfunctions of the enterprise and establish its strengths and weaknesses and finally during the design phase of the future system (TO BE).

The use of a generic model corresponding to a precise activity sector appears as pertinent. A new enterprise typology based on three criteria is established and a reference model according to each defined domain is being elaborated.

The first criterion chosen is the economic sector (Dossou and Mitchell 2013):

- The Primary sector is composed of activities linked to the natural resources exploitation such as agriculture, forestry; fishing and mining are included in the primary sector. It also includes all activities producing unprocessed raw materials.
- The Secondary sector corresponds to activities linked to the processing of raw materials from the primary sector in production and consumption goods are included together with the construction or manufacturing industry.
- The Tertiary sector contains service activities linked to that sector. We can find very wide types of activities such as commerce, administration, transport, financial and real estate activities, services to businesses and individuals, education, health and social action.

This sector decomposition is used by both economists and geographers. The sectors correspond to the three main economic sectors and are widely used.

The second criterion used is defined as size:

- Small companies correspond to enterprises with employees from 0 to 49
- Medium companies are composed of employees from 50 to 499
- Large companies correspond to enterprises with more than 500 employees.

For reducing the number of reference models to elaborate, only three sizes are defined.

The third criterion chosen is activity domain according to NAF code. A SIREN number is assigned to any French companies during their registration and this number is used for the lifetime of the company. As soon as the SIREN number is assigned, the company needs to select an APE code (Main Activity Practiced) which characterizes the activity. This code is used for the nomenclature of French activities (NAF).

The APE code is fundamental information for statistical data companies because all the rankings of firms by industry are based on it. The quality of the studies about the economic and the structural situation also depends on it.

We decided to only use the first level of the NAF for the primary and tertiary sectors. The explanation of this decision is because we are convinced that the performance models of these companies would be close to each other.

4 Test of the Defined Typology: Range of Enterprises

The elaboration of reference models for each domain has been undertaken. The process of elaboration is the same: acquisition of context, existing system modeling, analysis, design and finally proposition of reference model.

All the enterprises of Vendee are chosen as a study area. Indeed, there are more than seven thousand enterprises in Vendee, corresponding well to a quota of all enterprises in France for making a scientific study. The result of the study could easily be extended to enterprises nationwide in France, then throughout Europe. The quota chosen is really representative. So, Vendee enterprises represent for the study the global mathematical population. A meeting with the Chamber of Commerce allows to define how many enterprises would give an answer to a questionnaire sent to them for acquiring context and to make an enterprise classification. Because of the crisis and the difficult economic situation of enterprises, enterprises find it difficult to allocate the time to answer to this kind of questionnaire; only 10 % would surely give an answer. For the population, it means that 700 enterprises would be ready for giving us data for our analysis.

The proposition of an extension of GRAIMOD is in order to treat the data obtained and exploit it for elaborating reference models. GRAICARB, this extension, contains a data base with the questionnaire in which the responses will be studied. It allows to find good habits of enterprises by taking into account social, societal and environmental dimensions. The use of GRAIMOD for improving enterprise performance is now already efficient. Indeed, for an enterprise, the modeling, the analysis and the design phase are really well managed by the tool. The tool is also efficient for reducing lead time, choosing and implementing new tools in the enterprise, and implementing quality approach. But it is clear that the tool is less efficient in the management of improvement integrating carbon footprint reduction, social and societal dimensions, and respect of environmental norms. Then, GRAICARB will bring this efficiency, by focusing only on this criterion. The linear combination with the other performance criteria would be managed by GRAIMANAGER.

The ISO 26000 norm presents the main lines for all organizations wanting to assume the impacts of their decisions and activities. The societal responsibility is defined as the responsibility of an organization according to its decisions and activities towards society and the environment through an ethical and transparent approach which

- Contributes to the sustainable well-being including health and development of society.
- Takes into account all interested parties.
- Respects laws and is adapted to international norms.
- Is integrated to all the organization and is used in its relations.

For guiding the discussions between all the parties, the designers identified seven main questions:

- Environment
- Loyal practices
- Community and local development
- Organization governance
- Relations and working conditions
- Questions related to consumers
- Human rights.

Each of these questions is divided into action domains, explaining the main lines to be followed. An ISO 26000 approach respects three major steps:

- The realization of a diagnosis for defining action priority
- The deployment of identified actions
- The phase of account making.

The diagnosis management is based on:

- The definition of the perimeter of societal responsibility
- The identification of the parties concerned and their interest
- The review of the seven main questions.

The diagnosis serves to identify pertinent action domains on which the organization could be based for fixing its priorities. Then an auto-evaluation could be done according to requirements referential or an evaluation by an outside party could be done according to universal requirements referential.

The deployment implies the study of the organization values for defining strategy, objectives, resources and skill development. The results are followed by the regular review as part of a dynamic continuous improvement process.

The last step is destined to summarize what has been done and how to address what remains to be done.

The result of the questionnaire would allow to adjust the enterprise typology elaborated. It would show us the enterprises which changed their economic model by integrating other dimensions than cost, quality and lead time. For instance, some of them are green, ecological, virtuous, showing solidarity, ethical, or responsible enterprises. The study allows to focus on this kind of enterprises and to valorize them (Fig. 3). The objective is to show those not having chosen to follow this way the real advantages of such an approach and for those who have already started to help them become even more efficient.

Fig. 3 Collection of best practices

GRAICARB would be used for doing carbon management, implementing ISO 26000, managing ergonomic practices, managing how to make employees happy with their job and more efficient for the enterprise. The societal aspect could also be managed. The tool would also serve for improving local authorities.

To conclude, the study extended to the department of Vendee has been undertaken (Fig. 4). The first step was finished in February 2013. Then the typology elaborated will be adjusted and reference models developed for each new class. The next step will be modeling of the enterprises which will start using specific models in order to improve the reference models defined. Simultaneously, the elaboration of another economic model more adapted to the actual context and able to protect enterprises in the future will be developed.

5 New Economic and Sustainable Model

Existing economic models, including the capitalist one which many consider to be behind the actual world crisis, have shown their limitations. It is necessary to think about a new model that will be adapted to the actual world context and more efficient.

DATA ANALYSIS: 802 Enterprises

Criterion 1 - economic sector

Sector	Results
Primary	4
Secondary	528
Tertiary	270
	802

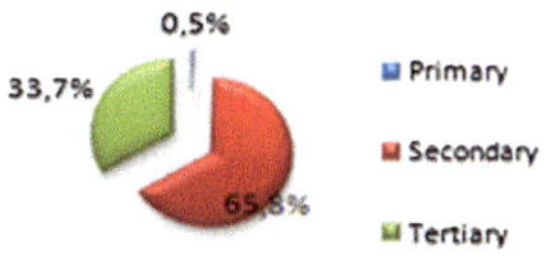

I- Primary sector

In the primary sector, enterprises don't exceed 500 employees.

Size	Results
PE	3
ME	1
GE	0
	4

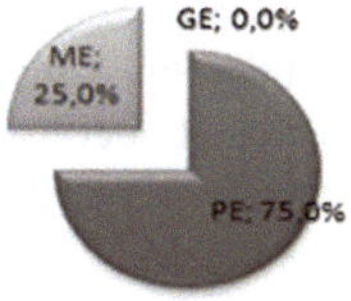

Activity

Activity code	Activity	Number
ASP	Agriculture	1
IE	Extraction Industry	3
		4

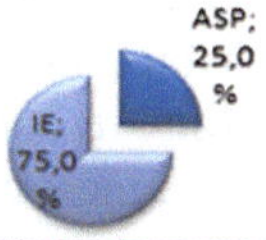

II- Secondary Sector

Size	Number
Small E	169
Medium E	283
Large E	76
	528

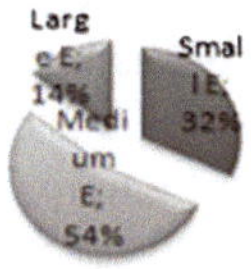

III - Activity: Small Enterprises

Activity Code	Activity	Number
C	CONSTRUCTION	29
IA	Automotive Industry	8
IAFB	Agribusiness, Food and Drink	11
IBFM	Wood Industry	7
ICFPCPM	Chemical Industry and non metallic manufacturing	14
IFPIEO	Electronic, Computer and Optic Industry	25
IMFPM	Metallic Industry	49
IPCI	Paper Industry	3
IRIME	Technical services (machines installation etc.)	19
ITHCC	Clothing manufaturers	4
	Small Enterprises	169

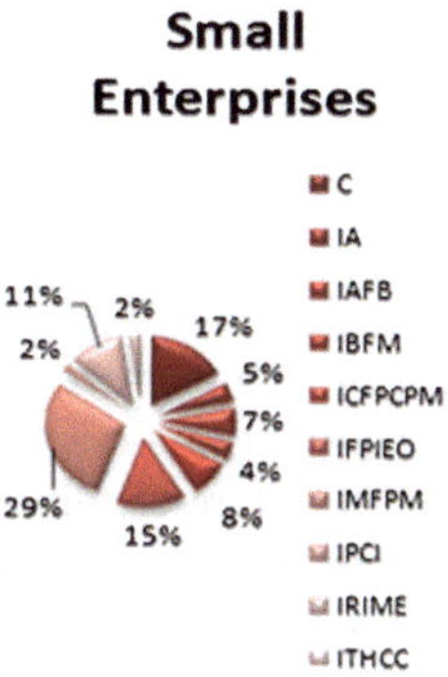

Fig. 4 Example of typology structure

The objective is to develop a model with a complete break from a purely liberal one. This model has to take into account the reason of the actual crisis and parameters such as the environment, society, social view, and to offer an optimization in terms of a balanced model.

The particularity of the new proposed model would be to think globally about local organization, an enterprise, a department or a country. It means that all parameters would have to be integrated in the definition of the model without giving priority for one criterion as it was in the old capitalist model.

For instance, for energy independence, it would be to think about how to implement a global energy vision according to a department or town. For the department of Vendee an audit would be done for analyzing in detail the existing system and then detect inconsistencies, and deduce points to improve. The next step would be to define a new specific Vendee model in accordance with other best models used in the world according to energy optimization but with the specificity and the identity of Vendee. The result would be a combination of best energy transition habits, integrating biomass and waste energy, or solar energy, wind energy, thermal resource energy, and so on. The model would be according to the parameters the most balanced possible. For validating this model, a demonstrator would be designed, elaborated and exploited on a new ICAM school for example. The model obtained and validated only by focusing on one aspect would be improved by introducing the other aspects in order to define an optimum.

This approach proposed for the energy self-sufficiency problem, would be generalized to all the previous questions in order to build reference models adapted to specific problems of enterprises and local government.

To conclude we could say that the department of Vendee department will serve as an example which will be studied completely and statistically with a scientific approach for defining a new economic and sustainable model.

6 Conclusion

Enterprise modeling is used for preparing enterprises for the post-crisis period by reorganizing them according to the evolution of society and globalization rules. It is important to show them how to integrate different new parameters in order to adapt to future conditions. The objective to make enterprises sustainable needs to redefine an optimum including environmental, social and societal dimensions in addition to the main performance criteria (QCD). GRAI Methodology supported by GRAIMOD a new software tool, allows to improve the global performance of Enterprises and particularly SMEs. A focus is made on GRAISSE and GRAI-CARB dedicated for managing carbon footprint and social, societal and environmental aspects in enterprise improvement.

In this chapter, the process of reference models elaboration is presented. An enterprise typology is also proposed with the objective to give the most appropriate model to each enterprise which desires to improve its performance. The new economic model is being elaborated and will be presented in detail in future papers. The proximity of ICAM Vendee to enterprises will allow not only to enrich the case base of GRAIMOD by doing modeling of them but also to improve the reference models and the typology elaborated.

References

Colin J (2002) From logistics operations control to SCM (De la maitrise des opérations logistiques au supply chain management). Gestion 2000, Janvier Février, pp 59–75

Doumeingts G, Ducq Y (1999) Enterprise modelling techniques to improve enterprise performances. In: The international conference of concurrent enterprising: ICE'99, The Hague, The Netherlands

Dossou PE, Mitchell P (2009) How quality management could improve the supply chain performance of SMES. FAIM 2009, Middlesbrough, United Kingdom

Dossou PE, Mitchell P (2010) Carbon reduction: new criteria of supply chain performance. FAIM 2010, Oakland, East Bay, USA

Dossou PE, Mitchell P (2013) Elaboration of reference models for improving enterprise performance. In: Advances in sustainable and competitive manufacturing systems. Lecture notes in mechanical engineering, Springer, Switzerland, pp 899–910

Rad AMM (2006) The impact of organizational culture on the successful implementation of total quality management. TQM Mag 18(6):606–625

Shamah AMR (2009) Going green in supply chain management as a prospective improvement. FAIM 2009, Middlesbrough, United Kingdom

Suh NP (1990) The principles of design. Oxford University Press, Oxford

Yahia ME, Mahmod R, Sulaiman N, Ahmad F (2000) Rough neural expert system. J Expert Syst Appl 18(2):87–99

Energy Audit Methodology and Energy Savings Plan in the Nautical Industry

Gilles Dedeban, Philip Mitchell and Paul-Eric Dossou

Abstract Despite technical and organizational progress, energy weighs heavily in the fixed costs for industrial enterprises. With the depletion of fossil resources and the inexorable rise of the resulting price, we are now at a crucial time when saving strategies and "best consumption" will be crucial for companies, especially in a period of recession where industrial plants have become improperly sized. In this article, the process of industrial energy audit based on reference standards and best practices is broken down into three main stages: (1) Developing a global energy balance of the company. (2) Identifying and modeling the main consumption, quantifying the potential savings. (3) Defining the actions and investment plan necessary to achieve these savings Through the example discussed, it is described how the energy management system has been implemented and what the results were after one year in the manufacturing sector of sailboats and powerboats of BENETEAU Group (11 production sites in France and 1 in Poland).

Keywords Energy management · Sailboat and powerboat manufacturing · Energy reduction

1 Introduction

The European Union is now confronted by an economic crisis and its consequences. In this context, enterprises have to pay attention to their organization in order to improve their performance. Indeed, enterprise modelling could represent

G. Dedeban (✉) · P. Mitchell · P.-E. Dossou
ICAM VENDEE, 28 Boulevard d'Angleterre, 85000 La Roche-Sur-Yon, France
e-mail: Gilles.dedeban@icam.fr

P. Mitchell
e-mail: philip.mitchell@icam.fr

P.-E. Dossou
e-mail: paul-eric.dossou@icam.fr

P. Golinska (ed.), *Logistics Operations, Supply Chain Management and Sustainability*, EcoProduction, DOI: 10.1007/978-3-319-07287-6_30,

for them a solution for resisting the crisis and preparing themselves to be competitive at the end of this crisis. The main criteria of performance improvement are not only quality, cost and lead time (QCD) but also a criterion associated to social, societal and environmental dimensions.

GRAI Methodology is one of the three main methodologies (with PERA and CIMOSA) used for improving enterprise supply chain performance. GRAIMOD is a tool for supporting this methodology. This chapter obviously includes the environmental dimension by focusing on energy reduction parameters according to ISO 14000 and ISO 26000 norms.

Despite technical and organizational progress, energy weighs heavily in the fixed costs for industrial enterprises. With the depletion of fossil resources and the inexorable rise of the resulting price, we are now at a crucial time when saving strategies and "best consumption" will be crucial for companies, especially in a period of recession where industrial plants have become improperly sized.

In this chapter guidelines are given on how to perform an industrial energy audit and to develop the approach used for improving and managing the performance of French boat builder Beneteau.

2 Methodology Used

The main criteria of performance are cost, lead time and quality. It is clear that there is a strong relation between these criteria. For instance, the improvement of quality facilitates the respect of lead time and reduces the cost of the product. But this improvement of quality also implies nowadays the respect of environmental demands. The new international context imposes upon the enterprises the need to adapt for being sustainable and green. In France, enterprises recognize the necessity of this change but have not been able to make changes because of the cost related to environmental operations. The integration of a criterion representing sustainable development (reduction in carbon, energy and waste, environmental, social and societal improvement) is necessary for defining the new optimum for enterprises and showing them how their performance will be improved. The methodology used is characterized by continuous improvement and represented in Fig. 1.

The approach was built relying on standard references:

- BP X30-120 of AFNOR,
- ISO 14001 and 50001,
- « Manuel CAP SPBI » (internal reference manual).

An Industrial Energy Audit consists of 3 stages:

1. Developing a global energy balance of the company.
   This preliminary analysis is to collect the data of the company, to compare the energy performance of known references and prepare a first assessment of sources of potential savings.

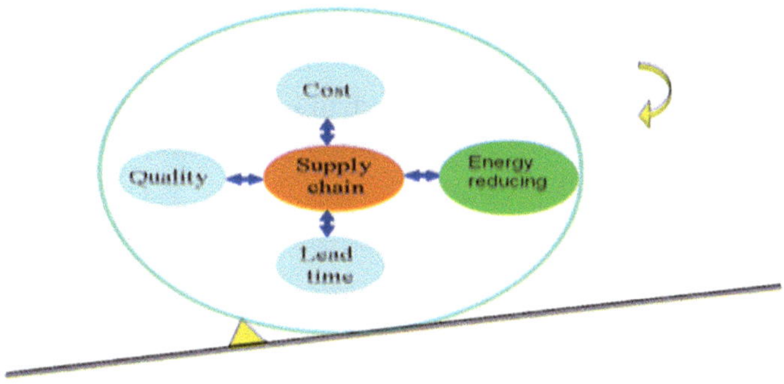

Fig. 1 Enterprise supply chain improvement criteria

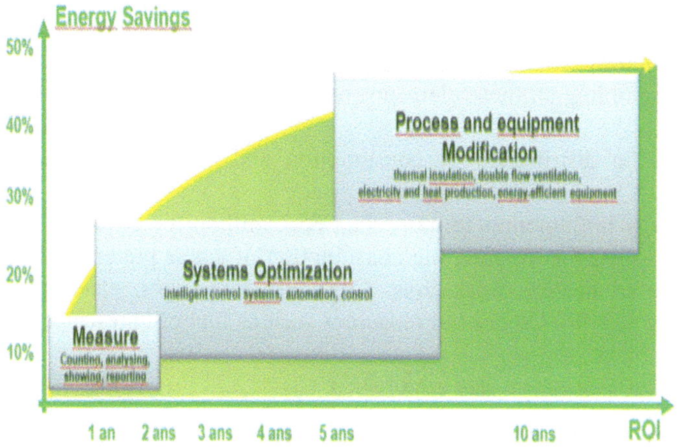

Fig. 2 Energy savings and return on investment

This is to identify simple actions with very fast return on investment and those requiring more in-depth study of profitability.

2. Quantifying the main potential savings.
   It is carried out by measurement campaigns, modeling and earnings estimates for prioritizing and selecting the most effective actions.
3. Developing the managerial organization and the necessary investment plan to achieve the expected savings.
   Up to 10 % energy savings can be obtained simply by raising awareness and monitoring. Going beyond this figure requires improvements in manufacturing processes and certain investment (Fig. 2).

3 Application

The project presented took place from March to July 2012 and entailed performing energy audits of the manufacture of sailboats and powerboats in the BENETEAU Group.

It focused on 11 production sites in France and 1 in Poland.

3.1 Energy Balance

The first step of the study was to collect data on consumption and production of relevant central services (BJ Technology) but also on the different sites (technical coordinators and site managers).

My period is the last full fiscal year, September 2010–August 2011.

Information to bring to the analysis were (Table 1):

After assembling all these elements, an initial energy balance of the group was undertaken.

Overall, the distribution of consumption and energy costs are as follows (Fig. 3):

The energy bill amounts to more than €5.2 million for 2010–2011.

If we consider the volume, the consumption of gas is predominant (62 %) for the purpose of heating workshops. Considering the cost, electricity takes the largest share (53 %). Other used resources such as fuel oil, propane, water remain marginal.

This initial assessment has established:

- A ranking of sites based on their consumption
- Consumption ratios depending on the surface area and volume of plants

Energy ratios reveal:

- Three workshops at net power consumption: woodworking, boating, maintenance/logistics for an average of 102 kWh/m^2
- Greater energy intensity for gas consumption in relation to age of sites.

3.2 Areas of Savings

To identify potential sources of economy and allow a realistic estimate of earnings, a modeling tool using EXCEL concerning heat losses in buildings was developed and electrical measurement campaigns were carried out on different sites.

Table 1 Data versus suppliers

Data	Suppliers
Costs power, gas, water, oil	Purchasing, remote reading
Plan production site and produced tons (excluding keel)	Information systems
Summary of energy meters electricity, gas, water	Energy officers
Programming workshops temperatures and ventilation outside temperature	
Daily outside temperature readings	
Heating, lighting and production equipments	
Hours, staff teams, product flows	Technical coordinator
Oil consumption	Energy officers
Resin and gelcoat consumption per site	Management control
Direct hours per site per activity and job training hours	Management control
Heat losses in buildings	BERI 21 (property manager of factories)

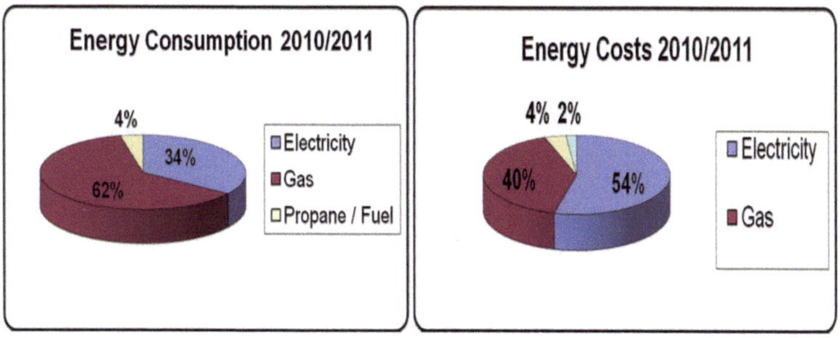

Fig. 3 Energy consumption and energy cost

3.2.1 Building Modeling

For a given site, the files are of 2 types:

- *Dth bâtiment:* file listing of heat gains and losses related to the structure and process of manufacture (1 file per building modeled)
  Information are here:

  site name, type and name of building exercise.
  building volume, type of heating and gases
  surface nature of the walls and components (windows, doors and gates)
  area, type of roof and components (domes, vents and windows)
  floor space, heat transfer between walls, roof and ridge
  max. exhaust air and time settings of the extraction rate
  dimension of roller shutter doors, frequency and time open daily
  hours of attendance and average occupancy of the building
  installed lighting power and hours of operation

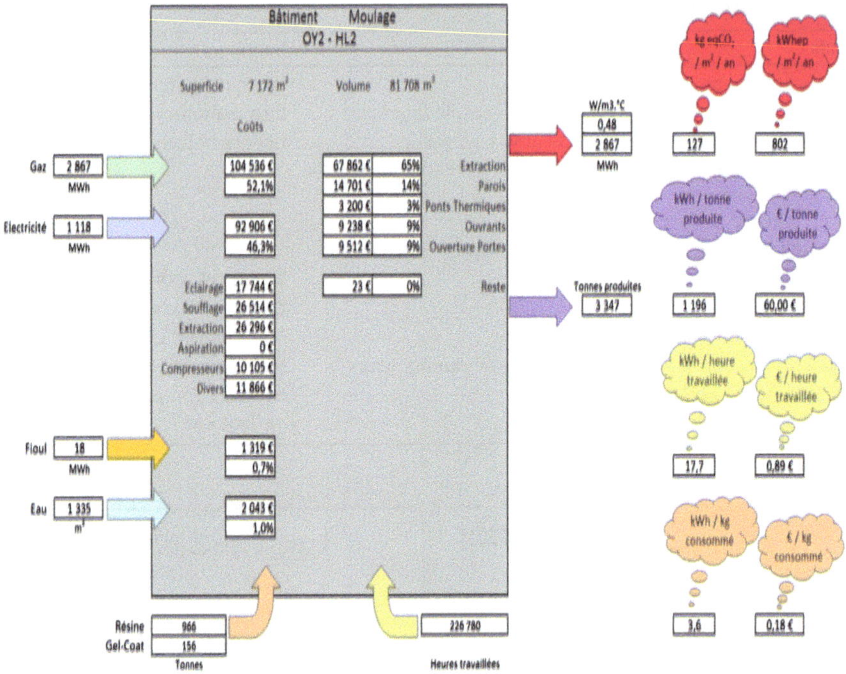

Fig. 4 Indicators

time slots and temperature levels, monthly degree-days for a heating temperature of 8–20 °C, number of days worked per month
price per kWh of natural gas and propane

- BilanE site: Balance modeling file (5 sheets per building)
  - BilanTh sheet: copy of the results sheet (first page) of *D*th
  - Data sheet: Specific data consumption and production for the site or the building concerned (with any weights)
  - Results sheet: automatically calculates:

    Environmental indicators

kg eq.CO2/m^2	$(0.09{*}kWh_{elec} + 0.28{*}kWh_{gaz} + 0.31kWh_{oil})/Area$
kWh_{EP} /m^2	$(2.58{*}kWh_{elec} + kWh_{gaz})/Area$

    Specific indicators of consumption and energy cost per consumed raw material, hours worked and tonnes produced (excluding keel) (Fig. 4)

  - Indicator sheet
    The distribution of energy costs, the cumulative energy consumption and specific energy indicators are displayed here as Green/Red graphics (Fig. 5).

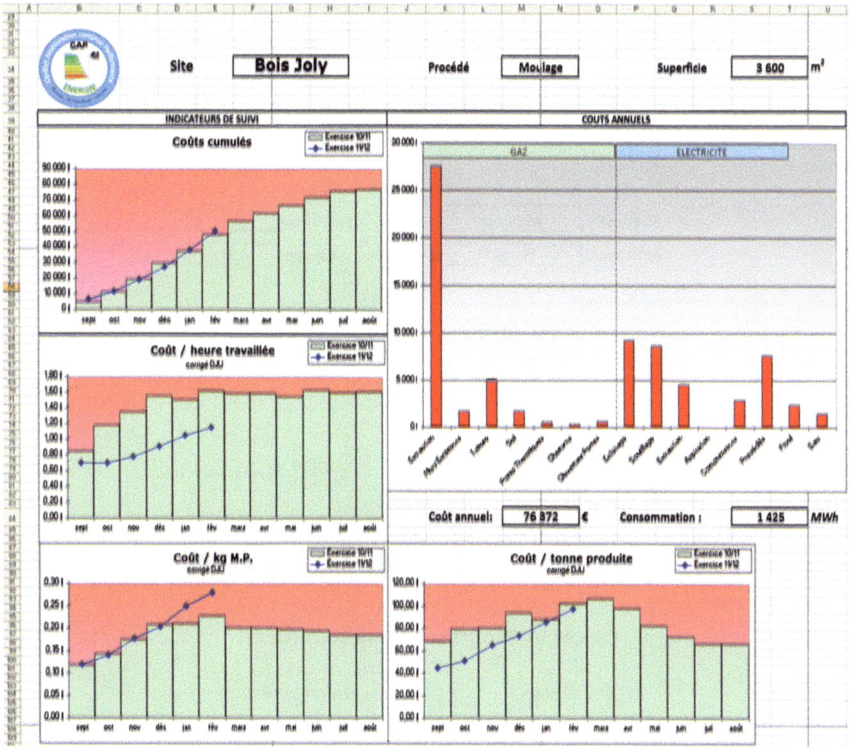

Fig. 5 The distribution of energy costs, the cumulative energy consumption and specific energy indicators. Data from year n – 1 as *green bar*, data from the current year in the form of *points*

Visualization of energy costs distribution helps guide the choice of energy saving measures to implement.

– Display sheet
Intended for display in the area of LEAN communication, this poster shows the Energy Performance Indicator of the building, the target to reach and the action plan implemented to improve the current level (Fig. 6).

3.2.2 Measurement Campaign

The electrical measurements are done with a qualified electrician.

Two types of complementary measures are useful:

1. Point measurements with clamp
They measure at a given instant (choose a time when the activity is in full swing) the proportions of consumed energy by electric departure (global per workshop, general day (or unscheduled), general night (or prior) upstream main

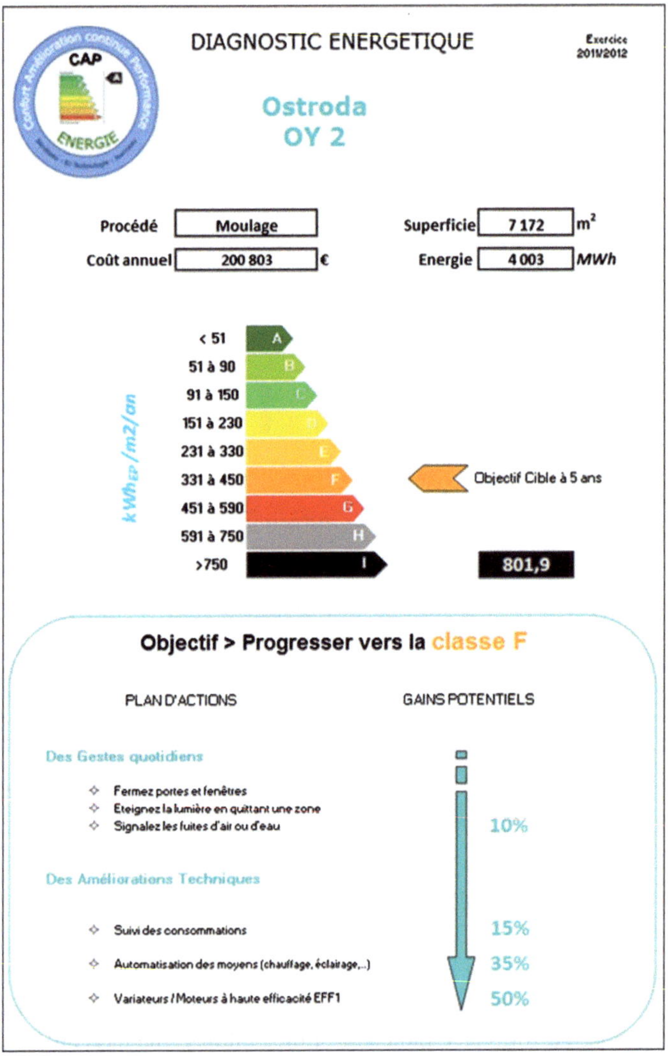

Fig. 6 Energy performance indicator of the building

lighting, blowing, extraction, starts, compressors, central vacuum, hydraulic circuits group (Fig. 7).

2. Measurements with network analyzer Chauvin Arnoux CA 8334 (software Qualistar View v2.6) (Fig. 8).

These measures can more accurately assess the consumption of electrical energy by observing a period of 24 h or even a part of the circuit (e.g. lighting departure.) or equipment (compressor) (Fig. 9).

	A	B	C	D	E	F
1		Pa (kW)	H. fonctionnement			compresseur réparti
2	Moulage	170,90	20	3418	45,00%	50,92%
3	Ebarbage	129,51	12	1554	20,46%	23,15%
4	Montage	108,79	16	1741	22,91%	25,93%
5	Compresseur	55,21	16	883	11,63%	
6						
7		**Total Mesuré**		**Energie calculée**	100,00%	100,00%
8		464,42		7596		
9						
10			**Relevé EDF**	**7589,50**		
11						

Fig. 7 Excel view

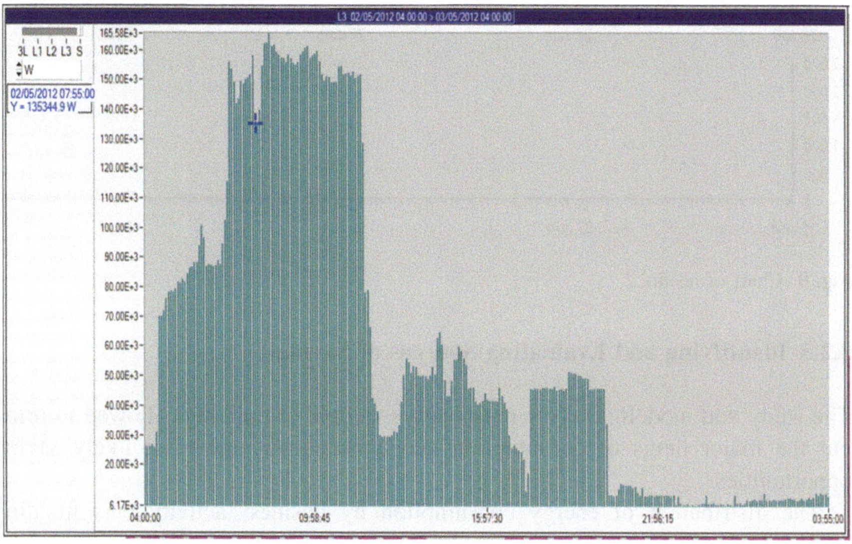

Fig. 8 Chart of results 1

Based on actual consumption and the utilization rate of equipment observed, they can adjust the theoretical estimates of consumption (e.g. annual consumption of a compressor).

In addition to verification, these measurements can be compared with data from meter reading and/or Central Technical Monitoring (CTM) available (Fig. 10).

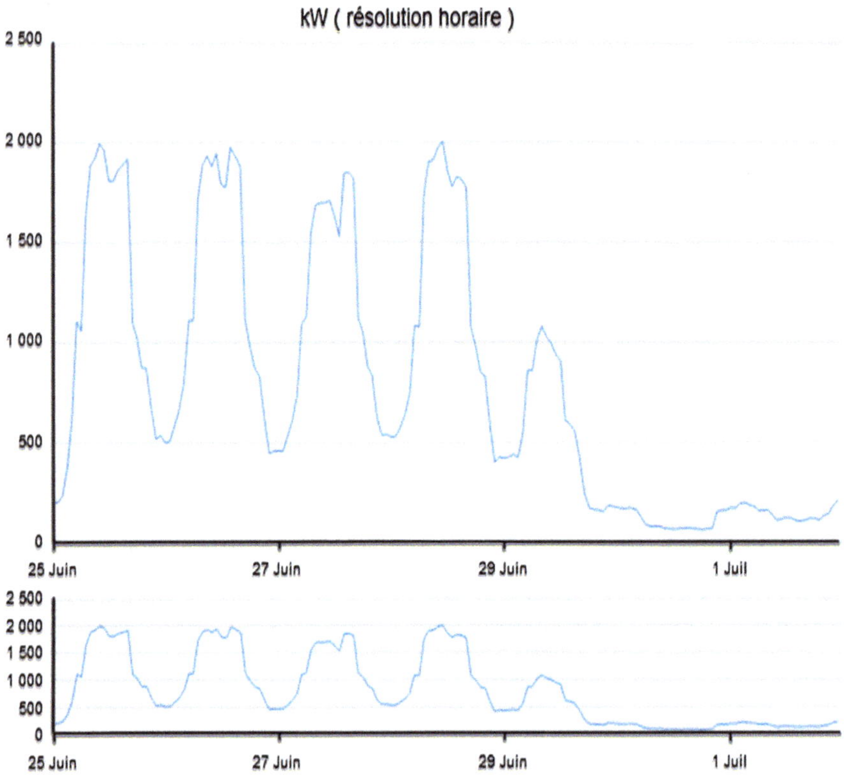

Fig. 9 Chart of results 2

3.2.3 Identifying and Evaluating Sources of Savings

The study and modeling of a representative sample of buildings allowed to bring out the major items of consumption and consequently the most likely saving opportunities.

The distribution of energy consumption by business activity puts molding operations at the forefront of energy-intensive activities (Fig. 11):

For the molding workshop, the utility distribution shows clearly that air extraction followed by lighting and production equipment (hydraulic, heaters, vacuum pumps…) constitute the main costs and then building heat loss (mainly roofs and openings) (Fig. 12).

3.2.4 Estimated Potential Savings

This economic approach was conducted in two stages:

- an investment plan for 1 year achieved for 2012–2013,
- an overall assessment of savings achievable in the longer term site by site.

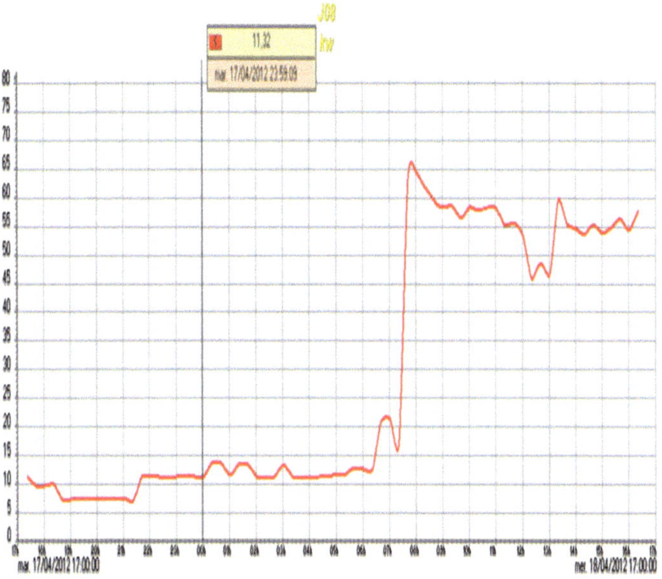

Fig. 10 Chart of results 3

Fig. 11 Energy cost

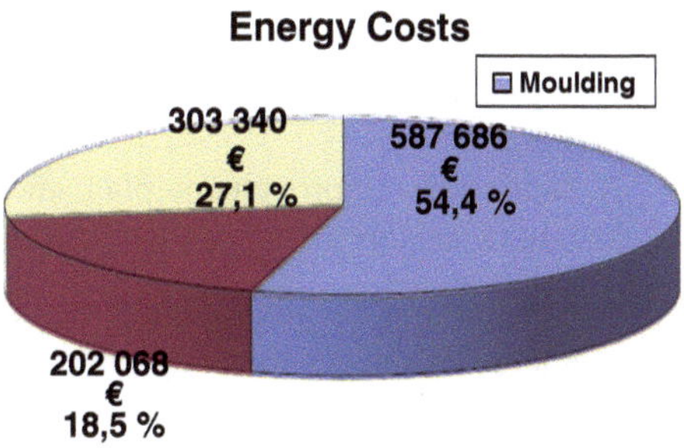

For the coming year, from investment requests from sites, the cost of the work required was calculated and actions were highlighted with a return of less than 1 year.

This document served to aid business decision-makers in the development of the budget in early July. This concerned implementing an Energy Metering plan and Central Technical Monitoring development on some sites.

For an provisional investment of k€ 145, savings of k€ 267 can already be generated in the first year.

For subsequent financial years, a table of estimated gains according to type of operation and site has been established with a reasonable target of 30 % savings.

For the rest, it still remains to validate the profitability of different projects to be implemented site by site and this will depend on quotes given for the work given by suppliers.

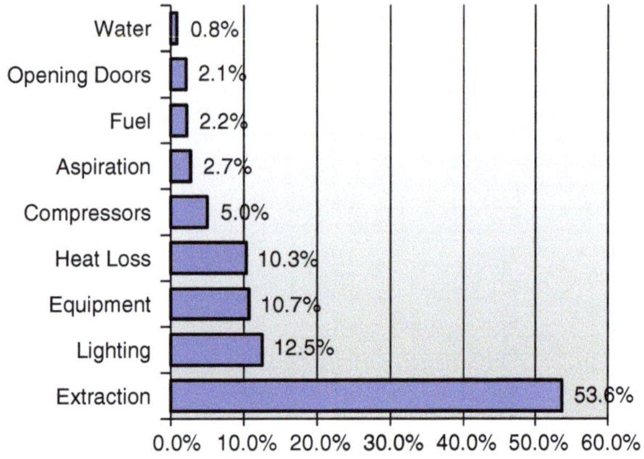

Fig. 12 Percentage of energy

3.2.5 Energy Saving Plan

To sustain the approach taken, an action plan has been undertaken and incorporated into the continuous improvement process. It is in the TOP 5 of the improvement actions under the Environmental Management Plan and is declined at each site with an annual target of 10 % savings.

It is built according to two axes:

- An awareness plan and monitoring at the sites: according to feedback, 10 % of savings can be obtained at no extra cost by good daily practice and without any investment.
- A continuous improvement plan and methodological support at the central maintenance service:

This should begin with the implementation of a metering plan within the group to automatically collect consumption data in order to refine the analysis and enable early identification of any drifts.

4 Conclusion

The 1st year, 246 improvement actions started focusing on the following areas:

- Education and training, implementation of an Action Plan for becoming a LEAN Site.
- Heating optimization and installation of ceiling fans.
- Electric metering plan.
- Reduction in energy anomalies at the weekend.
- Negotiation and optimization of gas and electricity contracts.

Savings achieved are in the order of 600 k€ (−12 %) for 2012–2013.

The level of savings corresponds very favourably to investment. Indeed, the final investment was much lower than initially thought at 120 k€ and the action plans undertaken should enable BENETEAU to make further gains, both economic and environmental.

References

1. ISO 50001 (2011) AFNOR, June 2011
2. BP X30-120 (2006) AFNOR, March 2006
3. 49 exemples de bonnes pratiques énergétiques en entreprise ADEME, November 2011
4. Rapport sur le développement industriel 2011 L'efficience énergétique industrielle pour une création durable de richesses ONUDI, November 2011
5. La consommation d'électricité dans l'industrie, Les cahiers de GLOBAL CHANCE - N° 27 - January 2010
6. Reference Document on Best Available Techniques (2009) European Commission, Feb 2009
7. Diagnostic de performance Energétique – Guide de Recommandation, Ministère du Logement, March 2009
8. Mener à bien un projet d'efficacité énergétique dans l'industrie, Gimélec, April 2008
9. Dossou PE, Mitchell P (2013) Elaboration of reference models for improving enterprise performance. In: Advances in sustainable and competitive manufacturing systems. Lecture notes in mechanical engineering. Springer, Berlin, pp 899–910

Part IV
Optimization of the Location Problems, the Inventory Management and the Vehicle Routing Problems

Strategic Inventory Positioning for MTO Manufacturing Using ASR Lead Time

Suk-Chul Rim, Jingjing Jiang and Chan Ju Lee

Abstract In most Make-To-Order manufacturing, work-in-process (WIP) inventory is usually piled up at almost every station in the factory in order to quickly meet the urgent request from the immediate downstream station. Depending on the station network configuration and lead time at each station, some of the WIP inventories do not contribute to reducing the manufacturing lead time of the final product at all. Therefore, it is important to identify the minimum set of stations to hold WIP inventory such that the total inventory holding cost is minimized, while the required due date for the final product is met. In this study, we present a model to determine the optimal position and quantity of WIP inventory for a given bill of material using the actively synchronized replenishment (ASR) lead time; and present a solution procedure using genetic algorithm.

Keywords Strategic inventory positioning · ASR lead time · WIP inventory, genetic algorithm

1 Introduction

Materials requirement planning (MRP) has been studied and implemented since Joseph Orlicky developed MRP in 1964. MRP is used to insure that materials are available to meet manufacturing and market requirements. MRP uses a forecast or master production schedule (MPS) and bill of material (BOM) as input data to determine when and how many items to produce or purchase to meet the MPS. MRP has been widely used in most manufacturing industry as a tool to generate production schedule and purchasing order.

S.-C. Rim (✉) · J. Jiang · C. J. Lee
Department of Industrial Engineering, Ajou University, Suwon, 443-749, South Korea
e-mail: scrim@ajou.ac.kr

P. Golinska (ed.), *Logistics Operations, Supply Chain Management and Sustainability*, EcoProduction, DOI: 10.1007/978-3-319-07287-6_31,

MRP has been widely and successfully used in most manufacturing firms in its first two decades of introduction. However, since the forecast accuracy is getting lower due to the rapid changes of technology and wider customer choices, MRP has been widely used in most manufacturing industry as a tool to generate production schedule and purchasing order.

Calculations from this forecast often are inconsistent with actual market demand. In 2011, Ptak and Smith introduce a new type of MRP called Demand Driven MRP (DDMRP) in the book of "Orlicky's Material Requirement Planning". DDMRP replaces the previous convention of safety stock with strategically replenished positions. DDMRP is an innovative multi-echelon pull methodology to plan inventories and materials. It enables a company to build more closely to actual market requirements and promotes better and quicker decisions and actions at the planning and execution. In this chapter we address the problem of inventory positioning that is determination of the BOM node where inventory should be held to minimize holding costs and satisfy the given limited lead time of the end product.

Normally, we devote to solving two questions of inventory planning: how much inventory to be hold, and when to reorder in MRP. However, in DDMRP, the focus is where to position the inventory. Because the "how much" question is meaningless until we can answer the "where" question correctly. Certainly, DDMRP has its own logical and effective approach to answer the "how much inventory we should hold" question. Choosing the best inventory locations can solve problems of unsatisfactory inventory performance and delivery performance. Meanwhile, the opportunity to satisfy a certain lead time of the end product that customers require becomes bigger.

1.1 ASRLT

We should mention a new lead time from the earlier shape of the DDMRP technology. Actively synchronized replenishment (ASR) lead time is defined as the longest unprotected or unbuffered sequence in the BOM for a particular parent (Ptak and Smith 2011). To make manufacturing lead time (MLT) a realistic planning input, all events in the BOM should be stocked so that they are always available. To make cumulative lead time (CLT) a realistic planning input means that no events on the longest path are stocked. So we can see MLT and CLT exist under two extremes. Compared with MLT and CLT, ASRLT is much more realistic and variable.

Figure 1 shows that the MLT of Part 101, the end product of this BOM, is 2 days, given three components of level 2 are all available. The longest leg to calculate the CLT is 26 days from 101 through 201, 301, 402, to 501P.

Suppose the five colored nodes 202, 301, 304, 401, and 402 of the BOM have stocks. Then the ASR lead time of the end product is $2 + 6 = 8$ days. If part 203 additionally holds inventory, the ASR lead time of part 101 becomes

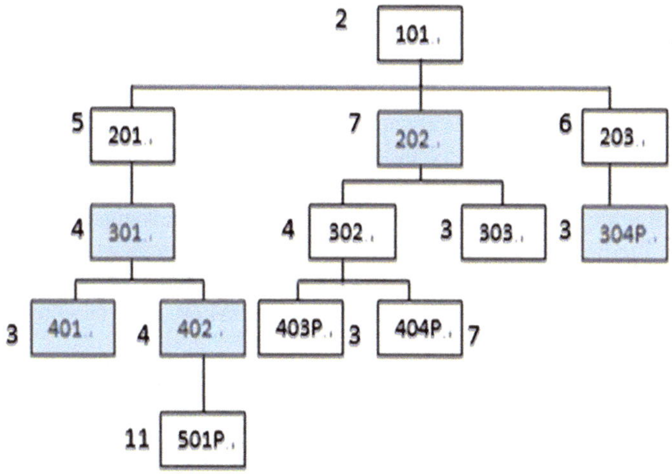

Fig. 1 An example of BOMs

Fig. 2 The meaning of each buffer zone

2 + 5 = 7 days. The more nodes we allocate the inventory at, the shorter (or equal) the ASR lead time will be, obviously with higher total inventory cost. Therefore the problem we address in this chapter is to determine the optimal set of nodes in the BOM to hold WIP inventory to minimize the total inventory cost, while satisfying the customer's lead time requirement of the end product.

1.2 Buffer Profiles

In order to calculate the holding cost, we also have to determine "*how much*". Based on several factors, different materials and parts behave differently. However, many also behave very similarly so that parts and materials are grouped into like "*buffer profiles*". Buffer profiles make a unique top level and zone definition for each part. Four key factors including item type, supply and demand variability, lead time and minimum order quantity form the various groups. There are five color-coded zones that comprise the total buffer. Just as shown Fig. 2, light blue displays an out-of-stock position; green represents an inventory position that requires no action; yellow represents a part that has entered its re-build zone; red represents a part that is in jeopardy and has two subzones including red zone base and red zone safety; dark red represents a stock out.

2 Literature Review

Many researchers have made contributions to the inventory positioning problem. Whybark and Yang present a carefully controlled simulation experiment to determine where to position inventory to get the best service level (Whybark and Yang 1996). Kaminsky and Kaya present an effective approach for determining locations at which to store inventory in the supply chain, sequencing specific jobs at specific facilities, and quoting lead times, so that system-wide costs are minimized, quoted lead times are relatively short, and quoted lead times are typically met (Kaminsky and Kaya 2008). However, DDMRP is a new topic, so few researches related DDMRP have been given.

Our paper is to determine the optimal inventory position based on the replenishment model introduced in DDMRP. Most previous literatures to determine the inventory position addressed the optimization problem to increase service fill rate, minimizing total inventory holding cost against a facility or full supply chains. The objective of increasing service fill rate and minimizing total inventory holding cost in our paper is the same as previous studies but the method to calculate the lead time and the inventory holding cost is totally different.

A number of papers studied on the inventory positioning problems based on Simpson model (Simpson 1958). Simpson model suggested that "*all or nothing*" policy through base-stock model in order to find the points in serial line systems whether inventory should be or inventory should not be. He stated that the service time is independent variable and the total inventory holding cost is the objective function. Hereby when calculating the total inventory cost, he regarded the safety stock for time t period as average inventory. The time t period subtracts the service time from the supply time. That supply time is the lead-time committed by the upstream stage plus processing lead-time of the current stage, and the service time is the lead-time to commit to the downstream stage. The safety stock is defined that multiply standard deviation of the demand for time t period by safety factor.

Graves and Willems studied to extend Simpson model to assembly, distribution and spanning tree network structures (Graves and Willems 1996: Graves and Willems 2000; Graves and Willems 2008). They have used the periodic review base-stock-system model, assuming no capacity constraints. The stated service time in the chapters is as decision variable and the objective function is to minimize the total inventory holding cost and then DP algorithm is proposed. When calculating the total inventory holding cost, they solved it just as what Simpson did.

Lesnaia followed the framework of Graves and Willems (2000) and applied to the supply chain of a manufacturing firm. The difference is the assumption of stochastic model as service time mode unlike deterministic model of Graves and Willems service time mode (Graves and Willems 2000). They proposed general-network algorithm to solve the safety stock placement problem and determine the candidates of optimal service times for the path.

Magnanti et al. applied to the environments including production/assembly stages for components in acyclic supply chain network that is not a spanning tree

(Magnanti et al. 2006). The decision variables are guaranteed service delivery time and inbound replenishment lead time, and the objective function is to minimize the total inventory cost in all stages of the network. They developed the efficient MIP formulation and used a successive piecewise linear approximation approach.

Kaminsky and Kaya designed effective heuristic for inventory positioning, order sequencing, and short and reliable due-date quotation for the supply chain (Kaminsky and Kaya 2008). They assumed the environment nodes of convergence and divergence in supply chains, and employed the base-stock-system. They defined two expected times as decision variables, one expected time is the safety stock of components stored at a facility j that are received from upstream facility k, the other expected time is the inventory of finished goods at facility j. They defined the objective function with the value that sum average inventory cost, average lead-time cost, and average tardy cost in a facility and then suggested LP model for the solution.

Inderfurth (1991), Inderfurth and Minner (1998) have extended the Simpson model to divergent systems and convergent systems in the network. These papers used a periodic review base-stock control policy and defined the minimizing inventory cost as the objective function and a service time as decision variables, using the simple dynamic program to optimize. They adopted the same framework with the deterministic service times that are quoted by the upstream nodes to the downstream.

And there are studies to apply the concept of postponement to determination of inventory placement in a supply chain. Georage et al. used the postponement concept to address the problem to determine supply chain node for holding inventory. This idea was initially introduced by Alderson (1950) and then has been studied by Bucklin (1965), Lee et al. (1993), Davis (1993), Feitzinger and Lee (1997), Ernst and Kamrad (2000).

Our paper proposes GA method to determine the inventory position in BOM to minimize the total inventory cost with fulfillment of service time that is required to be satisfied with customers. We define the total inventory cost as our objective function and the strategic inventory position as our decision variable, which is different from the service time stated in previous studies. In previous literatures, the placement of inventory depended on whether service time, which supply to adjacently downstream stage from each stage, is zero or full lead-time. However in our paper, the placement of inventory is determined by ASRLT satisfying the lead-time required by a customer. Unlike existing papers using base-stock-system, our paper calculates average inventory based on the buffer profile and then reflects specifically both deviation of demand and deviation of supply. This is different from assuming safety stock as average inventory in previous chapter. This is the first study to calculate average inventory and to determine the inventory position with this kind of method.

3 The Strategic Inventory Positioning Problem with DDMRP

In this section, we will present a mathematical model to determine the locations among the stations in a factory which hold WIP inventory in order to meet the customer lead time (also called service time, see (Simpson 1958), while minimizing the total inventory holding cost for a particular product. We name this problem as the strategic inventory positioning (SIP) problem. We will use the following notations to describe the SIP problem:

$part_{i,j}$ jth part counted from the left in the ith level of the BOM,
$t_{i,j}$ processing time of $part_{i,j}$,
$a_{i,j}$ ASR lead time of $part_{i,j}$,
$r_{i,j}$ required quantity of $part_{i,j}$ to make one immediate parent part,
$n_{i,j}$ required quantity of $part_{i,j}$ to make unit end product,
$p_{i,j}$ the number counted from the left in the BOM of the immediate parent part of $part_{i,j}$ (e.g. in Fig. 1, $p_{3,2} = 2$, $p_{3,4} = 3$),
$s_{i,j}$ the set of numbers counted from the left in the BOM of the immediate child of $part_{i,j}$ (e.g.in Fig. 1, $s_{2,2} = \{2, 3\}$, $s_{3,2} = \{3, 4\}$),
$adu_{i,j}$ average daily usage of $part_{i,j}$,
$c_{i,j}$ annual inventory cost of $part_{i,j}$
$ai_{i,j}$ average inventory quantity of $part_{i,j}$,
$v_{i,j}$ unit price of $part_{i,j}$,
$ltp_{i,j}$ percentage usage of $part_{i,j}$ over ASR lead time,
$vp_{i,j}$ percentage of Red Zone Base of $part_{i,j}$ that Red Zone Safety accounts for:
h annual inventory holding cost rate,
u average daily usage of the end product,
Q lead time for the end product requested by the customer (i.e., service time),
$y_{i,j}$ yellow zone quantity of $part_{i,j}$,
$g_{i,j}$ green zone quantity of $part_{i,j}$,
$rb_{i,j}$ red zone base quantity of $part_{i,j}$,
$rs_{i,j}$ red zone safety quantity of $part_{i,j}$,

Decision variables:

$$x_{i,j} = \begin{cases} 1 & \textit{if } part_{i,j} \textit{ has inventory} \\ 0 & \textit{if } part_{i,j} \textit{ has no inventory} \end{cases} \tag{1}$$

$$\mathrm{TC} = \sum_{\forall i} \sum_{\forall j} c_{i,j} \cdot x_{i,j} \tag{2}$$

Table 1 Recommended impact ranges for green and red zone base

Long lead time	20–40 % usage over LT
Medium lead time	41–60 % usage over LT
Short lead time	61–100 % usage over LT

Table 2 Recommended impact ranges for red zone safety

High variability	60–100 % red zone base
Medium variability	41–60 % red zone base
Low variability	20–40 % red zone base

$$c_{i,j} = v_{i,j} \cdot h \cdot ai_{i,j} \tag{3}$$

Ptak and Smith define the buffer profiles as the families or groups of parts for which it make sense to devise a set of rules, guidelines, and procedures that can be applied the same way to all members of a given buffer profile (Ptak and Smith 2011). They propose the following frame work: in DDMRP, the on-hand inventory position averages in the lower half of the yellow zone so the average inventory quantity can be calculated to be the total quantity of the red zone plus half quantity of the green zone, as Eq. 4. The yellow zone for all buffer profiles is usually set to be equal to the usage over ASR lead time, as Eq. 5.

$$ai_{i,j} = 0.5g_{i,j} + rb_{i,j} + rs_{i,j} \tag{4}$$

$$y_{i,j} = adu_{i,j} \cdot a_{i,j} \tag{5}$$

The lead time category has a direct impact on the size of the green and red zone bases. The longer the lead time of the part, the higher should be the inventory of the part. Table 1 describes the recommended ranges of impact for the green and red zone bases for each lead time category. We do not consider the minimum order quantity (MOQ) in this chapter. Then the red zone base and green zone can be expressed as Eq. 6 and 7, respectively.

$$rb_{i,j} = ltp_{i,j} \cdot y_{i,j} \tag{6}$$

$$g_{i,j} = ltp_{i,j} \cdot y_{i,j} \tag{7}$$

The variability category will size the red zone safety portion of the total red zone. Red zone safety is an expression of a percentage of the red zone base. Table 2 shows how red zone safety is sized. Red zone safety can be shown as:

$$rs_{i,j} = vp_{i,j} \cdot rb_{i,j} \tag{8}$$

In the BOM, we can express

$$n_{i,j} = r_{i,j} \cdot n_{i-1,p_{i,j}} \tag{9}$$

$$adu_{i,j} = u \cdot n_{i,j} \tag{10}$$

$$a_{i,j} = \begin{cases} \underset{m \in s_{i,j}}{Max}[(1 - x_{i+1,m})a_{i+1,m} + t_{i,j}], & if\ s_{i,j} \neq \varphi \\ a_{i,j} = t_{i,j} \quad if\ s_{i,j} = \varphi : end\,node \end{cases} \tag{11}$$

We want to find a set X* = $\{X_{i,j}\}$ which minimizes the total cost, given the ASR lead time of the end product is no longer than the given service time Q. Now the SIP problem is mathematically formulated as follows:

$$\mathrm{TC} = \sum_{\forall i} \sum_{\forall j} v_{i,j} \cdot h \cdot a_{i,j} \cdot x_{i,j} \tag{11}$$

$$\begin{aligned} &st. \\ &ai_{i,j} = (1.5 + vp_{i,j}) \cdot ltp_{i,j} \cdot adu_{i,j} \cdot a_{i,j} \end{aligned} \tag{12}$$

$$adu_{i,j} = u \cdot n_{i,j} \tag{13}$$

$$n_{i,j} = r_{i,j} \cdot n_{i-1,p_{i,j}} \tag{14}$$

$$a_{i,j} = \begin{cases} \underset{m \in s_{i,j}}{Max}[(1 - x_{i+1,m})a_{i+1,m} + t_{i,j}], & if\ s_{i,j} \neq \varphi \\ a_{i,j} = t_{i,j}, \quad if\ s_{i,j} = \varphi : end\,node \end{cases} \tag{15}$$

$$a_{1,1} \leq Q \tag{16}$$

The formulation given in Eq. 16 is a non-linear programming (NLP) model so that the model is hard enough to be solved. Accordingly, a heuristic search algorithm is required to solve the model. In this study, a genetic algorithm (GA) is used to solve the model. In the next section, we will present the GA to efficiently solve the problem given in Eq. 16.

4 GA-Based Solution Procedure

Genetic algorithm (GA) has been widely used for obtaining good solution to various optimization problems for the last three decades. GA is a class of search techniques inspired from the biological process of evolution by means of natural selection. GA combines the genetic operators such as selection, crossover, and mutation with the goal of finding near-optimal solution for the problem. GA stops searching for the optimal solution when a specific termination condition is met. An initial population is created. Each individual in the population is called chromosome, representing a particular solution to the problem. Chromosome is made up of a collection of genes. Chromosomes evolve through successive iterations, called generations, using the genetic operators. During each generation, each of the chromosomes is evaluated using the fitness function.

4.1 Genetic Representation

For the SIP problem for the product of n parts in the BOM, we define a chromosome as an array of n binary genes, each of which represents whether the corresponding part in the BOM holds inventory. To form the initial population of size m, chromosomes are randomly generated until m valid chromosomes are collected. A chromosome is a valid one if the ASR lead time of the end product is no longer than the lead time requested by the customer (i.e., service time, Q).

4.2 Initial Population

In this step, a collection of chromosomes is randomly generated. Once a chromosome is generated, the validity of the chromosome is tested according to the ASRLT of the end product. If a certain lead time of the end product cannot be satisfied, the chromosome generated will be abandoned and a new chromosome will be generated until the population size that we set is attained.

4.3 Evaluation

When GA is applied for an optimization problem, a fitness value needs to be assigned for a chromosome. When we maximize the objective function, the value of the objective function can be used as fitness value. However, we want to minimize the objective function, so here, we set that the fitness value of the chromosome equals to a big enough number minus the total cost generated from the chromosome. The fitness value of each chromosome is employed to selection operation.

4.4 Selection Operation

Selection in GA is to select from the population a couple of chromosomes to generate child chromosomes. Selection weeds out the poor chromosomes and keeps the good ones. Several selection methods, such as roulette wheel, tournament, ranking, and elitist are discussed in Michalewicz (1996) and Goldberg (1989).

In this chapter, we will use the roulette wheel selection method because it provides higher chance for the good chromosomes to be selected as parents of the next generation. However, since using the roulette wheel selection method only does not guarantee the survival of the current best solution, we also use elitism to explicitly guarantee the best chromosome to remain in the population.

After each generation, solutions are tested for validity considering if satisfying the constraint of lead time of the end product. If the constraint is satisfied, the corresponding chromosome will immigrate to the next population; otherwise the solution will be abandoned. Then the new generation will continue until a sufficient number of chromosomes are produced.

4.5 Crossover

In the crossover process, we randomly select a pair of parent chromosomes from the population with probability proportional to its fitness function value so that 'better' chromosomes will have higher chance to be parents. In this chapter, a single-point crossover operator is used. Figure 3 shows a graphical representation of the crossover operation using the crossover point at the 10th gene.

4.6 Mutation

Mutation is the second operation of the GA for exploring better solution. A random number between 0 and 1 is generated for each gene of the two child chromosomes generated by the crossover operation. If the random number generated for the k-th gene (r_k) of a chromosome is less than a predetermined mutation probability Pm, then the gene mutates, that is, the binary value of the *k*th gene (b_k) changes from 0 to 1 or 1 to 0. Figure 4 shows a mutation operation with Pm set to be 0.10

4.7 Termination Condition

In order to terminate the process of GA, termination conditions must be given. In this chapter, we stop the algorithm after thirty generations.

In summary, the steps of the GA in this chapter are as follows:

1. Set the parameters Pc = 0.8, Pm = 0.10 and the population size N = 500.
2. N arbitrary chromosomes are randomly generated to form the initial population.
3. Evaluate the fitness function of all chromosomes in the population.
4. Select two chromosomes to crossover and mutate by roulette wheel selection and elitist method.

1	0	0	0	1	0	0	1	0	**0**	**1**	**0**	**1**
									↕	↕	↕	↕
0	1	1	1	1	0	0	1	0	**1**	**0**	**0**	**1**

Fig. 3 An example of crossover operation with crossover point at 10th gene

b_k	1	0	**0**	0	1	0	**0**	1	0	1	1	0	1
r_k	0.13	0.7	**0.09**	0.3	0.45	0.6	**0.05**	0.6	0.55	0.32	0.11	0.47	0.36
							↓						
	1	0	**1**	0	1	0	**1**	1	0	1	1	0	1

Fig. 4 An example of mutation operation with mutation probability 0.10

5. Operate crossover to each pair of chromosomes with probability Pc.
6. Operate mutation to each chromosome with probability Pm.
7. Check the validity of the new chromosomes and continue to generate until a sufficient population size is reached. Then replace the current population by the resulting new population.
8. Evaluate the objective function
9. If the termination condition is met, stop the system. Otherwise, go to step 4.

5 Computational Results

In order to demonstrate the application of the proposed methodology and to evaluate its performance, in this section, two numerical examples are given.

5.1 Small Instances

The example problem has 13 parts in the BOM given in Fig. 1. Table 3 shows the numbers of parameters. Here, we assume that every part has the same variability and the percentage *vp* is set as 50 % for all parts. The percentage (*ltp*) is 30, 50 and 80 % for long, medium, and short lead time, respectively. In most cases, the crossover rate is typically set from 0.7 to 0.9 and the mutation rate is set from 0.05 to 0.15. Here, we set the crossover rate Pc as 0.8 and the mutation rate Pm as 0.1. For small instances, we set the population size N as 50. The iteration will stop after 30 generations.

Table 3 Data of parameters for small instances

Part	1	2	3	4	5	6	7	8	9	10	11	12	13
t	2	5	7	6	4	4	3	3	3	4	3	7	11
r	1	1	1	3	2	5	1	1	1	1	2	4	6
p	0	1	1	1	1	2	2	3	1	1	2	2	2
v	700	120	250	100	65	80	110	40	24	16	21	27	9
Q	10												
h	0.25												
vp	0.5												
ltp	Long lead time (26 +days)					30 %							
	Medium lead time (11–25 days)					50 %							
	Short lead time (1–10 days)					80 %							

Table 4 The experimental results

Pc	Pm	N	Average ranking among 2880
0.8	0.1	500	9

Table 5 The top 10 min TC obtained from enumeration

Ranking	TC
1	1654.6
2	1749
3	1753
4	1761.6
5	1786.6
6	1856
7	1860
8	1878.6
9	1881
10	1882.8

In order to show the effectiveness of the proposed GA method in solving SIP problem, the results obtained from GA is compared with those from enumeration. 2,880 feasible solutions that are ranked from the minimum to maximum total cost generated from those feasible solutions are obtained using enumeration. When we get the solution from GA, the ranking in the enumeration will be matched in the list of feasible solutions. Here, we run GA for 20 times and get the average ranking. The results are summarized in Table 4. Error ($TC_{GA} - TC_{min}/TC_{min}$) is not employed to test the effectiveness of the GA here (TCGA represents the total cost of the best solution obtained from GA, TC_{min} is the minimum cost from enumeration), because the total cost of the feasible solution is not consecutive. Table 5 shows the top 10 minimum total cost obtained from the enumeration.

Table 6 Data of parameters form medium-sized problem

Part	t	r	p	v
1	2	1	0	1200
2	5	1	1	900
3	7	1	1	890
4	6	1	1	910
5	4	2	1	860
6	4	3	1	880
7	3	2	1	800
8	3	1	1	780
9	3	2	2	790
10	4	3	2	750
11	3	2	3	690
12	5	3	3	660
13	5	4	4	600
14	4	3	4	640
15	2	1	4	700
16	4	3	1	660
17	6	2	1	580
18	3	1	2	550
19	3	1	3	570
20	5	2	3	600
21	7	3	4	540
22	6	2	4	510
23	3	2	4	590
24	4	1	5	400
25	3	1	6	490
26	3	1	7	480
27	5	2	7	520
28	4	2	8	400
29	3	1	9	480
30	2	3	9	490
31	5	1	1	460
32	4	1	1	500
33	3	3	3	470
34	3	2	3	300
35	5	2	5	380
36	4	2	8	330
37	3	3	8	340
38	4	1	12	350
39	6	1	15	360
40	4	3	15	310
41	4	2	2	330
42	4	2	3	290
43	3	2	3	270
44	3	2	5	260
45	2	1	7	270

(continued)

Table 6 (continued)

Part	t	r	p	v
46	5	1	7	270
47	3	3	9	100
48	2	2	9	110
49	3	4	5	50
50	5	2	7	30

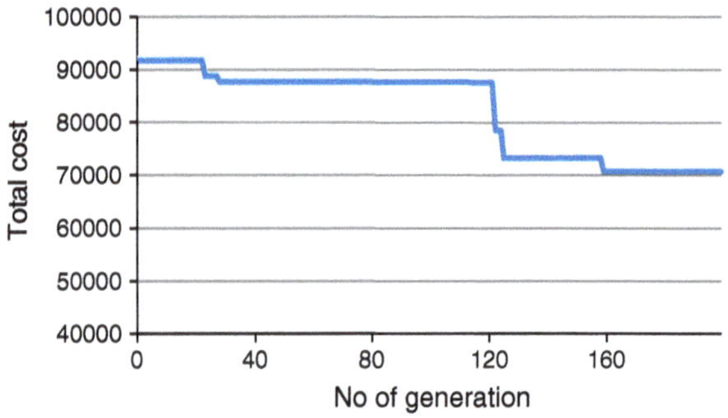

Fig. 5 The convergence path (N = 50)

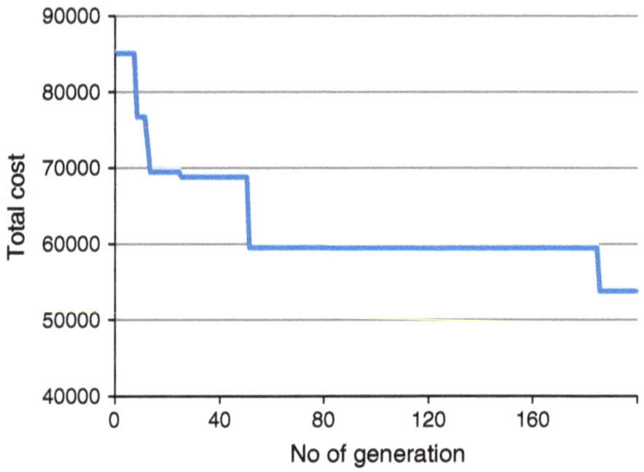

Fig. 6 The convergence path (N = 80)

5.2 Medium-Sized Problem

We can evaluate the effectiveness of GA to solve the SIP problem with small instances compared with the result obtained from enumeration. However, GA is used to solve the problem with big size instances that enumeration cannot solve. Here we have 50 parts in our BOM, which means that there are 2^{50} solutions. We cannot enumerate it using computer. Table 6 lists the parameter for medium-sized problem in which the number of *vp, lvp, h, Pc, Pm* is the same with those in small

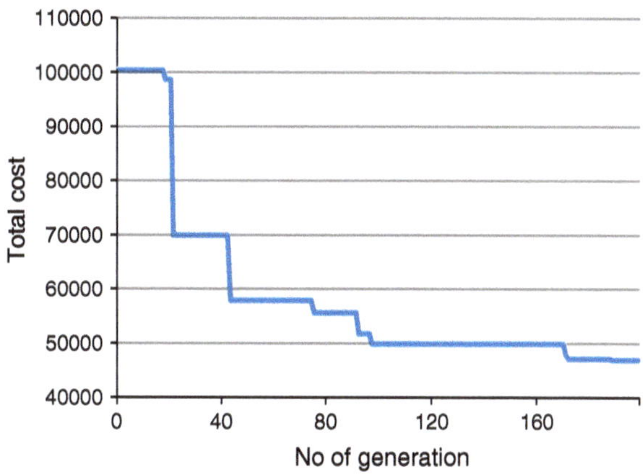

Fig. 7 The convergence path (N = 120)

instances and Q here is set as 14. The system will be stopped after 200 generations. When N is selected 50, 80 and 120, the convergence path graphs are presented in Figs. 5, 6 and 7, respectively.

6 Conclusion

Demand-driven MRP is a new type of MRP introduced by Ptak and Smith. In DDMPR, a new and more realistic type of lead time called ASRLT is proposed. Focus on DDMRP is changed from how much and when to position the inventory to where to position the inventory. In this research, a strategic inventory positioning with DDMRP model was developed. However, it is cannot be solved using mathematical method because it is a non-linear problem. Enumeration is limited when solving the problem with larger parts in the BOM due to the unreasonable computational time, so GA is proposed to solve this problem. Results from the enumeration are used to test the effectiveness of the GA method when solving the small instance problem. For medium sized problem and big sized problem, we just can use GA to solve it.

References

Alderson W (1950) Marketing efficiency and the principle of postponement. Cost Profit Outlook 3:15–18

Bucklin LP (1965) Postponement, speculation and the structure of distribution channels. J Mark Res 2(2):26–31

Davis T (1993) Effective supply chain management. Sloan Manag Rev 34(4):35–46

Dellaert D, Jeunet J, Jonard N (2000) A genetic algorithm to solve the general multi-level lot-sizing problem with time-varying costs. Int J Prod Econ 68:241–257

Ernst R, Kamrad B (2000) Evaluation of supply chain structures through modularization and postponement. Eur J Oper Res 124(3):495–510
Feitzinger E, Lee HL (1997) Mass customization at Hewlett-Packard: the power of postponement. Harvard Business Rev 75(1):116–121
Goldberg D (1989) Genetic algorithms in search: optimization and machine learning. Addison-Wesley, Reading
Graves SC, Lesnaia E (2004) Optimizing safety stock placement in general network supply chains. In: Singapore-MIT alliance annual symposium conference proceedings, Singapore**
Graves FC, Willems SP (2000) Optimizing strategic safety stock placement in supply chains. Manufact Serv Oper Manage 2(1):68–83
Graves FC, Willems SP (2008) Strategic inventory placement in supply chains. Manufact Serv Oper Manage 10(2):278–287
Inderfurth K (1991) Safety stock optimization in multi-stage inventory systems. Int J Prod Econ 24:103–113
Inderfurth K, Minner S (1998) Safety stocks in multi-stage inventory systems under different service measures. Eur J Oper Res 106:57–73
Kaminsky P, Kaya O (2008) Inventory positioning, scheduling and lead-time quotation in supply chains. Int J Prod Econ 114:276–293
Lee HL, Billington C, Carter B (1993) Hewlett-Packard gains control of inventory and service through design for localization. Interfaces 23(4):1–11
Loannou G, Prastacos GP, Skinntzi G (2004) Inventory positioning in multiple product supply chains. Ann Oper Res 126:195–213
Magnanti TL, Shen ZM, Shu J, Simchi-Levi D, Teo CP (2006) Inventory placement in acyclic supply chain networks. Oper Res Lett 34:228–238
Michalewicz Z (1996) Genetic algorithms + data structures = evolution programs, 3rd ed. Berlin, Springer
Ptak C, Smith C (2011) Orlicky's material requirements planning, 3rd edn. McGraw-Hill Publishers, Springer
Simpson KF Jr (1958) In-process inventories. Oper Res 6(6):863–873
Whybark DC, Yang S (1996) Positioning inventory in distribution system. Int J Prod Econ 45:271–278
Graves SC, Willems SP (1996) strategic safety stock placement in supply chains. In: Proceedings of the 1996 MSOM conference, Hanover, NH

Improving Efficiency of a Process in Warehouse with RFID: A Case Study of Consumer Product Manufacturer

Natanaree Sooksaksun and Sriyos Sudsertsin

Abstract This warehouse has two main problems: long cycle time of receiving process and poor accuracy of inventory location. Therefore, the objective of this research is to improve the process in the warehouse by using radio frequency identification (RFID) technology. There are four steps in this research. First of all, the current of the receiving process is studied. Second is to give the guideline for RFID application. Third is to implement of RFID in this warehouse. The last one is to compare the result of the current process and the proposed process. The passive ultra high frequency (UHF) RFID is selected for used in this warehouse. The RFID readers are attached on forklift trucks, the warehouse entrance and loading dock area. The system used RFID tags in two forms: one used to identify pallets and another to indicate the locations of shelves on which pallets are stored. After using RFID in receiving process, the results showed that the cycle time decreases from 66 min to 47 min which is down to 28.79 %. Moreover, accuracy of inventory location increased from 72.8 % to 100 %.

Keywords RFID · Warehouse · Receiving process

1 Introduction

Warehouse management is one significant activity of supply chain management. Operating in the warehouse involves several processing activities: receiving, storage, picking and shipping. The efficiency of a warehouse depends on how efficiently it can perform the activities. Especially, the receiving process where is

N. Sooksaksun (✉) · S. Sudsertsin
Faculty of Industrial and Technology Management, Department of Industrial Management, King Mongkut's University of Technology North Bangkok, Prachinburi Campus, Prachinburi, Thailand
e-mail: natanaree.s@fitm.kmutnb.ac.th

P. Golinska (ed.), *Logistics Operations, Supply Chain Management and Sustainability*, EcoProduction, DOI: 10.1007/978-3-319-07287-6_32,

the beginning of all activities in warehouse. Receiving the wrong products or putting products in incorrect locations can result in errors just as easily as can picking the wrong item.

Radio Frequency Identification (RFID) is technology which uses radio frequency to identify objects and transfer data by the wireless non-contact and it can be automatically tracked and traced on each product item or pallet by using the RFID tag. RFID technology consists of an antenna and a transceiver, or reader and a transponder, or tag. When the RFID tag passes through the field of the scanning antenna, it detects the activation signal from the antenna and it transmits the information on its microchip to be picked up by the scanning antenna. RFID tags can be read in a wide variety of circumstances, where barcodes or other optically read technologies are useless and it is developed for collecting and tagging data will help managing warehouse data more effectively.

In warehouse of consumer product manufacturer, the complexity of warehouse management depends on the number Storage Keeping Unit (SKU), quantities of each SKU and the number of orders received and shipping. This warehouse has two main problems: poor accuracy of inventory location and long cycle time of receiving process. It is necessary to improve warehouse management efficiency and reduce error rates. Therefore, the objective of this research is to improve the process in the warehouse by using RFID technology.

The remainder of this chapter is organized as follows. In Sect. 2, the related literatures are reviewed. The methodology of this research is explained in Sect. 3. In Sect. 4, the results of the methodology are shown. The last one is conclusion section.

2 Literature Review

In the past decades, RFID technologies have attached considerable attentions (Sarac et al. 2008). Currently, RFID is an importance technology for revolutionizing a wide range of applications including supply chain management (James et al. 2013). RFID can apply in different parts of supply chain such as warehouse management, transportation management, production scheduling, order management, inventory management and asset management systems (Banks et al. 2007).

The RFID system consists of three components: reader, tag and host computer (Sulaiman et al. 2012). The tag is the part that collects real time data and then transmits that data via radio waves. The tags usually have two parts, a small chip and an antenna. Information is stored and processed by the chip while the antenna is used to receive and transmit information. RFID tags can be either passive or active tag. An active tag has a small battery on board and is activated when in the presence of an RFID reader. A passive tag is cheaper and smaller because it has no battery. The information is read by an RFID reader when a tag passes by it. The reader can track the tag's movement in real time and pass its digital identity and other relevant information to a computer system.

There are many researches in field of RFID. For example, Spekman and Sweeney (2006) presented a comprehensive overview of RFID technology (Spekman and Sweeney 2006). The goal is to provide insights regarding the implementation and use of RFID by focusing on its advantages and its problem. Wang et al. (2010) proposed a digital warehouse management system in the tobacco industry based on RFID technology. Zhu et al. (2012) provided an overview of the current state of RFID application in different industries and its impact on business operations. Dwivedi et al. (2013) studied the factors affecting the use of RFID systems and user satisfaction in a library context by empirically testing relevant constructs from DeLone and McLean's is success model. Liu et al. (2013) viewed the RFID technology as an operation strategy to ensure food safety, and proposes three pricing decision models according to different operation strategies and market situations. Fan et al. (2014) considered the situation of a retailer subject to inventory inaccuracies stemming from shrinkage problems. They apply a newsvendor model to analyze how to reduce inventory shrinkage problems by deploying RFID.

3 Methodology

The methodology of this research has 4 steps.

- Study an existing warehouse
  This step studies the general information of an existing warehouse such as receiving process by using 3G principles. The first G is Genba or the actual place. The second G is Genbutsu or the actual thing. The last one is Genjitsu or Genshou or the actual situation. Moreover, the cycle time of receiving process and the accuracy of inventory location are collected.
- Study how to apply RFID
  This step studies how to apply RFID. The fundamental level of RFID is studied. For example, what is RFID? How do the parts fit? After that, the RFID system of this warehouse is designed.
- Implement RFID in the warehouse
  In this step, the RFID system that is designed is implemented in the warehouse. After that, the system is tested.
- Compare before and after implement RFID
  After the RFID system is used, the cycle time of receiving process and the accuracy of inventory location are collected. Next, the data are compared between before and after using RFID.

4 Results

- The general information of an existing warehouse

This factory is a consumer product manufacturer. The finished goods are sent to the warehouse. The distance between the factory and the warehouse is approximate 4 km. The finished goods are transported from the factory to warehouse by truck which contains 10 pallets per trip. Figure 1 shows the existing of receiving process.

The information of the receiving process is not real-time that mostly used paper-based. Moreover, this warehouse has two main problems: long cycle time of receiving process and poor accuracy of inventory location. The cycle time of receiving process (10 pallets per trip) is 66 min per trip. The accuracy of inventory location is 72.8 %. Therefore, the RFID system is selected to improve warehouse management efficiency and reduce error rates.

- The guideline for RFID application

The concept of RFID system is shown in Fig. 2.

The passive ultra high frequency (UHF) RFID base on EPC class 1 Gen1 standard is selected for this warehouse. The advantage is inexpensive price and not require power source to operate. For the RFID tags, two forms are selected to use. First form used to identify pallets, called "*pallet tag*" (see Fig. 3a). Another used to indicate the locations of shelves on which pallets are stored, called "location tag" (Fig. 3b).

For RFID Reader, there are three forms. The first form is the desktop reader (Fig. 4a) that is used in production line to generate data to RFID tag before attach to finished goods pallet. The second form (Fig. 4b) is the reader which mounted on forklift trucks with antenna and touch screen computer. This reader can read both pallet tag and location tag. The last one (Fig. 4c) is RFID reader and antenna which mounted on the warehouse entrance and loading dock. It can be automatically traced on each finished goods pallet to be moving pass.

- The RFID implementation in warehouse

Integrating RFID technology into warehouse process, based on the analysis of the current processes, the receiving process in Fig. 1 is changed after using RFID to be the receiving process in Fig. 5.

From the Fig. 5, after the finished good is produced, production data is linked to the RFID tags which attach to each pallet. Warehouse staff will know the number of pallet and prepare space of storage in warehouse. The first gate reader where is installed between end of production line and warehouse can be automatically traced on each finished goods pallet to be moving pass to specify the destination warehouse. The second gate reader where is installed at shipping area can be traced on each pallet to be moving pass and specify automatically the location in the destination warehouse and drop pallets into the truck. When transport pallets to the destination warehouse, RFID tag which attach to each pallets is read by forklift

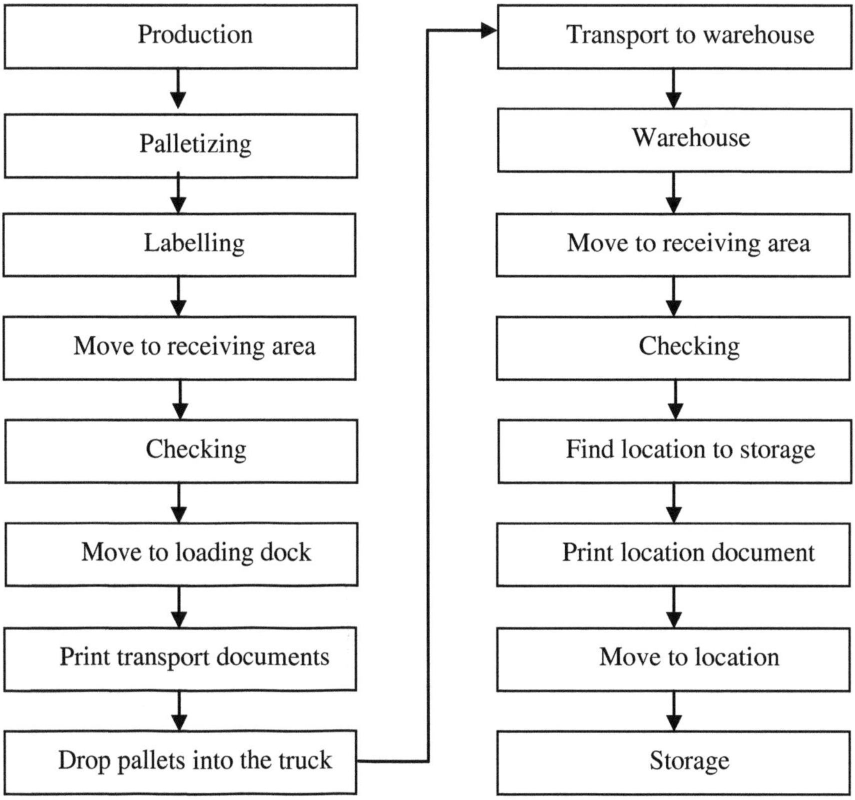

Fig. 1 The existing of the receiving process

reader and show the location to be stored on computer monitor which mounted on the forklift truck. The forklift truck moves pallet to the location and stores while location tag is read to verify by forklift reader.

- The result which compare before and after implement RFID

After implement RFID, the data is collected. The results are shown in Table 1.

From the Table 1, the implement RFID in the receiving process can help to solve the problems. The average total cycle time of receiving process is decreased by 28.8 %. Moreover, the accuracy of inventory location is 100 %.

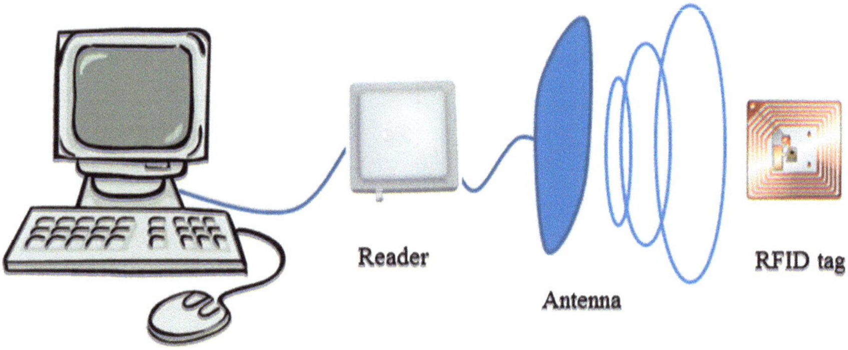

Fig. 2 RFID system

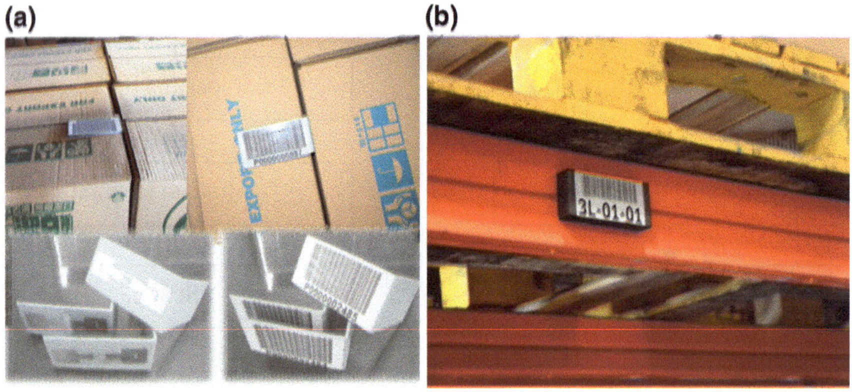

Fig. 3 RFID tags

5 Conclusion and Recommend

The objective of this research is to improve the receiving process in the warehouse by using RFID technology. This warehouse has two significant problems. The first problem is an inaccuracy of inventory location that the accuracy of inventory location is 72.8 %. The second problem is long cycle time of receiving process. The average total cycle time of receiving process is 66 min per trip while one trip is 10 pallets. After the RFID system is implemented, the accuracy of inventory location is improved from 78.2 % to 100 %. Moreover, the cycle time is decreased by 28.8 % or 19 min.

However, this research focuses only receiving process in this warehouse while the RFID can use in the other processes in the warehouse. Therefore, this warehouse has gap for improve in the other processes such as shipping process.

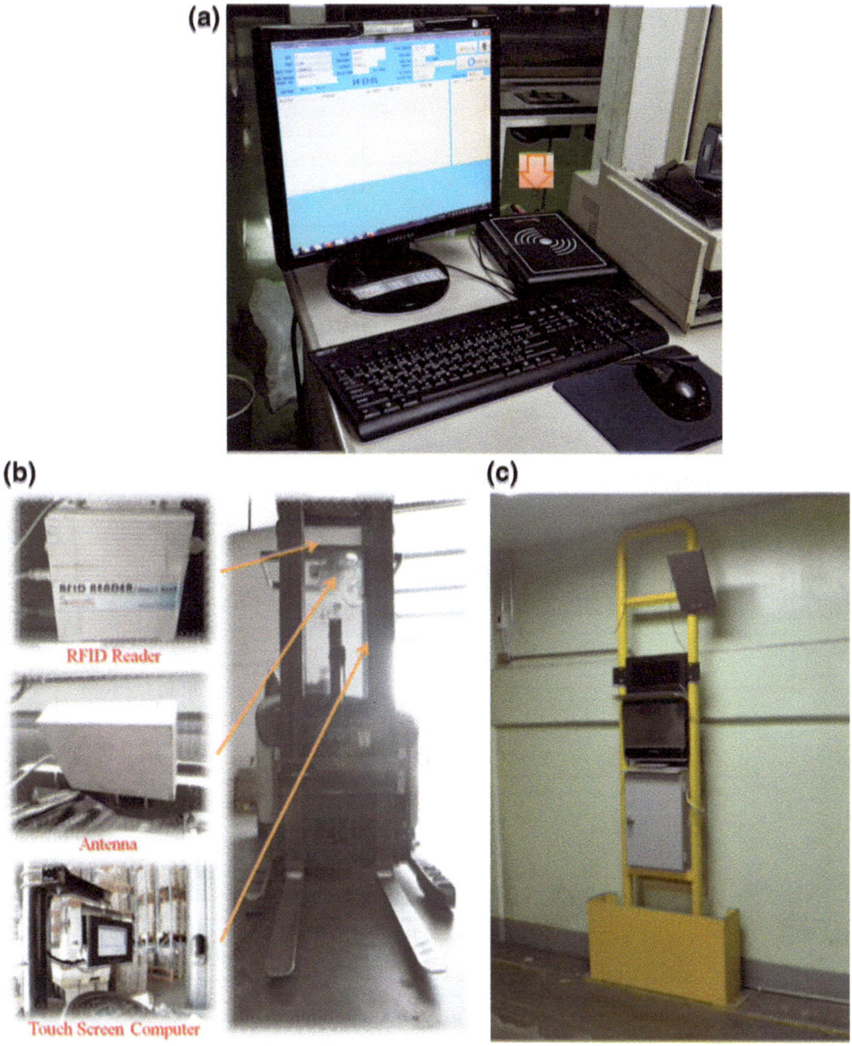

Fig. 4 RFID reader

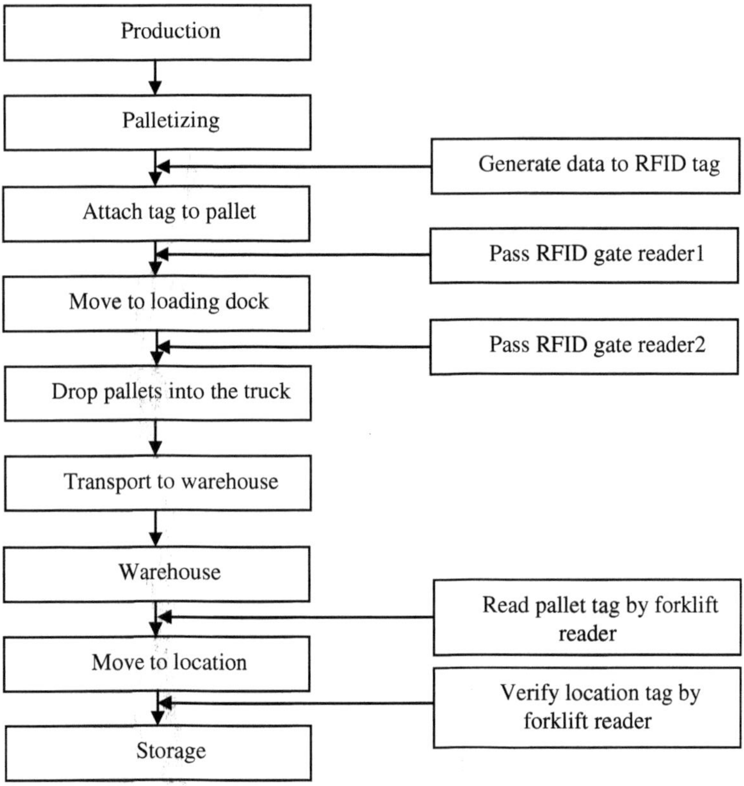

Fig. 5 The receiving process after using RFID

Table 1 Compare before and after implement RFID

Items	Before implement RFID	After implement RFID	Difference
The average total cycle time of receiving process (10 pallets/trip)	66 min per trip	47 min per trip	28.8 %
The accuracy of inventory location	72.8%	100 %	27.2 %

References

Banks J, Hanny D, Pachano MA, Thompson LG (2007) RFID applied. Wiley, New York

Dwivedi YK, Kapoor KK, Williams MD, Williams J (2013) RFID systems in libraries: an empirical examination of factors affecting system use and user satisfaction. J Inf Manag 33(2):367–377

Fan T, Chang X, Gu C, Yi J, Deng S (2014) Benefits of RFID technology for reducing inventory shrinkage. Int J Prod Econ 147:659–665

James CC, Chen-Huan C, PoTsang BH (2013) Supply chain management with lean production and RFID application: a case study. Expert Syst Appl 40:3389–3397

Liu S, Zhang D, Zhang R, Liu B (2013) Analysis on RFID operation strategies of organic food retailer. Food Control 33:461–466

Sarac A, Absi N, Dauzère-Perès S (2008) A simulation approach to evaluate the impact of introducing RFID technologies in a three-level supply chain. In: Proceedings of the 2008 winter simulation conference, pp 2741–2749

Spekman RE, Sweeney PJ II (2006) RFID: from concept to implementation. Int J Phys Distrib Logistics Manag 36(1):736–753

Sulaiman S, Umar UA, Tang SH, Fatchurrohman N (2012) Application of radio frequency identification (RFID) in manufacturing in Malaysia. Procedia Eng 50:697–706

Wang H, Chen S, Xie Y (2010) An RFID-based digital warehouse management system in the tobacco industry: a case study. Int J Prod Res 48(9):2513–2548

Zhu X, Mukhopadhyay SK, Kurata H (2012) A review of RFID technology and its managerial applications in different industries. J Eng Tech Manage 29(1):152–167

Model of Forklift Truck Work Efficiency in Logistic Warehouse System

Paweł Zając

Abstract This chapter deals with the modelling of forklift truck work efficiency in logistic warehouse systems. So far (Bangsow 2012; Bednarz et al. 2011; Makris et al. 2006; Ören 2011) truck work efficiency has been modelled with only kinetic energy taken into account. In this chapter, mathematical formulas supplementing the existing energy efficiency model with the potential energy connected with the lifting and lowering of the unit load in storage processes are derived. The derived kinematic formulas and energy relations enable the proper modelling and simulation of forklift truck work efficiency. Work efficiency is highly important considering the wastage of the energy supplied to the logistic system for handling unit loads.

Keywords Forklift truck modelling · Energy consumption by logistic warehouse systems

1 Introduction

According to (Zając 2011b), the term *productivity* means a ratio of the input to the effects. At the technological level, energy efficiency is the dependence between the consumed energy and the effect produced by this energy, often referred as an *energy service* (e.g. the number of kilometres a vehicle has travelled consuming a litre of a given fuel). In order to increase energy efficiency either less energy must be consumed to achieve the same level of services or the same energy consumption level must be maintained while providing a larger number of services. According to Energy Efficiency Act of 15 April 2011, energy efficiency is a ratio of the size of the

P. Zając (✉)
Faculty of Mechanical Engineering, Wrocław University of Technology,
ul. I. Łukasiewicza 7/9, 50-371 Wroclaw, Poland
e-mail: pawel.zajac@pwr.wroc.pl

P. Golinska (ed.), *Logistics Operations, Supply Chain Management and Sustainability*, EcoProduction, DOI: 10.1007/978-3-319-07287-6_33,

usable effect produced by a given object, piece of equipment or a system in their typical operating conditions to the amount of energy consumed by this object, piece of equipment or system in order to obtain this effect Tompkins et al. (2010).

In analysis (Bangsow 2010; Zając 2010a) one can read that the benefits from energy efficiency can be measured by its economic value. At the national economy level this amounts to a dependence between GDP and energy consumption, referred to as GDP energy consumption (Homburg 2007).

Poland consumes twice as much energy for producing a Euro national unit income than the average for the 27 EU member countries, and over three times as much than the most energy efficient economy of Denmark. There has already been much progress in energy consumption reduction in Poland—since 2005 the reduction has been nearly twofold, but the distance to the average for the 27 countries (not to mention the most developed and most energy-efficient countries) is still large (European Energy Efficiency 2012; Eurostat; Kemme 2013).

In the case of Logistic Warehouse Systems, efficiency is a ratio of the energy supplied to the materials handling equipment as well as to the lighting, heating, cooling, automatic identification and EDI subsystems and the energy necessary to handle the passage of the unit load through the warehouse (Fig. 1). The supplied energy is not only consumed, but also recovered through proper technical solutions and stored to be reused for useful work. The operation of the warehouse subsystems entails losses stemming from the efficiency of the energy conversions (Homburg 2007; Kwaśniowski et al. 2011). The efficiency of, for example, diesel-powered equipment (such as a forklift truck) is 0.36 while that of electric-powered equipment amounts to 0.92 Zając (2011c).

Geothermal energy and solar energy are seldom used in LWS because of the high installation costs and little interest in these technologies from project owners.

Knowledge of the energy intensity of the vertical movement of the forklift truck is important for determining the energy consumption of logistic storage system. Control the movement of energy quotient ratio, expressed quotient: horizontal to vertical can be used for the proper use of a forklift truck in warehouse work. Energy intensity ratio of the vertical motion can serve as a supplement to the method of "frequency of downloads" (*ABC*) resulting populate the sockets of the rack from the lower levels. Knowledge of the energy raising is important when deciding on the recovery of energy from lifting the forklift Zając (2011a).

2 Description of Forklift Truck Efficiency Model

In the case of static storage, when a pallet has been put in a storage place, its energy balance does not change. This model does not take into account the energy consumption connected with:

- the lighting of the facility,
- the energy receivers in the office (e.g. an electric kettle),

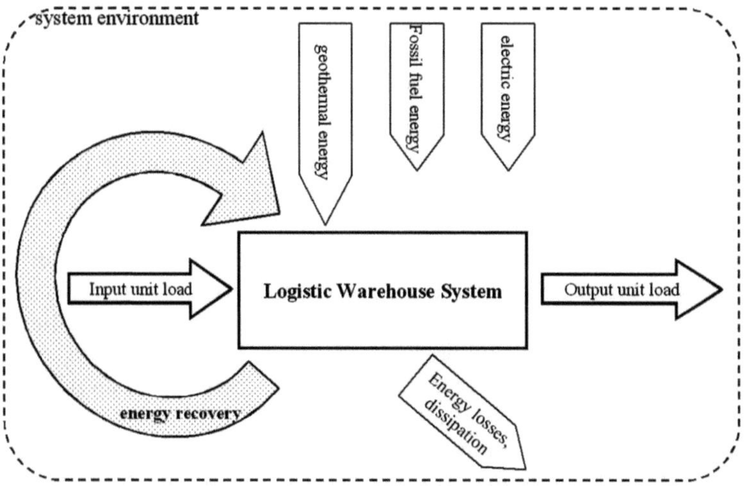

Fig. 1 Different kinds of energy supplied to LWS (personal research). (Bujak and Zając 2012)

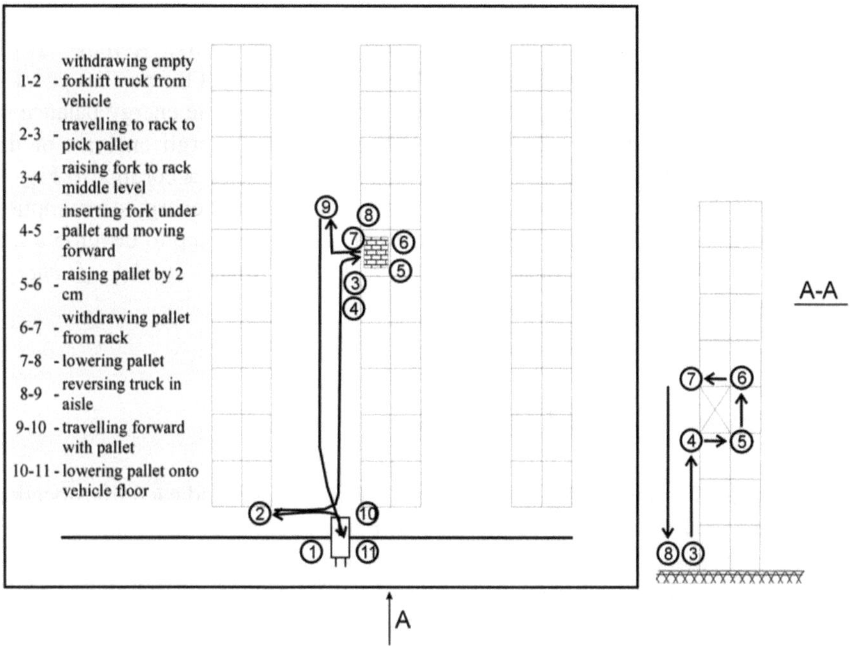

Fig. 2 Forklift truck route diagram for typical terminal (personal research)

- thermal insulation efficiency (e.g. the insulation of the warehouse buildings, the use of controllers in heat sources such as heaters),
- the degree of insolation of the facility, etc.

Let us assume that a forklift truck handles palletized loads transporting them from a rack in the middle of the warehouse to a vehicle waiting at the ramp, as shown in Fig. 1, where digits in circles represent the major points in forklift truck work (Fig. 2).

Each of the above operations: {1–2}, {2–3}, {8–9}, {9–10} includes starting, travelling and braking. The duration of each of the operations depends on the distance covered and the technical specifications of the truck (its travelling speed).

Detailed calculations should precisely determine the speed profile, the starting and braking time, the forces on the wheels and the energy spent on the operations (Bednarz et al. 2011; Law and Kelton 2000).

A methodology for modelling energy consumption by the forklift truck (whose working cycle is shown in Fig. 3) is presented in Fig. 4. This forklift truck work efficiency model does not take into account the energy (and so the energy efficiency) connected with raising and lowering the unit load. A concept of modelling forklift truck efficiency, with the potential energy connected with the operation of the forklift truck mast carriage taken into account, is presented in the next section.

In the chapters (Bangsow 2010, 2012; Homburg 2007; Kemme 2013; Kwaśniowski et al. 2011; Makris et al. 2006)—are the work of the past 10 years—do not take into account the potential energy component in the energy balance of the magazine. The author of (Zając 2010b) proposed a exergii analysis of the energy balance of potential and kinetic energy and the other according to Fig. 1, and proved that the change in potential energy impact on the energy consumption of forklift truck work (Zając 2010b) This article is the next step to develop a full analysis of the energy consumption of a forklift truck—*model of energy consumption.*

3 Energy of Vertical Translocation

It is assumed that the warehouse floor is the reference level and each translocation of the unit load by a device in the vertical plane changes the amount of energy spent on the passage of the load through LWS, increasing the amount of consumed energy by the energy spent on the successive translocations of the pallet in the warehouse or decreasing it by the recovered (e.g. by putting the unit load on the rack) energy. Finally, when the pallet leaves the warehouse, one can calculate how much energy was needed to handle the pallet by the successive devices. The physical source $E_p = m \cdot g \cdot h$ is potential energy (1), where: efficiency of the lifting $\eta_p = 0,8$ (Tompkins et al. 2010) and power $P = \left(Q_{Unit} + Q_{fork}\right) \cdot V_{lifting}$ and ΔS is pitch of the fork.

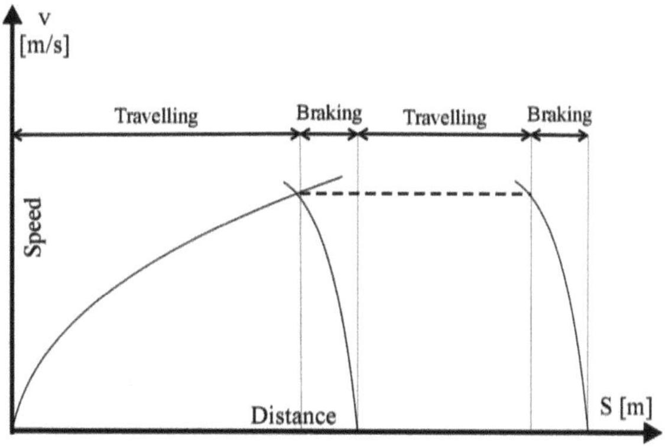

Fig. 3 Hypothetical course of forklift truck work. (Logistics Cost and Service 2007)

$$\eta_p \cdot P \cdot A_c \cdot \Delta S = m \cdot g \cdot h \qquad (1)$$

The potential energy is equal to the work which needs to be performed in order to translocate the pallet with the load in the vertical. The translocation of loads in the vertical in LWS is carried out by the lifting system which is used to grip and raise/lower the unit load in the vertical. This mechanism in forklift trucks (as well as in storage and retrieval machines) consists of a telescopic frame, a carriage, a carriage raising mechanism and a frame tilting mechanism. The telescopic frame is made up of two main parts: fixed outer frame 8 and moving inner frame 9 (Fig. 5) which moves upwards and downwards relative to the outer frame. The fixed outer frame of the lifting mechanism is rotationally attached by means of two axles 23 to the body frame brackets between the driving mast wheels. Two brackets with rollers 11 guiding the inner slidable frame are fixed from the inside to the top ends of the vertical guides. Slides 7, constituting rails for the moving frame rollers, are welded from the inside to the clamps of the vertical guides.

Moving frame 9 is made up of two guides made of channel bars, connected together with a crossbar at the top. In the lower part of each of the guides, two rollers 6 are mounted on roller bearings and one roller 4 is screwed to the channel bar. Guiding rollers 6 keep the moving frame in the plane of the truck longitudinal axis and prevent it from oblique orientation while rollers 11 and 4, located on the fixed frame and the moving frame, keep the moving frame in a plane parallel to the axis of the driving wheels as well as prevent it from tilting.

Carriage l, for attaching working fittings and for gripping loads, consists of two longitudinal members connected together with crossbars. Rollers 2 and 5 of the carriage, mounted on its longitudinal members, roll on the guides welded from the inside to the clamps of the vertical guides of the moving frame. A fork is attached by means of fixtures 17 and 18 to the carriage crossbars, to which protective frame 15 is screwed.

The carriage lifting mechanism consists of plate link chains 14, bracket 12 with guiding rollers and single-acting plunger-type hydraulic lifting actuator 13. The plate link chains run on guiding rollers mounted on the hubs of the bracket on the

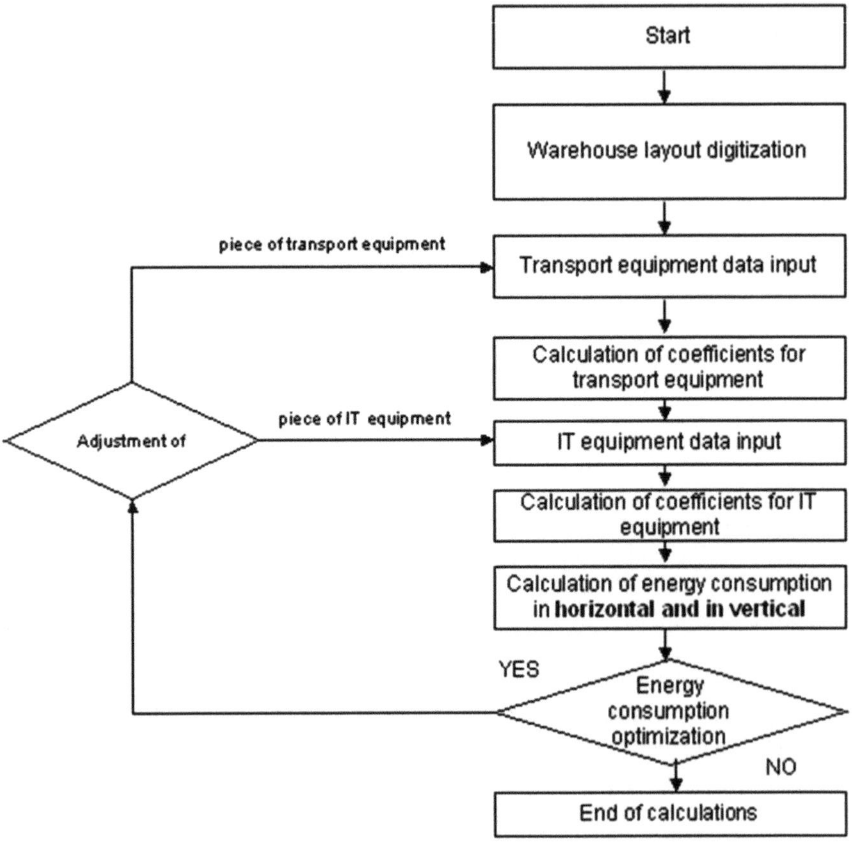

Fig. 4 Energy consumption method algorithm (personal research)

rolling bearings. Some ends 3 of the chains are fixed to the carriage while the other ends are fixed to two-armed levelling lever 19 mounted on axle 20 in the lifting cylinder head whereby the two chains are uniformly loaded when the carriage tilts.

The carriage moves down under the weight of the load and the moving parts of the lifting mechanism while the plunger shifting inside the cylinder forces oil from the latter through a throttling orifice to a tank. The lower end of the actuator is attached via a yoke to the lower crossbar of the fixed frame. The working liquid (e.g. spindle oil) from a distributor is supplied through flexible conduit 22 and throttle valve 24 to the lifting actuator cylinder.

The frame tilting mechanism consists of two double-acting piston-type tilt actuators 21. The lugs of the tilt cylinders are rotationally attached to the body frame brackets while the piston rods with tips screwed on them are attached to the brackets of the fixed frame of the lifting mechanism. From the hydraulic distributor the working liquid is directed via stubs to the upper or lower space of the cylinders, moving the pistons with the piston rods in them and tilting the telescopic frame forward or backwards.

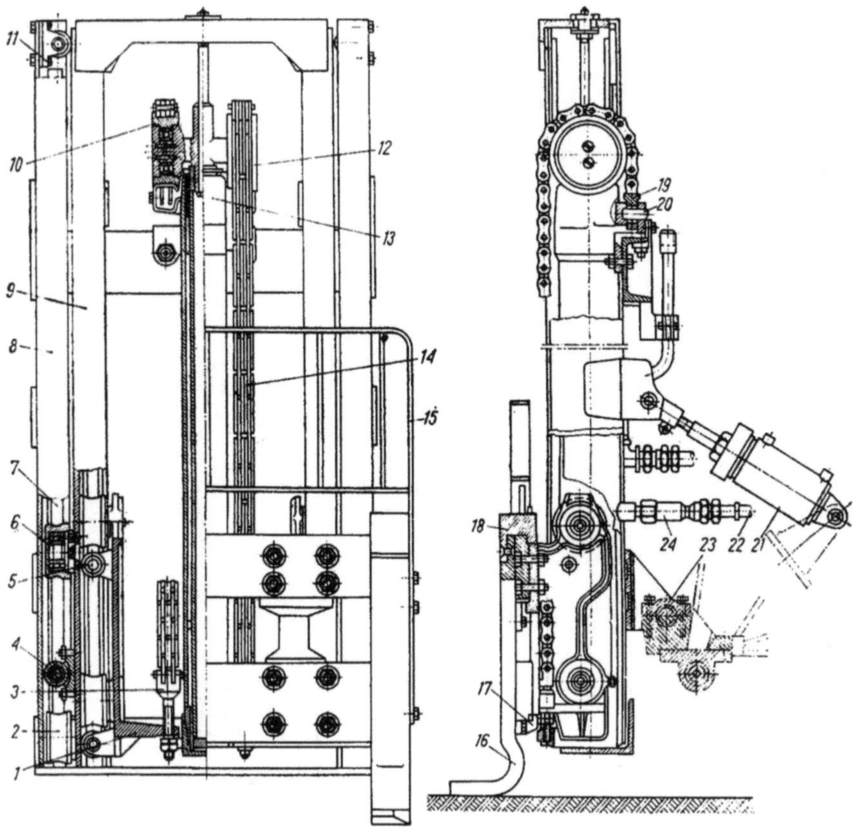

Fig. 5 Schematic of forklift truck lifting system

The hydraulic system (Fig. 7) consists of a double-acting hydraulic vane pump, a hydraulic distributor with a pressure reducing valve, hydraulic actuators (cylinders), a working liquid tank and conduits. The system operates as follows. Oil is poured through inlet 3 with a filter into the tank. Pump 1 sucks in oil from the tank through flexible conduit 2 and then pumps it via conduit 13 into slide valve 12. From the slide valve the oil flows via conduit 11 to single-acting lifting cylinder 14, then via conduits 6 and 7 to tilt cylinders 5 and via conduits 8 and 9 to cylinders (actuators) 10 of the working fittings. From the working fitting cylinders (actuators) the oil returns via the distributor and conduit 4. The forklift truck steering mechanism consists of a controller pedal, a hydraulic brake control pedal, a hand-brake lever, a travel range control lever, a steering column, an ignition switch, and a lever controlling the working fitting, tilt and mast lifting cylinders (actuators) (Fig. 6).

The lifting mechanism frame can be tilted forward by $3 \div 5°$ and backwards by $10 \div 15°$.

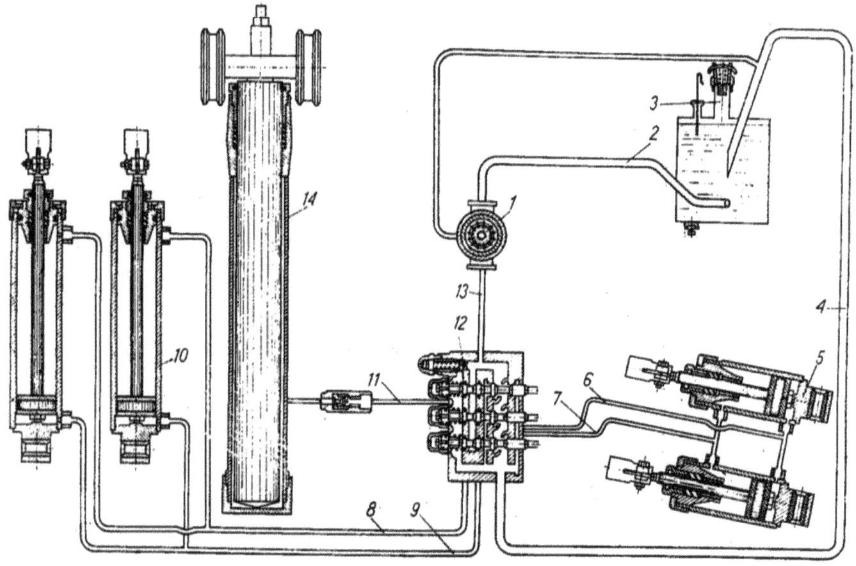

Fig. 6 Schematic of hydraulic system of forklift truck

Fig. 7 Schematic used for calculating forklift truck work

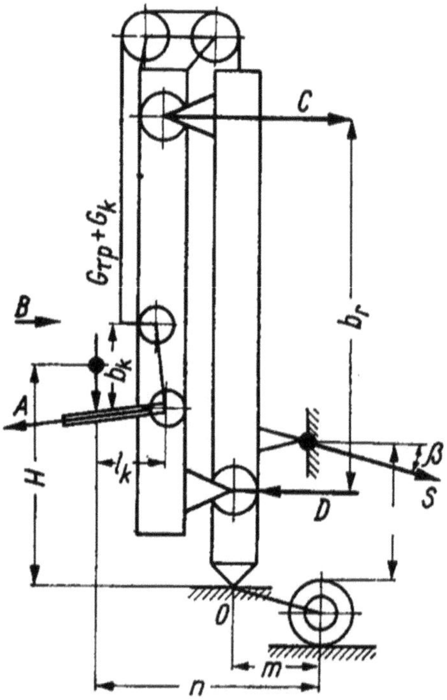

The power of the hydraulic drive in forklift trucks whose hydraulic actuators are supplied by a single pump, is calculated for the maximum force of the lifting mechanism actuator since it is impossible to combine lifting with the work of the hydraulic actuators for the other operations.

The necessary pump motor power is:

$$P = \frac{\sum W \cdot \upsilon}{2 \cdot 1000 \eta_h \cdot \eta_p} \tag{2}$$

where

- $\sum W$—the work of lifting the carriage with the fork, taking into account the friction in the guides of the carriage and the moving frame [N],
- υ—load raising speed [m/s],
- η_h—actuator efficiency (0.8 is assumed),
- η_p—pump efficiency (0.8 is assumed).

The lifting work is defined by formula (3).

$$\sum W = \frac{2(G_l + G_k)}{\eta_b} + G_r + W_t \tag{3}$$

where

- G_l—the force of gravity (weight) of the load [N],
- G_k—the force of gravity of the carriage and the working fittings [N],
- G_r—the force of gravity of the inner frame [N],
- W_t—the work of the friction forces during the movement of the carriage and the moving frame of the lifting mechanism [N].

In order to calculate the frictional resistance one should determine the support reaction to carriage rollers A and B (Fig. 5) and moving frame rollers C and D:

$$A = B = (G_l + G_k)\frac{l_k}{b_k} \tag{4}$$

$$C = D = (G_l + G_k + G_r)\frac{l_r}{b_r} \tag{5}$$

where

- l_r, l_k—the distance from the centre of gravity of respectively the load with the carriage and the carriage with the moving frame to the load gripping point (l_r can be assumed to be equal to l_k),
- b_r, b_k—the spacing of the bumper rollers of respectively the carriage and the moving frame.

Then

$$W_t = \left[2(G_l + G_k)\frac{l_k}{b_k}\left(\frac{fd}{D} + \frac{2\mu}{D}\right) + 2(G_l + G_k + G_r)\frac{l_p}{b_p}\left(\frac{fd_1}{D_1} + \frac{2\mu}{D_1}\right)\right]k[\mathrm{N}] \quad (6)$$

where

- d and d_l—the diameters of the hubs of the frame carriage bumper rollers,
- D and D_l—the diameters of the bumper rollers of the carriage and the frame,
- f—the coefficient of friction in the hubs,
- μ—the coefficient of friction of the rolling of the rollers on the guides,
- k—a coefficient representing other additional resistances (the type of wheels and their bearing system, $1.1 \div 1.3$ is assumed).

Force S on the frame tilt actuator piston rod is calculated from the equation of the moments of the forces acting on the frame relative to point 0 (i.e. the point of hinged fixation), as it tilts forwards at angle α at the upper position of the moving frame and the carriage with the load.

$$S = \frac{(G_1 + G_k + 0.75G_{rg})H_{\max}tg\alpha + G_l(n-m)}{h\cos\beta}[N] \quad (7)$$

where

- G_{rg}—the force of gravity (weight) of the main frame, including the tilt actuator,
- 0.75 and 0.25—coefficients approximately indicating the location of the centres of gravity of the frames,
- $H_{\max.}$—the maximum carriage-lifting height.

The dependence between the actuator piston rod force and the hydraulic actuator parameters can be expressed by formula (7) when the piston rod is inside the cylinder and by formula (8) when the piston rod is not inside the cylinder.

$$S = \frac{\pi\left(D_c^2 - d_t^2\right)p\eta_m}{4} \quad (8)$$

$$S = \frac{\pi D_c^2}{4}p\eta_m \quad (9)$$

where

- D_c—the cylinder diameter [cm],
- d_t—the piston rod diameter [cm],
- p—pressure [N/cm^2],
- η_m—the actuator efficiency (0.95 is assumed).

The working liquid consumption when the piston is inside the cylinder, taking into account losses, amounts to:

$$V = \frac{\pi\left(D_c^2 - d_t^2\right)v_t}{4\eta_c}\left[\mathrm{cm^3}/_{\mathrm{min}}\right] \tag{10}$$

$$V = \frac{\pi D_c^2 v_t}{4\eta_c}\left[\mathrm{cm^3}/_{\mathrm{min}}\right] \tag{11}$$

where

- v_t—the piston speed [cm/min/],
- η_c—the volumetric actuator efficiency (approximately equal to 1 when rubber and leather seals are used).

Forklift trucks incorporate hydraulic pumps with a delivery of about 80 [l/min] at about 1,700 [rpm] and a working pressure of 1,000 [kN/m^2] for mast raising/lowering and forwards/backwards tilt.

In order to translocate the unit load in the vertical in LWS, energy must be supplied to the latter. In the case of a forklift truck or a storage and retrieval machine, taking out the unit load from a given layer larger than a unit requires energy for the raising of the empty fork carriage while the lowering of the carriage necessitates the introduction of a force braking the carriage with the unit load, ensuring the safe lowering of the latter to the required level. Depending on the size of the forklift truck, the carriage weighs 100–300 [kg].

In such cases, technical systems making it possible to recover the remaining part of the potential energy of the forklift carriage and the unit load are used. Thanks to energy recovery LWS energy consumption and the energy consumption index of unit load passage through LSW are reduced.

Therefore pallet lifts and S/R machines (in some warehouse zones) are introduced into LSWs, for which the power demand can be defined by respectively formula (11) and (12).

$$N_s = \frac{kG_l v}{1000\eta}[kW] \tag{12}$$

$$N_u = \frac{\sum W \cdot v}{1000\eta}[kW] \tag{13}$$

where

- G_l—the force of gravity of the load [N],
- V—the carriage (platform) lifting speed [m/s],
- η—the efficiency of the device mechanisms,
- k—a coefficient representing the influence of the counterweight, amounting to 04–06, when there is no counterweight, $k = 1$,
- $\sum W$—the S/R machine driving resistance on the track [N], defined as for a runway truck.

The test calculations were performed for an organization chart of Fig. 2, for the carriage of the EGF-220 (detailed www.jh-online.pl).

$$\sum_t = \left[2(G_l + G_k)\cdot\frac{l_k}{b_k}\left(\frac{fd}{D}+\frac{2\mu}{D}\right) + 2(G_l + G_k + G_r)\frac{l_p}{b_p}\left(\frac{fd_1}{D_1}+\frac{2\mu}{D_1}\right)\right]\cdot k =$$

$$\begin{bmatrix} 2(19620+2943)\cdot\dfrac{500}{500}\left(\dfrac{0.09\cdot 40}{100}+\dfrac{2\cdot 0.85}{100}\right)+ \\ 2(19620+2943+1471.5)\cdot\dfrac{500}{500}\left(\dfrac{0.09\cdot 40}{100}+\dfrac{2\cdot 0.85}{100}\right) \end{bmatrix}\cdot 1.2 = 5927[N] \tag{14}$$

$$\sum W = \frac{2(G_l + G_k)}{\eta_b} + G_r + W_t = \frac{2(19620+2943)}{0.8} + 1471.5 + 5927 = 63806[N] \tag{15}$$

The calculation results indicate the following conclusions:

- Energy intensity fork lift without load is approximately 30 % less than the energy consumption of its lifting of the load.
- The energy needed for lowering the load is approximately 50 ÷ 70 % less energy than lifting the load,
- The energy needed to leave the carriage without charge is 50 % greater than the energy needed for lowering the load.

4 Conclusion

In the literature on the subject the efficiency of the forklift truck is modelled without taking into account the potential energy resulting from the vertical translocation of the unit load. The lifting capacity of the forklift truck used in logistic warehouse systems ranges from 700 kg to as much as 5 tons. It seems that if the potential energy stemming from the weight of the pallet with the load were taken into account in forklift truck work efficiency calculations and simulations, the results of the latter would be more reliable. In this chapter, relations supplementing the existing forklift truck work efficiency model have been derived. Owing to them it will be possible to improve the model and to verify the obtained results with the results of in situ experiments. The relevant research is being conducted at Wrocław University of Technology.

References

Bangsow S (ed) (2012) Use cases of discrete event simulation appliance and research. Springer, Heidelberg

Bangsow S (2010) Manufacturing simulation with plant simulation and simtalk usage and programming with examples and solutions. Springer, Heidelberg

Bednarz ŁJ, Górski A, Jasieńko J, Rusiński E (2011) Simulations and analyses of arched brick structures. Autom Constr 20(7):741–754

Bujak A, Zając P (2012) Can the increasing of energy consumption of information interchange be a factor that reduces the total energy consumption of a logistic warehouse system? In: Mikulski J (ed) Communications in computer and information science, vol. 329, Springer, Heidelberg, pp 199–210, ISSN 1865-0929

European Energy Efficiency (2012) Analysis of ODYSSEE indicators, Feb 2012

Eurostat http://epp.eurostat.ec.europa.eu/portal/page/portal/eurostat/home/, http://epp.eurostat.ec.europa.eu/portal/page/portal/energy/data/database

Homburg C (2007) Betriebswirtschaftslehre als emprirische Wissenschaft—Bestandsaufnahme und Empfehlungen. Zeitschrift für betriebswirtschaftliche Forschung 56(7):27–60

Kemme N (2013) Design and operation of automated container storage system. Springer, Heidelberg

Kwaśniowski S, Zając M, Zając P (2011) Telematic problems of unmanned vehicles positioning at container terminals and warehouses. In: Mikulski J (ed) Communications in Computer and Information Science, vol 104, Springer, Heidelberg, pp 391–399, ISSN 1865-0929

Law AM, Kelton WD (2000) Simulation modelling and analysis, 3rd edn. McGraw Hill, Boston

Logistics Cost and Service (2007) Establish, Inc Herbert W Davis and Company, http://www.establishinc.com/pdfs/2007_CSCMP_Presentation.pdf Accessed Jun 30 2013

Makris PA, Makri AP, Provatidis CG (2006) Energy-saving methodology for material handling application, Appl Energy 83: 1249–1257 www.sciencedirect.com

Ören T (2011) The many facets of simulation through a collection of about 100 definitions, quarterly society for modeling & simulation magazine. Soc Model Simul Int 2: 142–151

Tompkins J, White J, Bozer Y, Tanchoco J (2010) Plant simulation product description, In: Facilities planning, 4th edn, Wiley, New York UGS Corp 2006

Zając P (2011a) Can the raising of energy consumption of information interchange be a factor that reduces the total energy consumption of logistic warehouse system?, In: Chlebus E (ed) International Conference "Production Engineering 2011", Wrocław, 30 June–1July 2011, pp 79–89

Zając P (2010) Computer control of AGV in the warehouse and on the terminal's. In: Fertsch M (ed) Innovative and intelligent manufacturing systems, Publishing House of Poznan University of Technology, Poznań, Poland, pp 273–284

Zając P (2011b) The choice of parameters of logistic warehouse system, with taking the energy into consideration. In: Grzybowska K, Golińska P (eds) Selected logistics problems and solutions. Publishing House of Poznan University of Technology, Poznań, Poland, pp 107–120

Zając P (2011c) The idea of the model of evaluation of logistics warehouse systems with taking their energy consumption under consideration. Arch Civil Mech Eng 11(2):479–492

Zając P (2010) Transport-storage system optimization in terms of exergy. In: Bruzzone A (ed) Proceeedings of the 13th international conference on harbor maritime multimodal logistics modeling and simulation, HMS 2010, Fes, Morocco, Oct 2010, Marseille: Laboratoire des Sciences de l'Information et des Systèmes 2010, pp 143–148

The Integration of Environmental Foot-Printing Strategies to the Capacitated Warehouse Location Problem with Risk Pooling

Noura Al Dhaheri, Maria Polo Alvez and Shin Ju-Young

Abstract In recent decades, academic and corporate interest in sustainable supply chain management has increased substantially due to international levitation efforts toward minimizing the greenhouse gas emissions. These international determinations to combat climate change have encouraged countries to put mechanisms in place ranging from taxes, permits and voluntary incentives to required regulatory policies, which have created a challenge for companies of all sizes when it comes to managing their supply chain. To establish the field further, this research aims to integrate environmental foot-printing strategies to efficient supply chain design. This can be done using a two-fold approach. The first is to consider the closed-loop supply chain theory in designing the supply chain and the second is to incorporate carbon trading schema into the costs aspects of the design. To manage a supply chain that is not only efficient, but encourages sustainability we propose a capacitated warehouse location problem with risk pooling (CLMRP), which experiences stochastic demand for a single product, to be the backbone for our closed-loop model with carbon trading schema. This green tactical-operational level scheme is formulated as a mixed integer nonlinear mathematical program and solved using the General Algebraic Modeling System (GAMS). The computational results and the sensitivity analysis are provided to verify our design.

Keywords Sustainable supply chain management · Closed-loop supply chain · Capacitated-warehouse location problem · Risk pooling

N. Al Dhaheri (✉) · M. P. Alvez · S. Ju-Young
Department of Engineering Systems and Management, Masdar Institute of Science and Technology, Abu Dhabi, UAE
e-mail: naldhaheri@masdar.ac.ae

P. Golinska (ed.), *Logistics Operations, Supply Chain Management and Sustainability*, EcoProduction, DOI: 10.1007/978-3-319-07287-6_34,

1 Introduction

In recent years increasing environmental concerns have led companies to undertake measures to integrate sustainable options into their supply chain networks. Treating these two concepts simultaneously helps not only to reduce the environmental economic impact of supply chains but also to decrease costs by using resources more efficiently and minimizing emission. To begin with, extending the forward supply chain to a reverse supply chain allows incorporating returns processes. Product acquisition from the end-users, reverse logistics, sorting, remanufacturing and recycling are some of the activities that can be included in the closed-loop supply chain (Guide et al. 2003), thus making it more sustainable. Furthermore, consideration of carbon taxing or of a carbon trading scheme can promote lower emissions, as companies aim to reduce these costs and through this they establish greener or more cost-effective supply chains.

Another important contribution in this direction is the joint consideration of location and inventory decisions. Even though supply chain management has traditionally considered these decisions separately, recently there is an increasing interest in combining the strategic level decision of location with the tactical level decision of inventory. This is due to the fact that the simultaneous study of these two aspects leads to improved management of logistics networks, as it allows for a more efficient operation of the supply chain, minimizing transportation, carrying inventory and fixed costs (Ozsen et al. 2008). Finally, when addressing problems that involve retailers facing stochastic demand, risk pooling can be considered, in order to reduce the effect of demand variability by decreasing safety stock and therefore reducing average inventory (Berman et al. 2010).

In this chapter we address the capacitated closed-loop facility location problem with risk pooling (CLMRP) and extend the model proposed by Diabat et al. by considering the CO_2 emissions generated along the forward and the reversed supply chain. The CLMRP encompasses the location and inventory management problems. This optimization problem minimizes the total cost incurred along the supply chain, including transportation, inventory, and location fixed costs. The model provides the optimal warehouse locations, order quantity, and reordering point. In order to assess the impact of a carbon market trading framework on the closed-loop supply chain strategic design and total cost, we conduct a sensitivity analysis using the commercial software GAMS.

In the following sections, we will first provide an overview of the literature addressing joint location and inventory models, closed-loop supply chains as well as the application carbon trading schemes in the supply chain. In the next section, the characteristics of the proposed model are outlined and the mathematical formulation is presented. The following section describes the sensitivity analysis performed, as well as the computational results derived from this. Finally, we conclude our work with important findings and suggest possible future work.

2 Literature Review

In this section we highlight the current literature that address the integrated location-inventory decisions as well as the incorporation of sustainability in the supply chain, both by implementing closed-loop supply chains as well as by applying carbon trading schemes. Al Zaabi et al. analyzed different barriers to the successful incorporation of a green supply chain (Al Zaabi et al. 2013).

As far as location and inventory models are concerned, most literature on the supply chain field has studied them independently. It was not until recently that the joint location-inventory problems gained attention from researchers. In the early work of (Barahona and Jensen 1998), the authors study the design of a distribution network for computer spare parts. They present an integer programming model for plant location considering inventory costs and solve it using the Dantzig-Wolfe decomposition. Diabat et al. present an integrated location-inventory model regarding a single plant distributing to multiple retailers (Diabat et al. 2009). As the retailers face stochastic demand risk pooling is taken into consideration and the authors implement a genetic algorithm to solve the problem. Diabat et al. evaluate a multi-echelon joint inventory-location model which is formulated as a nonlinear mixed-integer program, and is solved using a Lagrangian Relaxation-based approach (Diabat et al. 2013) and in later work Diabat et al. use a new Lagrangian Relaxation-based heuristic capable of solving larger size problems (Diabat et al. 2014). Al Dhaheri and Diabat model a capacitated inventory-location problem with multiple products and risk pooling considering carbon emissions (Al Dhaheri and Diabat 2011). Finally Tung et al. formulate an inventory routing problem to minimize both inventory and transportation costs and solve it by using a column generation-based approach (Le et al. 2013).

In terms of developing a closed-loop supply chain, two distinct and yet related fields are evolving in this area, reversed logistics and closed-loop supply chain (Dekker et al. 2004). While regular supply chain research considers the movement of a product from supplier to end-user, the reversed logistics considers the collection and recovery of used product from the end-user to the producer. Integrating the former with the later concepts led to the evolution of the closed-loop supply chain (Voß and Woodruff 2006). In this regard, certain notable papers are presented. Abdallah et al. present an un-capacitated closed-loop model that integrates the facility location and inventory problems in both the forward and reverse supply chain (Abdallah et al. 2012), while Amin and Zhang design a general closed-loop supply chain network based on a three-stage model which includes evaluation, network configuration (Amin and Zhang 2013), and selection and order allocation. Özceylan and Paksoy propose a closed-loop supply chain network that includes both forward and reverse flows with multi-periods and multi-parts (Özceylan and Paksoy 2013). They formulate a mixed integer mathematical model to find the optimal values of transportation amounts of manufactured and disassembled products in a closed-loop supply chain while determining the location of plants and retailers.

Carbon trading has been dealt with by Fahimnia et al., in which the authors develop a linear programming model to assess the impact of carbon pricing on a closed-loop supply chain (Fahimnia et al. 2013). Diabat et al. consider remanufacturing in a multi-echelon multi-commodity facility location problem for a closed-loop supply chain (Diabat et al. 2013). They implement a carbon cap scheme according to which a cost is incurred if the company's emissions exceed the cap, but on the other hand it allows for profit if the emissions are below the cap. Diabat and Simchi-Levi formulate a mixed-integer program to optimize the strategy that companies should follow in order to meet their carbon cap (Diabat and Simchi-Levi 2009).

The above review shows that most of the literature has addressed the integrated supply chain, closed-loop supply chain and carbon emissions trading problems separately. The innovation involved in this chapter lies in the joint analysis of these three problems. It implements a carbon trading scheme into a closed-loop supply chain network that considers the integrated capacitated inventory-location model with risk pooling.

3 Problem Description and Formulation

The CLSC problem under study is illustrated by Fig. 1. In the forward supply chain, a single product is produced by a single manufacturer either from raw material or from recycled material, and then travels to set end-users (e) via set of capacitated warehouses (ω). In the reverse supply chain, the end-of-life (EOL) products are collected by the capacitated collection centers (c), where they are inspected and separated into non-recyclable and recyclable items. The non-recyclable items have to be sent to disposal centers (d) of known capacity to be disposed. Whereas, the recyclable items are shipped to the recycling centers (r) to be processed into recycled material. Then there recycled materials are sent to the manufacturer to be reused to produce products from recycled material. This problem can be formulated using MINLP that aims to determine the tactical-operational supply chain planning decisions so as to minimize the overall supply chain cost and the economic impact of emissions on the supply chain design.

3.1 Assumptions

Assumptions in the proposed model include the following:

- Number, location and capacity of warehouses (ω), collection centers (c), disposal centers (d) and recycling centers (r) are known.
- Number and location of end-users (e) (i.e. customer zones or retailers) are known.
- The daily demand of end-users (e) is assumed to be uncorrelated over time and across end-users (e) and follows a Poisson process, which implies that the

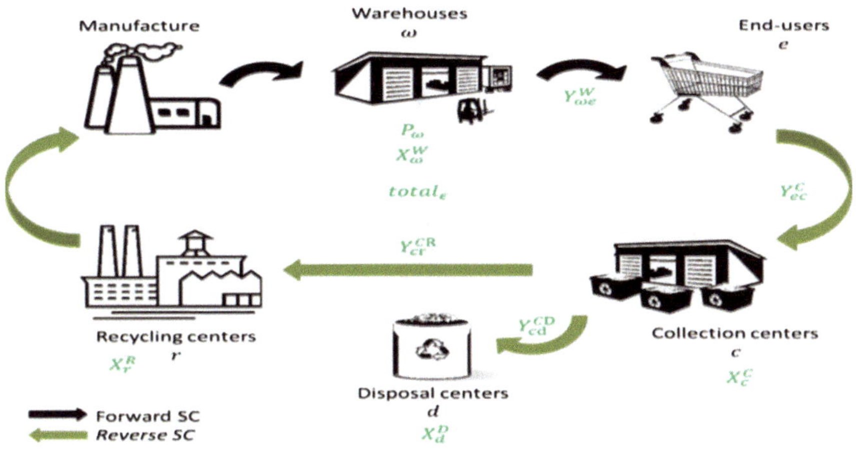

Fig. 1 Decision variables in the CLSC under investigation

variance of daily demands σ_2^e and the mean μ_e are equal for each end-user (e) (Al Dhaheri and Diabat 2011).

- To meet the demand of each end-user (e) the warehouse (ω) holds working inventory and a safety stock (Al Dhaheri and Diabat 2011).
- The quantity of returned EOL from each end-user (e) is assumed to be a fraction of the daily demand $(\breve{\mu}_e = \kappa\mu_e)$.
- There is a fixed fraction of the total EOL collected to be disposed (δ^x) and the remaining part will be recycled.
- The selling price of carbon emission, which equals the penalty of exceeding the emission cap for the overall supply chain, is known.

Following the notations provided in the Appendix section, the formulation of the model is presented in this section. First we show the objective function divided into two main parts: the CLSC cost, which comprises both forward and reverse cost elements, and the emissions trading. Then the suggested constraints are grouped into three groups to simplify understanding their relations.

3.2 Formulation of the Objective Function

The objective is to minimize the overall CLSC cost of opening and operating the facilities and distributing the products, as well as to minimize the emissions cost, which can be turned into profits gains if not exceeded the imposed emissions cap.

The overall cost of CLSC can be decomposed further into 11 cost elements. First we specify the cost of forward supply chain in Eqs. 1–4. After that the cost of reverse supply chain is expressed in Eqs. 5–8. Finally the emission trading is expressed in Eqs. 9–11.

3.2.1 The Cost of the Forward Supply Chain

$$\text{cost}_1 = \underbrace{\sum_{W} o_{\omega}^{W} X_{\omega}^{W}}_{\text{fixed cost of } \omega} \tag{1}$$

$$\text{cost}_2 = \underbrace{\sum_{W} \tau_{\omega} \sum_{E} \frac{\xi Y_{\omega e}^{W} \mu_e}{P_{\omega}}}_{\text{fixed cost of shipping orders}} + \underbrace{\sum_{W} \sum_{E} \tau_{\omega}^{v} \xi Y_{\omega e}^{W} \mu_e}_{\text{cost unit shipped to } \omega} + \underbrace{\sum_{W} \tau_{\omega e}^{vW} \sum_{E} \xi Y_{\omega e}^{W} \mu_e}_{\text{cost per unit shipped to } e} \tag{2}$$

$$\text{cost}_3 = \underbrace{\sum_{W} \frac{h_{\omega}^{W} P_{\omega}}{2}}_{\text{average cost holding } P_{\omega}} + \underbrace{\sum_{W} h_{\omega}^{W} z_{\alpha} \sqrt{\sum_{E} Y^{W} \mu_e l_{\omega}}}_{\text{holding stock cost}} + \underbrace{\sum_{W} g_{\omega} \sum_{E} \frac{\xi Y_{\omega e}^{W} \mu_e}{P_{\omega}}}_{\text{ordering cost}} \tag{3}$$

The total cost of the forward supply chain can be calculated by

$$\text{cost}_{\text{FSC}} = \text{cost}_1 + \text{cost}_2 + \text{cost}_3 \tag{4}$$

3.2.2 The Cost of the Reverse Supply Chain

$$\text{cost}_4 = \underbrace{\sum_{C} o_{c}^{C} X_{c}^{C}}_{\text{cost of opening c}} + \underbrace{\sum_{D} o_{d}^{D} X_{d}^{D}}_{\text{cost of opening } d} + \underbrace{\sum_{R} o_{r}^{R} X_{r}^{R}}_{\text{cost of opening } r} \tag{5}$$

$$\text{cost}_5 = \underbrace{\sum_{C} \rho_{c}^{C} \sum_{E} \xi Y_{ec}^{C} \breve{\mu}_e}_{\text{cost of collecting EOL}} + \underbrace{\sum_{D} \rho_{d}^{x} \sum_{C} Y_{cd}^{D} \delta^{x} \sum_{E} \xi Y_{ec}^{C} \breve{\mu}_e}_{\text{cost of opening disposing}\}} + \underbrace{\sum_{R} \rho_{r}^{R} \sum_{C} Y_{cr}^{R} (1-\delta^{x}) \sum_{E} \xi Y_{ec}^{C} \breve{\mu}_e}_{\text{cost of recycling}} \tag{6}$$

$$\begin{aligned} \text{cost}_6 = & \underbrace{\sum_{C} \sum_{E} \tau_{ec}^{E} \xi \breve{\mu}_e Y_{ec}^{C}}_{\text{per unit cost (puc) of EOL from e to c}} + \sum_{D} \underbrace{\sum_{C} \tau_{cd}^{X} Y_{cd}^{D} \delta^{x} \sum_{E} \xi \breve{\mu}_e Y_{ec}^{C}}_{\text{puc of products from c to d}} \\ & + \sum_{R} \sum_{C} \underbrace{\tau_{cr}^{R} Y_{cr}^{R} (1 - \delta^{x}) \sum_{E} \xi \breve{\mu}_e Y_{ec}^{C}}_{\text{puc of recyclable from c to r}} + \sum_{R} \tau_{r}^{RM} \underbrace{\sum_{C} Y_{cr}^{R} \eta (1 - \delta^{x}) \sum_{E} \xi \breve{\mu}_e Y_{ec}^{C}}_{\text{puc of recycled from r to plant}} \end{aligned} \tag{7}$$

The total cost of the reverse supply chain can be calculated by:

$$\mathrm{cost}_{\mathrm{FSC}} = \mathrm{cost}_4 + \mathrm{cost}_5 + \mathrm{cost}_6 \tag{8}$$

3.2.3 The Emissions Trading Scheme in the CLSC

$$\mathrm{cost}_{\mathrm{emission}} = \underbrace{\varepsilon(E^t + E^{\circ} - Cap_{\varepsilon})}_{\text{carbon emissions trading}} \tag{9}$$

$$E^{\circ} = \underbrace{\varepsilon_{fe}\Big(\sum_{W} X_{\omega}^{W} C_{\omega}^{W} v^{p} + \sum_{C} X_{c}^{C} C_{c}^{C} v^{EOL} + \sum_{D} X_{d}^{D} C_{d}^{D} v^{D} + \sum_{R} X_{r}^{R} C_{r}^{R} v^{R}\Big)^{1+\oplus}}_{CO_2\ emission\ \text{from warehouses, collection, disposal and recycling centers}} \tag{10}$$

$$\begin{aligned} E^t = \varepsilon_{ft} & \underbrace{(\mathrm{m}^{\mathrm{P}} \sum_{W} d_{\omega} \sum_{E} \xi Y_{\omega c}^{W} \mu_e)}_{\text{from shipping products to } \omega} + \underbrace{(\mathrm{m}^{\mathrm{P}} \sum_{W} \sum_{E} d_{\omega e}^{WE} \xi Y_{\omega e}^{W} \mu_e)}_{\text{from shipping products to } e} + \underbrace{(\mathrm{m}^{\mathrm{EOL}} \sum_{E} \sum_{C} d_{cr}^{EC} \xi \breve{\mu}_e Y_{ec}^{C})}_{\text{from shipping to c}} \\ & + \underbrace{(\mathrm{m}^{\mathrm{D}} \sum_{C} \sum_{D} d_{cd}^{CD} Y_{cd}^{D} \delta^{x} \sum_{E} \xi \breve{\mu}_e Y_{ec}^{C})}_{\text{of shipping nonrecyclable products c to } d} + \underbrace{(\mathrm{m}^{\mathrm{R}} \sum_{C} \sum_{R} d_{cr}^{CR} Y_{cR}^{R} (1 - \delta^{x}) \sum_{E} \xi \breve{\mu}_e Y_{ec}^{C})}_{\text{of shipping recyclable products from c to } r} \\ & + \underbrace{(\mathrm{m}^{\mathrm{RR}} \sum_{C} \sum_{R} d_{r}^{R} Y_{cr}^{R} \eta (1 - \delta^{x}) \sum_{E} \xi \breve{\mu}_e Y_{ec}^{C})}_{\text{from shipping products from r to plant}} \end{aligned} \tag{11}$$

3.3 *Formulation of the Constraints*

The constraints are grouped into three interrelated sets: the first set comprises the constraints of the forward supply chain (Eqs. 12–17), the second comprises the reverse supply chain (Eqs. 18–25) and the third is for restrictions on the decision variables (Eqs. 26–32).

3.3.1 Constraints of the Forward Supply Chain

$$\underbrace{\sum_{W} Y_{\omega e}^{W} = 1}_{\text{single supplier } \omega \text{ to e}}, \quad \forall e \in \mathrm{E} \tag{12}$$

$$\underbrace{Y^W_{\omega e} \leq X^W_{\omega} 1}_{\text{assignment to open warehouse}}, \quad \forall \omega \in \mathrm{W}; \forall \mathrm{e} \in \mathrm{E} \tag{13}$$

$$(\mathrm{P}_\omega + z_\alpha \underbrace{\sqrt{\sum_E l_\omega Y^W_{\omega e} \mu_e} + l_\omega \sum_E Y^W_{\omega e} \mu_e \leq C^W_\omega}_{\text{warehouse capacity}}, \quad \forall \omega \in \mathrm{W} \tag{14}$$

$$P_{EOQ\omega} = \underbrace{\sqrt{\frac{2(g^N_\omega + \tau_\omega) \sum_E \xi \mu_e}{h^W_\omega}}}_{\text{optimal economical ordering quantity}}, \quad \forall \omega \in \mathrm{W} \tag{15}$$

$$P_{capacity_\omega} = \underbrace{C^W_\omega - [z_\alpha - z_p] \sqrt{\sum_E Y^W_{\omega e} \mu_e l_\omega}}_{\text{optimal capacity ordering quantity}}, \quad \forall \omega \in \mathrm{W} \tag{16}$$

$$P_\omega = \underbrace{\min(P_{EOQ\omega}, P_{capacity_\omega})}_{\text{optimal quantity}}, \quad \forall \omega \in \mathrm{W} \tag{17}$$

.

3.3.2 Constraints of the Reversed Supply Chain

$$\underbrace{Y^C_{ec} \leq X^C_c}_{\text{collection center assignment to open collection center}}, \quad \forall \mathrm{e} \in \mathrm{E}; \forall \mathrm{c} \in \mathrm{C} \tag{18}$$

$$\underbrace{Y^R_{cr} \leq X^R_r}_{\text{recycling center assignment to open recycling center}}, \quad \forall \mathrm{r} \in \mathrm{R}; \forall \mathrm{c} \in \mathrm{C} \tag{19}$$

$$\underbrace{\sum_E \xi \breve{\mu}_e Y^C_{ec} \leq X^C_c C^C_c}_{\text{collection center capacity}}, \quad \forall \mathrm{c} \in \mathrm{C} \tag{20}$$

$$\underbrace{\sum_C Y^C_{ec} = 1}_{\text{single assignment e to c}}, \quad \forall \mathrm{e} \in \mathrm{E} \tag{21}$$

$$\underbrace{\sum_C \delta^x Y^D_{cd} \sum_E \xi \breve{\mu}_e Y^C_{ec} \leq X^D_d C^D_d}_{\text{disposal center capacity}}, \quad \forall \mathrm{d} \in \mathrm{D} \tag{22}$$

$$\underbrace{\breve{\mu}_e = \kappa \mu_e}_{\text{fraction of returned products}}, \quad \forall \mathrm{e} \in \mathrm{E} \tag{23}$$

$$\mathrm{v^R}\sum_{C} Y^R_{cr}(1-\delta^x)\sum_{E}\xi\breve{\mu}_e Y^C_{ec} + \underbrace{\mathrm{v^{RM}}\sum_{C} Y^R_{cr}\eta(1-\delta^x)}_{\text{recycling center capacity}}\sum_{E}\xi\breve{\mu}_e Y^C_{ec} \le C^R_r X^R_r, \quad \forall \mathrm{r} \in \mathrm{R} \tag{24}$$

$$\underbrace{\sum_{E}\sum_{C}\xi\breve{\mu}_e Y^C_{ec}}_{\text{total returned EOL}} = \underbrace{\sum_{C}\sum_{D} Y^D_{cd}\delta^x \sum_{E}\xi\breve{\mu}_e Y^C_{ec}}_{\text{total disposed products}} + \underbrace{\sum_{C}\sum_{R} Y^R_{cr}(1-\delta^x)\sum_{E}\xi\breve{\mu}_e Y^C_{ec})}_{\text{total recycling products}} \tag{25}$$

3.3.3 Restrictions on Decision Variables

$$X^W_\omega \in \{0,1\} \quad \forall \omega \in \mathrm{W} \tag{26}$$

$$Y^W_{\omega e} \in \{0,1\} \quad \forall \omega \in \mathrm{W};\ \forall \mathrm{e} \in \mathrm{E} \tag{27}$$

$$Y^C_{ec} \in \{0,1\} \quad \forall c \in \mathrm{C};\ \forall \mathrm{e} \in \mathrm{E} \tag{28}$$

$$X^C_c \in \{0,1\} \quad \forall c \in C \tag{29}$$

$$X^D_d \in \{0,1\} \quad \forall d \in \mathrm{D} \tag{30}$$

$$X^R_r \in \{0,1\} \quad \forall \mathrm{r} \in \mathrm{R} \tag{31}$$

$$P_\omega \ge 0 \quad \forall \omega \in \mathrm{W} \tag{32}$$

4 Computational Results and Sensitivity Analysis

In this section we analyze the effect that the carbon trading scheme and the recycling policy have on the supply chain tactical and operational decisions. The sensitivity analysis was performed using GAMS.

A 49-city dataset obtained from Daskin (1995) was used to validate the model. However, due to the computation limitation of the software in use, no optimal or near optimal solution was found. As a result, only the 5 cities of the data set were selected in order to carry out the analysis. Each city represents a potential location for a warehouse and an end-user, as well as for a collection center, disposal center and recycling center. The parameters used to test this model were determined as follows.

The mean of the daily demand of each of the end-users is computed as the total demand of each city divided by the total number of days in 1 year, which is

Table 1 The values of main system parameters

$\tau_{\omega} = \$5$	$z_{\alpha} = 1.96$
$Cap_{\varepsilon} = 10{,}000$ tons	$z_p = 0$
$l_{\omega} = 1$ day	$m^R = 2$ kg
$m^{RR} = 1$ kg	

assumed to be 250 days. The mean of the daily return of EOL from each end-user is known and its value is set to 0.5 (50 %). The maximum capacity and holding cost of each warehouse are assumed to be proportional to its fixed cost, while the maximum capacity of each collection, disposal and recycling center is determined as a proportion of the maximum capacity of the warehouse in the same city. That is 0.5, 0.25 and 0.25 respectively. The opening costs of the warehouses are considered to be fifty percent of their fixed costs, which are given in the dataset. In addition, the opening costs of the collection, recycling and disposal centers are obtained by dividing the opening cost of the warehouse in the same city by 2. The variable shipment costs are proportional to the distance between the two locations. The values chosen for the main parameters are shown in Table 1.

4.1 Effect of CO_2 Price on Supply Chain Total Cost

For this analysis we examined two scenarios. In the first scenario the CO_2 emissions cost is not considered, while in the second one it is. We varied the CO_2 price and analyzed its effect on the supply chain cost components. That is, on the transportation cost in both the forward and reverse supply chains as well as the costs derived from operating the warehouses and collecting, recycling and disposal centers.

4.2 Effect of CO_2 Cap on the Total Emission Produced

As shown from Fig. 2, the total cost of the CLSC decreases when the CO_2 price increases. This could be attributed to the trading mechanism proposed: the higher the carbon price, the more incentivized are the CLSC agents to cut emissions and boost profits by selling carbon emission rights. However, when the price is set too high, the system reaches its maximum possible emission reduction and is forced to endure the high costs of emission.

When we look at total emission numbers, for a fixed carbon price and return fraction, tighter emission cap will lead to lower actual carbon emissions. Furthermore, the breakdown of CO_2 emissions show an increase in the share of Forward supply chain emissions as the cap is tightened as shown in Fig. 3. This is due to the fact that forward supply emissions are more likely to stay fixed because

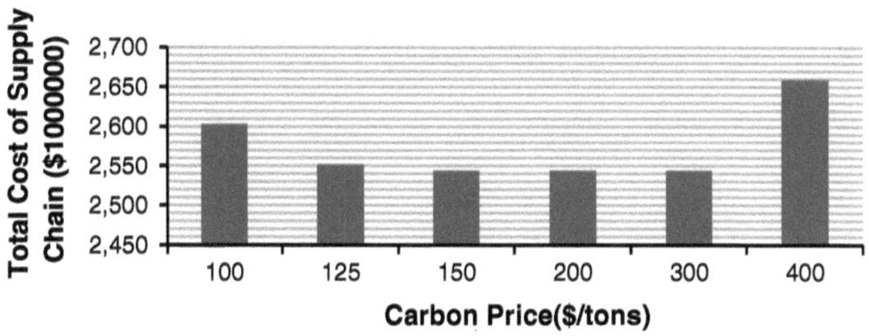

Fig. 2 Effect of carbon price on the supply chain total cost

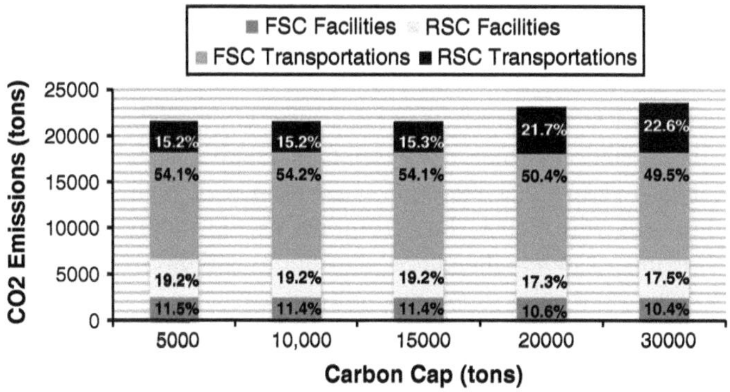

Fig. 3 Effect of carbon emissions cap on actual carbon emissions

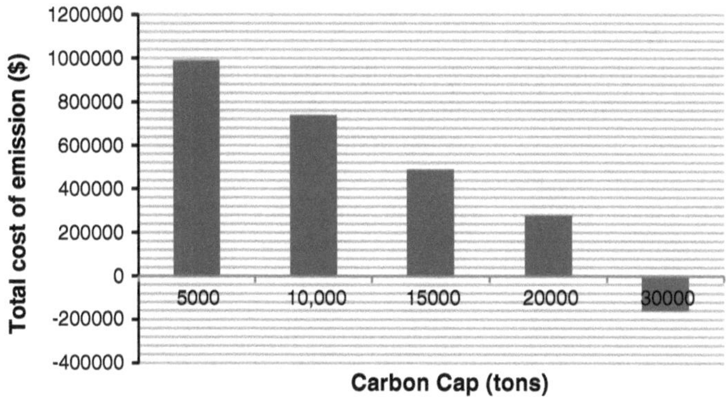

Fig. 4 Carbon cap effect on carbon emissions cost

they depend more on external economic factors (such as EOQ, storage costs, ordering cost etc.) and less on the cap itself. Meanwhile, RSC Transportation emissions do vary with the Cap. As a result, when the total cap drops, FSC emissions will emissions will vary little (which increases their relative share) while

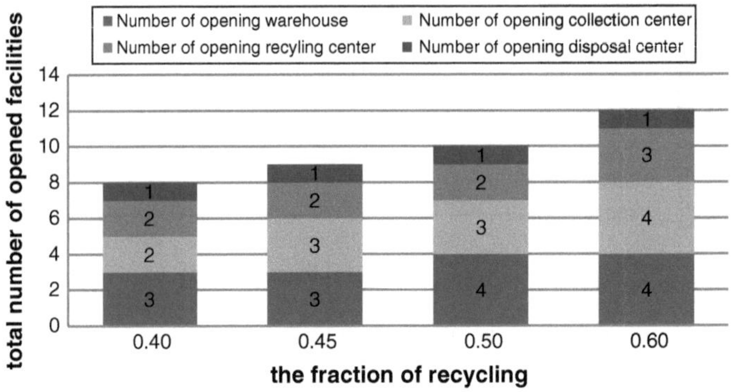

Fig. 5 Effect of recycling fraction on opening facilities

the more variable RSC Transportation emissions will bear the greatest part of the total emission reduction (which decreases their relative share).

Finally, Fig. 4 shows that emission costs drop with higher carbon caps. The more flexible caps will reduce the constraints on the supply chain's profit optimization allowing for more efficient SC allocations and offering the opportunity to generate revenue from sales of carbon emission rights.

4.3 Effect of Proportion of Returned Products

The recycling policy stipulates the proportion of products that must be returned by the end-users. We set the CO_2 emissions cap to 27,000 tons and varied the value of recycling policy in order to study its effect on the supply chain. Figure 5 shows that, as expected, the system opens more facilities in the reverse supply chain when the fraction increases to accommodate for the increasing returns.

We also studied the effect of the recycling fraction on total emissions and overall CLSC cost. Similarly, we found that a higher return fraction will lead to both higher emissions and overall CLSC costs.

5 Conclusion

Recently, the integration of environmental strategies into the design of supply chain has been receiving heightened attention, both practically and theoretically. This is happening as a result of the dual external effect of growing government regulation and escalating public demands for environmental accountability. In addition, companies have an internal self-interest in cost-effective supply chain design. This chapter proposed one way to consider such an integration taking into consideration more efficient inventory policies. We developed a mixed integer nonlinear programing model to model the capacitated—inventory location problem with risk pooling within the framework of a closed loop supply chain with carbon trading. The impacts of carbon trading on both a standard forward SC and a CLSC were evaluated. Parametric analysis was showed over several carbon prices and emission limits under real case scenarios. The development of such optimization models is predominantly valuable for companies, researchers and policy makers. Several important results from the primary analysis have been illustrated. First, we confirmed that, although not intuitive on the short term, tighter carbon emission caps will reduce the overall system's emissions, even after accounting for all trading opportunities. Second, we showed that under the carbon trading scheme, firms will take advantage of higher carbon prices by cutting down on emissions increasing their profit (up to a certain limit). Finally, we showed that the imposition of higher recycling fractions will lead to increases in the number of reverse facilities and total CLSC costs, but also in total emissions.

We studied the implications of some significant factors such as the rate at which products are returned, and the capacity of reverse operations, the carbon prices and cap. Further research could investigate the inventory policy in the revers supply chain and the effect of virgin raw material and recycled material in both the production cost and emission. The opportunities of including different transportation modes and routes can also be considered in future work.

A.1 Appendix

A.1.1 Notation

Sets

Set of potential warehouses	W indexed by ω
Set of end-users	E indexed by e
Set of potential collection centers	C indexed by c
Set of potential disposal centers	D indexed by d
Set of potential recycling centers	R indexed by r

Parameters

Fixed Cost of Opening and Operating

o_ω^W	Fixed cost of opening and operating warehouse (ω) ($)
o_c^C	Cost of opening and operating collection center (c) ($)
o_d^D	Cost of opening and operating disposal center (d) ($)
o_r^R	Cost of opening and operating recycling center (r) ($)

Revers Supply Chain Processes Cost

ρ_c^C Cost of collecting, inspecting and separating a unit of EOL ($/unit)
ρ_d^X Cost of disposing a unit of non-recyclable EOL ($/unit)
ρ_r^R Cost of recycling a unit recyclable EOL ($/unit)

Mean of Daily Demand and Daily Returns

μ_e Mean daily demand of products at end-user (e) (unit/day)
σ_2^e Variance daily demand at end-user (e) (unit^2/day)
$\breve{\mu}_e$ Mean of daily returns of EOL at end-user (e) (unit/day)

Transportation Cost (Fixed and Variables)

τ_ω Fixed cost of shipping an order to warehouse (ω) ($)
τ_ω^v Cost per unit shipped from manufacturer to warehouse (ω) ($/unit)
$\tau_{\omega e}^{vW}$ Cost per unit shipped from warehouse (ω) to end-user (e) ($/unit)
τ_{ce}^E Cost per unit shipped from end-user (e) to collection center (c) ($/unit)
τ_{cd}^X Cost per unit shipped from collection center (c) to disposal center (d) ($/unit)
τ_{cr}^R Cost per unit shipped from collection center (c) to recycling center (r) ($/unit)
τ_r^{RM} Cost per unit shipped from recycling center (r) to manufacturer ($/unit)

Holding Cost and Ordering Cost

h_ω^W Annual holding cost per item of products at warehouse (ω) ($)
g_ω Fixed cost of ordering from manufacturer by warehouse (ω) ($)

Capacities

C_ω^W Max capacity of warehouse (ω) for products (unit)
C_c^C Max capacity of collection center (c) for recyclable EOL and non-recyclable EOL (unit)
C_d^D Max capacity of disposal center (d) (unit)
C_r^R Maximum capacity of recycling center (r) in (m^3)

Fractions of Recycling, Disposing, Remanufacturing

δ^x Fraction of EOL that must be disposed
κ Fraction of daily returns
η Fraction of recyclable EOL that will be transformed into recycled material (%)
v^P The volume of a unit of product (m^3/unit)
v^{EOL} The volume of a unit of EOL (m^3/unit)
v^D The volume of a unit of disposal product (m^3/unit)
v^R The volume of a unit of recyclable EOL (m^3/unit)

Emissions

Cap_{ε}	Maximum emission cap on the overall CLSC
E	Emission trading price
m^P	Weight per unit of product (tons)
m^{EOL}	Weight per unit of EOL (tons)
m^D	Weight per unit of disposed material (tons)
m^R	Weight per unit of recyclable EOL (tons)
m^{RR}	Weight per unit of recycled (tons)
d_{ω}	Distance from manufacturer to warehouse (ω) (km)
$d^{WE}_{\omega e}$	Distance from warehouse (ω) to end-user (e) (km)
d^{EC}_{ec}	Distance from end-user (e) to collection center (c) (km)
d^{CD}_{r} cd	Distance from collection center (c) to disposal center (d) (km)
d^{CR}_{cr}	Distance from m collection center (c) to recycling center (r) (km)
d^{R}_{r}	Distance from recycling center (r) to manufacturer (km)
ε_{ft}	Average CO_2-emission factor per ton-km (g CO_2 per tons-km /1,000,000)
ε_{fe}	Average CO_2-emission factor per KWh (tons CO_2/KWh) ϕ was removed, only 1 factor
Θ	Coefficient for the electric consumption in location (ω), (c), (d) and (r)

General System Parameters

ξ	Number of days per year (250)
α	The probability of that stock outs occur
p	The probability that the inventory accumulation at DC exceeds the capacity of that DC
z_{α}	Standard normal deviation such that $\Pr ob(z \le z_{\alpha}) = \alpha$
z_p	Standard normal deviation such that $\Pr\ ob(z \le z_p) = p$
l_{ω}	Lead time for product orders to be shipped from manufacturer to warehouse (ω) (unit time)

A.1.2 Decision Variables

Binary Variables

$$X^{W}_{\omega} = \begin{cases} 1 & \text{if we opened a warehouse } \omega \\ 0 & \text{otherwise} \end{cases}$$

$$Y^{W}_{\omega e} = \begin{cases} 1 & \text{if we assigned a warehouse } \omega \text{ to serve end-user e} \\ 0 & \text{otherwise} \end{cases}$$

$X_c^C = \begin{cases} 1 & \text{if we opened a collection center c} \\ 0 & \text{otherwise} \end{cases}$

$X_d^D = \begin{cases} 1 & \text{if we opened a disposal center d} \\ 0 & \text{otherwise} \end{cases}$

$X_r^R = \begin{cases} 1 & \text{if we opened a recycling center r} \\ 0 & \text{otherwise} \end{cases}$

$Y_{ce}^C = \begin{cases} 1 & \text{if we assigned a collection center c to serve end-user e} \\ 0 & \text{otherwise} \end{cases}$

$Y_{cd}^D = \begin{cases} 1 & \text{if we assigned a disposal center d to collection center c} \\ 0 & \text{otherwise} \end{cases}$

$Y_{cr}^R = \begin{cases} 1 & \text{if we assigned a recycling center r to collection center c} \\ 0 & \text{otherwise} \end{cases}$

Continues Variables

$P\omega$	Reordered quantity of warehouse (ω) (unit)
$PEOQ_\omega$	Reordered quantity of warehouse (ω) based on Economic Order Quantity (unit)
$Pcapacity_\omega$	Reordered quantity of warehouse (ω) based on capacity of warehouse (unit)
E°	Amount of CO_2 emitted from locating and operating facilities (tons CO_2 emissions)
E^t	Amount of CO_2 emitted from transportation (tons CO_2 emissions)
$total_\varepsilon$	Total emission produced from the CLSC

References

Abdallah T, Diabat A, Simchi-Levi D (2012) Sustainable supply chain design: a closed-loop formulation and sensitivity analysis. Prod Plan Control 23(2–3):120–133. http://www.tandfonline.com/doi/abs/10.1080/09537287.2011.591622. Accessed 11 Nov 2013

Fahimnia B et al (2013) The impact of carbon pricing on a closed-loop supply chain: an Australian case study. J Clean Prod 59:210–225. http://linkinghub.elsevier.com/retrieve/pii/S0959652613004599. Accessed 15 Nov 2013

Al Dhaheri N, Diabat A (2011) An integrated supply chain problem with environmental considerations. In: Advances in sustainable manufacturing. Springer, Berlin, pp 383–388

Al Zaabi S, Al Dhaheri N, Diabat A (2013) Analysis of interaction between the barriers for the implementation of sustainable supply chain management. Int J Adv Manuf Technol 68(1–4):895–905

Amin SH, Zhang G (2013) A three-stage model for closed-loop supply chain configuration under uncertainty. Int J Prod Res 51(5):1405–1425. http://www.tandfonline.com/doi/abs/10.1080/00207543.2012.693643. Accessed 11 Nov 2013

Barahona F, Jensen D (1998) Plant location with minimum inventory. J Math Program 83(3):101–111

Berman O, Krass D, Tajbakhsh MM (2010) On the benefits of risk pooling in inventory management. Prod Oper Manag 20(1):57–71

Daskin MS (1995) Network and discrete location: Models, algorithms, and applications. Wiley, New York

Dekker R, Fleischmann M, Inderfurth K (eds) (2004) Reverse logistics: quantitative models for closed-loop supply chains. Springer, Berlin

Diabat A, Simchi-Levi D (2009) A carbon-capped supply chain network problem. In: IEEE international conference on Industrial engineering and engineering management, 2009 (IEEM 2009), pp 523–527

Diabat A, Aouam T, Ozsen L (2009) An evolutionary programming approach for solving the capacitated facility location problem with risk pooling. Int J Appl Decis Sci 2(4):389–405

Diabat A et al (2013a) Strategic closed-loop facility location problem with carbon market trading. IEEE Trans Eng Manag 60(2):398–408

Diabat A, Richard J-P, Codrington CW (2013b) A lagrangian relaxation approach to simultaneous strategic and tactical planning in supply chain design. Ann Oper Res 203(1):55–80

Diabat A, Olga B, Dima N (2014) An improved lagrangian relaxation-based heuristic for a joint location-inventory problem. Comput Oper Res

Guide VDR, Harrison TP, Van Wassenhove LN (2003) The challenge of closed-loop supply chains. Interfaces 33(6):3–6

Le T, Diabat A, Richard J-P, Yih Y (2013) A column generation-based heuristic algorithm for an inventory routing problem with perishable goods. Optim Lett 7(7):1481–1502

Özceylan E, Paksoy T (2013) A mixed integer programming model for a closed-loop supply-chain network. Int J Prod Res 51(3):718–734. http://www.tandfonline.com/doi/abs/10.1080/00207543.2012.661090. Accessed 11 Nov 2013

Ozsen L, Coullard CR, Daskin MS (2008) Capacitated warehouse location model with risk pooling. Naval Res Logist 55(4):295–312

Voß S, Woodruff DL (2006) Introduction to computational optimization models for production planning in a supply chain. Springer, New York

Storage Location Assignment Considering Three-Axis Traveling Distance: A Mathematical Model

Chompoonoot Kasemset and Pongsakorn Meesuk

Abstract This research aimed to present mathematical model for warehouse storage assignment to optimize total traveling distance. In this study, traveling distance was a three-axis traveling distance considering horizontal and vertical distance. The objective function of the model was to minimize the total traveling distance in order-picking process. To formulate the objective function, parameter of warehouse layout as width of aisle and storage row, length and height of storage block were included. The proposed model was tested by solving the numerical example of 3 products with 48 storage blocks and 3-level racks using Lingo software. The results showed that the total distance was 623.59 m. Product 2 had the highest ratio of number of pickings and storage blocks needed comparing among three products, thus this product was assign to the locations near I/O point. The results can ensure that the proposed model was able to give a valid solution and practical to apply for the real case problem in further study.

Keywords Storage location assignment · Three-axis distance · Mathematical model

1 Introduction

Storage location assignment (SLAP) in warehouse management is a problem related to assigning products to storage locations that aims to satisfy required performance measures (i.e. total transportation distance, total transfer time, storage utilization, etc.). Total traveling distance is one performance measure that many research works try to minimize to gain the effectiveness of warehouse operations.

C. Kasemset (✉) · P. Meesuk
Department of Industrial Engineering, Chiang Mai University, Chiang Mai, Thailand
e-mail: chompoonoot.kasemset@cmu.ac.th

P. Golinska (ed.), *Logistics Operations, Supply Chain Management and Sustainability*, EcoProduction, DOI: 10.1007/978-3-319-07287-6_35,

The objective of this study is to propose the mathematical model for storage location assignment considering three-axis distance. The objective function of this mathematical model is formulated based on width of picking aisle, width of storage row, length and height of storage block. A numerical example is tested and solved via Lingo software to verify and validate the model.

2 Literature Reviews

A classification of warehouse management problems were classified in (van den Berg and Zijm 1999). Storage location assignment was considered as one task in warehouse management problems related to the assignment of incoming stock to storage locations. The location of products was affected to the effectiveness of basic warehouse processes (Rouwenhorst et al. 2000), classified as receiving, storage, order picking and shipping, in terms of processing time and traveling distance.

Many policies were proposed for storage assignment addressed in (de Koster et al. 2007). Family grouping policy was the policy that considers the relations between products. Similar products were located in the same region of the storage area. As in (Kasemset and Rinkham 2012), family grouping policy was applied to improve the storage area of one camera and lense manufacturer in Thailand. Product model and part-customer were used to group all parts before assigned to each location considering the total distance.

The class-based storage policy distributed products, based on their demand rates, among a number of classes and reserved a region within the storage area for each class. The mathematical of class-based policy with meta-heuristics algorithm was proposed in (Sooksaksun et al. 2012) in order to find the near-optimal warehouse location assignment for each class of products when the problem formulated as non-linear formulation.

The dedicated storage policy attempted to reduce the mean travel times for storage/retrieval by storing products with high demand at locations that were easily accessible. This policy was close to the proposed model in this study that try to assign products to minimize total traveling distance when considering the product demand (number of picking). The total traveling distance in this study included two-axis on vertical and one-axis on horizontal distances. The detail of the proposed model was addressed in the following section.

3 Proposed Mathematical Model

In this section, the proposed mathematical model for storage assignment using three-axis traveling distance was introduced based on Fig. 1.

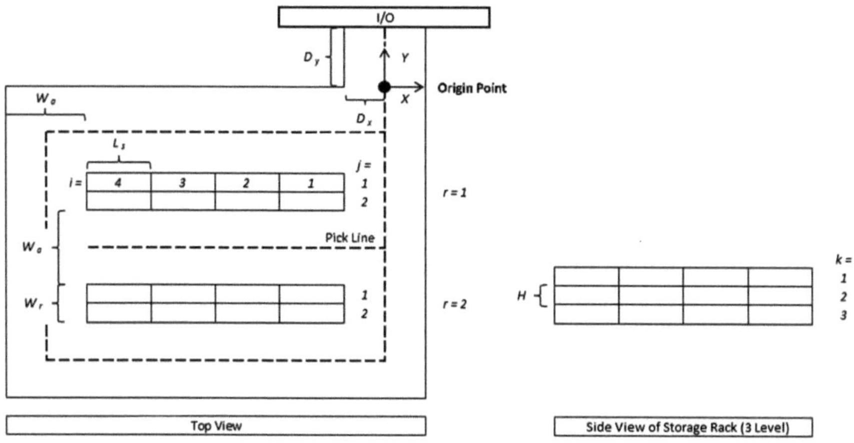

Fig. 1 Storage area layout and notation (48 storage blocks with 3 levels)

3.1 Assumptions

To formulate the mathematical model, the following assumptions were considered:

1. I/O point was located at one corner of warehouse.
2. Warehouse layout was symmetric.
3. The number of storage blocks (N) was limited.
4. One storage block can be used for only one product.
5. Traveling distances were measured along the aisle centerline and centerline of storage block.
6. Horizontal traveling distance considered by the picker moved along the aisle floor and vertical distance considered by height of rack shelf.

3.2 Mathematical Model

Indices:

- p = Product type (1, 2, 3, …, n)
- i = Position of storage block in each storage rack (1, 2, 3, …, m)
- j = Number of storage rack (1, 2, 3, …, q)
- k = Level of storage block (1, 2, 3, …, k)
- r = Number of storage row (1, 2, 3, …, r)

Decision Variable:

$$X_{\mathrm{pijkr}} = \begin{cases} 1, \text{ if product } p \text{ is assigned to storage position } i \text{ rack } j \text{ level } k \text{ row } r \\ 0, \text{ otherwise.} \end{cases} \tag{1}$$

Parameters:

- D_x = Distance from I/O point to origin point along X-axis
- D_y = Distance from I/O point to origin point along Y-axis
- W_a = Width of aisle
- W_r = Width of storage row (equal to two times of storage block width)
- L_s = Length of storage block
- H = Height of storage block
- T_p = Number of picking for product p
- S_p = Number of storage block needed for product p
- I = Order of storage block position i
- J = Order of storage rack j
- K = Order of level k
- R = Order of row r

Objective:
Minimize Z (Total traveling distance):

$$\sum_{p=1}^{n}\sum_{i=1}^{m}\sum_{j=1}^{q}\sum_{k=1}^{k}\sum_{r=1}^{r}\left(\frac{Tp}{Sp}\right)X_{pijkr}D_{ijkr} \tag{2}$$

Subject to:

$$\mathrm{D_{ijkr}} = TD_x + TD_y + TD_z \tag{3}$$

$$TD_x = D_x + (I - 0.5)L_s \tag{4}$$

$$TD_y = D_y + (0.5JW_a) + [(J-1)(W_r + 0.5W_a)] + [(W_r + W_a)(R-1)] \tag{5}$$

$$TD_z = H(K-1) \tag{6}$$

$$\sum_{p=1}^{n} X_{pijkr} \leq 1, \ \forall i,j,k,r \tag{7}$$

$$\sum_{i=1}^{m}\sum_{j=1}^{q}\sum_{k=1}^{k}\sum_{r=1}^{r} X_{pijkr} = S_p, \ \forall p \tag{8}$$

X_{pijkr} are binary variables

The decision variables in this problem were binaries. The objective function in Eq. 2 was to minimize the total traveling distance considering the horizontal and vertical distance. Equations 3–6 illustrated how to calculate total traveling distance when considering three-axis. Equation 7 was the constraint for assigning one or no product in one storage unit. Equation 8 was to ensure that each product stored in the storage units as equal to the number of storage needed.

4 Numerical Example

There were 48 storage blocks for 3 product types. The storage area layout was presented previously as Fig. 1. The parameters used in this example were shown in Table 1.

4.1 Lingo Model

To test the proposed mathematical model, the model was applied to solve the test problem using Lingo software. Lingo model for this example was presented as Fig. 2.

Lingo model in Fig. 2 was separated in three parts. The first part was how to define the used set and set members (variables and parameters). The second part was the proposed mathematical model in form of Lingo code starting from (van den Berg and Zijm 1999) objective function to minimize the total traveling distance (referred to Eq. 2), (Rouwenhorst et al. 2000) the way to calculate 3-axis distance (referred to Eqs. 3–6), (de Koster et al. 2007) referred to Eq. 7, (Kasemset and Rinkham 2012) referred to Eq. 8 and (Sooksaksun et al. 2012) was to defined decision variables as binary variables. The third part was input data for the numerical example.

4.2 Results

The solution report from Lingo showed that the minimum total traveling distance was 623.59 m and this solution was global optimum (shown in Fig. 3). Distance of each storage position was presented as in Fig. 4 (Left) and product assignment from the optimal results was presented as in Fig. 4 (Right).

The results of product assignment were concluded as in Table 2.

From Table 2, product type 2 had the highest portion of $T_p/S_p,$ so this product was located at the storage blocks close to the I/O point with the distance range

Table 1 Parameter Used in the Test Problem

	Parameter	Value	Parameter	Value
Product	S_1	17 blocks	T_1	15 times
	S_2	14 blocks	T_2	20 times
	S_3	7 blocks	T_3	5 times
Layout	D_x	2 m	W_a	4 m
	D_y	2 m	W_r	3 m
	L_s	2 m	H	1.7 m

```
Model:
Sets:
!ideices, parameters,decision variables;
Product/1..3/:S,T;
Position/1..4/:Q ;
Rack/1..2/:M ;
Level/1..3/:C ;
Row/1..2/:N ;
Links1(Product,Position,Rack,Level,Row): X;
Links2(Position,Rack,Level,Row): D;
Endsets                                                                  Part 1

!Math Model;
!Objective Minimize Total Traveling Distance;
Min = @sum(Links1(p,i,j,k,r): (T(p)/S(p)*X(p,i,j,k,r)*D(i,j,k,r)));    [1]

!This is how the total distance is calculated;                           [2]
@For(Position(i):@For(Rack(j):@For(Level(k):@For(Row(r):D(i,j,k,r)
= (Distx + ((Q(i)-0.5)*L))+(Disty +(0.5*M(j)*WA)+((M(j)-1)*(WR+(0.5*WA)))+((WR+WA)*(N(r)-
1)))+(H*(C(k)-1)))))));

!Constraints;
@For(Position(i):@For(Rack(j):@For(Level(k):@For(Row(r):@sum(product(p):X(p,i,j,k,r))<=
1)))));                                                                  [3]
@For(product(p):@sum(Links1(p,i,j,k,r):X(p,i,j,k,r))=S(p));        [4]
@FOR(Links1:@BIN(X));                                              [5]       Part 2

!Here is data;
Data:
S= 17 14 7;
T= 15 20 5;
Distx = 2;
Disty = 2;
H = 1.7;
L = 2;
WA = 4;
WR = 3;
Q = 1 2 3 4;
M = 1 2;
N = 1 2;
C = 1 2 3;
Enddata                                                                  Part 3

End
```

Fig. 2 Lingo model for the numerical example

7.0–15.7 m. Product type 1 and 3 were assigned to the storage with distance range 15.7–20 and 21.4–23.4 m, respectively because product type 1 had the portion of T_p/S_p greater than product type 3.

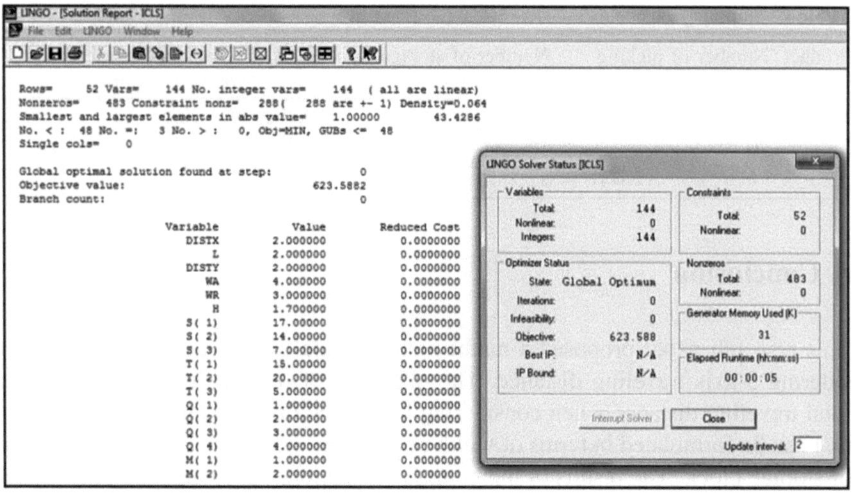

Fig. 3 Solution from Lingo

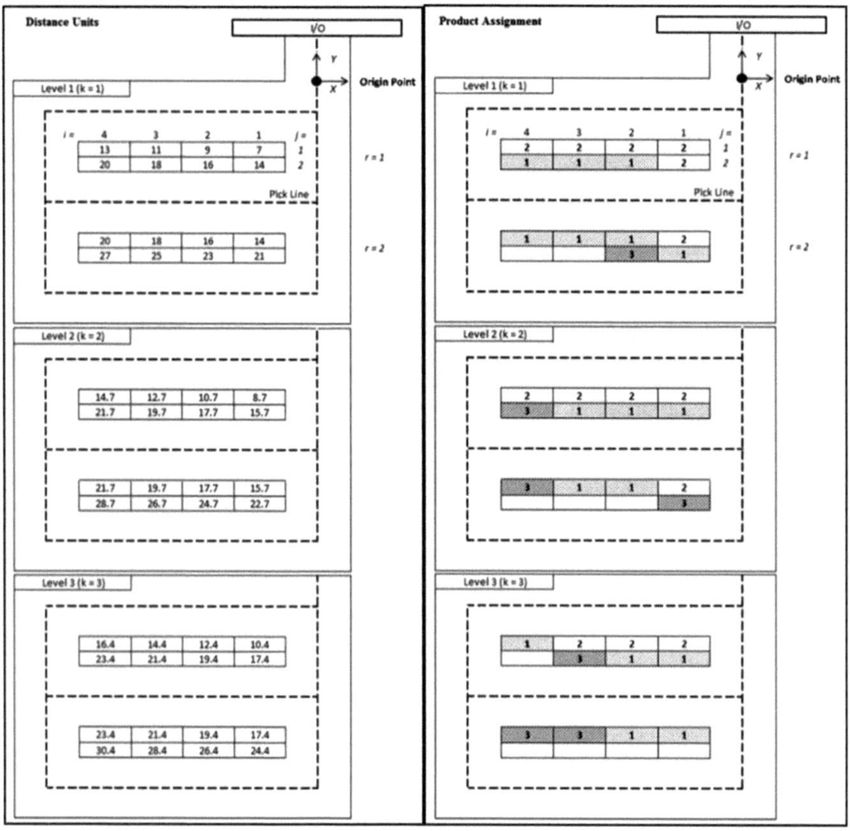

Fig. 4 The total distance considering 3-axis for each storage block (*Left*) and optimal product assignment (*Right*)

Table 2 Summary of results

Product	Number of picking time (T_p)	Number of storage needed (S_p)	T_p/S_p	Distance of assigned storage position
1	15	17	0.882	15.7–20.0
2	20	14	1.429	7.0–15.7
3	5	7	0.714	21.4–23.4

5 Conclusion

This research paper proposed a mathematical model for storage assignment considering 3-axis traveling distance. The objective of this model was to minimize total traveling distance when considering horizontal distance and vertical distance that can be formulated in terms of width of aisle and storage row, length and height of storage block. The results of the numerical example were valid and this verified that the logic of the proposed model was correct. Thus, the further study is to apply this proposed model for the real case that the size of the problem will be increased.

References

de Koster R, Le-Duc T, Roodbergen KHJ (2007) Design and control of warehouse order picking: a literature review. Eur J Oper Res 182:481–501

Kasemset C, Rinkham C (2012) Warehouse storage assignment: the case study of camera and lense manufacturer. In: Proceeding of 2011 IEEE international conference on industrial engineering and engineering management (IEEM), Singapore, pp 1108–1112. doi:10.1109/IEEM.2011.6118087

Rouwenhorst B, Reuter B, Stockrahm V, van Houtum GJ, Mantel RJ, Zijm WHM (2000) Warehouse design and control: framework and literature review. Eur J Oper Res 122:515–533

Sooksaksun N, Kachitvichyanukul V, Gong D (2012) A class-based storage warehouse design using a particle swarm optimization algorithm. Int J Oper Res 13:219–237

van den Berg JP, Zijm WHM (1999) Models for warehouse management: classification and examples. Int J Prod Econ 59:519–528

Solving a Multi-objective, Source and Stage Location-Allocation Problem Using Differential Evolution

Rapeepan Pitakaso and Thongpoon Thongdee

Abstract The purpose of this research is to develop an algorithm using Differential evolution algorithm in order to explore solutions of a Multi-Objective, Source ans Stage Location-Allocation Problem (MOSS-LAP). The development process was started from the design of standard DE and extended it into modified DE which the recombination process has been modified by using the advantages of CR constant randomization by determining two random values. The result obtained from this improvement came up with three solutions consisting of the initial solution, the best solution and the final solution then continued developing using the advantages of PSO to support the recombination process. Its outstanding qualification is the acceleration which helps in finding the best solution faster. With the new algorithm, it determines only the value of CR then selects the initial solution or the final solution to be used in the calculation according to the acceleration formula of PSO. When applying these three developed algorithms in solving problems of the case study, the result showed that DE-PSO gave better solution and took times in solving problems faster than the standard DE and the modified DE either small, medium or large problem. DE-PSO was 6.5 % better than the standard DE and 2.8 % better than the modified DE.

Keywords Location allocation problem · Meta-heuristics · Differential evolution · Multi-objective optimization

R. Pitakaso (✉) · T. Thongdee
Department of Industrial Engineering, Faculty of Engineering,
Ubon Ratchathani University, Ubon Ratchathani 34190, Thailand
e-mail: enrapepi@mail2.ubu.ac.th

T. Thongdee
e-mail: thongpoonthongdee@gmail.com

P. Golinska (ed.), *Logistics Operations, Supply Chain Management and Sustainability*, EcoProduction, DOI: 10.1007/978-3-319-07287-6_36,

1 Introduction

This research has studied on location selection using multi-stage approach and multiple sources of raw material in finding the opportunity to open the ethanol plant whether the raw material used in production is bagasse or cassava pulp, or even use both as raw materials in the same plant. Initially, Buddadee et al. (2008) studied the location selection of ethanol plant using bagasse by studying the effect of economic value and environmental impact. It found that the use of bagasse was cost effective and helped reduce the environmental impact in terms of reducing greenhouse gas emissions to the atmosphere. Later, Nanthasamroeng et al. (2008) studied the solutions of the problem on selecting the location of ethanol plant produced from bagasse using multi-objectives approach. In which, there were four specific characteristics for identifying the problems on location selection including (1) the customer's location is already fixed (2) the plant that is needed to locate the location itself (3) the location where the customer and factory are located (4) the distance between the factory and the customer location, Jozefowiez et al. (2008). In addition, the study is also considered the two-level location selection which is the two-level location routing problem.

This research has studied the same problem with Nanthasamroeng et al. (2008) but the form of problem has been extended to the multi-source of raw materials which means it takes into account the source of raw materials used to produce ethanol whether it is bagasse and cassava pulp. This problem cannot be solved by optimization software because of the limitation in the size of the experiment. When increasing the size of the source of raw materials and the potential locations for the production plant of ethanol, it found that the program was unable to solve the problem due to the memory backup of applications was insufficient to the complexity of increasing problems. Therefore, it is necessary to develop the meta-heuristic process to help in finding the best solution and to take the least time in the experiment with DE process.

2 Literature Reviews

A Summary of Meta-Heuristic Algorithms There has been many studies that proposed meta-heuristic algorithms that have been widely used.

1. Ant Colony Optimization (ACO), Doerner et al. (2007) is a methodology that mimics the behavior of ant colony foraging for foods. When determining the food foraging behavior of ants in numerical analysis, this methodology will give appropriate results for a specific area, but there are limitations of the application in terms of the determination of restriction. If the function has many values of the Local Optima and they are equal or similar, it may not find the Global Optima, which is similar to the behavior of ants when finding foods in the same quantities in several areas, many ants will swarm the foods only in

the area that has been found first instead of swarming with the same quantities, but if it has the optimum rounds of searching then it will find the Global Optima as well. It is therefore a methodology that should be very cautious in using the parameters.

2. Tabu Search (TS), Lin and Kwok (2006), Drezner et al. (2006), Stummer et al. (2004), Caballero et al. (2007) and Uno and Katagiri (2008) is a methodology similar to Simulated Annealing which is a methodology of learning experience. In which, Tabu Search has applied a better way in finding answers by recognizing and protecting the traditional values that are more worse in the Tabu List, which will help improve the next rounds of searching until it reaches the acceptable threshold. However, this methodology relies on the memory of the computer quite a lot. As a result, it makes the work more complex.
3. Particle Sworn Optimization (PSO), Thongpoon and Repeepan (2013) is a stochastic based methodology for determining the suitability which has been inspired by the behavior of dispersed particle as a group such as a group of birds that are flying the herd. PSO consists of a group of particles that are in motion in multiple dimensions, which are the way to find answers of the solutions based on the actual number of individual particles. The position of velocity in vectors is stored in the memory compared to the neighbor particles, and then selects to study the particles with potential velocity and the leading vectors moving in new directions until getting the Global Solution Vector.
4. Genetic Algorithms (GAs), Xu et al. (2008), Doerner et al. (2009), Cantarella and Vitetta (2006), and Leung (2007) is a stochastic based methodology for determining the suitability with the genetic hypothesis that mimics the Natural Evolution, based on the concept of the selection of species by means of natural selection in which it can be well applied to the numerical analysis in finding the answers of the solutions that are complex with many variables and conditions.
5. Differential Evolution (DE), Medaglia et al. (2009), and Raisanen and Whitaker (2005) is a stochastic based methodology for determining the suitability and is the random base Global Search Space which randomly finds comprehensive answers with the genetic hypothesis same as Gas, but it has distinctive advantage as it has less complexity of the methodology structure and makes more generalizations. It can also use the Floating Point Real Number in the calculation without the need to convert the decision variables into the binary numeral systems so that these are the main reasons that make the DE methodology fast with the robustness in finding answers more than other methodologies.

3 Research Methods

This research is divided into three phases which are the preparation phase, the algorithm design phase and the after-operation phase.

Firstly, do a literature review on meta-heuristic process relating to location routing problem using multi-stage approach, multi-source approach and multi-objective approach. Secondly, define the status of the problems by determining the mathematical model in the form of Integer Linear Programming as shown in Fig. 1 with the aim to find the location of ethanol plant with the lowest operating costs, the minimal environmental impact, the lowest security risk and the highest satisfaction of the surrounding societies and communities. Thirdly, gather information to be used in the case study on the ethanol production from bagasse and cassava pulp in the Northeast. Fourthly, do the methodological design of algorithms for solving problems with DE and test the results. Finally, summarize and discuss the results.

3.1 Mathematical Models for Selecting the Plant Location

All mathematical models for this research refer to the study of Thongdee and Pitakaso (2012) as "Solving a multi-objective, source and stage location-allocation problem: the case study on the ethanol production from bagasse and cassava pulp in the northeastern Thailand".

3.2 A Case Study of a Bagasse and Cassava Pulp Ethanol Plant in Northeastern Thailand

16 sugar plants and 46 starch plants in the Northeast have been designated as the source of raw materials which are spread out in different provinces as shown in Fig. 2 according to the research by the Thongdee and Pitakaso (2012) which was a study on solutions to prove whether the mathematical model is correct or not. Based on this research, 5 starch plants and 5 sugar plants were chosen using the criteria on location selection with a score and ranking, and then chose five plants with the maximum scores to be used in the experiments.

The location selection and the vehicle allocation in the case study on ethanol plant in the northeastern area of Thailand was the problem solving for the three objectives as follows; (1) Economic objectives; the cost reduction on transportation and plant construction, (2) Environmental objectives; the reduction of greenhouse gas emissions from all processes of production and transportation and, (3) The safety risk. This minimizes the risk that could happen to people who live in the transportation area if the leakage happens.

3.3 Research Resulting by DE

Thongdee and Pitakaso (2012) studied the same problem but it was tested by LINGO 11.0 software package which had a limitation on time constraint. Each case must be taken at least 30 h and if we need a large volume of sugars and starch, the trial would take a very long time to run or might not even be able to use the LINGO.

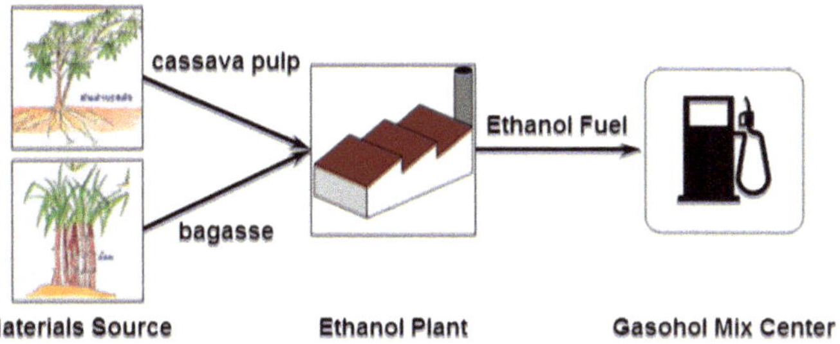

Fig. 1 The supply chain of the ethanol production for mixing to produce the gasohol

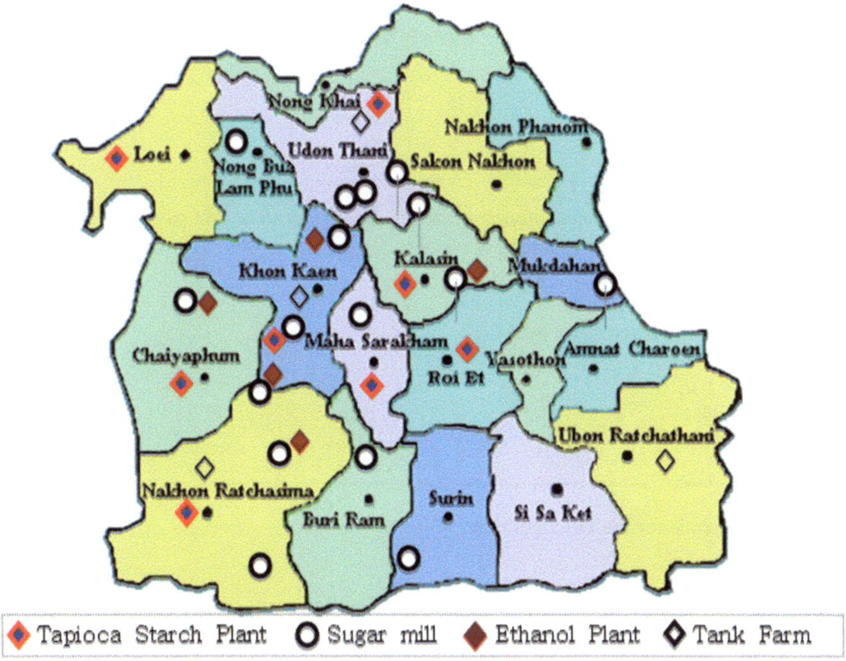

Fig. 2 Location of sugar plants, starch plants, ethanol plant and the gasohol mixing plant in north-eastern

For a better performance on testing, the test shall use higher level which is the heuristic and meta-heuristic method, and the form of algorithm design for the trial. Therefore, DE was developed to solve that problem and compared to LINGO 11.0.

3.3.1 DE Standard

For the process of developing the algorithm and how to calculate the location of ethanol plants using raw materials from bagasse and cassava pulp with the lowest operating costs, minimal environmental impact and the lowest security risk, the

```
Pseudo-code for the DE algorithm
Begin
    Set CR = 0.88, F = 2 and NP = 10
    G = 0
    Generate the initial population X_i,g ∀i, i=1,...,NP
    Evaluate the fitness (function value) for each individual f(X_i,g) ∀i, i=1,...,NP
    While stopping criterion (100 Loop) is not satisfied Do
        For i=1 to NP Do
            Select uniform randomly r1 ≠ r2 ≠ r3 ≠ i
            j_rand = randint [1,M_1,M_2]
            For j=1 to [M_1,M_2] Do
                If (rand_j [0,1] < CR or j=j_rand) Then
                    U_j,i,g+1 = X_j,r1,g + F(X_j,r2,g - X_j,r3,g)
                Else
                    U_j,i,g+1 = X_j,i,g
                End If
            End For
        End For
        For i=1 to NP Do
            Evaluate the fitness (function value) for each individual f(Ui,g+1) ∀i, i=1,...,NP
            If (f(U_i,g+1) is better than or equal to f(X_i,g)) Then
                Replace X_i,g+1 with U_i,g+1
            Else
                Replace X_i,g+1 with X_i,g
            End If
        End For
        G = G+1
    End While
End
```

Fig. 3 Pseudo code for the DE algorithm

researcher has proposed the approach for the development of meta-heuristic for the use in determining the location according to the pseudo code for using in the writing of C++ program as shown in Fig. 3 and the flow chart in Fig. 4.

Beginning Solution Process

The researcher has applied the beginning solution process which is inspired by the Genetic Algorithm (GA) by determining the number of chromosomes particle to be equal to the number of potential ethanol plant. The crossing particles in chromosome will be the parts of sugar plant, starch plant and oil storage which are the horizontal cells while the ethanol plant is the vertical cells.

Fig. 4 Development flow chart for DE algorithm

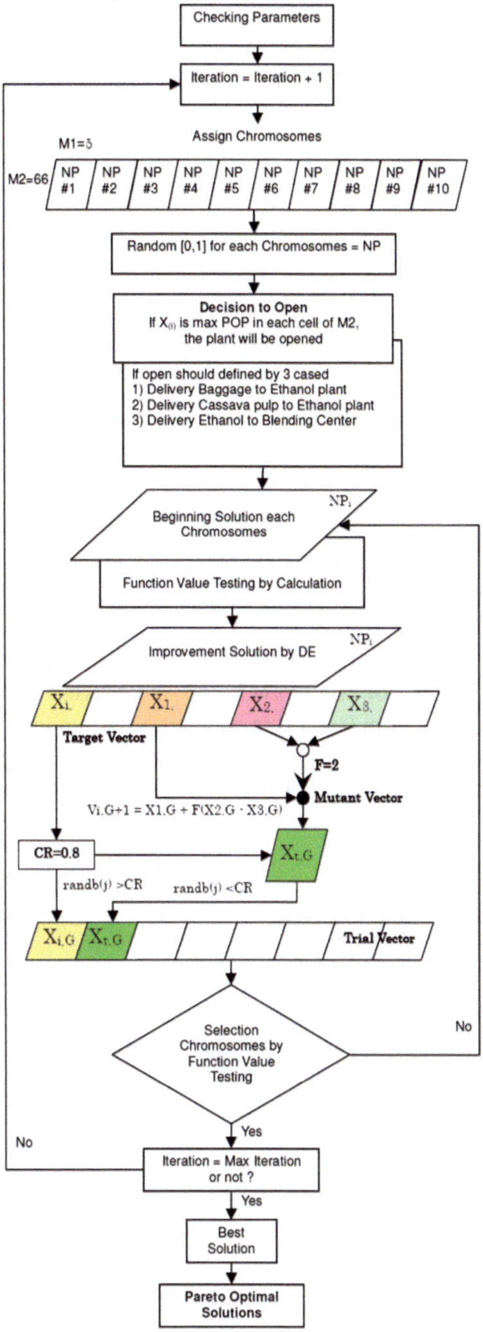

Table 1 The beginning solution of DE algorithm

	j1	j2	j3	j4	j5
i1	0.47	0.41	0.14	0.70	0.67
i2	0.21	0.12	0.17	0.95	0.06
i3	0.37	0.00	0.37	0.02	0.12
i4	0.94	0.59	0.53	0.52	0.83
i5	0.68	0.15	0.57	0.82	0.85
t1	0.47	0.93	0.15	0.89	0.58
t2	0.80	0.81	0.65	0.36	0.82
t3	0.89	0.22	0.30	0.60	1.00
t4	0.64	0.06	0.82	0.87	0.43
t5	0.19	0.78	0.65	0.36	0.98
k1	0.54	0.57	0.02	0.70	0.40
k2	0.61	0.35	0.28	0.92	0.08
k3	0.37	0.25	0.30	0.59	0.79
k4	0.97	0.93	0.30	0.16	0.49

For finding the beginning solution of the algorithm, it started from the determination of initial chromosome and generated a random number with values ranging from 0.0–1.0 for each array in the chromosome through working procedure according to RBS rules, which started from the creation of array with the number of column M1 = 5 and crossed with M2 = 66 then generated a random value from 0.0–1.0 in each array. When getting the value of each array then began to compare; any array that has the highest value in each row (M2) will be elected to the operation or transferred to each other which is the decision whether to open the ethanol plant or not (Table 1).

The Mutation Process

Starting with the evolution of mutation by gathering the target vectors to start producing the mutant vector in each of the target vector which is Xi,g, which was a vector in g generation. On the other hand, the mutation vector was Vi,g which was created according to the following equation.

$$V_{i,j} = X_{r1,g} + F\left(X_{r2,g} - X_{r3,g}\right) \qquad (1)$$

It is noted that Xr1, Xr2, Xr3 were randomly selected from the vector population which was different from the target vector Xi,g while F was the factor that controlled the degree of difference in victor between Xr2 and Xr3 then added to the standard vector which was Xr1 (Table 2).

Table 2 The mutation solution of DE algorithm

	j1	j2	j3	j4	j5
i1	1.01	−0.71	0.19	1.10	0.01
i2	0.12	−1.42	1.95	0.64	−0.02
i3	−0.36	0.70	0.15	−0.47	0.13
i4	1.06	0.60	−0.09	0.30	0.00
i5	−0.16	−0.35	0.50	1.17	0.63
t1	2.03	−0.56	0.76	1.11	−0.07
t2	1.13	1.38	−0.28	0.39	0.51
t3	0.71	−0.38	−0.49	0.83	−0.17
t4	−0.89	−0.03	1.69	0.45	0.80
t5	0.45	1.38	−0.60	1.93	1.90
k1	1.66	−0.80	0.62	0.43	−0.65
k2	0.75	−0.92	1.95	−0.14	−0.58
k3	0.26	−0.33	−0.09	1.43	0.66
k4	2.23	1.19	−0.35	−0.80	−0.85

The Crossover Genetic Process

This is the evolution resulted from the crossover between Xi,g and Vi,g in the creation of experimental vector which was Ui,g. It was derived from chromosome of the beginning solution and chromosome of the mutation solution. The principle is to randomly select a number from 0 to 1 for a set of chromosomes in the crossover process hereinafter referred to the Trial Vector which the random selection was similar to the determination of the set of beginning solution.

$$u_{j,i,g} = \begin{cases} v_{j,i,g}, & \text{if } u_j \leq c_r \text{ or } j = j_u \\ x_{j,i,g}, & \text{otherwise} \end{cases} \tag{2}$$

When Cr (Crossover Rate) is the constant [0, 1]

$$\mathrm{i} = 1, 2, \ldots\ldots \mathrm{N};\ \mathrm{j} = 1, 2, \ldots\ldots \mathrm{D}$$

By the Eq. (2), it is explained that when generating the random value to each array in the chromosome which was a number from 0 to 1, if the value is less than or equal to (Cr) Crossover Rate then select the value of the vector obtained from the mutant vector, otherwise choose the value obtained from the target vector.

For the sample of calculation, assuming that Cr = 0.8 then make random numbers and we have got 0.45, and if we assume that the position of the cell is i1: j1, so we must choose the answer which is 1.01. Similar to the trial vector, if the position is i1: j3, if making a random number, supposing we have got a random number 0.97, so we must choose the number 0.14 as an answer of the trial vector. All the details are shown in Table 3.

Table 3 The recombination solution of DE algorithm

	j1	j2	j3	j4	j5
i1	1.01	0.41	0.14	1.10	0.01
i2	0.12	−1.42	0.17	0.64	−0.02
i3	−0.36	0.00	0.15	−0.47	0.13
i4	0.94	0.60	0.53	0.52	0.83
i5	−0.16	−0.35	0.50	1.17	0.85
t1	2.03	−0.56	0.76	1.11	−0.07
t2	1.13	1.38	0.65	0.39	0.51
t3	0.71	0.22	−0.49	0.83	−0.17
t4	−0.89	−0.03	1.69	0.45	0.80
t5	0.45	0.78	−0.60	1.93	1.90
k1	0.54	−0.80	0.62	0.43	−0.65
k2	0.75	0.35	1.95	−0.14	−0.58
k3	0.37	−0.33	−0.09	1.43	0.66
k4	2.23	0.93	−0.35	−0.80	−0.85

Selection

The solution selection process is the selection which any vector that gives a better solution will be chosen. The selection process has defined each vector at Xi,g and trial vector is Ui,g which is used for comparison of the vector to the target equation and the survival as the next generation and the vector Xi,g + 1 will be selected based on the criteria set forth as follows.

$$x_{i,g+1} = \begin{cases} U_{i,g}, & \text{if } f\left(U_{i,g}\right) \leq f\left(x_{i,g}\right) \\ x_{i,g}, & \text{otherwise} \end{cases} \tag{3}$$

3.3.2 DE Modified (MODDE)

Modified DE is the improvement in the mutation process. The F constant is designed for self-adjustment upon the population which has three levels consisting of high, medium and low level. Then F constant will be adjusted to be at medium level upon the optimized solution of population.

Referring to the Eq. (2), it is simply fundamental mutate process (Crossover or Recombination) which has not yet been updated to better performance. Therefore, the researcher has then adjusted the process by allowing CR to be the value adjusted by it self (SELF ADJUSTED) as well as the F and is inspired by the combination process of GA species which usually is normally the data exchange of different chromosomes but DE does not use it. Therefore, it is an advantage in the development process of the mixed breed DE as shown in Eq. (4).

Table 4 The recombination solution of MODDE algorithm

	j1	j2	j3	j4	j5
Random number					
i1	0.45	0.93	0.97	0.06	0.74
	j1	j2	j3	j4	j5
Trial vector by recombination as MODDE					
i1	1.01	0.41	0.14	1.10	0.67

$$U_{j,i,g} = V_{j,i,g} + A(V_{gbest}) + B(V_{ibest}) + C(V_{friend}) \quad (4)$$

When

uj (Ramdom) > CR1 + CR2; X*ij* (Using initial solution)
uj (Ramdom) > CR1; GbestX*ij* (Using best solution)
uj (Random) < CR1; U*ij* (Using latest solution)

The sample of calculations, if we assume that CR1 = 0.4, CR2 = 0.5, and then make random numbers. Assuming that we have got 0.45 which is greater than CR1, and if the sample is at the position of the cell i1: j1, so we must choose the best answer which is the number 1.01. Likewise, referring to all answers for trial vector, if we make a random and we have got 0.97 which is greater than CR1 + CR2, so we must choose the number 0.14 which is derived from the beginning answer of trial vector (Table 4).

3.3.3 DE Particle Swarm (DE-PSO)

Improve the mutant process (Crossover or Recombination) as well as the MODDE approach but added the advantage of DE by applying the good behavior of PSO to be used in the development which is PSO. It will accelerate in finding the solution while DE will find the solutions in broad ways. And then PSO will find the best solution in that group. In addition, the researcher has also improved this process by the CR is designed to be adjustable by itself (SELF ADJUSTED) as well as the F, so it is an advantage in the development process of a mixed breed of DE as shown in the Eq. (5).

$$U_{j,i,g} = (X_{j,i} \text{ or } U_{j,i}) + V_{j,i,g} \quad (5)$$

Calculate acceleration from

$$Vj, i, g = (W1Xj, i) + (W2PbestXj, i - Xj, i) + (W3GbestXj, i - Xj, i) \text{ or } Uj, i$$

when

Table 5 The recombination solution of DE-PSO algorithm

	j1	j2	j3	j4	j5
Random number					
i1	0.45	0.93	0.97	0.06	0.74
	j1	j2	j3	j4	j5
The number to selected to PSO calculation					
i1	2.01	−4.17	0.66	2.21	−0.02
	j1	j2	j3	j4	j5
Trial vector by recombination as DE-PSO					
i1	1.01	0.41	0.14	1.10	0.01

uj (Random) > CR; Xi,j (using initial solution)
uj (Random) < CR; Ui,j (using latest solution)

The sample of calculations, if we assume that CR = 0.8, and then make numbers randomly. Assuming that we have got 0.45 which is less than CR, and if gives an example at the position of the cell i1: j1, so we have to pick the best answer which is 1.01 to be used according to the formula of PSO and will become the answer to the trial vector as well. If the position is i1: j3, if the number is random. Assuming that we have got 0.97 which is greater than CR, so we must choose 0.14 which comes from the beginning solution to be calculated by using the formula of PSO and becomes the solution of trial vector, and if we defines W1 = 2, W2 = 3, W3 = 4, so the result would be as shown in Table 5.

3.4 The Algorithm Testing Result

The researcher has tested the algorithm by designing the test and found that the solution has decreased the function value. This is in accordance with the objective of the research that intends to find the minimal value. The calculation result and the performance test of algorithm by differential evolution and the comparison of the performance between LINGO 11.0 and DE algorithm is shown in Table 6.

3.4.1 Small Problem (I5-T5-J5-K4)

This is the test by selecting five sugar plants and five starch plants using a factor analysis of plant location using rating plan and selected plants with the top 5 scores which is shown in the research by Thongdee and Pitakaso (2012) then tested with the Lingo V.11.0 software which was able to find the global solution, but it takes a long time in the trial which lasted 27 h 29 min. The results obtained from the experiment are shown in Table 6.

Table 6 The comparison of performance between LINGO 11.0 with DE algorithm

Case Study (problem)	Objective value (MB)		Time (sec)		Comparison (%)	
	Lingo	DE	Lingo	DE	Value	Time
I5-T5-J5-K4	42,914[a]	43,685[d1]	50,400[a]	80.831[d1]	1.8	−99.84
		43,339[d2]		65.520[d2]	1.0	−99.87
		43,086[d3]		55.440[d3]	0.4	−99.89
I10-J26-K5-T4	4.968 e + 006[b]	5.288 e + 006[d1]	9,700[b]	110.949[d1]	6.45	−98.86
		5.129 e + 006[d2]		107.659[d2]	3.25	−98.89
		5.032 e + 006[d3]		105.064[d3]	1.28	−98.92
I16-T46-J5-K4	5.850 e + 006[c]	6.381 e + 006[d1]	16,988[c]	133.066[d1]	9.1	−99.22
		6.142 e + 006[d2]		129.797[d2]	5.0	−99.24
		5.967 e + 006[d3]		114.688[d3]	2.0	−99.32

[a] Lingo test by global optimization (runtime 27:29:40, hh:mm:ss)
[b] Lingo test by lower bound (runtime 2:41:40, hh:mm:ss)
[c] Lingo test by lower bound (runtime 4:43:08, hh:mm:ss)
[d] DE test by Dev-C++ Program ([d1] DE STD, [d2] MODDE, [d3] DE-PSO)

Then the research has developed the algorithm to test and compare to the results obtained from the program by developing the algorithms with DE-PSO approach, which the result was a lot similar to the program. The function value is different from the program by 0.4 % but the time of the trials were 99.89 % faster than.

3.4.2 Medium Problem (I10-T26-J5-K4)

This is the test by the selection of 10 sugar plants and 26 starch plants from a factor analysis of plant location using rating plan and selected the plants with the highest scores in the first 10 and 26 ranks respectively, then tested by Lingo V.11.0 software programs. In which, the results was not be able to find the global solution due to the problem is larger and more complex which the program was unable to test. However, it was tested by the disposal of certain conditions. It was the removal for the program to be able to test which is called Lower Bound Test, but it takes a long time to experiment as well. The test lasted 2 h 41 min while the answer obtained was not the best solution (Objective Bound). the results obtained are shown in Table 6.

3.4.3 Large Problem (I16-T46-J5-K4)

This is the test by using all the plants operating in the Northeast, 16 sugar plants and 46 starch plants were tested by conducting the experiment with Lingo V.11.0 software programs. The results were not able to find the global solution due to it was a large problem and very complicated and cannot be tested by the program. However, it was tested by the disposal of certain conditions. It was the removal of restrictions in order for the program to be able to test which it is called the Lower

Bound Test, but it takes a long time as well. It took a total of 4 h 43 min at the response that the solution was not the best solution (Objective Bound). The results obtained are shown in Table 6.

Then the researcher has developed the algorithm to test and compare the results of the program. The algorithm developed in the first phase was the standard differential evolution as well as a small problem. The results of the answers were not that good which the solution was differ by 9.1 %, so we have designed the algorithms by modification of the crossover of differential evolution step which provides the solution to a higher cost of approximately 5 %, but it was 99 %, faster than Lingo. Therefore, it is needed to increase the advantages of DE in the next research by taking a good behavior of PSO to be used in development which is the acceleration in finding the solution. DE will look for solutions in broad ways and PSO will find the best solution in that group in order to get a better solution and faster. The results obtained from the design of algorithm with improved crossover step by PSO methodology will provide the cost of answers that are different by 2 %, but it takes 2 % faster than that Lingo 99 % which it is the time saving solutions. The results obtained are shown in Table 6.

3.5 Analyzing Multi-Objective Effect Using Pareto Front Technique

For finding the multi-objective optimization, there are two main methods used which are the weighted sum approach and Pareto-based approach. Since it is impossible to have only one answer to give the best of all objectives simultaneously for multi-objective problems, the groups of the ideal answer to this problem is the answer that is not dominated when comparing to all the answers. That solution can be obtained by means of the Pareto which is the principle of *Pareto Domination*.

This research has used the Pareto technique as well, by determining weights to each objective. The weight obtained has come from the random in each round of calculations. This principle used is shown in Fig. 5, then approach to the process of domination until getting various solutions. Then be graphed to determine the extent to which is no predominate. The results of the Pareto are shown in Fig. 6.

Referring Fig. 5, each loop of the experiments with many purposes, the algorithm has assigned a random weight equal to the number of objectives such as in there are 3 objectives in the case study, the algorithm will choose a random weight to three values (w1, w2, w3), then takes the weight to be calculated such as $w'1 = w1/(w1 + w2 + w3)$, which the values obtained from the calculation will be multiplied with the target function such as the objective number 1 ($f'1 = w'1*f1$). This principle will result in getting various solutions for many purposes. It is the only solution to determine weights of each objective while the weight obtained will be applied to each objective set out in the case study (Figs. 7, 8).

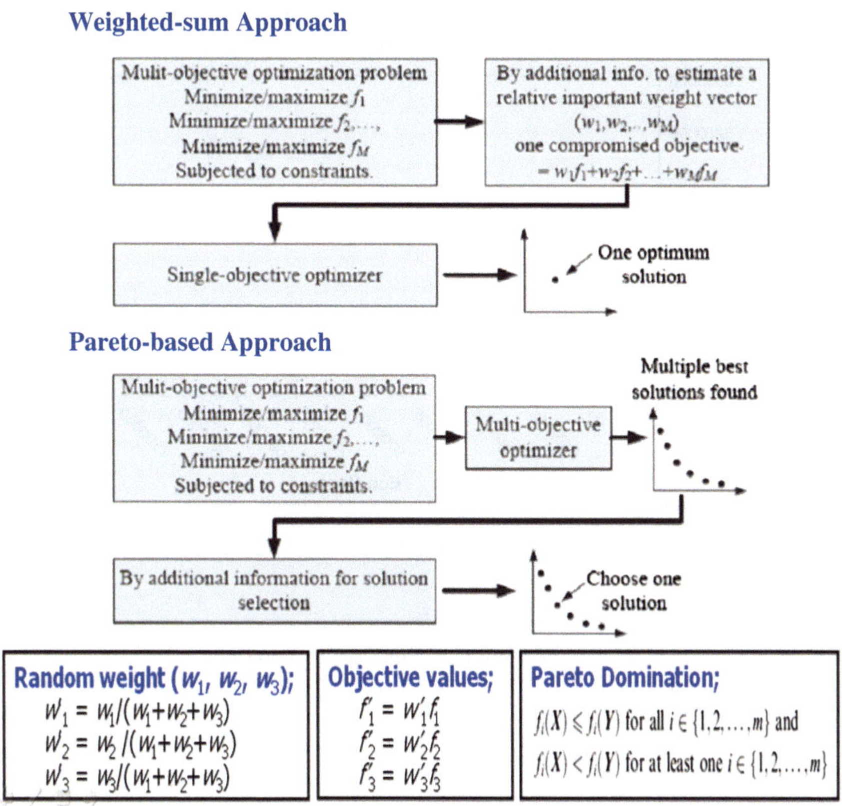

Fig. 5 Show the random weight of each objective function

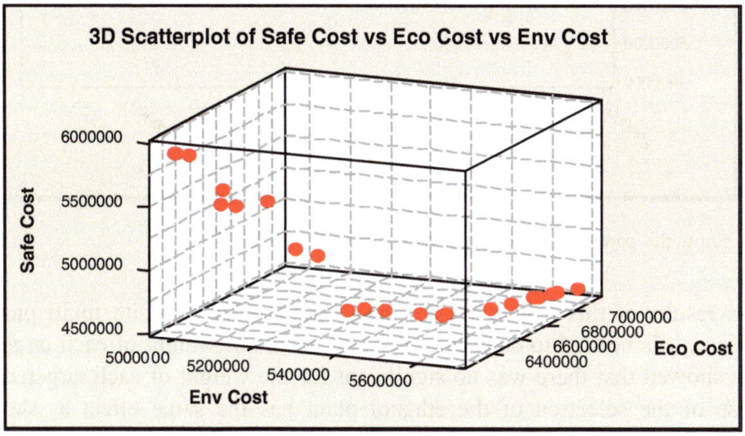

Fig. 6 Show the pareto front for multi-objectives

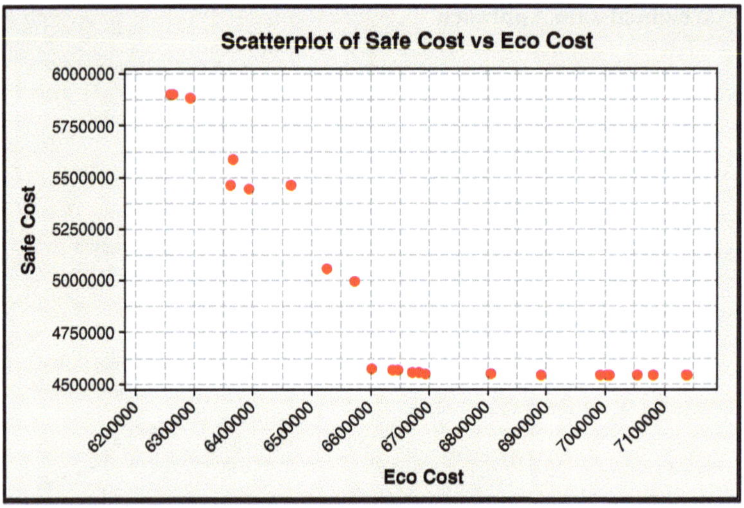

Fig. 7 Show the pareto front compare between ECO with SAFE

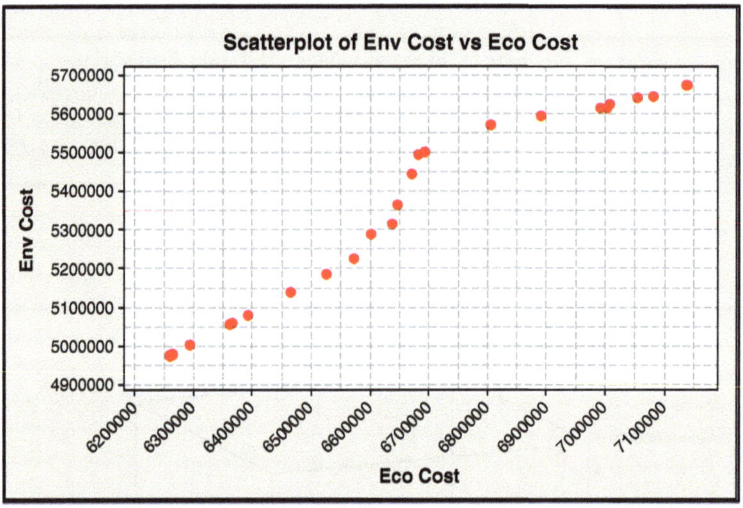

Fig. 8 Show the pareto front compare between ECO with ENV

The results of this research were the location selection with multi purposes. Therefore, it is needed to determine the importance or weight of each target. The results showed that there was no significant for the weight of each targeted value because of the selection of the ethanol plant has the same effect as shown in Table 7. However, referring to the objective values obtained from the calculation, it found that the targeted value in terms of security risks has affected the total cost as minimal. The targeted value that has maximum impact on the total cost is the

Table 7 Selection location of ethanol plant produced from bagasse and cassava pulp

Quantity material (ton)	Ethanol producing source	Ethanol receiving source	
		Quantity of ethanol receiving (litter)	Tank farm
329,557	VP starch (2000) Co., Ltd	2.38E + 07	Konkaen depot
145,824	Kaen Chalern Co., Ltd	1.08E + 07	Udonthani depot
164,474	PVD international Co., Ltd	1.28E + 07	Ubon Ratchathani depot
351,884	Sangaunwong industry Co., Ltd	2.49E + 07	Nakhon Ratchasima depot
274,002	CPS Starch Co., Ltd	5.22E + 06	Ubon Ratchathani depot
		1.42E + 07	Udonthani depot

objective of economics. Therefore, when considering the impact that has been resulted from the distance transport of raw materials to the ethanol plant and from the ethanol plant to the oil storage, therefore, the conclusion is that, the selection of plants for the production of ethanol is needed to consider the integrated transport system because the effect occurred depends on the solutions or the value of all the functions. In which, the distance of the transport has the greatest impact.

4 Summary

The results obtained from the differential evolution process by particle swarm with the methodology of PSO has given the solution with 2 % difference, but at a faster time than Lingo approximately 99 %, which saves time in solving problems. However, the DE-PSO is still not good enough because the solution obtained is higher than Lingo by approximately 2 %. Therefore, further research is required to find or add the strategic advantages of DE by applying the behavior of the mutation. The research that has been conducted is still using the standard approach, but there is as adjustment of F values automatically only. If it is improved by this new method and there is variety of selection of the strategies, the determination of vector to be used in the mutant process more 3 vectors or selecting the best vector to be calculated, it will provide a variety of solutions and makes it better and faster as well.

In addition to the development of the new algorithm in order to get the best solution with less time for the experiment, there is another interesting approach to the future development which is to apply the developed algorithms to the standards of problem solving that are closed to the problem of choosing the location and assignment of multi-step routes and multi-purposes such as a P-median, Location-allocation, location routing problem or the problem of selecting the location and assigning the one-step route and so on.

Acknowledgments Ubon Ratchathani University research team thanks to the University for the Knowledge of how to do research and the Energy Conservation Promotion Fund; the Department of Energy for the financial support of this research.

References

Medaglia AL, Villegas JG, Rodriguez-Coca DM (2009) Hybrid bi-objective evolutionary algorithms for the design of a hospital waste management network. J Heuristics 15:153–176

Buddadee Bancha, Wirojanagud Wanpen, Watts Daniel J, Pitakaso Rapeepan (2008) The development of multi-objective optimization model for excess bagasse utilization: A case study for Thailand. Environ Impact Assess Rev 28(6):380–391

Stummer C, Doerner K, Focke A, Heidenberger K (2004) Determining location and size of medical departments in a hospital network: a multiobjective decision support approach. Health Care Manage Sci 7:63–71

Lin CKY, Kwok RCW (2006) Multi-objective metaheuristics for a location-routing problem with multiple use of vehicles on real data and simulated data. Eur J Oper Res 175(3):1833–1849

Cantarella GE, Vitetta A (2006) The multi-criteria road network design problem in an urban area. Transportation 33:567–588

Jozefowiez N, Semet F, Talbi E (2008) Multi-objective vehicle routing problems. Eur J Oper Res 189:293–309

Xu J, Liu Q, Wang R (2008) A class of multi-objective supply chain networks optimal model under random fuzzy environment and its application to the industry of Chinese liquor. Inform Sci 178:2022–2043

Doerner KF, Gutjahr WJ, Nolz PC (2009) Multi-criteria location planning for public facilities in tsunami-prone coastal areas. OR Spectrum 31(3):651–678

Doerner K, Focke A, Gutjahr WJ (2007) Multicriteria tour planning for mobile healthcare facilities in a developing country. Eur J Oper Res 179:1078–1096

Raisanen L, Whitaker RM (2005) Comparison and evaluation of multiple objective genetic algorithms for the antenna placement problem. Mob Networks Appl 10:79–88

Nanthasamroeng N, Pitakaso R, Buddadee B (2008) A multi objectives model for multi-echelon location problem: application in ethanol plant location analysis in Thailand. In: Proceeding in international conference on intelligent manufacturing and logistics, Waseda University

Caballero R, Gonzalez M, Guerrero FM, Molina J, Paralera C (2007) Solving a multiobjective location routing problem with a metaheuristic based on tabu search: application to a real case in Andalusia. Eur J Oper Res 177:1751–1763

Leung SCH (2007) A non-linear goal programming model and solution method for the multi-objective trip distribution problem in transportation engineering. Optim Eng 8:277–298

Drezner T, Drezner Z, Salhi S (2006) A multi-objective heuristic approach for the casualty collection points location problem. J Oper Res Soc 57:727–734

Uno T, Katagiri H (2008) Single- and multi-objective defensive location problems on a network. Eur J Oper Res 188:76–84

Thongdee Thongpoon, Pitakaso Repeepan (2012) Solving a multi-objective, source and stage location-allocation problem: a case study of a bagasse and cassava pulp ethanol plant in northeastern Thailand. KKU Res J 17(1):71–87

Thongdee Thongpoon, Pitakaso Repeepan (2013) Solving a multi-objective, source and stage location-allocation problem using MODDE. KKU Res J 18(4):662–676

A Study on the Optimum Location of the Central Post Office in Bangkok: Applying the Travelling Salesman Problem

Keizo Wakabayashi, Akihiro Watanabe, Jun Toyotani, Kuninori Suzuki, Koichi Murata and Sarinya Sala-ngam

Abstract Many post offices can be seen in Thailand and most Thai people use post offices to send their packages. However customers do not get good service from post offices as expected and many problems can be found, such as the big problem of package lateness. Therefore, the post offices in Thailand are required to think of a new system of effective delivery to bring back reliability from customers. In order to find a solution of these problems, the central post office is focused on in this study and to achieve this study of purposes, the travelling salesman problem will be applied in this chapter.

Keywords Optimum relocation · Travelling salesman problem · Optimum total travelling distance · Cluster-first route-second method · Latitude/longitude

1 Introduction

In many logistics companies in Thailand, most Thai people use post offices to send their packages, because the cost is low and many post offices can be seen. Therefore, post offices can be found easier than other logistics companies. However, customers do not get good service from post offices. Many problems can be found, such as lateness on delivery and damaged or lost packages without the post offices being held responsible.

Obviously, the post offices in Thailand need to think of a new system of effective delivery to bring back reliability from customers, provide professional services to customers, and correspond to customer's needs. To find a solution of

K. Wakabayashi · A. Watanabe · J. Toyotani · K. Suzuki · K. Murata · S. Sala-ngam (✉)
Industrial Engineering and Management, College of Industrial Technology, Nihon University, 1-2-1 Izumi, Narashino, Chiba 275-8575, Japan
e-mail: Cisa14001@g.nihon-u.ac.jp

K. Wakabayashi
e-mail: Cisa12002@g.nihon-u.ac.jp

P. Golinska (ed.), *Logistics Operations, Supply Chain Management and Sustainability*, EcoProduction, DOI: 10.1007/978-3-319-07287-6_37,

these problems, the post offices in Thailand are required to create a technique in order to search for the best location for its central post office. Therefore, the central post office is focused on in this study. In appearance, the central post office where packages are compiled from post offices and then distributed to their target post offices would be in the middle of that area. However, the central post office will need to find the optimum relocation based on determining facts, such as information on the frequency of delivery and the distances between the central hub post office and its spoke post offices.

As noted above, when determining the optimum relocation on a map, if geographic information of post offices and delivery destinations can be defined with the Euclid space, the optimum relocation for the central post office will be the position on a map, which are the shortest total distances by the frequency of delivery and the Euclid distance.

The purpose of this study are to examine the efficiency of the present central post office in Bangkok and in order to find the optimum relocation for the central post office in order to solve the problems started above and to provide professional services to customers. In order to find the best location for the central post office, the information from the post offices, such as the distance between the other post offices and the central post office, the number of package are required. In recent years, according to the development of Information Technology, addresses have been able to be changed into position coordinates (latitude and longitude). Google and Google Maps are popularized generally. These web sites are used to find target positions and distances on a map. This study is centralized on the central post office and the post offices in Bangkok, Thailand, and is focused on finding the optimum relocation for the central post office. To achieve the purposes of this study, Google Maps and The Geocoding web site are used to find the target locations and change its addresses into position coordinates, and the average of these coordinates will be used to find the optimum relocation. Moreover, the solving the travelling salesman problem is applied to examine whether another place where becomes the best possible location than optimum location by used density function.

2 The Methods Used to Find the Optimum Relocation for the Central Post Office

First of all, the Monte Carlo method is said to be the easiest method in finding the optimum location for the central post office using the distances between and the location of the head post offices in certain areas. It is a method of asking for the latitude and longitude of some place in the specified area where post offices can be seen, and searching for the point that is calculated as the arrange distance between these locations and the central post office.

In addition, in order to solve this problem by consulting the travelling salesman problem, generally programming language such as C language is used to search for the optimum relocation automatically in comparison with the shortest possible overall round-trip. In this study, solving this simple problem and the optimum relocation for the central post office is found by shortening the overall round trip distances by the model of the cluster-first route-second method (Kubo 2007) which is one way of solving travelling salesman problem.

2.1 *Formulation of the Relocation Place by the Density Function*

In this study, the position of post offices is set to $(\bar{x}_j, \bar{y}_j)$ as $j = 1 \sim N$, and the total number of post offices is set to N. Moreover, as a method of finding the optimum relocation the density function model researched by Okabe and Suzuki (2008) will be consulted in this study. The density function of Okabe and Suzuki showed that if the density of post offices in scope is defined as $f(\bar{x}_j, \bar{y}_j)$, the total number of post offices in the whole region is determined by integrating the whole region using the following formula.

$$N = \iint f(x, y)dxdy \tag{1}$$

Incidentally, a certain specific point $(\bar{x}_j, \bar{y}_j)$, and the number of post offices in site area A is given by the following formula.

$$p_j = \iint_A f(\bar{x}_j, \bar{y}_j)dxdy \tag{2}$$

According to formula (2), when the point on a map of the central post office is set to *(x, y)*, the distance from the central post office to other post offices is given be the next formula.

$$d = \sqrt{(x - \bar{x}_j)^2 + (y - \bar{y}_j)^2} \tag{3}$$

Moreover, if all the distances from the central post office to other post offices are added, the total distance T is obtained. And the distance from the central post office of one time of delivery to another post office is shown by t_j. Therefore, T shows the sum total distance of all the circumferential delivery records.

$$t_j = d_j p_j = \sqrt{(x - \bar{x}_j)^2 + (y - \bar{y}_j)^2}\frac{1}{N} \tag{4}$$

$$T = \sum_{j=1}^{N} d_j p_j = \sum_{j=1}^{N} \sqrt{(x - \bar{x}_j)^2 + (y - \bar{y}_j)^2} \iint_A f(\bar{x}_j, \bar{y}_j dxdy \tag{5}$$

In each package delivery record, to add the number of times of delivery as one time per matter to all the operation records, the value of density function for one delivery record is 1/N. Moreover, when delivering packages to the same place, as m times in the unit of month or year, delivery is as follows.

$$p_j = \iint_A f(\bar{x}_j, \bar{y}_j)dxdy \cong \frac{1}{N} \times m \tag{6}$$

$$T = \sum_{j=1}^{N} d_j p_j = \sum_{j=1}^{N} \sqrt{(x - \bar{x}_j)^2 + (y - \bar{y}_j)^2}\frac{m}{N} \tag{7}$$

Next, in order to find the location of the central post office, what is needed is the shortest total distance from the central post office to other post offices. Namely, if the total distance can be defined as formula (7), it becomes possible to calculate the shortest value. Furthermore, it is possible to find the partial differentiated value if by using each coordinate system of space in formula (8) and (9), the conditional formula turns out to be 0.

$$\frac{\partial T}{\partial x} = 0 \tag{8}$$

$$\frac{\partial T}{\partial y} = 0 \tag{9}$$

And for this value to be the minimum value, the following condition needs to be fulfilled.

$$\frac{\partial^2 T}{\partial x^2} \rangle 0 \tag{10}$$

$$H = \frac{\partial^2 T}{\partial x^2}\frac{\partial^2 T}{\partial y^2} - \frac{\partial^2 T}{\partial x \partial y}\frac{\partial^2 T}{\partial y \partial x} \rangle 0 \tag{11}$$

Thus, the position of the optimum relocation for the central post office is determined as x, y by solving the constructed equation from above [formula (10) and (11)].

As mentioned before, to easily achieve the purposes of this study, Google Maps and the Geocoding web site are used to change post offices's addresses in Bangkok, Thailand into latitude and longitude data, and then find the optimum relocation for the central post office by calculating the average of those latitudes and longitudes. In order to find the optimum location, the coordinate value which makes total of distance from post offices to other post offices into the minimum needs to be calculated, even though distribution of the total distance like T at formula (7) is shown when making the position of the central post office change. Thus, if formulas

(10) and (11) are solved, and checked and the partial differentiating value is positive, the following formula will be obtained.

$$\frac{\partial T}{\partial x} = \sum_{j=1}^{N} (2x - 2x_j) = 0 \tag{12}$$

$$\frac{\partial T}{\partial y} = \sum_{j=1}^{N} (2y - 2y_j) = 0 \tag{13}$$

And the following formula is obtained by solving formula (12) and (13) in more detail.

$$x = \frac{\sum_{j=1}^{N} x_j}{N} \tag{14}$$

$$y = \frac{\sum_{j=1}^{N} y_j}{N} \tag{15}$$

Therefore, the position coordinate can be solved by formula (14) and (15). To find the optimum relocation for the central post office and by calculating the average of these position coordinates, this method is very effective in achieving these purposes, so the destiny function by Okabe and Suzuki is used in this study.

2.2 The Solving Method of the Travelling Salesman Problem

Many methods can be solved using the travelling salesman problem. However, to achieve these purposes, the Cluster-First Route-Second Method is adopted in this study. First, as a result of the density function method, the point that can be found by the density function method is called the Calculating Point. And then, that calculating point is made the center and about eight points are designed to arrange around it, and post offices are divided into some cluster using these points. On this occasion, the sum total of the treated number of packages which is contained in a cluster are required to make it not exceed the loading capacity of a track. In various methods for the division of a cluster, the area segmentation method is considered to be the best solution in the case of surrounding the central post office with all other post offices. Moreover, in methods of dividing area segmentation into various forms, as is determined by the study scope, the sectorial partitioning scheme is thought of to be the best.

Next, the search for the round way begins by arranging points and patrols the point by having added other post offices to post offices included in each cluster. The round ways are made to repeat rounding from nine arranging points starting from one point and then going to other post offices respectively included in a same cluster until finishing all the arranging points and patrolling post offices in all clusters. As a result of this, of the nine arranging points (including calculating point), the point that the total distance travelling around all of post offices is the shortest is considered to be the optimum relocation.

Therefore, in order to find the optimum relocation for the central post office, the density function by Okabe and Suzuki is used in this study. Not only that, but to confirm that the optimum location that is found by using the density function method of Okabe and Suzuki is really the best, the cluster-first route-second method is also applied to achieve the purposes in this study.

3 Methodology

3.1 Scope

In Fig. 1, Bangkok, Thailand is centered as scope on this study and the position of the central post office and post offices are investigated. As a result of this investigation, the present central post office (yellow point), can be found in one place and the post offices (blue point) can be found in twenty-five places.

3.2 Method of Obtaining Latitude and Longitude Information

The Geocoding web site is used to obtain the position coordinates of the central post office and the post offices in Bangkok automatically by changing the addresses to latitude and longitude. The most famous API web site is Google Maps, thus this web site is used to change address to latitude and longitude by a lot of people.

3.3 The Cluster-First Route-Second Method

First, the scope is divided like a form of sectorial partitioning scheme. And Google Maps is used to find the distances from each arranging point to the post offices by one place and the distances between the post offices with other post offices.

Fig. 1 The positions of the central post office and other post offices in Bangkok

Secondly, a list of distances ranging from shortest to longest is created. Finally, the optimum relocation can be found by using the total distances of round ways found in this list.

4 Results

The Table 1 shows latitude and longitude information of the post offices in Bangkok which changed from addresses to positions coordinate by the Geocoding web site.

From the density function of Okabe and Suzuki which is mentioned at Chap. The Impact Of Demographic Changes On Human Resources Management In European Supply Chains-Selected Aspects, the optimum location coordinate can be found simply by calculating the average of these latitude and longitude by using formulas (14) and (15), as following:

$$x = \frac{344.082568}{25} = 13.76330272$$
$$y = \frac{2514.036828}{25} = 100.5614731$$

Resulting from above, the coordinates of the optimum relocation would be at 13.76330272, 100. 5614731. And this position can be searched for on a map by using the Geocoding web site in Fig. 2.

Table 1 Latitude and longitude of the central post office and other post offices

No.	Post office	Latitude	Latitude
1	Central BKK post office	13.742445	100.518345
2	Jatujak post office	13.851296	100.581926
3	Bang khun thian post office	13.685206	100.445237
4	Luksi post office	13.882652	100.571219
5	Samsen nai post office	13.793772	100.54937
6	Dusit post office	13.783218	100.515266
7	Rat burana post office	13.668633	100.505228
8	Bangkok noi post office	13.75521	100.471638
9	Khlong chan post office	13.767207	100.650403
10	Ram inthra post office	13.873657	100.598514
11	Yannawa post and telegraph office	13.6986	100.494373
12	Phasi charoen post office	13.709188	100.385776
13	Bang sue post office	13.817278	100.532586
14	Bangna post office	13.668899	100.60478
15	Bang rak post office	13.728048	100.515235
16	Rong muang post office	13.744841	100.522643
17	On nut post office	13.714263	100.648354
18	Phlapphla chai post office	13.752623	100.514213
19	Ratchadamnoen post office	13.757871	100.499811
20	Lat krabang post office	13.743052	100.791318
21	Lat phrao postoffice	13.789881	100.603461
22	Nong chok post office	13.85243	100.860973
23	Phra khanong post office	13.714618	100.592444
24	Samre post office	13.715029	100.489509
25	Chorakhe bua post office	13.834006	100.660146
26	Taling chan post office	13.78109	100.432405
	Total	**344.082568**	**2514.036828**

Fig. 2 Result of the optimum location by using the density function of Okabe and Suzuki

Fig. 3 Eight points are arranged around calculating point

Table 2 Result of the total travelling distances from eight arranging points to all post offices

No.	Arranging points	Total travelling distances (Km)
1	First point	297.9
2	Second point	315.6
3	Third point	321.5
4	Fourth point	298.3
5	Fifth point (calculating point)	305.6
6	Sixth point	312.0
7	Seventh point	303.1
8	Eighth point	305.5
9	Ninth point	301.8

Next, to find the best possible relocation more than the optimum location by density function method, the optimum location by density function is centered, and eight points are arranged around it in Fig. 3.

Moreover, with solving the travelling salesman problem, the optimum relocation can be found using the shortest total distance of the travelling all the post offices by making the starting point move to arranged points and patrolling all post offices in every cluster, moving it from one point and repeating the process. In Table 2 shows a result of the travelling distances from arranging points to all post offices.

From the result of this, the optimum relocation for the central post office in Bangkok is the first arranging point in Fig. 4.

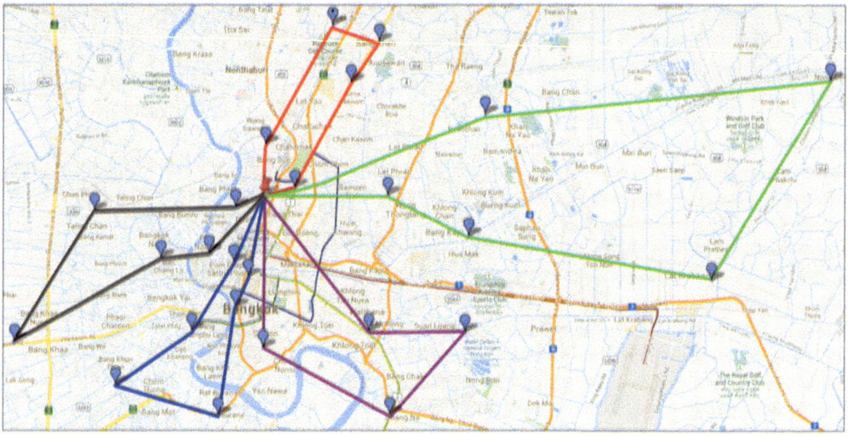

Fig. 4 The optimum round ways from the first arranging point

Fig. 5 The optimum relocation for the central post office in Bangkok, Thailand

Table 3 Distance from current the central post office and the optimum location by using the density function method and the cluster-first route-second method to all post offices

No.	Post offices	Distance from current the central post office	Distance from the optimum location by the density function	Distance from the optimum location by the cluster-first route second method
1	Current the central post office	0	8.1	6.5
2	The optimum location by the density function method	8.1	0	7
3	The optimum location by the cluster- first route-second method	6.5	7	0
4	Jatujak post office	16.9	11.6	9.5
5	Bang Khun Thian post office	17.6	23.8	18.3
6	Luksi post office	25.8	18.2	14.7
7	Samsen Nai post office	9.7	6.8	3.1
8	Dusit post office	6.2	7.1	3
9	Rat Burana post office	12.6	17.9	18
10	Bangkok Noi post office	9.4	13.1	10.4
11	Khlong Chan post office	19.6	13.4	15.9
12	Ram Inthra post office	23.8	15.3	13.2
13	Yannawa post and telegraph office	8.8	17.9	13.3
14	Phasi Charoen post office	22.5	29	22.2
15	Bang Sue post office	15.4	14.3	3.8
16	Bangna post office	18.2	17	20.3
17	Bang Rak post office	3.3	10.2	8.1
18	Rong Muang post office	0.9	7.1	5.7
19	On Nut post office	20.3	15.6	19.4
20	Phlapphla Chai post office	1.9	8.4	6.3
21	Ratchadamnoen post office	5.2	9.6	6.3
22	Lat Krabang post office	41.2	28.5	33
23	Lat Phrao post office	18.7	7.4	8.8
24	Nong Chok post office	50.3	40	44.5
25	Phra Khanong post office	12.5	8	15
26	Samre post office	7.0	15.1	10.9
27	Chorakhe Bua post office	31.8	20.3	19.7
28	Taling Chan post office	15.0	16.6	13.9
	Total	**429.1083333**	**407.3**	**370.8**

5 Conclusions

From this study, in order to find the optimum relocation for the central post office in Bangkok, Thailand, first the density function of Okabe and Suzuki is used by calculating the average of these latitude and longitude with using formulas (14) and (15), as following:

$$x = \frac{344.082568}{25} = 13.76330272$$
$$y = \frac{2514.036828}{25} = 100.5614731$$

From above, to find this location, the Japanese Geocode web site is used to find that location on a map. As a result from this, the optimum location which used the density function of Okabe and Suzuki can be seen in Fig. 5. Next to find the best possible relocation more than the optimum location by using the density function method, the cluster-first route-second method is used. A result of this can be solved by arranging eight points around the optimum location by using the function method and then find the shortest of the total distances from the arranging points to all post offices. A result of the optimum relocation by using the cluster-first route-second method can be seen in Fig. 5.

In Fig. 5, the yellow point is the present location of the central post office in Bangkok, the green point is the optimum relocation determined by using the travelling salesman problem, and the red point is the optimum relocation determined by using the density function of Okabe and Suzuki.

As a result, the optimum relocation for the central post office in Bangkok, Thailand can be found by using the density function of Okabe and Suzuki and the travelling salesman problem; the cluster-first rout-second method. Not only that but the total distances between the optimum location found by the travelling salesman problem, the optimum location found by the density function by Okabe and Suzuki and the present location of the central post office with the post offices are found and compared. As results from study on the travelling salesman problem, the total distance from the optimum location to all the post offices is improved by 13.8 % and from study on the density function by Okabe and Suzuki, the total distance from the optimum location to all the post offices is improved by 5.4 % with calculating the total distance from current the central post office and the optimum location by using the density function and the cluster-first route-second method to all post offices which are shown on the Table 3.

Therefore, from the result of this study, the green point is considered to be the optimum relocation for the central post office in Bangkok, Thailand. This place is Nakhon Chai Si road, Dusit, Bangkok. Many problems which are found in the post offices in Bangkok, Thailand will be solved and delivery will become more efficient if the central post office is relocated here.

References

Google Maps web site. https://www.maps.google.com
Japanese Geocoding Web site. https://www.geocoding.jp
Kubo M (2007) Applied mathematics of logistics. 1st edn, Kyoristu, Japan, pp 176–184
Okabe A, Suzuki A (2008) Mathematics of the optimum relocation. 9th edn, Asakura bookstore, Japan, (mathematics of people of today series), pp 52–79
Thailand post office web site. https://www.thailandpost.co.th
Cook WJ, Matsuura S (2013) In pursuit of the traveling salesman: mathematics at the limits of computation. 1st edn, Degu and Houeisya, Japan, pp 14

A Closed-Loop Capacitated Warehouse Location Model with Risk Pooling

Nabil Kenan, Marwa Attiya and Bedoor AlShebli

Abstract The irreversible environmental harm caused by various industrial operations, along with the newly enacted environmental regulations, shifts the focus of governments and corporations in the direction of sustainable supply chains rather than the traditional ones. One of the main ways this can be done is by introducing a reverse supply chain, in addition to the traditional forward supply chain, to create the so-called closed-loop supply chain. We extend a capacitated warehouse location model with risk pooling (CLMRP) by adding a reverse supply chain where recovered products are sorted and collected by collection centers (CCs) and sent to remanufacturing centers (RCs). The products are then either sent back to the forward supply chain or sent to recycling centers or disposal sites. The objective of the closed loop capacitated warehouse location model with risk pooling (CLCLMRP) is to minimize the sum of the fixed facility location, transportation, and inventory carrying costs for both the forward and reverse supply chains. The model will also determine distribution centers (DCs) and remanufacturing centers' (RCs) locations, shipment sizes and frequency from plant to DCs and RCs, working inventories and safety stock levels at DCs and RCs and assignment of retailers' demand to the DCs and returns to the RCs. Results showed that an increase in both shipping costs and inventory holding costs result in moving the supply chain towards a decentralized supply chain. An increase in both level of returns from the retailers to the CCs and from the CCs to the RCs leads to an increase in the total cost of the closed-loop supply chain.

N. Kenan (✉) · M. Attiya · B. AlShebli
Department of Engineering Systems and Management,
Masdar Institute of Science and Technology, Abu Dhabi, UAE
e-mail: nkenan@masdar.ac.ae

M. Attiya
e-mail: mattiya@masdar.ac.ae

B. AlShebli
e-mail: balshebli@masdar.ac.ae

P. Golinska (ed.), *Logistics Operations, Supply Chain Management and Sustainability*, EcoProduction, DOI: 10.1007/978-3-319-07287-6_38,

Keywords Capacitated · Closed-loop · Disposal · Recycling · Reverse supply chain · Supply chain

1 Introduction

Dealing with supply chains from a strategic point of view and an operational point of view separately, due to the nature of their time horizon (Ozsen et al. 2008), is now a thing of the past. Operational decisions such as inventory management are considered short term, while strategic decisions such as facility locations are considered long term. By looking at the big picture, firms realize that in order to achieve big cut downs on costs and develop an efficient supply chain, with customer satisfaction as a basis for success, optimizing the process as a whole has become a necessity. Therefore, there is clearly a need to integrate strategic level and operational level models in a way to facilitate solving the facility location, distribution, and inventory management problems simultaneously.

Irreversible harm is being done to the environment, and business practices are contributing towards a big portion of this harm. Stern (2006) predicts that an annual loss of 5 % in the GDP will occur if strong and early actions are not taken in that matter. This has shifted the focus of governments and businesses in the direction of sustainable practices away from the traditional business ones.

According to Brundtland (1987), in order to achieve sustainability, we need to meet the needs of the present without compromising the ability of the future generations to meet their own needs. One of the most important ways in making a supply chain more sustainable is introducing a reverse supply chain that focuses on product recollection and remanufacturing. Along with the forward supply chain, this forms a closed-loop supply chain.

Reverse logistics in broader perspective involves activities related to resource reduction, recycling, substitution, reuse of materials and disposal (Stock 1998). Also, it includes planning and control of product flow from the manufacturer till the recovery or disposal point (De Brito and Dekker 2004). The new emerging field of reverse logistics proves to be one of the most challenging research areas in terms of efficient design of a product recovery network. According to Abdallah et al. (2012), the efficient design of a product recovery network is one of the main challenges facing the new emerging field of reverse logistics.

In this paper, we extend the capacitated warehouse location model using risk pooling (CLMRP) introduced by Ozsen et al. (2008) and later solved using a genetic algorithm by Diabat et al. (2009). The extension will focus on developing a reverse logistics supply chain where the product will be collected from the end-user, remanufactured and reentered into the forward supply chain once again.

The extension of the model will be a closed-loop capacitated warehouse location model with risk pooling (CLCLMRP). In the model, we consider a supply chain network in which a single plant produces a single product and delivers it to a set of retailers, with uncertain demand, through a set of capacitated DCs. Since the demand of the retailers is uncertain, the DCs must keep safety stock of inventory in order to prevent stock-outs during lead times and maintain customer goodwill. Single sourcing is assumed, where each retailer is only assigned a single DC.

Returned products will be sent back to the forward supply chain through capacitated RCs. The objective of the RCs is to remanufacture the products and re-enter them into the forward supply chain or to disassemble the functioning parts and send them back to the plant. Single sourcing is also assumed in this case, where each retailer sends the returned products to a single RC.

The objective of the CLCLMRP is to minimize the sum of the fixed facility location, transportation, and inventory carrying costs for both the forward and reverse supply chains. The model will also determine the DC and RC locations, shipment sizes and frequency from plant to DC and RC, working inventories and safety stock levels at DCs and RCs and assignment of retailers' demand to the DCs and returns to the RCs.

2 Literature Review

Our supply chains literature is divided into three major parts: traditional forward location theory, reverse logistics theory, and risk pooling. Traditional forward location theory deals with determining the optimal number of allocated DCs and their locations. Erlebacher and Meller (2000) formulate a model that deals with DC locations based on the assumption that customer locations are continuously represented. Daskin et al. (2002) formulate a DC location model that takes into consideration working inventory and safety stock costs in addition to transportation costs. The model was formulated as a nonlinear integer programming model and a Lagrangian relaxation model is proposed as solution. Shen et al. (2003) consider a joint location-inventory problem with a single supplier involved and multiple retailers. In their model, some retailers serve as DCs to accommodate the stochastic demand involved in the problem. The lower bound from their model gives the optimal solution. Diabat et al. (2013c) solve a multi-echelon joint inventory-location model using a Lagrangian relaxation based approach. The model is a mixed integer nonlinear program and the efficiency of the algorithm was studied through a computational study. Le et al. (2013) develop a mathematical model for a specific inventory routing problem and solve it using a column generation approach and this model is later solved bu Diabat et al. (In press) using a hybrid tabu search based heuristic. They present computational experiments to show the effectiveness of their methodology. Al Zaabi et al. (2013) study the different barriers of implementing successful sustainable supply chain management. They study 13 different barriers found from different literature. Diabat et al. (2014) present a Lagrangian relaxation based heuristic to solve a multi-echelon joint inventory-location (MJIL) problem that makes location, order assignment, and inventory decisions simultaneously. The heuristic was capable of efficiently solving large-size instances of the problem. Their heuristic yields optimal or near-optimal solutions. Al Dhaheri and Diabat (2010) show that optimization can be used to reduce CO_2 emissions in different applications. Abdallah et al. (2013) integrate the installation of photovoltaic cells of different sizes on the rooftops of the different supply chain facilities and calculate the costs incurred per kWh.

Reverse logistics theory deals with the recollection of the products from the end user and delivering them to the RCs where they are either remanufactured or disassembled and entered once again into the forward supply chain. Abdallah et al. (2012) introduce an uncapacitated closed-loop location-inventory model. The model shows the interdependency between location-inventory decisions in the forward and reverse supply chain. They perform a sensitivity analysis and show that the value of the recovered products is important for a closed-loop network to be economically feasible. Diabat et al. (2013) introduce a closed-loop facility location problem with carbon emission constraints that comply with the Carbon Market. The study shows that procurement activities produce a major load of carbon emissions and hence remanufacturing can be introduced to decrease this cost and decrease the overall carbon footprint of the supply chain. Demirel and Gökçen (2008) developed a mixed integer mathematical model for a remanufacturing system in a closed-loop supply chain. The solution of the model provides optimal values for production and transportation quantities of both the remanufactured and manufactured products. Amin and Zhang (2013) built a closed-loop supply chain model with multiple plants, CCs, demand markets and products. They use a mixed integer linear program to minimize the total costs. Diabat et al. (2013b) consider a closed-loop location-inventory problem where each of the forward and reverse supply chains is single echelon and spare parts are manufactured at the RCs in the reverse supply chain and pushed back to the forward supply chain. Govindan et al. (2013) present an overview of using contracts as a coordination mechanism through a closed-loop supply chain by presenting an overview of contracting literature and classifying the different coordination contracts.

In order to accommodate for both location constraints and inventory constraints, many models have recently started using risk pooling in their work. Park et al. (2010) consider a single sourcing, forward supply chain with risk pooling. Their model is a nonlinear integer programming model that determines the locations where DCs should be opened and the assignment of DCs to suppliers and retailers to DCs to minimize transportation and inventory costs. Tavakkoli-Moghaddam et al. (2013) present a novel mathematical model that considers risk pooling in a three-level supply chain with single sourcing. Their model was formulated as a bi-objective stochastic mixed integer nonlinear programming model which aims to determine the locations of DCs, the assignment of retailers to DCs, the amount of products to be ordered and the inventory decisions in a sense that minimized total system costs and transportation times. Kumar and Tiwari (2013) consider a closed-loop supply chain model with risk pooling in their work. In order to benefit from the risk pooling they consider two cases, first when the retailers act independently and second when the DCs and retailers work jointly.

3 Closed Loop Capacitated Location Model with Risk Pooling

3.1 Previous Model Formulation

In their paper, Ozsen et al. (2008) work on the forward supply chain of the model using a *(Q, r)* model with Type *I* service and EOQ to determine the order quantity. Their CLMRP models the storage and movement of a single product from a single plant to a set *I* of retailers through a set *J* of candidate DCs. They assume direct shipments from the DCs to the retailers and that a retailer has to be served by a single DC. To introduce the CLMRP, they introduce the following notation.

I	the set of retailers, indexed by i
J	the set of DCs, indexed by j
μ_i	mean (per unit time) demand at retailer i, $\forall i \in I$
g_j	fixed cost per shipment from the plant to a DC located at location j, $\forall j \in J$
a_j	per-unit shipment cost from the plant to DC located at location j, $\forall j \in J$
f_j	total cost of establishing and operating a DC at location, j, $\forall j \in J$
d_{ij}	cost per unit shipped from DC j to retailer i, $\forall i \in I, j \in I$
Q_j	order quantity of the product from DC at location j to the plant. $\forall j \in J$
F_j	fixed cost of placing an order from distribution center j to the plant, $\forall j \in J$
C_j	available capacity at DC j, $\forall j \in J$
h	inventory holding cost per unit of commodity per unit of time at any warehouse
β	weight factor associated with the transportation cost, $\beta > 0$
θ	weight factor associated with inventory cost, $\theta > 0$
χ	number of days in a year
L_j	the lead time in days for deliveries from the plant to DC j, $\forall j \in J$
z_α	standard normal deviate such that $P(z \leq z_\alpha) = \alpha$

They define the following decision variables:

$$X_j = \begin{cases} 1, & \text{if we locate a DC at candidate site } j, \ \forall j \in J \\ 0, & \text{otherwise} \end{cases}$$

$$Y_{ij} = \begin{cases} 1, & \text{if demands at customer } i \text{ are assigned to a DC at candidate site } j, \ \forall i \in I, \forall j \in J \\ 0, & \text{otherwise} \end{cases}$$

They formulate the CLMRP as follows

Minimize

$$\sum_{j\in J}\left[f_jX_j+\beta\chi\sum_{i\in I}d_{ij}\mu_iY_{ij}+\theta hz_\alpha\sqrt{L_j}\sqrt{\sum_{i\in I}\mu_iY_{ij}}\right]$$
$$+\sum_{j\in J}\left[(F_j+\beta g_j)\frac{\chi\sum_{i\in I}\mu_iY_{ij}}{Q_j}+\beta\chi\sum_{i\in I}a_j\mu_iY_{ij}+\theta\frac{hQ_j}{2}\right] \quad (1)$$

Subject to

$$\sum_{j\in J}Y_{ij}=1,\ \forall i\in I \quad (2)$$

$$Y_{ij}-X_j\leq 0,\ \forall i\in I,\forall j\in J \quad (3)$$

$$Q_{capacity}+\left(z_\alpha\sqrt{L_j}\sqrt{\sum_{i\in I}\mu_iY_{ij}}+L_j\sum_{i\in I}\mu_iY_{ij}\right)\leq C_j,\ \forall j\in J \quad (4)$$

$$Q_j\geq 0,\ \forall j\in J \quad (5)$$

$$Y_{ij}\in\{0,1\},\ \forall i\in I,\forall j\in J \quad (6)$$

$$X_j\in\{0,1\},\ \forall j\in J \quad (7)$$

The objective function (1) sums the fixed cost of locating distribution centers, the DC-retailer transportation cost, the safety stock cost, and the working inventory cost. Constraints (2) require that each retailer is assigned to exactly one distribution center. Constraints (3) state that retailers can only be assigned to open DCs. (4) and (5) are constraints of the capacitated EOQ problem. (6) and (7) are standard binary constraints.

In order to make sure that the (uncertain) demand from the retailers is met, we add another two constraints and thus slightly modify the model in Ozsen et al. (2008) work. Constraints (8) ensure that the order quantity at each DC either equals the economic order quantity (EOQ) or does not exceed the capacity of the DC. Constraint (9) ensures that the demand from each DC is met no matter how many retailers are assigned to it.

$$Q_j=\min(Q_{EOQ,j},Q_{capacity}),\ \forall j\in J \quad (8)$$

$$\sum_{j\in J}\left(Q_j+\left(z_\alpha\sqrt{L_j}\sqrt{\sum_{i\in I}\mu_iY_{ij}}+L_j\sum_{i\in I}\mu_iY_{ij}\right)\right)X_j=\sum_{i\in I}\mu_i \quad (9)$$

3.2 Extension to the Previous Model

As discussed previously in Sect. 1, the extension will focus on adding a reverse supply chain with CCs and RCs to decrease the minimum overall carbon footprint of the entire chain. Figure 1 shows the flow of the product through the entire network.

M the set of CCs, indexed by m
K the set of RCs, indexed by k
S_k fixed annual cost of locating an RC at site $k, \forall k \in K$
F_m fixed annual cost of locating a CC at site $m, \forall m \in M$
δ fraction of average daily returns of expected average daily demand
d_{im} cost per unit shipped from retailer i to CC m, $\forall i \in I, \forall m \in M$
d_{mk} cost per unit shipped from CC m to RC k, $\forall m \in M, \forall k \in K$
H_k fixed cost for placing an order for sub-assembly at RC $k, \forall k \in K$
D_k expected annual demand of RC k for subassemblies, $\forall k \in K$
Q_k order quantity of sub-assembly at RC $k, \forall k \in K$
C_k capacity constraint of RC $k, \forall k \in K$
n_k fixed cost of shipping an order to RC $k, \forall k \in K$
b_k per unit shipping cost of sub-assemblies to RC $k, \forall k \in K$
$\overline{L}$ the lead time of delivering the aggregate sub-assembly to the RCs
σ_i^2 variance (per unit time) of demand at retailer $i, \forall i \in I$
p fraction of average daily returns shipped from CC m to RC k

We define the following decision variables:

$$W_k = \begin{cases} 1, & \text{if we locate a RC at candidate site } k, \forall k \in K \\ 0, & \text{otherwise} \end{cases}$$

$$T_m = \begin{cases} 1, & \text{if we locate a CC at candidate site } m, \forall m \in M \\ 0, & \text{otherwise} \end{cases}$$

$$Z_{im} = \begin{cases} 1, & \text{if returns from customer } i \text{ are assigned to a CC at} \\ & \text{candidate site } m, \forall i \in I, \forall m \in M \\ 0, & \text{otherwise} \end{cases}$$

$$Z_{mk} = \begin{cases} 1, & \text{if returns collected by CC at site } m \text{ are assigned to} \\ & \text{an RC at candidate site } k, \forall k \in K, \forall m \in M \\ 0, & \text{otherwise} \end{cases}$$

We formulate the reverse logistics part of the CLCLMRP as follows:

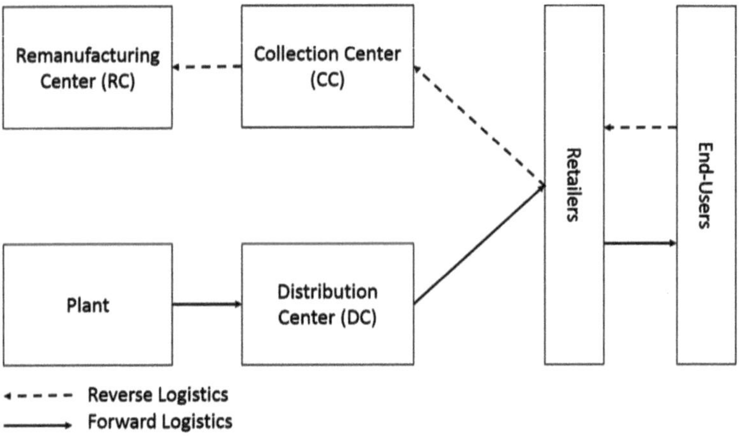

Fig. 1 Network diagram for the closed-loop supply chain

Minimize

$$\begin{aligned}&\sum_{k\in K} S_k W_k + \sum_{m\in M} F_m T_m + \beta \sum_{m\in M}\sum_{i\in I} \chi\delta\mu_i d_{im} Z_{im} + \beta \sum_{m\in M}\sum_{k\in K} p\chi\delta\mu_i d_{mk} Z_{mk} \\ &\quad + \sum_{k\in K}\left[\frac{H_k D_k}{Q_k} + \beta\left(\frac{n_k D_k}{Q_k} + b_k D_k\right) + \theta\frac{hQ_k}{2}\right] \\ &\quad + \left[\theta h z_\alpha \sum_{m\in M}\sqrt{\sum_{i\in I}\overline{L}\delta^2\sigma_i^2 Z_{im}}\right] \\ &\quad + \theta\sum_{m\in M}\sum_{i\in I} h\chi\delta\mu_i Z_{im} + \beta\sum_{i\in I} p\chi\delta\mu_i\end{aligned} \tag{10}$$

Subject to

$$\sum_{m\in M} Z_{im} = 1\,,\ \forall i \in I \tag{11}$$

$$\sum_{m\in M} Z_{mk} = 1,\ \forall k \in K \tag{12}$$

$$Z_{im} - T_m \leq 0,\ \forall i \in I, \forall m \in M \tag{13}$$

$$Z_{mk} - W_m \leq 0,\ \forall k \in K, \forall m \in M \tag{14}$$

$$Z_{im} \in \{0,1\},\ \forall i \in I, \forall m \in M \tag{15}$$

$$Z_{mk} \in \{0,1\},\ \forall m \in M, \forall k \in K \tag{16}$$

$$W_k \in \{0,1\}, \ \forall k \in K \tag{17}$$

$$T_m \in \{0,1\}, \ \forall m \in M \tag{18}$$

$$Q_k + z_\alpha \sum_{m \in M} \sqrt{\sum_{i \in I} \overline{L} \delta^2 \sigma_i^2 Z_{im}} \leq C_k, \ \forall k \in K \tag{19}$$

$$Q_k \geq 0, \ \forall k \in K \tag{20}$$

The first term in the objective function (10) is the annual fixed cost of locating RCs. The second term is the annual fixed cost of locating CCs. The third term is the cost of shipping all returned products from customers to CCs. The fourth term is the cost of shipping fraction p of returned products from CCs to RCs. The fifth term is the expected working inventory costs at RCs. The sixth term is the safety stock cost incurred at RCs. The seventh term is the holding cost incurred at CCs. The eighth term is the cost of shipping remanufactured products from RCs back to the plant.

Constraints (11) state that each retailer is allocated only one CC to ship the returned products to. Constraints (12) state that each CC can only ship the returned products to one RC. Constraints (13) state that the assignment of a retailer to a CC is not possible unless a CC is opened at site m.

Constraints (14) state that assignment of a CC to an RC is not possible unless an RC is opened at site k. Constraints (15) and (16) ensure single sourcing in the reverse supply chain where each retailer is allocated one CC, and each CC is allocated one RC. Constraints (17), (18) and (20) are standard integrality constraints. Constraints (19) ensure that the quantity of sub-assemblies ordered at RC k should not exceed the capacity of that RC.

4 Sensitivity Analysis

In this section, we summarize the effects of varying different parameters on the network of DCs, CCs, and RCs. The sensitivity analysis was done using GAMS. The instances were solved on an 8 GB RAM running a 2.6 GHz on the 64 bit version of the Windows 7 Professional operating system. We attempted a sensitivity analysis on a 5-city problem by considering only the first 5 cities from the 49-city dataset obtained from Daskin (1995). Each city represents a retailer, as well as a candidate DC, CC and RC location. We generated 5 instances each time to study the effect of varying the shipping cost factor (β), the inventory holding cost factor (θ), the level of returns factor (δ) and the level of shipping factor (p) on the decisions of opening DCs, CCs and RCs. The model coefficients used for the sensitivity analysis are shown in Table 1.

The fixed facility location costs of the DCs, CCs and RCs are assumed to be equal and are obtained by dividing the facility location costs in Daskin (1995) by

Table 1 Model coefficients for sensitivity analysis

Parameter	Description	Value
F_j, H_k	Fixed order cost at sites *j* and *k*	10
g_j, n_k	Fixed shipping cost at sites *j* and *k*	10
a_j, b_k, d_{im}, d_{mk}	Unit shipping cost from the plant to sites *j* and *k*	5
z_α	Service level parameter	1.96
h	Unit holding cost	1
χ	Days per year	365

100. The daily demand (μ_i) was obtained by dividing the population at each city in the dataset by 1,000. The standard deviation of the demand at each retailer (σ_i) was assumed randomly.

4.1 Effect of Shipping Costs

In order to study the effect of shipping costs, we set $\theta = 0.01$, $\delta = 0.5$, and $p = 0.5$, and varied the value of the shipping cost factor (β). The effect of shipping costs of the network of DCs, CCs and RCs are shown in Figs. 2 and 3. We observe that as the shipping costs increase, the number of DCs and RCs opened increases to mitigate the effect of the increase in shipping cost. The number of RCs is already at its maximum value (5) from the start because their capacities are low compared to those of the DCs and so the returned products from the retailers through the CCs to the RCs have to be spread among all the available DCs.

4.2 Effect of Inventory Costs

To study the effect of inventory costs, we set $\beta = 0.01$, $\delta = 0.5$ and $p = 0.5$, and varied the value of the inventory holding cost factor θ. The effects of inventory costs on the closed-loop network are shown in Figs. 4 and 5. The EOQ is inversely proportional to the inventory holding cost. This is why it is observed that as the inventory holding cost increases, the number of facilities opened increases in order to satisfy demand from the safety stock available at each location. With the increase in the number of facilities opened, the overall cost of the closed-loop supply chain increases which can be seen in Fig. 5. The increase is not linear because the opening of a new facility depends on whether the available facilities can still satisfy the demand or not. The number of RCs is at its maximum value (5) from the start because their capacities are low compared to those of the DCs and so the returned products from the retailers through the CCs to the RCs have to be spread among all the available DCs.

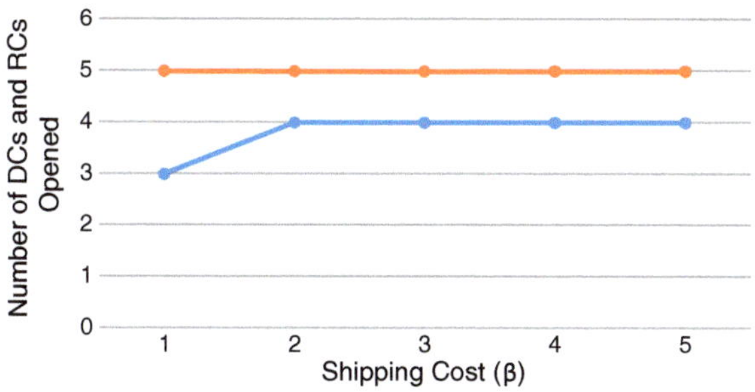

Fig. 2 Number of facilities open versus the shipping cost factor (β)

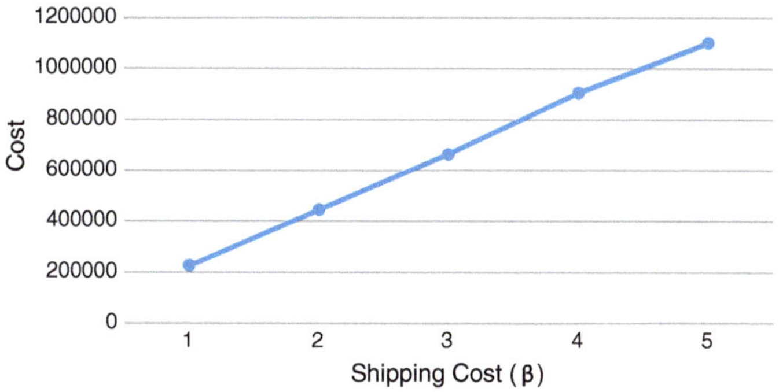

Fig. 3 Total cost of the closed-loop supply chain versus the shipping cost factor (β)

Fig. 4 Number of facilities open versus the inventory holding cost factor (θ)

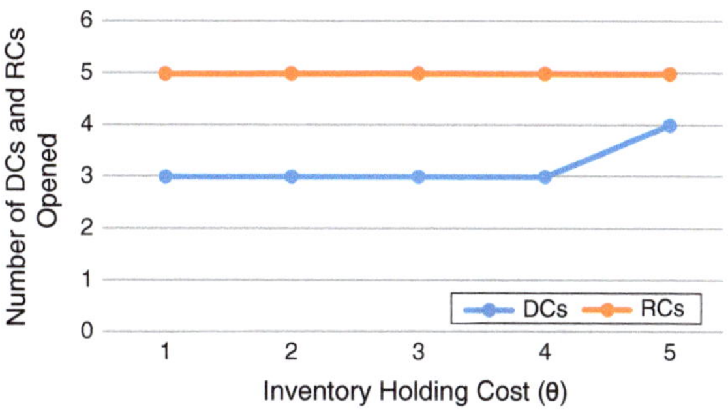

4.3 Effect of Level of Returns on the Closed-Loop Network

To study the effect of the level of returns, we set $\theta = 0.1$, $\beta = 0.02$, and varied the value of the level of returns factor (δ and p). The effects of the levels of return are shown in Figs. 6 and 7. We observe that as the level of returns increases, the total costs involved in the closed-loop supply chain increase. This is mostly because as

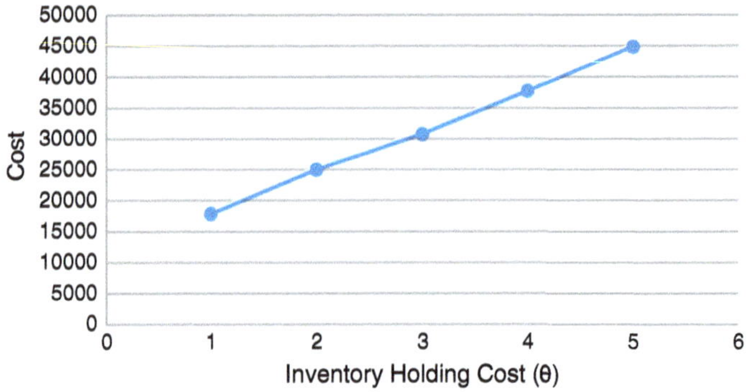

Fig. 5 Total cost of the closed-loop supply chain versus the shipping cost factor (θ)

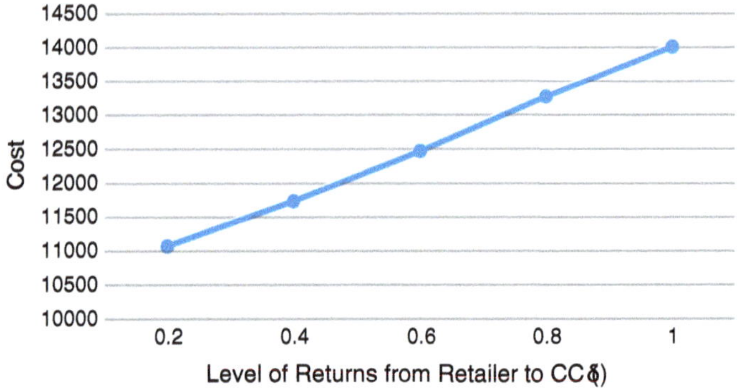

Fig. 6 Total cost of the closed-loop supply chain versus the Level of Return (δ)

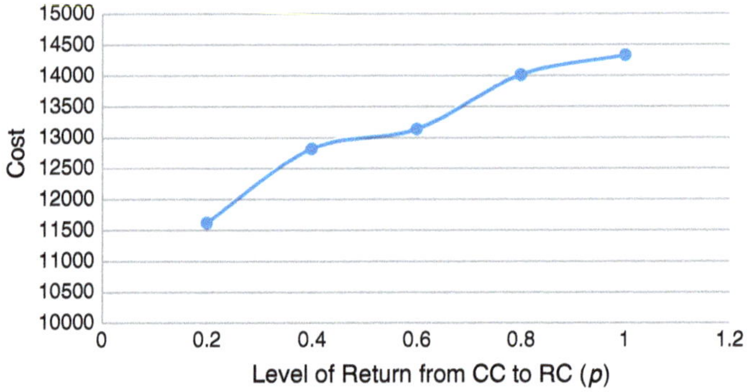

Fig. 7 Total cost of the closed-loop supply chain versus the Level of Return (p)

the number of returned products increases, the shipping costs involved increase. Also, the number of returned products are forcing more RCs to be opened and thus the cost of establishing and operating an RC becomes a part of the total cost.

5 Conclusions and Future Research

In this paper we have presented a formulation for a CLCLMRP problem, which integrates location and inventory decisions in both the forward and reverse supply chain. In order to integrate both the forward the reverse logistics CCs were established whose duty is to collect returned products from the suppliers, send them to the RCs which in turn remanufacture them and send them back to the forward supply chain.

Sensitivity analysis was used to the study the effects of the transportation costs, inventory holding costs and level of recovery. It was observed that an increase in both the inventory holding costs and the shipping costs resulted in the increase of the number of facilities opened and moved the supply chain towards a decentralized supply chain. A very important factor in deciding whether a facility should be opened at a given candidate site was the safety stock to be held at that site.

Our future research will focus on a number of additions to be made to the current work. As a start, the sensitivity analysis will be done beyond linear relationships. This will also include developing efficient algorithms to solve the CLCLMRP problem, considering capacities for the CCs in the loop, taking into consideration the fact that any order to be done from any DC can be satisfied by the plant and the RC. Another research that can have time invested in would be adding the cost of carbon emissions across the entire loop.

References

Abdallah T, Diabat A, Simchi-Levi D (2012) Sustainable supply chain design: a closed-loop formulation and sensitivity analysis. Prod Planning Control 23(2–3):120–133

Abdallah T, Diabat A, Rigter J (2013) Investigating the option of installing small scale PVs on facility rooftops in a green supply chain. Int J of Prod Eco 146(2):465–477

Al Dhaheri N, Diabat A (2010) A mathematical programming approach to reducing carbon dioxide emissions in the petroleum refining industry. In:Engineering Systems Management and Its Applications (ICESMA), 2010 Second International Conference on IEEE

Al Zaabi S, Al Dhaheri N, Diabat A (2013) Analysis of interaction between the barriers for the implementation of sustainable supply chain management. Int J Adv Manuf Technol 68(1–4):895–905

Amin SH, Zhang G (2013) A multi-objective facility location model for closed-loop supply chain network under uncertain demand and return. Appl Math Model 37(6):4165–4176

Brundtland G (1987) Our common future: Report of the 1987 World Commission on Environment and Development. Oxford University Press, Oxford

Daskin M (1995) Network and discrete location: models, algorithms and applications. Wiley, New York

Daskin MS, Coullard CR, Shen ZJM (2002) An inventory-location model: formulation, solution algorithm and computational results. Ann Oper Res 110(1–4):83–106

De Brito M P, Dekker R (2004) A framework for reverse logistics, Springer, Heidelberg

Demirel NÖ, Gökçen H (2008) A mixed integer programming model for remanufacturing in reverse logistics environment. Int J Adv Manufact Technol 39(11–12):1197–1206

Diabat A, Abdallah T, Al-Refaie A, Svetinovic D, Govindan K (2013a) Strategic closed-loop facility location problem with carbon market trading. IEEE Trans Eng Manag 60(2):398–408

Diabat A, Abdallah T, Henschel A (2013b) A closed-loop location-inventory problem with spare parts consideration. Comp Oper Res

Diabat A, Abdallah T, Le T (In press) A hybrid tabu search based heuristic for the periodic distribution inventory problem with perishable goods. Ann Oper Res

Diabat A, Aouam T, Ozsen L (2009) An evolutionary programming approach for solving the capacitated facility location problem with risk pooling. Int J Appl Decis Sci 2(4):389–405

Diabat A, Richard J-P, Codrington CW (2013c) A Lagrangian relaxation approach to simultaneous strategic and tactical planning in supply chain design. Ann Oper Res, 203(1): 55–80

Diabat A, Battaïa O, Nazzal D (2014) An improved lagrangian relaxation-based heuristic for a joint location-inventory problem. Comput Oper Res

Erlebacher SJ, Meller RD (2000) The interaction of location and inventory in designing distribution systems. IIE Trans 32(2):155–166

Govindan K, Popiuc MN, Diabat A (2013) Overview of coordination contracts within forward and reverse supply chains. J of Cleaner Prod 47:319–334

Kumar SK, Tiwari M (2013) Supply chain system design integrated with risk pooling. Comput Ind Eng 64(2):580–588

Le T, Diabat A, Richard J-P, Yih Y (2013) A column generation-based heuristic algorithm for an inventory routing problem with perishable goods. Optim Lett 7(7):1481–1502

Ozsen L, Coullard CR, Daskin MS (2008) Capacitated warehouse location model with risk pooling. Naval Res Logistics (NRL) 55(4):295–312

Park S, Lee TE, Sung CS (2010) A three-level supply chain network design model with risk-pooling and lead times. Transp Res Part E: Logistics Transp Rev 46(5):563–581

Shen ZJM, Coullard C, Daskin MS (2003) A joint location-inventory model. Transp Sci 37(1):40–55

Stern N (2006) The Great Britain treasury. 2006. The economics of climate change: the stern review, Cambridge University Press, Cambridge

Stock, J. R. (1998). Development and implementation of reverse logistics programs. In: proceedings of the annual conference , council of logistics management

Tavakkoli-Moghaddam R, Forouzanfar F, Ebrahimnejad S (2013) Incorporating location, routing, and inventory decisions in a bi-objective supply chain design problem with risk-pooling. J Ind Eng Int 9(1):1–6

A Joint Inventory-Location Model with CO_2 Emission Taken into Account in Design of a Green Supply Chain

Faisal Alkaabneh, Abdullah Kaya and Jasem AlHammadi

Abstract This chapter develops a green supply chain design model that incorporates the cost of carbon emissions into the objective function. The goal of the model is to simultaneously minimize logistics costs and the environmental cost of CO_2 emissions by strategically locating warehouses within the distribution network. A multi-echelon joint inventory-location model that simultaneously determines the location of warehouses and inventory policies at the warehouses and retailers is developed. The supply chain design model also integrates published experimental data to derive nonlinear concave expressions relating vehicle weight to CO_2 emissions as a transportation costs. The objective function determines the number of warehouses to establish, their location, the sets of retailers that are assigned to each warehouse, and the size and timing of orders for each facility to minimize the sum of inventory, shipping (including the CO_2 emissions effect), ordering, and location costs, while satisfying end-customer demand. The developed model is Mixed Integer Nonlinear model solved by using CONOPT solver in GAMS.

Keywords Joint-inventory location model · Green supply chain · Mixed integer nonlinear model

1 Introduction

As climate change concerns intensify, goal of decreasing carbon emissions due to human activities is recognized more important than ever both by public and private agents. Transportation of goods between the suppliers, warehouses and retailers is

F. Alkaabneh (✉) · A. Kaya · J. AlHammadi
Department of Engineering Systems & Management, Masdar Institute of Science & Technology, Abu Dhabi, UAE
e-mail: falkaabneh@masdar.ac.ae

P. Golinska (ed.), *Logistics Operations, Supply Chain Management and Sustainability*, EcoProduction, DOI: 10.1007/978-3-319-07287-6_39,

actually one of the activities highly responsible for carbon emissions (McKinnon 2007; Elhedhli and Merrick 2012). Efficient allocation of warehouses and inventory management can help to decrease the carbon emissions substantially. This study integrates two former studies to achieve an eco-friendly supply chain. The 'The Impact Of Demographic Changes On Human Resources Management In European Supply Chains-Selected Aspects' considers simultaneously the decision of locating warehouses and inventory policies to decrease the system wide costs of a supply chain (Diabat et al. 2013). We think that their decision of using economic order quantity (EOQ) for the inventory of retailers can yield good management of a supply chain. The 'Barriers of the Supply Chain Integration Process' incorporates carbon emissions into the objective of function (Elhedhli and Merrick 2012). They try to reduce carbon emissions and transportation costs at the same time by allocation of warehouses in a supply chain without consideration of inventory management. Hence, we adopt their approach of handling carbon emissions into our objective while considering both strategic facility and tactical inventory decisions at the same time. Our objective function minimizes the total location, inventory, ordering, shipping cost and carbon emissions simultaneously. This chapter is organized as follows. We discuss some important literature review in Sect. 2 which is about joint inventory-location models and handling of carbon emissions into the objective function. In Sect. 3 we introduce our description and formulation of the problem. In Sect. 4 we discuss our results after solving the problem by using GAMS. In the last section we deliver conclusions of our study and discuss potential work for the future.

2 Literature Review

With rapid expansion of the supply chain activities, the distribution networks expanded enormously in the global scale. The long traveling distances between nodes and warehouses leads to sharp increase in carbon footprint due to the emissions from vehicles. Thus, supply models have recently considered the carbon emission into account, where traditionally models focused mainly on minimizing fixed and operating cost (Elhedhli and Merrick 2012). Another study conducted a qualitative research to measure the performance of the green supply chain and its environmental counterparts, and they concluded through their findings that there are number of issue that still need to be addressed (Hervani et al. 2005). Merrick and Bookbinder developed nonlinear expressions that link the vehicle weight to the carbon dioxide (CO_2) emissions, where the average travel speed and the trip travel distance were important parameters (Merrick and Bookbinder 2010). Their finding indicated the importance of the facility locations for a green supply chain networks along with the inventory management.

Traditional inventory management decisions and facility location decisions are considered to be independent in supply chain network design whereas recent models integrate both decisions are simultaneously. The need for multi-echelon (location-inventory) models is increasing, raising more attention in taking both warehouse and retailers into consideration to minimize transportation costs. Nozick and Turnquist used multi product two-echelon inventory system to optimize the inventory location for individual products, and both location-inventory decisions were integrated (Nozick and Turnquist 2001). Miranda and Garrido proposed an integrated model of both inventory and location highlighting two warehouse capacity constraints (Miranda and Garrido 2006). One deals with the received lot size for orders and the other for stochastic bound on the inventory capacity. Shen and Qi studied a joint location-inventory model that aimed to minimize the cost of both of them through formulating a nonlinear programming model that later was solved by the proposed Lagrangian relaxation solution algorithm (Shen and Qi 2007). Moreover, they suggested a low-order polynomial algorithm can be solved by solving Lagrangian relaxation sub-problems in non-linear programming model. Rappold and Van Roo proposed a two-echelon, single item, service parts supple chain model with stochastic demand to solve integrated inventory-location and capacity investment problems (Rappold and Van Roo 2009). Their developed model helps in making efficient and feasible decisions of integrated facility location and multi-echelon inventory optimization. Diabat et al. have also studied a Multi-echelon Joint Inventory-Location (MJIL) problem that determines the location of warehouses and inventory policies simultaneously, and solved it by using Lagrangian-relaxation-based heuristic approach (Diabat et al. 2013).

As shown above, the work in the supply chain design and management is moving to new directions and insights that have not been taken into consideration before (Diabat et al. 2013a, b; Abdallah et al. 2012; Abdallah et al. 2013; Le et al. 2013; Zaabi et al. 2013; Diabat et al. 2009; Diabat et al. 2014; Al Dhaheri et al. 2011; Diabat et al. 2013c ; Diabat et al. (In press); Govindan et al. 2013). In the last decade, due to environmental and ecological responsibility, enterprises are trying to find innovative solutions to mitigate the CO_2 emissions due to human activities. One of those activities is the transportation system worldwide. Therefore, the logistics systems in the supply chain designs have to take into account the effects of CO_2 emissions to provide more sustainable supply chain designs and logistic systems that take emit as low as possible CO_2 emissions. Having this global new trend to reduce the CO_2 emissions, the motivation behind this chapter is to incorporate the CO_2 emissions cost as a transportation cost in the design of supply chain system as well as the consideration of the traditional decisions in the design of an efficient supply chain systems, specifically, the number of warehouses to establish, their location, the sets of retailers that are assigned to each warehouse, and the size and timing of orders for each facility to minimize the sum of inventory, shipping (including the CO_2 emissions effect), ordering, and location costs. An thus, the contribution of this chapter is to provide more insights in the design of supply chain system that incorporate the effect of CO_2 emissions as a cost in the objective function along with the traditional decisions that are mentioned earlier.Govindan et al. (2012) evaluated of performance measures and supply chain profit behavior under buyback, revenue sharing, quantity flexibility and advanced purchase discount contracts versus no coordination and wholesale price systems.

3 Problem Description and Reformulation

MJIL problem taking into considerations the cost of CO_2 emissions can be formulated as:

Minimalize

$$\sum_{j \in J} (f_j + \frac{\hat{k}_j}{\hat{T}_j}) X_j + \sum_{j \in J} \sum_{i \in I} (\frac{k_i}{T_{ij}} + b_{ij} + c_{ij} T_{ij} + e_{ij} \max\{\hat{T}_j, T_{ij}\}) Y_{ij} \quad (1)$$

Subject to

$$\sum_{j \in J} Y_{ij} \leq 1 \, for \; all \, i \in I \quad (2)$$

$$Y_{ij} \leq X_j for \; all \, i \in I, j \in J \quad (3)$$

$$Y_{ij}, X_j \in \{0, 1\} for \; all \, i \in I, j \in J \quad (4)$$

$$\sum_{j \in J} X_j = p \quad (5)$$

$$\hat{T}_j / T_{ij} = 2^{N_{ij}} for \; all \, i \in I, j \in J \quad (6)$$

$$N_{ij} \in Z for \; all \, i \in I, j \in J \quad (7)$$

$$\hat{T}_j, T_{ij} > 0 for \; all \, i \in I, j \in J \quad (8)$$

The objective function minimizes the total inventory, ordering, location, and shipping costs. Constraints 2 require that each retailer be assigned to exactly one warehouse. Constraints 3 prevent retailers from being assigned to warehouses that are closed. Constraints 4 require that the decision variables Y_{ij} and X_j are binary. Constraint 5 assures that the number of selected warehouses will be exactly as specified by the parameter p, which is the number of warehouses to locate. Finally, Constraints 6 through 8 require a power-of-two inventory policy at the retailers and the warehouses.

Let fj be the fixed cost of opening and operating warehouse j, let s_{ij} be the per-unit shipping cost from warehouse j to retailer i, and let $\hat{s}_j$ be the per-unit shipping cost from the plant to warehouse j. Let $\hat{T}_j$ be the time between orders placed by warehouse j to the plant, and let T_{ij} be the time between orders placed by retailer i to warehouse j (assuming that retailer i is served by warehouse j). Let $\hat{h}_j$, $\hat{o}_j$ be respectively the unit-inventory holding cost per unit time at warehouse j, and the fixed cost per order at warehouse j. Let h_i, o_i be respectively the unit-inventory holding cost per unit time at retailer i, and the fixed cost per order at retailer i. Let TRL is the truck load (i.e. the weight that the single truck can lift) and the dis_{ij} to be the distance between warehouse j and retailer i. And let COE the amount of CO_2 emitted from the truck in grams while transporting 1 ton of goods for 1 mile. Constraint 13 presents the holding cost.

Constraint 14 calculates the CO_2 cost in the model, where TCO_2 is the CO_2 emissions and PL is the truck payload. To model the effect of CO_2 emissions, we will assume that the amount of CO_2 emitted the supply chain network depends on the transportation system. Therefore, we modelled the cost of CO_2 based on the amount of CO_2 emitted from the trucks used to transport goods from the warehouses to the retailers and from the retailers to the warehouses. Different scenarios will be developed based on the truck type used on the transporting the goods (including the truck load) and cost weight of transportation (β_{trn}).

$$k_i = \beta_{inv} o_i t_b \tag{9}$$

$$\hat{k}_j = \beta_{inv} \hat{o}_j t_b \tag{10}$$

$$b_{ij} = \beta_{trn}(\hat{s}_j + s_{ij})(COE * dis_{ij}/TRL) d_i t_b \tag{11}$$

$$c_{ij} = \beta_{inv}(h_i - \hat{h}_j) d_i t_b \tag{12}$$

$$e_{ij} = 0.5 * \beta_{inv} \hat{h}_j d_i t_b \tag{13}$$

$$CO_2\_Cost = \beta_{trn} * \sum_i \sum_j dis_{ij} * d_i * COE * TCO_2/PL \tag{14}$$

where β_{inv} is the weighting factor for inventory and ordering costs, and is the weighting β_{trn} factor for inventory and ordering costs.

4 Computational Analysis

We tested our new formulation using GAMS version 24.0.2 using Dell workstation with Intel(R)Xeon(R) CPU and two E620 @2.40 GHz processors with Windows 7 operating system on the data set, derived from Daskin, a 49-node data set (48 continental US state capitals plus Washington, DC; 1990 census data) (Daskin 1995). The GAMS solver was specified to be CONOPT 3 version 3.15I. The parameters used in the computational analysis are:

Table 1 Emissions factor (g/ton-mile)

Truck type	Payload (tons)	CO_2
Type 8b	12.5	187
Type 5	3.1	230
Type 2b	1.6	289

$$\beta_{trn} \in \{0.01, 10\}$$
$$p \in \{3, 4, 5\}$$
$$\hat{o}_j = \$1.80$$
$$t_b = 60\,days$$
$$h_i = \$4.30$$
$$\hat{h}_j = \$2.60$$
$$o_i = \$2.10$$
$$d_i = \text{assumed by dividing the city population by 100}$$
$$COE = \$25/ton$$

Moreover, three different types of truck used for transporting the goods are used, each of this truck types has its own pay load and CO_2 emissions, and this is shown in Table 1 adapted from a previous study (Facanha and Horvath 2007):

The results are illustrated in Table 2. Column 1 in Table 2 gives the number of warehouses (WH) for the selected scenario, column 2 presents the scenario name and truck type and the value of β_{trn}, column 3 gives the number of warehouses found for the optimal configuration given the parameters for the scenario, the fourth column gives the CO_2 cost and the fifth column gives the total cost for the network including fixed costs of locating warehouses and transportation, ordering, and inventory costs.

Several interesting results are found from Table 1. For instance, the configuration changes with changing the truck type which clearly shows the effect of accounting the CO_2 cost with the capacity of the truck to transport the goods from the warehouses to the retailers. In scenario 1 the selected warehouses are located in AL, AR, and TX for the same number of warehouses and β_{trn} with truck type 8b. However the optimal warehouses locations are found in AR, ID, and NV for the truck type 5.

Another interesting finding is the effect of changing the CO_2 effect (i.e. β_{trn}). This is clearly illustrated by given the same truck system with the same number of warehouses to locate. As an example, for β_{trn} equals 0.01, three warehouses and truck type 8b the optimal warehouses are found to be AL, AR, and TX. For the same characteristics and changing β_{trn} to be 10, the optimal locations of warehouses is found to be AL, CA, and WY.

One of the aspects that should be discussed is the trade-off between the increase in the fixed costs due to increase in the number of warehouses and the decrease in transportation costs due to increase the proximity of the retailers to the warehouses.

Table 2 Computation results

# of WH	Scenario name and characteristics	WHs assigned	CO_2 cost	Total cost
3	Scenario 1 (Truck type 8b, $\beta_{trn} = 0.01$)	AL, AR, TX	8,188	19,300
4	Scenario 2 (Truck type 8b, $\beta_{trn} = 0.01$)	AL, AR, LA, TX	7,824	18,967
5	Scenario 3 (Truck type 8b, $\beta_{trn} = 0.01$)	AL, CA, LA, MO, TX	5,130	38,099
3	Scenario 4 (Truck type 5, $\beta_{trn} = 0.01$)	AR, ID, NV	17,565	34,498
4	Scenario 5 (Truck type 5, $\beta_{trn} = 0.01$)	AL, AZ, CA, TX	25,206	39,154
5	Scenario 6 (Truck type 5, $\beta_{trn} = 0.01$)	AR, CA, GA, IN, TX	19,901	36,723
3	Scenario 7 (Truck type 2, $\beta_{trn} = 0.01$)	IN, NV, NY	40,074	56,897
4	Scenario 8 (Truck type 2, $\beta_{trn} = 0.01$)	AL, AR, ME, UT	45,133	64,500
5	Scenario 9 (Truck type 2, $\beta_{trn} = 0.01$)	AR, AZ, CA, CO, CT	35,537	54,765
3	Scenario 10 (Truck type 8b, $\beta_{trn} = 10$)	AL, CA, WY	4,817,258	10,767,690
4	Scenario 11 (Truck type 8b, $\beta_{trn} = 10$)	AL, CO, DE, IA	3,588,616	9,769,759
5	Scenario 12 (Truck type 8b, $\beta_{trn} = 10$)	AL, CA, CO, DE, IL	2,289,513	8,275,691
3	Scenario 13 (Truck type 5, $\beta_{trn} = 10$)	AL, CA, WY	23,619,920	29,676,940
4	Scenario 14 (Truck type 5, $\beta_{trn} = 10$)	AR, CO, CT, FL	18,759,290	24,934,200
5	Scenario 15 (Truck type 5, $\beta_{trn} = 10$)	AL, AR, CA, CT, OR	13,264,010	19,386,070
3	Scenario 16 (Truck type 2, $\beta_{trn} = 10$)	AL, CA, WY	57,502,940	127,119,900
4	Scenario 17 (Truck type 2, $\beta_{trn} = 10$)	AL, CA, DC, IA	30,975,810	37,135,530
5	Scenario 18 (Truck type 2, $\beta_{trn} = 10$)	AL, AR, CA, CT, DC	28,984,830	35,166,790

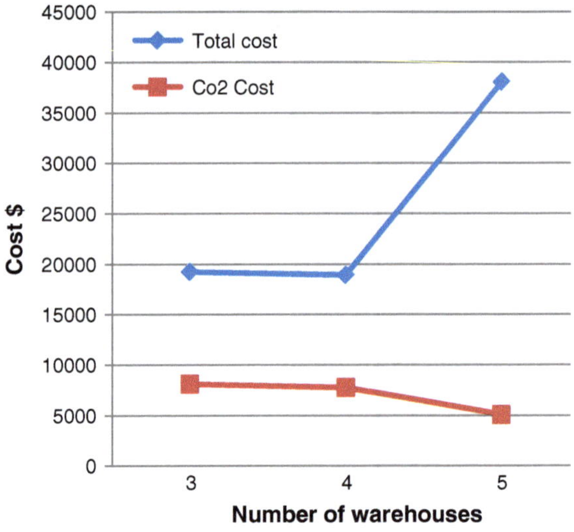

Fig. 1 Comparison between *total cost* and CO_2*cost* for $\beta_{trn} = 0.01$

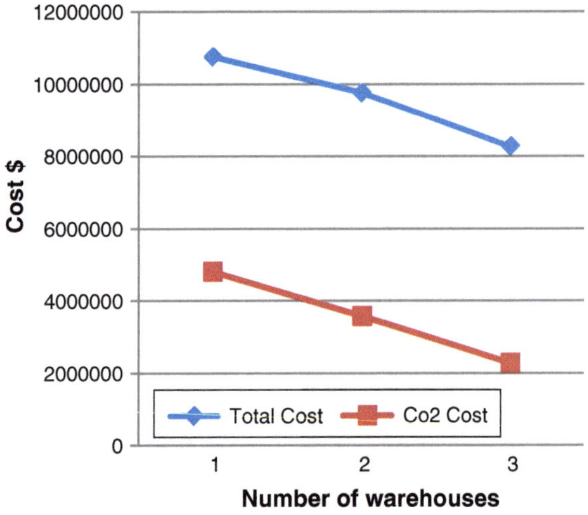

Fig. 2 Comparison between *total cost* and CO_2*cost* for $\beta_{trn} = 10$

Figure 1 illustrates this effect. In Fig. 1 the β_{trn} is fixed to be 0.01 with the truck type 8b, as shown the CO_2 is constantly decreasing as the number of warehouses to locate increases. The total cost is increasing because the fixed cost of locating warehouses exceeds the costs for CO_2 emissions. The same results are presented in Fig. 2 for the β_{trn} equals 10. It is easy to see the dramatic decrease in both the total network cost and the CO_2 cost. This is due to the fact that when β_{trn} is high the system favours to select warehouses based on their proximity to the retails even if the fixed costs of the warehouses is high.

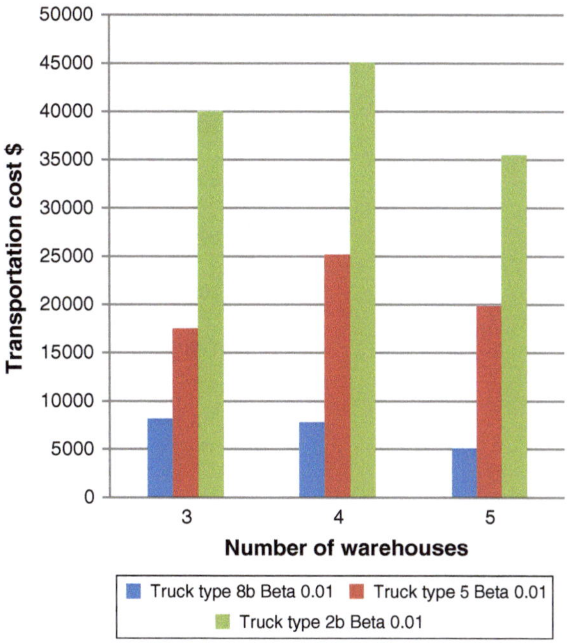

Fig. 3 Transportation cost for each truck type given β_{trn} equals 0.01

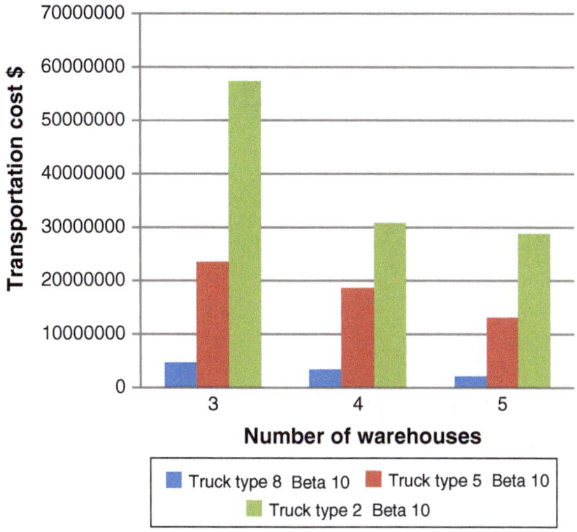

Fig. 4 Transportation cost for each truck type given β_{trn} equals 10

Regarding the truck type for transporting the goods it clearly as shown in Figs. 3 and 4, truck class 8b is the optimal truck system for transportation as it has the lowest transportation cost among the tested types.

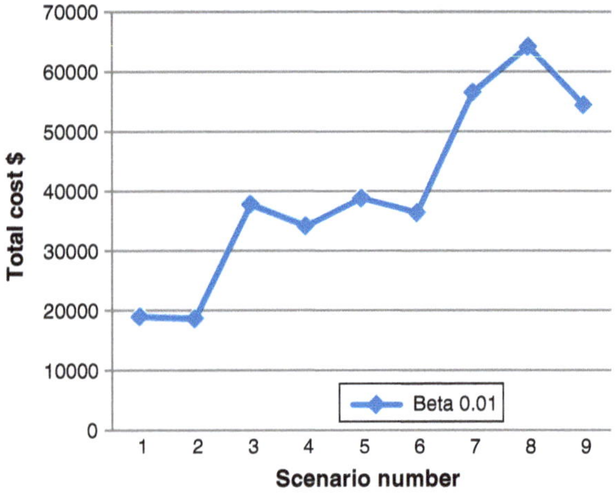

Fig. 5 Total system cost for β_{trn} equals 0.01

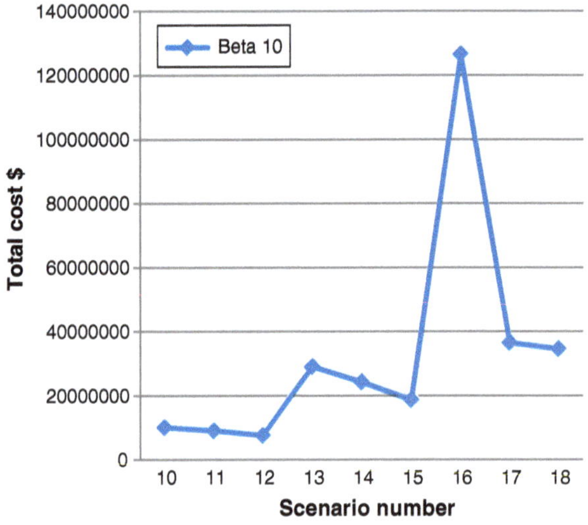

Fig. 6 Total system cost for β_{trn} equals 10

Finally the combined effect of the number of warehouses with the truck type for transportation for β_{trn} equals to 0.01 is shown in Fig. 5 while for β_{trn} equals 10 is shown in Fig. 6.

Based on the results shown in Fig. 5 and for β_{trn} equals 0.01, the minimum cost is achieved in scenario 2 which is to have 4 warehouses and truck class 8b for transportation, however for β_{trn} equals 10 the minimum cost is achieved in scenario 12 which is to have 5 warehouses and truck class 8b for transportation. This can be explained due to the huge weight given to the transportation while normal effect is given for the fixed costs of locating facilities as warehouses.

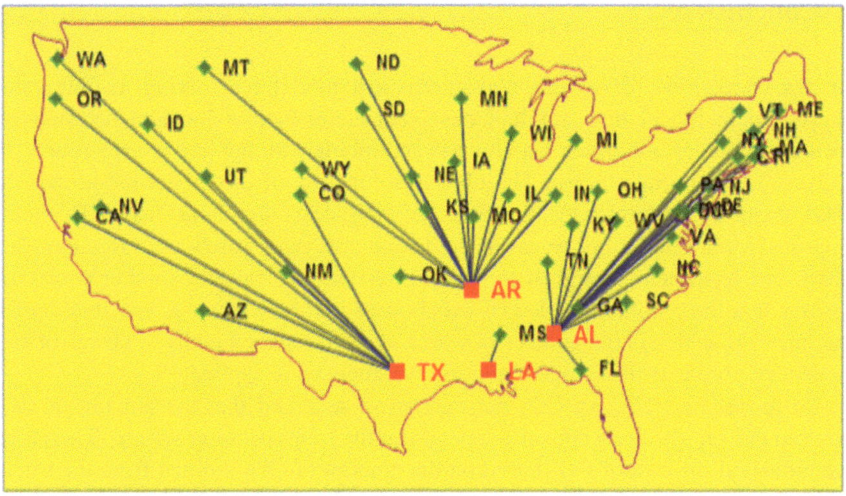

Fig. 7 Optimal configuration system cost for β_{trn} equals 0.01

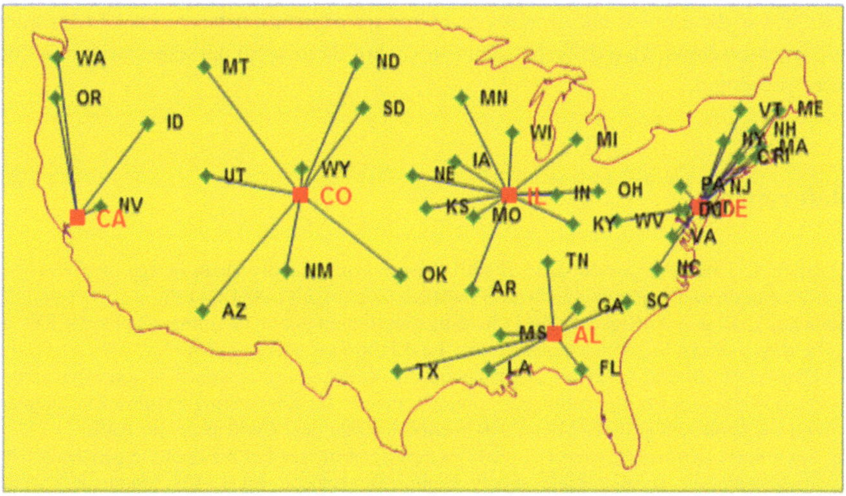

Fig. 8 Optimal configuration system cost for β_{trn} equals 10

According to the analysis presented above the optimal locations for warehouses and the retailers assigned to the warehouses is shown in Fig. 7 for β_{trn} equals 0.01 policy and β_{trn} equals 10 policy it is shown in Fig. 8.

5 Conclusions

In this chapter we developed a mathematical model for solving an integrated supply chain model that simultaneously considers facility location decisions as well as inventory policies at warehouses and retailers taking into consideration the CO_2 as a transportation cost. The objective function of the model minimizes the total inventory, ordering, location, and shipping costs simultaneously. The model imposes each retailer to be assigned to exactly one warehouse. Retailers are prevented from being assigned to warehouses that are closed. Another constraint assures that the number of selected warehouses will be exactly as specified by the parameter p, which is the number of warehouses to locate. Finally, a power-of-two inventory policy at the retailers and the warehouses is applied.

To model the effect of CO_2 emissions, we assumed that the amount of CO_2 emitted the supply chain network depends on the transportation system. Therefore, we modelled the cost of CO_2 based on the amount of CO_2 emitted from the trucks used to transport goods from the warehouses to the retailers and from the retailers to the warehouses. Different scenarios were developed based on the truck type used on the transporting the goods (including the truck load) and cost weight of transportation.

We observed that, the configuration of the network of warehouses and retailers change for different truck systems used for transportation based on CO_2 emitted and truck payload, also different scenarios were developed and comparison results were presented.

References

Abdallah T, Diabat A, Simchi-Levi D (2012) Sustainable supply chain design: a closed-loop formulation and sensitivity analysis. Prod Plan Control 23(2–3):120–133

Abdallah T, Diabat A, Rigter J (2013) Investigating the option of installing small scale PVs on facility rooftops in a green supply chain. Int J Prod Eco 146(2):465–477

Al Dhaheri N, Diabat A (2010) A mathematical programming approach to reducing carbon dioxide emissions in the petroleum refining industry. In: Engineering Systems Management and Its Applications (ICESMA), 2010 Second International Conference on, IEEE

Al Zaabi S, Al Dhaheri N, Diabat A (2013) Analysis of interaction between the barriers for the implementation of sustainable supply chain management. Int J Adv Manufact Technol 68(1–4):895–905

Daskin MS (1995) Network and discrete location: models, algorithms, and applications. Wiley, New York

Diabat A, Aouam T, Ozsen L (2009) An evolutionary programming approach for solving the capacitated facility location problem with risk pooling. Int J Appl Decis Sci 2(4):389–405

Diabat A, Richard J P, Codrington C W (2013a) A Lagrangian relaxation approach to simultaneous strategic and tactical planning in supply chain design. Annals of operation research 203(1):55–80

Diabat A, Abdallah T, Al-Refaie A, Svetinovic D, Govindan K (2013b) Strategic closed-loop facility location problem with carbon market trading. IEEE Trans Eng Manage 60(2):398–408

Diabat A, Abdallah T, Henschel A. (2013c) A closed-loop location-inventory problem with spare parts consideration. Comput Oper Res (in press)
Diabat A, Abdallah T, Le T (in press) A hybrid tabu search based heuristic for the periodic distribution inventory problem with perishable goods. Ann Oper Res (in press)
Diabat A, Battaïa O, Nazzal D (2014) An improved lagrangian relaxation-based heuristic for a joint location-inventory problem. Computers & Operations Research, in press.
Elhedhli S, Merrick R (2012) Green supply chain network design to reduce carbon emissions. Transp Res Part D: Transp Environ 17(5):370–379
Facanha C, Horvath A (2007) Evaluation of life-cycle air emission factors of freight transportation. Environ Sci Technol 41(20):7138–7144
Govindan K, Popiuc M. N., Diabat A (2013) Overview of coordination contracts within forward and reverse supply chains. J Clean Prod 47:319–334
Govindan K, Diabat A, Popiuc M. N (2012) Contract analysis: A performance measures and profit evaluation within two-echelon supply chains.Comput Indust Eng 63(1):58–74
Hervani AA, Helms MM, Sarkis J (2005) Performance measurement for green supply chain management. Benchmarking: Int J 12(4):330–353
Le T, Diabat A, Richard J-P, Yih Y (2013) A column generation-based heuristic algorithm for an inventory routing problem with perishable goods. Optim Lett 7(7):1481–1502
McKinnon AC (2007) CO_2 emissions from freight transport in the UK. Logistics Research Centre, Heriot-Watt University, Edinburgh
Merrick R, Bookbinder JH (2010) Environmental assessment of shipment release policies. Int J Phys Distrib Logistics Manage 40(10):748–762
Miranda PA, Garrido RA (2006) A simultaneous inventory control and facility location model with stochastic capacity constraints. Netw Spat Econ 6(1):39–53
Nozick LK, Turnquist MA (2001) A two-echelon inventory allocation and distribution center location analysis. Transp Res Part E: Logistics Transp Rev 37(6):425–441
Rappold JA, Van Roo BD (2009) Designing multi-echelon service parts networks with finite repair capacity. Eur J Oper Res 199(3):781–792
Shen ZJ, Qi L (2007) Incorporating inventory and routing costs in strategic location models. Eur J Oper Res 179(2):372–389

Analysis and Suggestion of an e-Commerce Logistics Solution: Effects of Introduction of Cloud Computing Based Warehouse Management System in Japan

Keizo Wakabayashi, Kuninori Suzuki, Akihiro Watanabe and Yutaka Karasawa

Abstract It has become increasingly important to have both visible and accurate real-time inventory management tools when managing merchandising/logistics in the clothing industry. A comprehensive, sophisticated, and company-wide inventory management system, which can integrate incoming inventory, logistics center inventory, store inventory and return goods inventory is required. The capability to deal with and analyze massive amount of sales data for thousands of stores, for each store, area, and brand in a short period of time, by unit control (e.g. color, size), is also a requirement. In order to meet these requirements to control and manage inventory, an expensive and heavily-featured warehouse management system (WMS) is being conventionally used today. However, an ever-more sophisticated clothing market, with a shortened fashion cycle and diversified trends and tastes, has increased the demand for more advanced and inexpensive IT-driven logistics services. The ability to promptly introduce and upgrade state-of-the-art IT logistics systems is essential, not only for establishing actual physical stores, but also for developing web-based clothing stores. Using case studies, this chapter will discuss the possibilities for a Cloud Computing based WMS and services in the fashion and apparel industry and business.

Keywords Cloud computing · Warehouse management system · Fashion logistics · Inventory management

K. Wakabayashi (✉) · K. Suzuki · A. Watanabe
Nihon University, Izumicho, Narashino, Chiba, Japan
e-mail: wakabayashi.keizou@nihon-u.ac.jp

K. Suzuki
e-mail: QWL02764@nifty.ne.jp

Y. Karasawa
Kanagawa University, Kanagawa-ku, Yokohama, Kanagawa, Japan

P. Golinska (ed.), *Logistics Operations, Supply Chain Management and Sustainability*, EcoProduction, DOI: 10.1007/978-3-319-07287-6_40,

1 Research Focus: Improvement of Logistics and Inventory Management in the Fashion Industry

In this chapter, we will cast a light on improvement of logistics and inventory management by introducing a Cloud Computing based WMS in the fashion industry; fashion markets has been defined as exhibiting typically the following characteristics short lifecycle, high volatility, low predictability, and high impulse purchase. The combined effect of these pressures clearly provides a challenge to logistics management (Christopher et al. 1998) and the increased attention placed on warehouse and fulfillment operation in the last few years a harvest of improvement in technology and functionality in warehouse management systems (Frazelle 2002, p. 270).

Apparel products tend to be delivered in the beginning of each season. The inventory for replenishment in the warehouse is required at least. It is sometimes difficult to procure special textiles or furs, and therefore additional orders cannot be done. Some products cannot be ordered the same way basic items are, by using logical inventory strategies and statistics. If many products are not sold even after price markdowns have been taken, the inventory will be transferred to the next year. When there is no profit, the inventory rests; on the other hand, when profit can be found, the inventory is appreciated.

Under these circumstances, it is necessary to be aware of the exact number of items that can be sold. This is the total number of real inventory items and incoming inventory items subtracted by the number of items scheduled to be shipped. These numbers need to be visualized in real time through actual inventory management operations in order to have the minimum levels of required inventory items at any particular point in time (Suzuki 2007, pp. 52–53).

In the apparel industry, the visualization of inventory in real time and accurate inventory management are both becoming increasingly important. A comprehensive, sophisticated, and company-wide inventory management system that can integrate incoming inventory, logistics center inventory, store inventory, and return goods inventory is required. The capability to manage and analyze massive amounts of sales data by unit control (e.g. color and size) and brand, for thousands of stores in a short period of time, is an important requirement.

In addition, the fashion retailing industry has various forms, such as on-line shops, department stores, outlets, etc., so their order information and treatment information will vary. Even the same products have different shipping styles when sold by different venues.

Concerning department stores, order forms, invoices, price and tags are different depending on the buying systems used. Especially in Japanese department stores, the process used for price tagging is different for each department store.

Regarding the outlets, some have a central buying system, while others may ship after assorting items for each location destination. Even the same shops' orders to the logistics centers may use different logistics centers depending on the items, which may be unique for the Japanese outlet stores.

In order to meet these requirements to control and manage inventory, an expensive and heavily featured warehouse management system is being conventionally used today. However, an increasingly sophisticated clothing market, combined with a shortened fashion cycle and diversified trends and tastes, has increased the demand for more advanced and inexpensive IT-driven logistics services.

As for well-being of employees in the fashion industry, there have been few systematic studies assessing Human Resource Management and Employee Satisfaction in logistics (Geringer et al. 2002, pp. 5–30). However, the solutions that have been presented so far tend to focus on the designing of better systems and operational process (Berglund et al. 1999, pp. 59–70).

Logistics still has to rely on human resources to a great extent in order to achieve better collaborations in supply chains and smooth operations in warehouses.

Managers in logistics centers may be looking only at the job territories of their own, while floor workers there may be unaware that the accumulation of mistakes and damages they make everyday is causing enormous logistical inefficiency and dissatisfaction of employees in the supply chains (Shinohara 2006, p. 3). Christopher (1992) also sees that the major reason for supply chain inefficiencies is the lack of co-ordination and linkage between various parties in the chain.

However, such problems can be solved by introducing a Cloud Computing based WMS; operational progress information in warehouses or logistics centers can be known in real time. In addition, the progress of each worker can be analyzed by processing data. The labor management function that evaluates each worker's efficiency can be included. The workers can be transferred from the section whose operation makes rapid progress to the section falling behind in the schedule, and their distribution can be corrected by remote control using a WMS system. The concrete number of completed operations, such as picking, checking, etc., can be shown in the warehousing site, to bring out the workers' highest sense of motivation and to enhance their well-being by providing them a sense of fulfillment concerning their work in logistics centers.

This chapter tries to clear view of logistical effectiveness of the Cloud Computing based WMS, referring to well-being of employees in the fashion industry.

In the following section, the effects of using a Cloud Computing based WMS are to be introduced and examined. On this basis, the case studies are presented in Sect. 4.

2 Previous Studies Around Fashion Logistics

Logistics studies have gathered interest of many scholars and experts and numerous theories have been developed (Shinohara 2006, p. 15).

Christopher defines logistics as a planning orientation and framework that seeks to create a single plan for the flow of product and information through a business, and supply chain management builds upon this framework and seek to achieve linkage and co-ordination between process of other entities in the pipeline, i.e.

suppliers and customers, and the organization itself (Christopher 2005, p. 4). On this basis, one goal of supply chain management might be to reduce or eliminate the buffers of inventory between organizations in a chain through the sharing of information on demand and current stock levels (Christopher 2005, p. 4).

Christopher points out that "*Fashion retailing, and manufacturing sector that supports it, are clearly highly dependent on an agile logistics capability*" (Christopher et al. 1998). Successful companies in fashion markets seem not just to be able to capture the imagination of the customer with their products, but are often characterized by their agility. They are finding that it is possible to make significant improvements by reducing the logistics lead time and capturing information sooner on actual customer demand (Christopher et al. 1998, p. 97).

This can be related with the concept of WMS that will be discussed in more detail later in this chapter. A Cloud Computing based WMS is a system that can fulfill the process from receiving to shipping in the logistics center while minimizing inventory level via internet in order to compete successfully in short lifecycle and volatile markets and to bring the products to market in the shortest possible timescale.

There have been many reports and studies on "*sweatshops*". Sweated labor was emerging in apparel production in many countries of the developing world and was reemerging even in some advanced countries (Rosen 2002). The well-being of employees in the factories has been seriously considered, pointed out the case of sweatshops.

However, not only in the factory, but also logistics work in warehousing should be improved because this is some of the most dangerous work in all of industry and offers a low quality of working life in many organizations (Frazelle 2002, p. 343). In most warehouses today, it is difficult to keep someone on-post for more than three month (Frazelle 2002, p. 344). However, introducing a Cloud Computing based WMS more human-friendly warehousing with well-being of the employees in the logistics center can be achieved. New technology can greatly increase worker productivity in logistics, too.

By examining the previous studies around fashion logistics, without doubt, a Cloud Computing based WMS can be useful in fashion markets to manage fashion logistics pipeline successfully, while their employees can be satisfied by using this information system.

On the basis of the previous studies, using the case studies, we will discuss the possibilities for a Cloud Computing based WMS and services in the fashion and apparel industry and business.

3 The Effects of Using a Warehouse Management System

True breakthroughs in logistics performance are achieved when new ways are identified to substitute information for inventory and work content (Frazelle 2002, p. 276). It is sure that the great improvements in logistics productivity were achieved when personal computers were mainstreamed, enabling logisticians real-time and broad access to demand, supply, inventory, and ship information (Frazelle 2002, p. 276).

The WMSs are powerful optimization tools in warehouse operations, logistical information sharing, inventory management, and the like. We can say that the main effects of using a Cloud Computing based WMS are follows:

1. Careful Information Management in Warehouses.
   By the installation of bar code systems, wireless LAN systems, or automatic identification systems, warehousing and shipping check and storage location management can be easily introduced. Goods and information can be managed as a whole (Suzuki 2004, pp. 102–110).
2. Supply-chain-wide and company-wide information sharing.
   For example, receiving and shipping information, and inventory information in a logistics department can be shared with different departments in the same company, such as the sales department. Also, this information and data can be shared outside the company by accessing the Internet link to the warehousing software (Suzuki 2004, pp. 58–61).
3. Standardization and Regularization for the Warehouse Operation.
   The functions for supporting the warehouse operation can help the working efficiency and its accuracy. The material flow in the warehouses will be smoother as the information on task instructions is made based on the information on orders. Also, because each item can be systematically located, even for someone other than a person in charge of replenishment, it is easier to confirm inventory and replenish items. Of course, there is a great contribution to building the working surroundings that are efficient and considerate of the workers' health (Suzuki 2004, pp. 102–108).
4. Control of Operational Progress Information.
   Operational progress information in warehouses or logistics centers can be known in real time. In addition, the progress of each worker can be analyzed by processing data. The labor management function that evaluates each worker's efficiency can be included. The workers can be transferred from the section whose operation makes rapid progress to the section falling behind in the schedule, and their distribution can be corrected by remote control using a WMS system. The concrete number of completed operations, such as picking, checking, etc., *can be shown in the warehousing site, to bring out the workers'* highest sense of motivation and to enhance their well-being by providing them a sense of fulfillment concerning their work in logistics centers (Frazelle 2002, pp. 270–271).
5. Enforcement of Inventory Management.
   By controlling and managing real-time inventory in logistics centers, surplus and safety inventories can be minimized. By avoiding stock-out items, the rate of goods delivery can be improved. Also, various data regarding usual and expected changes in inventory can be sent to order resources' computers. That is, automatic orders can be done (Frazelle 2002, pp. 91–144).

6. Reducing Operational Costs.
   Thanks to correct stock information, the costs and time involved in inventory or stocktaking can be reduced. Checking inventory by human hands, which is time-consuming, can be replaced by software that does this more quickly and accurately, thus reducing costs.
7. Improvement of Customer Service.
   Achieving a reduction in lead-time, improving delivery rates, and offering value-added services such as distribution processing, customer service levels can be eminently improved. The profit margin for each order can also be increased. All of these can enhance customers' satisfaction and sense of well-being (Frazelle 2002, pp. 70–90).

4 Case Study: Introduction of a Cloud Computing based WMS

Magaseek, which is a listed company on the *Mothers*, which is part of the Tokyo Stock Market, collaborates with popular female magazines to entice Web users to purchase their goods, operating one of the biggest fashion Web sites in Japan.

Magaseek grew rapidly, and the arrangement of its logistics system could not keep up with its growing orders. The company tried to deal with this problem in a conventional way, but it was also worried that IT—based logistics system development with a traditional WMS software company would take longer than *Magaseek* needed. Then *Magaseek* decided to utilize *Logizard-Plus*, a Cloud Computing based WMSs which had a good reputation in the apparel industry in Japan, and with the introduction of *Logizard-Plus*, *Magaseek* quickly achieved its timely and visible logistics system in real-time, and all logistical information and data can be shared outside the company by accessing the internet link to the *Logizard-Plus*. Supply-chain-wide and company-wide information sharing can be achieved (Suzuki and Endo 2008, pp. 124–125).

Major apparel companies in Japan develop many brands, some of which have sales in the range of 5,000–10,000 million yen or more. Logistics centers are operated in many cases on a single-brand basis as well as on a company-wide basis (Suzuki and Endo 2008, p. 126). However, it was not easy to operate on time with an IT system until recently.

For this solution, even the major Japanese companies began to introduce and install a Cloud Computing based WMS. By watching the trends of the fashion scene, which change quickly and suddenly, the companies can introduce a highly advanced merchandising and logistics system. The companies should build a highly advanced IT system that shows accurate alerts, and that is practical for checking the merchandising trends for each store and each product on a daily basis. The Cloud Computing based WMS is useful for achieving this purpose, because it can be easily introduced or eliminated with few initial investment costs.

Also, achieving a reduction in lead-time, improving delivery rates, and offering value-added services such as distribution processing, customer service levels can be eminently improved (Suzuki and Endo 2008, pp. 126–132).

5 Conclusion

We have looked at the possibilities for a Cloud Computing based WMS and come to conclusion that it can be good use of the characteristics in fashion markets such as short lifecycles, high volatility, low predictability, high impulse purchase (Christopher et al. 1998, p. 82). As the case studies are examined, the Cloud Computing based WMS such as *Logizard-Plus* is suitable to current paradigms of logistics management in fashion markets, and also effective in warehouse operations and inventory management.

References

Berglund M et al (1999) Third-party logistics: is there a future? Int J Logistics Manage 10:59–70
Christopher M (1992) Logistics and supply chain management, 2nd edn. Pitman Publishing, London
Christopher M (2005) Logistics and supply chain management, 3rd edn. Prentice Hall, London
Christopher M, Lowson B, Peck H (1998) "Fashion logistics and quick response", logistics and retail management. Kogan Page, London
Frazelle EH (2002) Supply chain strategy. McGraw-Hill, New York
Geringer JM et al (2002) In search of 'best practices' in international human resource management: research design and methodology. Hum Resour Manage 41:5–30
Rosen EI (2002) Making sweatshops. (University of California Press, Berkeley)
Shinohara M (2006) European and Japanese logistics paradigms. Maruzen Planet, Tokyo
Suzuki K (2004) Senryaku-warehouse-no-keyword. Faraokikaku, Tokyo
Suzuki K (2007) Ryutsuzaiko-no-kanrisakugen. Nikkan-kogyo-shimbumsha, Tokyo
Suzuki K, Endo H (2008) Shohinkanri-buturyukanri. Nikkan-kogyo-shimbun, Tokyo

A Simulated Annealing Heuristic for the Vehicle Routing Problem with Cross-docking

Vincent F. Yu, Parida Jewpanya and A. A. N. Perwira Redi

Abstract Cross-docking is a warehousing strategy that involves moving of products from pickup nodes to delivery nodes through cross-dock in a short amount of time. This study proposed a simulated annealing heuristic (SA) to solve the vehicle routing problem with cross-docking introduced by Lee et al. (2006). The objective of the problem is to determine the number of vehicles and the best vehicle routes that minimize the sum of the operational cost of vehicles and transportation cost. Computation results show that SA can obtain better results in terms of objective function value at a slightly longer computational time.

Keywords Cross-docking · Simulated annealing · Vehicle routing problem

1 Introduction

Cross-docking concept replaced the warehousing concept in the retail and trucking industries by immediately consolidating products from disparate manufacturers to the destinations in full-load of vehicle capacity and realizing economy of scale in outbound transportation. In the cross-docking distribution system, products are transported from manufacturers to the cross-docking center or depot. After

V. F. Yu (✉) · P. Jewpanya · A. A. N. Perwira Redi
Department of Industrial Management, National Taiwan University of Science and Technology, Taipei 106, Taiwan
e-mail: vincent@mail.ntust.edu.tw

P. Jewpanya
e-mail: parida.jewpanya@gmail.com

A. A. N. Perwira Redi
e-mail: wira.redi@gmail.com

P. Golinska (ed.), *Logistics Operations, Supply Chain Management and Sustainability*, EcoProduction, DOI: 10.1007/978-3-319-07287-6_41,

products arrive at the cross-docking center, it will be unloaded at the receiving dock, split into smaller lots and consolidated with other products according to customer orders. Then, products are moved to suitable locations on the shipping dock and are loaded into truck for transporting to customers (Boloori Arabani et al. 2009). In the system described above, several problems can be identified which include vehicle routing problem, truck scheduling problem, dock assignment problem, cross-docking center process problem and cross-docking distribution planning problem.

Vehicle routing problem (VRP) with cross-docking in the supply chain concerns about the determination of optimal set of route to be performed by fleet of vehicle to serve a given set of customer. Lee et al. (2006) studied the vehicle routing problem in cross-docking network with the pickup and delivery process. They pointed out that two key considerations in cross-docking operation are the simultaneous arrival of vehicles and consolidation of orders for deliveries. The products from various suppliers are consolidated according to customer demands and loaded to each vehicle in the cross-dock for delivery to customers. The objective was to find the number of vehicle and then find the optimal vehicle routing schedule in order to minimize the transportation cost. A tabu search algorithm was proposed to solve the problem. The results were compared with the results obtained by enumeration method.

Liao et al. (2010) also considered a vehicle routing problem for cross-docking in order to control the inventory flow. The objective of the problem is to determine the number of vehicles and the best routes that minimize the sum of the operational cost of vehicles and transportation cost. The problem is NP-hard so a new tabu search (TS) algorithm is proposed to solve this problem. The results compared favorably with those obtained by Lee et al. (2006) and showed that the new TS algorithm gives a better solution with faster computation time. Also, the average improvements are as high as 10–36 % for different sizes of problems.

The vehicle routing problem with cross-docking is studied subsequently in several papers with variant new concepts. For instance, Hasani-Goodarzi and Tavakkoli-Moghaddam (2012) studied the cross-docking problem that focus on the split vehicle routing in the cross-docking network (SVRP). Santos et al. (2013) considered the pickup and delivery problem with cross-docking (PDPCD). Due to the problem in VRP variant is NP-hard, several heuristics and meta-heuristics are proposed. For instance, Dondo and Cerdá (2013) studied the vehicle routing problem with cross-docking (VRPCD) and proposed the sweep heuristic algorithm, Lin et al. (2009) considered the SA algorithm to solve the truck and trailer routing problem (TTRP), also in Yu et al. (2010) with a capacitated location routing problem. Moreover, Mousavi and Tavakkoli-Moghaddam (2013) proposed a new algorithm to solve a cross-docking center location and a vehicle routing scheduling model. This technique base on a two-stage hybrid simulated annealing

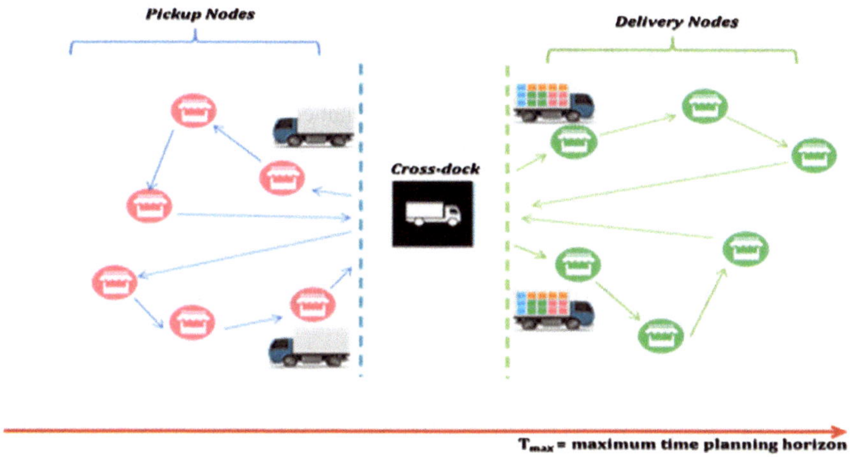

Fig. 1 The vehicle routing with cross-docking (VRPCD)

(HSA) with a tabu search (TS). In this chapter, the simulated annealing is proposed to solve this problem and a comparison of its performance with those of the TS algorithms proposed by Lee et al. (2006) and Liao et al. (2010) are presented.

2 Problem Description

This study considers the vehicle routing problem with cross-docking. It is assumed that all the vehicles are located in the cross-dock (0). All nodes n ($n = P + D$) that consists of pickup nodes (P) and delivery nodes (D) are allowed to visit only once. Moreover, the pickup vehicles start from the cross-dock to pick up products (p_i) at the manufacturers and arrive at the cross-dock simultaneously. Then, the delivery vehicles move to the customers to deliver products (d_j) after that return to the cross-dock after finishing delivery process.

Moreover, the number of vehicles utilized should be less than or equal to the number of available vehicles (m). The total quantity of pickup (p_i) should equal the quantity to be delivered (d_i). For each vehicle, the load on the pickup route and on the delivery route cannot exceed the capacity of the vehicle (Q). And all processes have to be finished within time horizon T (Fig. 1).

The objective of the problem is to determine the number of vehicles and the best vehicle routes as well as the arrival time of each vehicle so as to minimize the sum of the operational cost (C_k) of vehicles and the transportation cost (tc_{ij}).

$$Min\left(\sum_{i=0}^{n}\sum_{j=0}^{n}\sum_{k=1}^{m} tc_{ij} \cdot x_{0j}^{k}\right) \quad (1)$$

This problem is explained above was considered in Lee et al. (2006) and Liao et al. (2010). Since the problem is NP-hard, an efficient heuristic algorithm is needed to obtain good solutions in a reasonable amount of time. The SA algorithm was used to solve this problem and will be explained in Sect. 3. Then, the results will be compared with previous results obtained by Lee et al. (2006) and Liao et al. (2010) in Sect. 4.

3 The Proposed Simulated Annealing Algorithm for the VRPCD

Simulated annealing (SA) algorithm is a local search-based heuristic. SA was proposed by Metropolis et al. (1953) and popularized by Kirkpatrick et al. (1983). An important characteristic of this algorithm is its capability of escaping from being trapped into a local optimum by accepting, with a small probability, worse solutions during its iterations.

In this study, an SA heuristic is proposed to solve VRPCD. The algorithm starts with an initial solution method specially designed for the problem. To improve the solution, some predefined neighborhood operators are considered in this algorithm. The detail of the proposed SA for the VRPCD will be discussed in this section.

3.1 *Solution Representation*

The solution representation of VRPCD with P pickup nodes and D delivery nodes consists of a permutation of customers P (1, 2, …, P) from pickup nodes and a permutation of D customers ($P + 1$, $P + 2$, …, $P + D$) from delivery nodes. N_{dummy} zeros are used to separate routes, in addition to the time limit and vehicle capacity constraint. For problem with vehicle capacity Q, N_{dummy} is calculated as:

$$\left\lceil \sum_i \frac{d_i}{Q} \right\rceil \tag{2}$$

where d_i is the demand of customer i.

The representation method allows routes to terminate randomly by the dummy zeros, the route maximum travelling time or capacity constraints. This mechanism enables the algorithm to search more diversified areas in the search space. The solution representation ensures the solution will not violate any problem constraints. A solution representation example is shown in Fig. 2. There are 4 pickup and 6 delivery nodes in the problem. 2 vehicle routes are considered. The first

1	0	2	4	3	7	9	5	0	10	6	8
Pickup					Delivery						

Fig. 2 An example of solution representation of pickup and delivery nodes

route starts at cross-dock (0) and then goes to pickup node 1. Then, the vehicle goes back to cross-dock. For the second route, vehicle starts at the same cross-dock, then pick up products from pickup node 2, 4 and 3 respectively. After that, vehicle goes back to cross-dock.

3.2 Initial Solution

A greedy insertion strategy at each stage of route construction is used instead of a random generation strategy because it may give better candidate solutions for the problem. The important factors considered is effective utilization of vehicle capacity within the maximum traveling time. Both pickup and delivery solutions are created using the same initial solution procedure.

The initial solution starts by setting the current node at the cross-dock facility. After that, one out of three nearest un-routed customers from the current node is randomly selected to be the inserted into the solution. This approach taking account of well know nearest neighborhood method with some modification in order to keep diversity of solution generated.

After finishing the selection process, two feasibility checks are performed for each customer insertion. First, vehicle capacity is checked. If inserting a customer into a route violates vehicle capacity, the route is terminated and the customer is assigned to a new vehicle. Second, the maximum traveling time constraint is checked. If the vehicle arrives later than the latest traveling time of a route, the same treatment is employed. If none of the constraints is violated, the current customer is assigned to the current vehicle and the current node is changed to the last customers inserted. The procedure is terminated when all customers are assigned to a vehicle. Finally, N_{dummy} zeros is randomly inserted to the solution. The utilization of N_{dummy} zeros enable this solution representation to search larger solution space that have possibility to gain better result.

3.3 Simulated Annealing Heuristic

The proposed SA heuristic starts with an initial solution constructed by a greedy procedure which is explained in the previous section. The current solution is set to be the initial solution. Then, the algorithm takes a new solution from the predefined neighbourhood of the current solution. The neighbourhood structure used in the proposed SA heuristic features five types of commonly used moves: swap, reversion and N-insertion with $N = 1, 2$ and 3 representing the number of nodes to be inserted.

Six parameters are used in the SA: I_{iter}, T_0, K, $N_{non\text{-}improving}$ and α. I_{iter} denotes the number of iterations at a particular temperature. T_0 represents the initial temperature, while T_F is the final temperature which SA heuristic is terminated. K is the Boltzmann constant used in the probability function to determine whether to accept a worse solution or not. $N_{non\text{-}improving}$ is the maximum allowable number of temperature reductions during which the best objective function value has not been improved. α is the coefficient controlling the cooling process.

As initial condition the current temperature T is set to be the same as initial temperature T_0. The current best solution X_{best} and the best objective function F_{best} value obtained so far are set to be X and *obj (X)*, respectively. To improve the solution, the next solution Y is generated from neighbourhood $N(X)$ and its objective function value is evaluated. Let Δ denote the difference between *obj (X)* and *obj (Y)*, that is $\Delta = obj\ (Y) - obj\ (X)$. The probability of replacing X with Y, given that $\Delta > 0$, is exp $(-\Delta/KT)$. This is accomplished by generating a random number $r \in Uniform\ [0, 1]$ and replacing the solution X with Y if $r < exp\ (-\Delta/KT)$. Meanwhile, *if* $\Delta \leq 0$, the probability of replacing X with Y is *1*. X_{best} records the best solution found so far as the algorithm progresses.

The current temperature T is decreased after running I_{iter} iterations since the previous temperature decrease, according to the formula $T \leftarrow \alpha T$, where $0 < \alpha < 1$.

The algorithm is terminated when the current temperature T is lower than T_F or the current best solution X_{best} is not improved in $N_{non\text{-}improving}$ consecutive temperature reductions. After the end of SA procedure, a local search procedure which sequentially performs 2-Opt, swap and insertion is used to improve the current best solution. Following the termination of the SA procedure, the number of vehicles and vehicle routes can be derived from X_{best}.

4 Computation Results

The proposed SA heuristic algorithm is coded in C++ and performed on a PC with a 3.4 GHz processor and 4 GB of RAM, under the Windows 7 operating system.

The proposed SA heuristic is evaluated against the TS algorithms of (Liao et al. 2010) and Lee et al. (2006). The dataset generated by Lee et al. (2006) and Liao et al. (2010) consists of 3 types of problems. These three problem sets consist of 10, 20, and 30 nodes respectively. According to Lee et al. (2006), the dataset was generated based on the parameter values given in Table 1. The planning horizon, T was assumed to be 960 min and the operational cost of vehicle was 1,000 for all problems. Transportation cost and transportation time were generated asymmetrically (Table 2).

Table 1 Parameter values

	10 nodes	30 nodes	50 nodes
p + d	10	30	50
m	10	20	30
T	960	960	960
Q	70	150	150
tc_{ij}	U(20,200)	U(20,200)	U(20,200)
P_i, D_i	U(5,50)	U(5,20)	U(5,30)
No. of suppliers	4	7	12
No. of retailers	6	23	38

Table 2 Computational time of Lee et al. (2006), Liao et al. (2010) and the proposed SA

Instance	Lee et al. TS			Liao et al. TS			Proposed SA		
	10	30	50	10	30	50	10	30	50
1	1.52	3.00	5.62	0.22	0.37	0.49	1.91	2.80	7.49
2	1.74	3.55	5.73	0.23	0.16	0.64	1.78	2.90	6.62
3	2.37	4.32	5.80	0.19	0.43	0.30	2.44	3.00	7.10
4	1.60	2.09	5.87	0.27	0.23	0.44	1.75	3.00	6.89
5	2.28	2.26	7.28	0.21	0.39	0.61	1.75	3.50	7.09
6	1.82	2.10	5.38	0.03	0.22	0.56	1.81	3.10	7.47
7	2.80	2.61	7.65	0.01	0.11	0.55	2.77	3.00	6.11
8	1.85	3.00	9.88	0.04	0.16	0.42	2.21	3.30	6.91
9	2.04	3.10	5.54	0.25	0.16	0.38	1.75	2.80	7.11
10	1.82	2.38	5.77	0.36	0.20	0.52	2.05	2.60	7.10
11	1.80	3.03	5.37	0.00	0.29	0.37	1.93	3.00	7.15
12	1.72	2.64	4.46	0.24	0.16	0.17	1.76	2.90	6.75
13	1.54	2.92	4.62	0.24	0.15	0.16	1.88	2.90	6.32
14	1.53	2.76	5.44	0.00	0.18	0.53	1.69	3.60	7.38
15	1.61	3.06	6.31	0.41	0.38	0.55	1.75	3.10	7.97
16	2.05	3.15	5.18	0.03	0.28	0.30	2.05	3.10	6.95
17	2.26	2.14	6.93	0.06	0.34	0.44	2.60	3.00	6.72
18	1.74	1.70	6.25	0.19	0.28	0.53	1.88	2.90	7.07
19	2.21	2.77	5.68	0.03	0.36	0.28	1.92	3.10	7.93
20	2.55	2.72	4.79	0.00	0.24	0.36	2.32	2.50	7.56
Average	1.94	2.77	5.98	0.15	0.25	0.43	2.00	3.00	7.10

4.1 Parameter Setting

Parameter selection may influence the quality of the computational results. Thus, to determine the appropriate parameter values for SA, an experimental study is conducted. The following combinations of the parameter values were tested:

- I_{iter} : 1500, 3000
- T_0 : 100, 200
- T_F : 0.01, 0.001
- α : 0.99, 0.993
- $N_{non\text{-}improving}$: 1,000,000

The results indicated that the best solution quality was obtained by setting $I_{iter} = 1500$, $T_O = 100$, $T_F = 0.001$, $\alpha = 0.933$, $N_{non\text{-}improving} = 1{,}000{,}000$. Therefore, these parameter values were used for further experiments in this study.

4.2 Result Comparison

Performance analysis of approximation algorithm such as simulated annealing concentrates on solution quality, computational time and robustness. Therefore algorithm performance was measured in terms of the percentage gap between solution objective values and previous best known solution objective values with consideration of CPU time and average performance of the algorithm.

Table 3 presents the comparison of our algorithm with the TS algorithm proposed by Liao et al. (2010) and Lee et al. (2006). For each instance, 10 replications were made and the average was recorded. Liao et al. (2010) indicated that their TS algorithm yield superior results with less computational time than the TS algorithm of Lee et al. (2006). Thus, the GAP is calculated as the percentage difference in objective value between the TS algorithm of Liao et al. (2010) and the proposed SA heuristic. Comparative results in Table 3 show that the average improvement from Liao et al. (2010) is 2.9, 1.0 and 5.8 % for the three problem sets, respectively. As shown in Table 2, the computational time difference between the TS algorithm of Liao et al. (2010) and the proposed SA heuristic is, on average, 2.00, 3.00 and 7.10 s for the three problem sets, respectively.

Table 3 Total cost comparison of Lee et al. (2006), Liao et al. (2010) and the proposed SA

Instance	Lee et al. TS			Liao et al. TS			Proposed SA			Percentage gap		
	10	30	50	10	30	50	10	30	50	10	30	50
1	7,571.4	12,366.7	24,284.6	6,847.6	7,692.9	20,704.6	6,953.0	7,550.2	19,804.5	1.5	−1.9	−4.3
2	7,103.7	14,173.0	23,435.6	6,816.8	7,787.2	20,816.8	6,741.0	7,832.7	18,248.1	−1.1	0.6	−12.3
3	9,993.5	13,836.8	23,449.4	9,615.6	7,893.6	19,612.2	9,269.0	7,747.4	18,133.9	−3.6	−1.9	−7.5
4	8,338.0	10,995.4	23,471.1	7,289.7	7,792.2	19,549.0	7,255.0	7,677.4	19,083.6	−0.5	−1.5	−2.4
5	8,709.9	11,757.8	23,406.2	6,599.0	7,224.8	20,448.0	6,524.0	7,579.7	18,877.3	−1.1	4.9	−7.7
6	9,143.5	11,027.7	24,026.6	9,324.6	7,245.9	21,212.0	7,613.0	7,053.0	19,783.0	−18.4	−2.7	−6.7
7	12,721.2	11,899.2	24,190.0	12,083.0	8,206.9	20,640.2	11,990.0	7,720.1	19,690.0	−0.8	−5.9	−4.6
8	9,275.7	12,825.5	23,158.9	8,719.6	7,880.9	20,664.1	8,158.0	7,709.8	18,939.2	−6.4	−2.2	−8.3
9	8,096.5	12,718.6	23,594.7	7,362.2	8,157.3	18,920.0	7,120.0	7,882.5	18,510.6	−3.3	−3.4	−2.2
10	7,044.8	11,794.7	23,530.5	6,204.5	7,924.7	20,384.2	6,056.0	7,734.9	19,607.3	−2.4	−2.4	−3.8
11	8,051.8	12,094.9	23,371.7	7,635.3	7,452.6	19,941.6	7,434.0	7,721.7	18,675.8	−2.6	3.6	−6.3
12	8,661.0	12,132.5	21,082.8	7,867.2	8,320.0	17,258.4	7,800.0	7,899.8	17,550.3	−0.9	−5.1	1.7
13	7,370.2	13,223.4	21,610.7	7,097.9	8,222.7	17,829.9	6,934.0	7,863.7	18,039.2	−2.3	−4.4	1.2
14	7,132.3	12,413.9	23,397.9	5,208.0	8,211.7	19,845.2	4,704.0	8,141.1	18,252.5	−9.7	−0.9	−8.0
15	7,563.4	12,521.4	24,041.9	7,103.2	8,144.6	21,863.0	7,088.0	7,941.6	19,803.8	−0.2	−2.5	−9.4
16	9,983.6	12,044.4	22,893.4	8,768.7	7,451.7	20,144.2	8,616.0	7,901.9	18,808.3	−1.7	6.0	−6.6
17	9,538.1	12,699.4	22,950.4	9,003.0	8,086.2	20,093.3	9,003.0	8,055.0	18,713.8	0.0	−0.4	−6.9
18	8,057.4	11,001.4	24,358.2	6,887.5	7,576.0	20,244.8	6,911.0	7,798.3	18,579.7	0.3	2.9	−8.2
19	9,042.6	12,724.4	25,068.7	7,123.0	7,871.2	19,955.0	7,051.0	7,964.3	18,453.2	−1.0	1.2	−7.5
20	10,478.0	12,357.7	23,232.1	10,471.0	7,883.7	19,267.7	10,004.0	7,522.4	18,167.3	−4.5	−4.6	−5.7
Average	8,693.8	12,330.4	23,427.8	7,901.4	7,851.3	19,969.7	7,661.2	7,764.9	18,786.1	−2.9	−1.0	−5.8

5 Conclusion

The vehicle routing problem for cross-docking network involves various decisions to control the product distribution flow from manufacturers to cross-dock and cross-dock to customers. This study proposed an SA algorithm to solve the VRPCD. Moreover, in the objective function, the transportation costs that are commonly incurred in the above activities are also included. The proposed algorithm is compared with the algorithms of Lee et al. (2006) and Liao et al. (2010).

In summary, the proposed SA heuristic obtained 50 solutions with the same or better results reported in previous research out of 60 instances in the three benchmark problems. Although the required CPU time is longer than that required by the approach of Liao et al. (2010), it is still reasonable. Furthermore, the comparison shows that the proposed algorithm can improve the average objective value reported in previous research. The results indicate the effectiveness and robustness of the SA heuristic in solving the VRPCD.

References

Boloori Arabani AR, Fatemi Ghomi SMT, Zandieh M (2009) A multi-criteria cross-docking scheduling with just-in-time approach. Int J Adv Manuf Technol 49(5–8):741–756

Dondo R, Cerdá J (2013) A sweep-heuristic based formulation for the vehicle routing problem with cross-docking. Comput Chem Eng 48:293–311

Hasani-Goodarzi A, Tavakkoli-Moghaddam R (2012) Capacitated vehicle routing problem for multi-product cross- docking with split deliveries and pickups. Procedia Soc Behav Sci 62:1360–1365

Kirkpatrick S, Gelatt DG Jr, Vecchi MP (1983) Optimization by simmulated annealing. Science 220(4598):671–680

Lee YH, Jung JW, Lee KM (2006) Vehicle routing scheduling for cross-docking in the supply chain. Comput Ind Eng 51(2):247–256

Liao C-J, Lin Y, Shih SC (2010) Vehicle routing with cross-docking in the supply chain. Expert Syst Appl 37(10):6868–6873

Lin S-W, Yu VF, Chou S-Y (2009) Solving the truck and trailer routing problem based on a simulated annealing heuristic. Comput Oper Res 36(5):1683–1692

Metropolis N, Rosenbluth AW, Rosenbluth MN, Teller AH, Teller E (1953) Equation of state calculations by fast computing machines. J Chem Phys 21(6):1087

Mousavi SM, Tavakkoli-Moghaddam R (2013) A hybrid simulated annealing algorithm for location and routing scheduling problems with cross-docking in the supply chain. J Manuf Syst 32(2):335–347

Santos FA, Mateus GR, da Cunha AS (2013) The pickup and delivery problem with cross-docking. Comput Oper Res 40(4):1085–1093

Yu VF, Lin S-W, Lee W, Ting C-J (2010) A simulated annealing heuristic for the capacitated location routing problem. Comput Ind Eng 58(2):288–299

Pollution-Inventory Routing Problem with Perishable Goods

Ahmed Al Shamsi, Ammar Al Raisi and Muhammad Aftab

Abstract A major aspect in inventory routing problems (IRP) is the CO_2 emissions from vehicles and its impact on the environment. Vehicle emissions depend on various factors such as vehicle's weight, traveling speed, etc. This chapter presents the Pollution-Inventory Routing Problem with perishable goods (PPIRP), which is an extension of the Inventory Routing Problem with perishable goods (PIRP) that account not just for the transportation and inventory costs, but also for the cost due to greenhouse emissions. We develop a mathematical model for the PPIRP and solve it using GAMS. Through extensive computational experiments, we evaluate the impact of vehicle emissions on the results and find that our model achieves approximately 61 % reduction in carbon emissions and a 23 % decrease in empty vehicle trips compared to PIRP model. The objective of this chapter is to shed light on the trade-off between emission costs and total costs, and make the model more environmental friendly with a potential for pareto-improvements.

Keywords Inventory routing · Vehicle routing · Perishable goods · Greenhouse gas emissions · Energy consumption

A. Al Shamsi (✉) · A. Al Raisi · M. Aftab
Masdar Institute of Science and Technology, Abu Dhabi, UAE
e-mail: aahalshamsi@masdar.ac.ae

A. Al Raisi
e-mail: basalraisi@masdar.ac.ae

M. Aftab
e-mail: cmuhaftab@masdar.ac.ae

P. Golinska (ed.), *Logistics Operations, Supply Chain Management and Sustainability*, EcoProduction, DOI: 10.1007/978-3-319-07287-6_42,

1 Introduction

One of the main reasons of global warming is carbon emissions. It is important that companies make a precise assessment of their carbon emissions because industrial emissions represent a large portion of the global carbon emissions. In particular, carbon emissions from transport activities are significantly contributing to the overall carbon emissions by a company. Therefore, transport emission reduction is critically important and will have a significant impact on the carbon footprint of a company. Govindan et al. (2012) evaluated of performance measures and supply chain profit behavior under buyback, revenue sharing, quantity flexibility and advanced purchase discount contracts versus no coordination and wholesale price systems. Recently some studies considered the consideration of CO_2 emissions in the design of supply chain, Abdallah et al. (2012); Diabat et al. (2013a); Diabat et al. (2014); Al Dhaheri and Diabat (2010); Diabat (2013b); Govindan et al. (2013); Abdallah et al. (2013); Le et al. (2013); Diabat et al. (in press); Diabat et al. (2009); Diabat et al. (2013c); Al Zaabi et al. (2013). Some of these studies considered developing solution algorithms to tackle the developed mathematical programs such as Diabat et al. (2013a) and Diabat et al. (2014)

Basically a Vehicle Routing Problem (VRP) deals with routing decisions only. In a VRP, vehicles leave the depot based on customers' demand on a given day, and the company allocates the orders for each vehicle and decides feasible routes. A feasible route in VRP is defined as a sub-tour for which the sum of demands of customers on the root is less than the vehicle capacity. Inventory Routing Problems (IRPs), on the other hand, deal with both routing and inventory decisions in the same model. A feasible route in IRP starts from the depot, visits a subset of customers at most one time, and then returns to the depot. Moreover, there are no customer orders in IRP scenario, but the company chooses when and how much to deliver to the costumers.

The Inventory-routing problem with perishable goods PIRP is an extension of the classic IRP where it takes the perishable goods into consideration. The quality of the perishable goods is affected by many factors but the main factor is temperature. Perishable good typically have a shelf-life and value attached to them. The value of perishable goods either remains constant or decrease with time. After exceeding shelf-life, perishable goods have no value and get disposed (Le et al. 2013). Perishable goods need to be delivered in a timely manner using temperature controlled vehicles (Chaug-Ing et al. 2007). These vehicles consume more fuel than regular vehicles and also emit more CO_2 to the atmosphere. These vehicles are bounded by many factors that decide the amount of CO_2 they emit e.g., the load of the vehicle and vehicle speed etc. By studying these factors closely in transportation processes will enhance different ways to reduce CO_2 emissions.

In this chapter, we extended the Inventory Routing Problem with perishable goods (PIRP) model to include the cost of CO_2 emissions due to transportation. Our extended Pollution Inventory Routing Problem (PPIRP) model is similar to the model developed in (Le et al. 2013) but with cost of CO_2 emissions incorporated into the objective function. The notation used in this chapter is chosen so as to be as close as possible to the notation used in (Le et al. 2013). The objective of our model is to reduce carbon emissions of freight transport operations. The model assumed the perishable goods have a fixed lifetime after which it will be disposed. The CO_2 emissions are calculated based on the vehicle load and distance. The model solves the problem by choosing the best route and the amount of goods to deliver to each customer.

The rest of the chapter is organized as follows. In Sect. 2, we provide a review of the relevant literature. We present the model and formulations of Pollution Inventory Routing Problem (PPIRP) in Sect. 3. In Sect. 4, we evaluate our model through extensive simulations. Finally, we summarize and discuss several future extensions in Sect. 4.

2 Literature Review

Inventory Routing Problems (IRPs) is an area of research where the earliest contributions appeared in the eighties and is steadily growing ever since. Initial work in this area was motivated by real-world applications e.g. companies departing from Retailer Managed Inventory (RMI) to Vendor Managed Inventory (VMI) technique in supply chain management. Bertazzi and Speranza (2012, 2013) provide detailed reviews articles on various techniques employed to solve different types of IRPs.

Inventory control for perishable products has also gained considerable attention due to its widespread existence in the real-life. In Wang and Li (2012), the authors present a scheme that aims to improve the performance of supply chains with perishable goods through a pricing mechanism where retailers take the pricing decisions based on dynamic identification of the product shelf life. Information about product shelf life is accurately captured through modern tracking and monitoring techniques. They evaluate their proposed scheme by modeling the retailer's profit under different pricing policies. Le et al. (2013) integrate inventory and transportation models for perishable goods. They consider a deterministic multi-period IRP for perishable goods (PIRP) where perishable goods are assumed to have a fixed shelf-life. They formulate their problem as a linear, mixed integer program with the objective of minimizing the sum of transportation and inventory costs while satisfying the additional constraint of perishability. They develop a column generation-based heuristic algorithm that exploits the problem structure to find a near optimal solution. Through numerical evaluation they demonstrate the effectiveness of their proposed heuristic algorithm. They also argue that their solution approach can be extended to other types of inventory routing problems.

Recently, many studies have explored sustainability in supply chains because carbon emissions from supply chain activities are significantly contributing to global warming. Diabat and Simchi-Levi (2009) propose a model for supply chain management that takes environmental performance into account by introducing carbon emission constraints. They formulate their model as mixed-integer program (MIP) that captures the impact of different carbon emission caps on the supply chain cost. Their model aims to help companies find an optimal strategy to meet their carbon cap. They carry out a computational study to show the viability of their model and argue that their findings can help supply chain managers keep their companies competitive under various scenarios. Ericsson et al. (2010) develop a system that aims to reduce CO_2 emissions through a navigational aid that helps drivers determine the optimal routes based on the lowest fuel consumption rather than cost or time. To evaluate the performance of their framework, they equipped more than 200 cars with a device that recorded driving behavior and pattern of their system among other parameters. They show that sizeable reduction in fuel consumption can be achieved by adopting routes that optimize fuel consumption, as compared to adopting routes that minimize cost or time. The authors in Diabat et al. (2013) present a model that incorporates the impact of carbon emissions into the design of a closed-loop supply chain. They formulate the problem as mixed integer linear program (MILP) and show that it is possible to reduce the overall carbon footprint of the supply chain through remanufacturing. Their analysis shows that the actual reduction in carbon emission depends upon supplier decision and cost of carbon emissions.

Figliozzi (2010) presents a vehicle routing problem which includes minimization of emissions and fuel consumption as part of the objective function. The author takes into account vehicle costs, distance traveled, route duration and emissions to develop a special algorithm that solves the problem. The author demonstrates that emissions can be significantly reduced with only slight increase in routing costs by taking routes that are not necessarily optimal with respect to cost. Bektaş and Laporte (2011) integrate CO_2 emissions from transportation into a Vehicle Routing Problem (VRP). They first calculate emissions based on the type of vehicle, speed of vehicle and the load factor. Then, they convert the emissions into costs and include it in the objective function. They perform computational experiments to shed light on the tradeoff between CO_2 emissions and vehicle load, speed, and total cost. They also show, with the help of an example, that the cost of the optimal route is not necessarily the one that minimizes energy demand.

Treitl et al. (2012) are among the first to incorporate environmental aspects into an Inventory Routing Problem. They extend the IRP model to also include carbon emissions and apply it to a real-world case study from the petrochemical industry. Their work focuses on the analysis of transport activities where the results from detailed transport analysis are compared with the results from previous studies. In the previous studies, the main focus was on facility location decisions rather than on transportation decisions. Their results indicate the importance of detailed transport process analysis and suggest a potential for achieving pareto-improvements by showing that at least one considered aspect is improved while another aspect is (at least) not deteriorated. They report reduction in total kilometers driven as well as empty driven kilometers, while significantly increasing the average load factor, thus leading to improvements in economic as well as environmental performance.

Table 1 Parameters for emission calculation

Empty vehicle weight (w, in tons)	25
Acceleration of vehicle (a, in m/s^2)	0
Gravitational constant (g, in m/s^2)	9.81
Road angle (h, in degrees)	0
Coefficient of rolling resistance (C_{ro})	0.01
Coefficient of drag (C_d)	0.77
Frontal surface (A, in m^2)	6.5
Air density at ground level at 20° C (ρ, in kg/m^3)	1.2
Conversion factor for vehicle energy consumption (cf_v)	6.94E-08
Cost of CO_2 emissions per kg (in dollars)	0.093

Our research differs from prior contributions in that we not only integrate environmental issues into an Inventory Routing Problem but also assume that goods are perishable. The model we use is based on the model proposed in (Le et al. 2013) and (Treitl et al. 2012).

3 Problem Description and Reformulation

We extended the model proposed in Le et al. (2013) by incorporating the total CO_2 emissions cost in the objective function. Let P_{rt} represent the amount of energy (in Joule) that is necessary for the vehicle to traverse the route r in period t. Then P_{rt} can be calculated based on the vehicle speed (v_r,) the length of the route (l_r,) the empty vehicle weight (w), load of the vehicle k on route r in period t (q_{tij}). Calculation of P_{rt} also depends on several other parameters as explained in Table 1. First we calculate the energy consumed by vehicle and then convert it to CO_2 emissions (in kilograms) using the conversion factor (cf_v, KJ to kg CO_2). Our calculations for emissions and fuel usage are based on Bektaş and Laporte (2011) and Barth and Boriboonsomsin (2009). In addition, the total amount of emissions (E_{tot}, in kilograms) must not exceed an upper emission limit (E_{cap}).

We introduce the following additional notation:

- N Set of customers $N = 1,\ldots,|N|$
- V Set of nodes $V = \{0\} \cup N$, where node 0 represents the depot
- T Set of time periods $T = 1,\ldots,|T|$
- K Set of homogeneous vehicles $K = 1,\ldots,|K|$
- R Set of all feasible routes
- C Vehicle capacity
- τ_{max} Maximum shelf-life
- d_{it} Demand of customer $i \in N$ in time period $t = 1,\ldots, T,\ldots,T+\tau_{max} - 1$
- u_{it} Upper bound inventory level at customer $i \in N$ in time period $t \in T$,

$$u_{it} = (\sum\nolimits_{\tau st+\tau_{max}} d_{i\tau}) \quad (1)$$

h_{it} Inventory holding cost of customer $i \in N$ in time period t
I_{i0} Inventory level at customer $i \in N$ at the beginning of time period $t = 1$
l_{ij} Length of arc (i, j)

$$a_{ir} = \begin{cases} 1 & \textit{if route } r \in \text{R visits custpmer i} \in \text{N} \\ 0 & \text{otherwise} \end{cases} \quad (2)$$

c_r Transportation cost of route r ∈ R
c_E Cost of CO_2 emissions in kg

The decision variables of the model are defined as follows:
I_{it} Inventory level of customer $i \in N$ at the end of time period $t \in T$

$$\theta_{rt} = \begin{cases} 1 & \textit{if route r is selected in time period t} \in T \\ 0 & \text{otherwise} \end{cases} \quad (3)$$

a_{irt} Quantity delivered to customer $i \in N$ by route $r \in R$ in time period $t \in T$ in tons
q_{tij} Load of the vehicle k on route $r \in R$ in period t

The optimal solution of PPIRP is obtained by solving the following mixed integer nonlinear programming model (MINLP):
Objective Function

$$Z^* = \min \sum_{t \in T} \left(\sum_{r \in R} c_r \theta_{rt} + \sum_{i \in N} h_{it} I_{it} \right) + c_E E_{tot} \quad (4)$$

Subject to

$$\sum_{r \in R} \alpha_{ir} \theta_{rt} \leq 1, \ \forall i \in N, \, t \in T \quad (5)$$

$$\sum_{i \in N} \alpha_{ir} a_{irt} \leq C\theta_{rt}, \ \forall r \in R, \, t \in T \quad (6)$$

$$I_{it-1} + \sum_{r \in R} \alpha_{ir} a_{irt} = d_{it} + I_{it}, \ \forall i \in N, \, t \in T \quad (7)$$

$$I_{it} \leq u_{it}, \ \forall i \in N, \, t \in T \tag{8}$$

$$\sum_{r \in R} \theta_{rt} \leq |K|, \ \forall t \in T \tag{9}$$

$$P_{rt} = (a + g(sin\vartheta_r + C_{ro}cos\vartheta_r))(w\theta_{rt}l_r + q_{tij}l_{ij}) + (0.5C_d A\rho)v_r^2 l_r \ \forall r \in R, \, t \in T, \, i \in N, \, j \in V \tag{10}$$

$$E_{tot} = cf_v \sum_{r \in R} \sum_{t \in T} P_{rt} \tag{11}$$

$$E_{tot} \leq E_{cap} \tag{12}$$

$$\theta_{rt} \in [0, 1], \ \forall r \in R, \, t \in T \tag{13}$$

$$a_{irt}, I_{it} \geq 0, \ \forall i \in N, r \in R, \, t \in T \tag{14}$$

The objective function (4) represents the minimization of the sum of transportation cost, inventory cost and CO_2 emissions cost. The first term in the objective function represents the transportation cost, the second term represents inventory holding cost, while the third term represents the CO_2 emissions cost. Constraints (5) ensure that a customer is visited at most once per time period. Constraints (6) require that the vehicle capacity be respected. Constraints (7) are inventory balance equations that relate customer demands, incoming deliveries and inventories. Constraints (8) guarantee that a customer never has an inventory level that is greater than the total demands in the next ($\tau_{max} - 1$) consecutive time periods. This, in turn, imposes the constraint that perishable goods will never be discarded. Constraints (9) require the maximum number of different routes selected in a time period to be less than the number of vehicles. Equation (10) determine the energy (in Kilo-Joule) that is required for a vehicle *k* to traverse arc (*i, j*) in time period *t*. Equation (11) calculates the total amount of CO_2 emissions caused by transport and warehousing activities. Constraint (12) assures that the total amount of emissions does not exceed a certain emission cap. Lastly, Constraints (13) require the variables θ_{rt} to be binary and Constraints (14) require that inventory levels and quantities of goods delivered to customers be non-negative. In our model, non-negative inventory levels guarantee that no stock-outs occur at any customer during the planning time horizon.

4 Computational Analysis

In this section, we present the results of the simulations that we ran (using LINDO solver in GAMS) in order to evaluate the performance of our model. We provide a description of our experimental setup, simulation results, and discussion on the results.

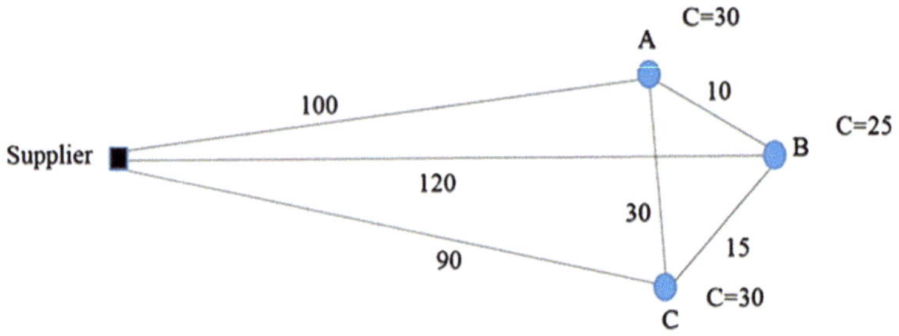

Fig. 1 Graphical representation of the problem

4.1 Simulation Setup

We consider the setting of discrete time with finite planning horizon, where the length of planning horizon is 9 days. There is one supplier and three customers, where the supplier is responsible for restocking the supplies of all three customers, see Fig. 1. The goods are delivered at the beginning of time period before the demand occurs. The supplier has only one truck with a capacity of 20 tons and travelling costs of two dollars per kilometer. The truck is assumed to travel at a constant speed. The inventory cost for customer A, B and C is 3$, 1$, and 1$ per ton per time period respectively. The inventory level of each customer never exceeds the total demand in next 3 days ($\tau_{\max} - 1$) at any time period, where the maximum shelf-life is 4 days ($\tau_{\max} = 4$). Initially (i.e. at $t = 0$), the inventory level for each customer is assumed to be full. Figure 1 shows the maximum inventory capacity (C) for the three customers, the distances (in km) between the supplier and each customer as well as the distances between customers. In this chapter, we use marginal CO_2 damage costs ($93/ton) as the cost of emissions. According to (Bektaş and Laporte 2011), an analysis of 133 estimations, gathered from 28 published studies, concludes the marginal CO_2 damage costs to be $93/ton and can be as higher as $350/ton.

4.2 Results and Discussion

First, we ran the simulation with the proposed model without imposing the carbon emission cap [(i.e. by ignoring constraint (Eq.12)] and setting the cost of emissions to zero, so that emission will not affect the routing decisions. Then, we introduced a positive carbon emission cost and a positive emission cap in our model, repeated the simulation, and compared the results. The results are displayed in Table 2, which shows the customer demand, inventory level and the quantity delivered by truck to each customer for each time period. In the table, the columns under the PIRP heading correspond to the scenario with zero emission costs and no emission

Table 2 Analysis of the simulation data

Time		PIRP setting			PPIRP setting		
		A	B	C	A	B	C
0	Initial Inventory	25	18	30	25	18	30
1	Delivery	0	5	0	0	0	0
	Demand	−6	−5	−9	−6	−5	−9
	Inventory	19	18	21	19	13	21
2	Delivery	0	0	0	0	0	0
	Demand	−6	−5	−8	−6	−5	−8
	Inventory	13	13	13	13	8	13
3	Delivery	0	0	0	0	0	0
	Demand	−8	−3	−6	−8	−3	−6
	Inventory	5	10	7	5	5	7
4	Delivery	0	0	19	0	19	0
	Demand	−5	−3	−7	−5	−3	−7
	Inventory	0	17	19	0	21	0
5	Delivery	18	0	0	6	0	8
	Demand	−6	−7	−8	−6	−7	−8
	Inventory	12	0	11	0	14	0
6	Delivery	0	18	0	7	0	10
	Demand	−7	−5	−8	−7	−5	−8
	Inventory	5	13	3	0	9	2
7	Delivery	0	0	20	5	0	15
	Demand	−5	−5	−7	−5	−5	−7
	Inventory	0	8	16	0	4	10
8	Delivery	14	0	0	14	0	6
	Demand	−6	−4	−9	−6	−4	−9
	Inventory	8	4	7	8	0	7
9	Delivery	0	0	0	0	4	0
	Demand	−8	−4	−7	−8	−4	−7
	Inventory	0	0	0	0	0	0

cap, while the columns under the PPIRP heading show the simulation results when carbon costs and emission cap were imposed.

Table 3 compares the total costs and carbon emission between the two cases. The CO_2 emissions given in Table 3 are the emissions due to transportation, which depend on the load of the truck, the speed of the truck, and the total distance travelled both loaded and empty. Our model is able to decrease the empty trips per kilometer (less empty trips means more efficient) by 26.2 % and reduce the overall carbon emissions by approximately 60 % with only a slight increase in total costs and distance travelled. Moreover, it can be seen the imposing emission constraints not only alters the route choice but also the delivered quantities and the load of the vehicle.

Table 3 Simulation results

	PIRP model	PPIRP model	Change %
Number of deliveries for all customers	6	10	60.0
Number of trips	6	6	0.0
Distance traveled (km)	1240	1320	6.5
Load factor (%)	78.3	78.3	0.0
Empty trips (%)	45.2	33.3	−26.2
Inventory and transportation cost ($)	2436	2509	3.0
Total CO_2 emissions (kg)	234.68	91.035	−61.2
Total costs ($)	2458	2517	2.4

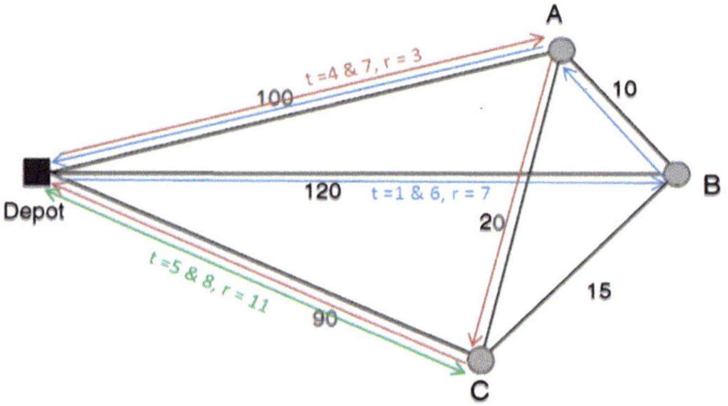

Fig. 2 Feasible routes of the PIRP model at time period *t*

PIRP model forces to minimize the total distance traveled by the truck and the inventory for each time period because the objective function of PIRP model minimizes the transportation cost and inventory cost. On other hand, PPIRP model tries to minimize the total cost of CO_2 emissions in addition to the previous two costs. Thus, the PPIRP model will try to first deliver the higher loads to the corresponding customers, as the CO_2 emissions are a function of overall vehicle load including the load of goods. Consequently, the PPIRP model might sometimes choose the longer routes in order to unload the heavier loads first.

Figures 2 and 3 show the routes that each model follows to deliver the perishable goods to each customer (A, B and C). We can see that both model used the same number of routes (i.e. 6 routes); however the second model (PPIRP) used different routes in terms of traversal order from the first model as can be seen. The second model (PPIRP) selects the routes that will deliver the heavier loads first so that it can visit more customers in each trip, thus minimizing the empty trips.

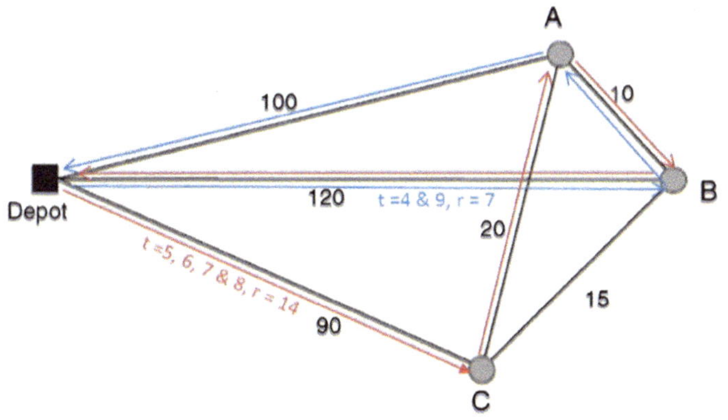

Fig. 3 Represents the feasible routes of the PPIRP model at time period *t*

5 Conclusion

Our model (PPIRP) successfully decreases the CO_2 emissions approximately by 61 % as compared to the original PIRP model. Lowering CO_2 emissions result in a slight increase in the total costs (about 2.4 %) because the model delivers the heavier loads first so that more customers can be visited in each trip and empty trips can be decreased. As a consequence, sometimes longer routes are selected, increasing the total distance traveled by 6.5 %.

Our model could be could be modified and applied to the reduction of CO_2 emissions in bigger problems with more vehicles, and longer time periods. CO_2 emissions from the warehousing activities could also be added to the model. In addition, instead of using a constant vehicle speed, the speed can be a variable with an upper and a lower limit, so the model will select the speed with minimum CO_2 emissions keeping in view the timing constraints and the drivers comfort.

Acknowledgments The authors wish to express their appreciation to Dr. Ali Diabat, Associate Professor—Engineering Systems and Management at Masdar Institute, for his help and support.

References

Abdallah T, Diabat A, Rigter J (2013) Investigating the option of installing small scale PVs on facility rooftops in a green supply chain. Int J Prod Econ 146(2):465–477

Abdallah Tarek, Diabat Ali, Simchi-Levi David (2012) Sustainable supply chain design: a closed-loop formulation and sensitivity analysis. Prod Plan Control 23(2–3):120–133

Al Dhaheri N, Diabat A (2010) A mathematical programming approach to reducing carbon dioxide emissions in the petroleum refining industry. In: Engineering Systems Management and Its Applications (ICESMA), 2010 Second International Conference on, IEEE

Al Zaabi S, Al Dhaheri N, Diabat A (2013) Analysis of interaction between the barriers for the implementation of sustainable supply chain management. Int J Adv Manuf Technol 68(1–4):895–905

Barth M, Boriboonsomsin K (2009) Energy and emissions impacts of a freeway-based dynamic eco- driving system. Trans Res D Transp Environ 14(6):400–410

Bektaş T, Laporte G (2011) The pollution-routing problem. Transp Res Part B: Method 45(8):1232–1250

Bertazzi L, Speranza MG (2012) Inventory routing problems: an introduction. EURO J Transp Logistics 1(4):307–326

Bertazzi L, Speranza MG (2013) Inventory routing problems with multiple customers. EURO J Transp Logistics 2:1–21

Chaug-Ing H, Sheng-Feng H, Hui-Chieh L (2007) Vehicle routing problem with time-windows for perishable food delivery. J Food Eng 80(2):465–475. ISSN 0260-8774

Diabat Ali, Aouam Tarik, Ozsen Leyla (2009) An evolutionary programming approach for solving the capacitated facility location problem with risk pooling. Int J Appl Decis Sci 2(4):389–405

Diabat A, Simchi-Levi D (2009). A carbon-capped supply chain network problem. In: IEEE International Conference on Industrial Engineering and Engineering Management, pp. 523–527

Diabat Ali, Richard Jean-Philippe, Codrington Craig W (2013a) A Lagrangian relaxation approach to simultaneous strategic and tactical planning in supply chain design. Ann Oper Res 203(1):55–80

Diabat A, Abdallah T, Al-Refaie A, Svetinovic D, Govindan K (2013b) Strategic closed-loop facility location problem with carbon market trading. IEEE Trans Eng Manage 60(2):398–408

Diabat A, Abdallah T, Henschel A (2013c) A closed-loop location-inventory problem with spare parts consideration. Comput Opers Res

Diabat A, Abdallah T, Le T (2014) A hybrid tabu search based heuristic for the periodic distribution inventory problem with perishable goods. Ann Oper Res (in press)

Diabat A, Olga B, Dima Naz (2014) An improved lagrangian relaxation-based heuristic for a joint location-inventory problem. Comput Oper Res

Ericsson E, Larsson H, Brundell-Freij K (2006) Optimizing route choice for lowest fuel consumption–Potential effects of a new driver support tool. Transp Res Part C: Emerg Technol 14(6):369–383

Figliozzi M (2010) Vehicle routing problem for emissions minimization. Transp Res Rec: J Transp Res Board 2197(1):1–7

Govindan K, Diabat A, Popiuc M. N (2012) Contract analysis: A performance measures and profit evaluation within two-echelon supply chains. Comput Ind Eng 63(1):58–74

Govindan K, Popiuc M. N., Diabat A (2013) Overview of coordination contracts within forward and reverse supply chains. J Clean Prod 47:319–334

Le T, Diabat A, Richard J. P., Yih Y. (2013) A column generation-based heuristic algorithm for an inventory routing problem with perishable goods. Optim Lett 7(7):1481–1502

Treitl S, Nolz PC, Jammernegg W (2012) Incorporating environmental aspects in an inventory routing problem. A case study from the petrochemical industry. Flex Serv Manuf J 26:1–27

Wang X, Li D (2012) A dynamic product quality evaluation based pricing model for perishable food supply chains. Omega 40(6):906–917

A Meta-heuristic Approach for VRP with Simultaneous Pickup and Delivery Incorporated with Ton-Kilo Basis Saving Method

Yoshiaki Shimizu and Tatsuhiko Sakaguchi

Abstract Under growing concerns with sustainable society, green or low carbon logistic optimization is becoming a keen interest to provide a plausible solution aiming at qualified service in global and competitive distribution system. As a key technology for such deployment, this chapter proposes a hybrid method of simultaneous pickup and delivery VRP aiming at rational framework available for real world applications. In its general procedure, the initial solution is derived from a modified saving method that consider the cost accounting known as Weber model or a bi-linear model of distance and weight. Then, it is updated by a modified tabu search to improve the tentative solution as much as possible. The idea is further extended to a non-linear or generalized Weber model. Numerical experiments are taken place to validate effectiveness of the proposed method through comparison.

Keywords Simultaneous pickup and delivery VRP · Hybrid meta-heuristic approach · Weber model · Modified saving method

1 Introduction

Under growing concerns with sustainable society, green or low carbon logistic optimization is becoming a keen interest to provide a fruitful solution aiming at qualified service in global and competitive distribution system. As a key technology for such deployment, we have engaged in the practical studies on vehicle routing problems (VRP) noticing that the transportation cost and/or CO_2 emission actually depend not only distance but also loading weight. This bi-linear cost accounting (unit cost times loading weight times traveling distance) is known as

Y. Shimizu (✉) · T. Sakaguchi
Toyohashi University of Technology, 1-1 Hibarigaoka, Tenpaku-cho, Toyohashi 441-8580, Japan
e-mail: shimizu@me.tut.ac.jp

P. Golinska (ed.), *Logistics Operations, Supply Chain Management and Sustainability*, EcoProduction, DOI: 10.1007/978-3-319-07287-6_43,

the Weber model and applied popularly to the conventional location/allocation problems belonging to a strategic level problem. Regarding operational problems like VRP, however, this idea has never been considered previously. Giving a general solution procedure in terms of a modified saving method for the Weber model, we developed a hybrid method so that we can practically cope with various types of VRP, i.e., delivery, direct pickup, and drop by pickup both for single depot and multi-depot problems (Shimizu 2011a, b). Against this, it is almost impossible to cope with those problems by conventional mathematical programming approaches. This chapter extends such our idea to a simultaneous pickup and delivery VRP (VRPSPD), and propose a novel saving-based hybrid method that is amenable for various real world applications.

The rest of the chapter is organized as follows. In Sect. 2, we briefly describe problem statements with a review of the related studies. Giving problem formulation in Sect. 3, we explain the proposed method in Sect. 4. Section 5 shows numerical experiments carried out to validate the effectiveness of our method. Finally, we give some conclusions in Sect. 6.

2 Problem Statement and Review of the Related Studies

Let us consider such a logistic network composed of single depot and multiple customers as illustrated in Fig. 1. Thereat, every vehicle starting from the depot visits its client customers via circular route, and return to the original depot so that a certain objective function can be optimized under some prescribed constraints. This generic problem has been studied popularly as VRP. It is a well-known combinatorial optimization problem, which minimizes the total distance traveled by a fleet of delivery vehicles under various constraints.

One of the recent studies on VRP is an extension from the generic customer demand satisfaction and vehicle payload limit. The other concerns have been paid to multi-depot problem (Chen et al. 2005) and multi-objective formulation (Jozefowiez et al. 2008). As a special variant of VRP, many researches are recently interested in VRP with pickup and delivery from certain aspects. They are classified into the following three categories (Nagy and Salhi 2005).

- Delivery first, Pickup second (VRPB): pickup only after delivered,
- Mixed Pickup and Delivery (MVRP): delivery and pickup in any sequence along the routes,
- Simultaneous Pickup and Delivery (VRPSPD): simultaneous delivery and pickup.

In this study, we concern VPRSPD since this is the most practical and general compared with the others. The VRPSPD refers to MVRP when either of pickup demand or delivery demand is placed at each customer.

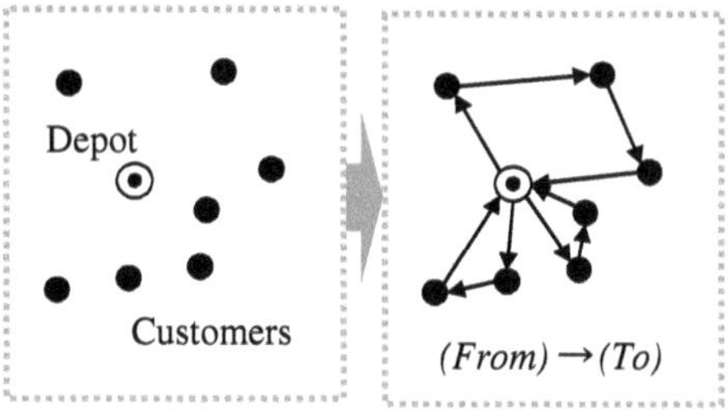

Fig. 1 Logistic network design problem concerned here

The first work in this area is studied by Min (1989) who proposed a cluster-first route-second approach and made a case study in public library A universal mix integer programming mathematic model is formulated associated with time windows and travelling distance constraints by Cao and Lai (2007) and it is solved using an improved genetic algorithm (IGA). Xie et al. (2007) gave an inserting criterion based on travelling distance. Ai and Kachitvichyanukul (2009) applied a particle swarm optimization (PSO) with multiple social structures. Subramanian et al. (2010) applied a parallel technique using multi-start heuristic. It is based on a variable neighbourhood descent procedure with a random neighbourhood ordering in an iterated local search (ILS) framework. Catay (2010) proposed an ant colony algorithm employing a new saving-based visibility function and pheromone updating procedure. Being interested in cost saving and environmental protection, Lai and Cao (2010) solved the problem with time windows by an improved differential evolution. A new meta-heuristic method called bacterial foraging optimization algorithm (BFOA) is applied by Hezer and Kara (2011) under the condition of total distance travelled. Capacitated vehicle routing problem (CVRP) is studied on the branch-and-cut algorithm by Subramanian et al. (2011). Noticing the importance of VRPSPD in reverse logistics, Jun and Kim (2012) proposed a sweep-based route method to generate better initial solution that could be improved by the following procedure associated with inter and intra routes. Goksal et al. (2013) used PSO under the variable neighbourhood descent algorithm (VND). Table 1 summarizes the VRPSPD related works mentioned above, describing their main contributions and/or approaches.

Due to difficulty of solution, however, only small problems formulated by no more than 400 customers are solved in numerical experiments to validate effectiveness of the respective method. Moreover, they have never adopted the transportation cost accounting by the Weber model. Real world applications of VRPSPD are frequently encountered in the distribution system of bottled drinks, groceries, LPG tanks, laundry service of hotels, some reverse logistics, etc. Since in these cases, we cannot ignore the weight expect for a few exceptions, the present approach has a practical significance compared with the previous studies mentioned above.

Table 1 Related studies associated with VRPSPD

Work	Year	Contributions and/or approach
Min	1989	First work, a cluster-first route-second approach
Cao and Lai	2007	MIP model, with time windows and traveling distance constraints, improved genetic algorithm
Xie, Qiu and Zhang	2007	Inserting criterion based method on traveling distance
Ai and Kachitvichyanukul	2009	PSO with multiple social structures
Subramanian et al.	2010	Parallel approach, a variable neighborhood descent in local search
Catay	2010	Ant colony algorithm employing a new saving-based visibility function and pheromone updating procedure
Lai and Cao	2010	Time windows, improved differential evolution
Subramanian et al.	2011	Branch-and-cut algorithm
Hezer and Kara	2011	Bacterial foraging optimization algorithm
Jun and Kim	2012	Sweep-based route method to generate better initial solution
Goksal, Karaoglan and Altiparmak	2013	Particle swarm optimization, variable neighborhood descent algorithm

3 Problem Formulation

The Weber model for VRPSPD is formulated by the following combinatorial optimization problem.

(p.1) min

$$\sum_{v\in V}\sum_{m\in M}\sum_{m'\in M} c_v d_{m'm}(g_{mm'y} + w_v) z_{mm'v} + \sum_{v\in V} F_v y_v$$

Subject to

$$\sum_{m\in M} z_{kmv} \le 1, \quad \forall k \in K;\ \forall v \in V \tag{1}$$

$$\sum_{m'\in M} z_{mm'v} - \sum_{m'\in M} z_{m'mv} = 0,\ \forall m \in M;\ \forall v \in V \tag{2}$$

$$\sum_{j\in J} z_{jj'v} = 0,\ \forall j' \in J;\ \forall v \in V \tag{3}$$

$$\sum_{v\in V}\sum_{k\in K} g_{jkv} z_{jkv} \le U_j x_j,\ \forall j \in J \tag{4}$$

$$g_{mm'v} \leq W_v z_{mm'v}, \ \forall m, \ m' \in M; \ \forall v \in V \tag{5}$$

$$\sum_{m \in M} \sum_{m' \in M} z_{mm'v} \leq L y_v, \ \forall v \in V \tag{6}$$

$$\sum_{k \in K} g_{kjv} = \sum_{k \in K} \sum_{k' \in K} p_k z_{kk'v}, \ \forall j \in J; \ \forall v \in V \tag{7}$$

$$\sum_{v \in V} \sum_{m \in M} g_{mkv} - \sum_{v \in V} \sum_{m \in M} g_{kmv} = q_k - p_k, \ \forall k \in K \tag{8}$$

$$\sum_{v \in V} g_{mm'v} = \sum_{v \in V} (u_{mm'v} + s_{mm'v}), \ \forall m, \ m' \in M \tag{9}$$

$$\sum_{v \in V} \sum_{m \in M} u_{kmv} - \sum_{v \in V} \sum_{m \in M} u_{mkv} = p_k, \ \forall k \in K \tag{10}$$

$$\sum_{v \in V} \sum_{m \in M} s_{kmv} - \sum_{v \in V} \sum_{m \in M} s_{mkv} = q_k, \ \forall k \in K \tag{11}$$

$$\sum_{j \in J} \sum_{k \in K} z_{jkv} = y_v, \ \forall v \in V \tag{12}$$

$$\sum_{j \in J} \sum_{k \in K} z_{kjv} = y_v, \ \forall v \in V \tag{13}$$

$$\sum_{m \in \Omega} \sum_{m' \in \Omega} z_{mm'v} \leq |\Omega| - 1, \ \forall \Omega \subseteq M \backslash \{1\}, \ |\Omega| \geq 2, \ \forall v \in V \tag{14}$$

$$x_j \in \{0, \ 1\}, \ \forall j \in J; \ y_v \in \{0, \ 1\}, \ \forall v \in V$$
$$z_{mm'v} \in \{0, \ 1\}, \ \forall m, \ m' \in M; \ \forall v \in V$$
$$g_{mm'v} \geq 0, \forall m, \ m' \in M; \forall v \in V$$

Notations of this problem are shown below.

Variables

- $g_{mm'v}$ [ton]: total load of vehicle v on the path from $m \in M$ to $m' \in M$
- $s_{mm'v}$ [ton]: delivery load of vehicle v on the path from $m \in M$ to $m' \in M$
- $u_{mm'v}$ [ton]: pickup load of vehicle v on the path from $m \in M$ to $m' \in M$
- $x_j = 1$ if candidate depot j is opened; otherwise 0
- $y_v = 1$ if vehicle v is used; otherwise 0
- $z_{mm'v} = 1$ if vehicle v travels on the path from $m \in M$ to $m' \in M$; otherwise 0

Parameters

- c_v [cost unit/ton/km]: transportation cost per unit load per unit distance of vehicle v

- p_k [ton]: pickup demand of customer k
- q_k [ton]: delivery demand of customer k
- $d_{mm'}$ [km]: path distance between $m \in M$ and $m' \in M$
- F_v [cost unit]: fixed charge for working vehicle v
- L [-]: auxiliary constant (large integer number)
- w_v [ton]: unladen weight of vehicle v
- U_j [ton]: maximum capacity of depot j
- W_v [ton]: maximum capacity of vehicle v

Index set

- J: depot; K: customer; V: vehicle; $M = J \cup K$, Ω: sub-tour

Here, objective function is composed of routing transportation cost and fixed charges of vehicles. On the other hand, each constraint means as follows: each vehicle cannot visit the customer twice (Eq. 1); coming in vehicle must leave out (Eq. 2); making a path between depots is avoided (Eq. 3); upper holding capacity at depot and upper bound load capacity for vehicle are given by Eq. 4 and Eq. 5, respectively; vehicle must travel on a certain path (Eq. 6); load of the returned vehicle must be equal to the total pickup loads (Eq. 7); Equation 8 and Eq.9 are material balances at each depot and path, respectively; difference from the foregoing visit is described by Eq. 10 and Eq. 11 for pickup and delivery, respectively; each vehicle leaves only one depot and return there (Eq. 12 and Eq. 13); Equation 14 gives a sub-tour elimination condition. Integrality conditions and positive conditions are imposed on the respective variables.

It is well known that this kind of problem belongs to an NP-hard class, and becomes extremely difficult to obtain a rigid optimal solution for real-world size problems. Hence, it is meaningful for those applications to provide a practical method that can derive a near optimum solution with reasonable computational efforts. Actually, we propose a hybrid method composed of the ordinary and meta heuristic methods.

4 Hybrid Approach for Practical Solution

4.1 Transportation Cost Accounting

As a key component of the hybrid method mentioned above, we applied the modified saving method whose cost accounting relies on the bilinear model known as the Weber model (Ton-Kilo basis). It is described as $c_v d_{ij}(w_v + q_j)$. Here w_v and q_j are unladen weight of vehicle v and demand of customer j as weight, respectively. In order to evaluate the cost from a more flexible viewpoint, the above formula is possible to extend to the generalized Weber model descried as $c_v \gamma d_{ij}^{\alpha}(w_v + q_j)^{\beta}$ where α and β denote the elastic coefficients for the distance and weight, respectively and γ is a constant.

Moreover, for practical evaluation of economy, it is suitable to account the fixed-charge of working vehicle F_v besides transportation cost. Eventually, we can evaluate the total cost TC by the following equation.

$$TC = \sum_{i=1}^{|V|} TR_i + |V| \cdot F_v \tag{15}$$

where

TR_i denotes circular transportation cost of root i, and $|V|$ total number of roots (necessary vehicle).

4.2 Proposed Method

To practically work with the above problem, we used the procedure outlined below. In this procedure, we apply two major components termed modified saving method and modified tabu search in a consecutive manner as follows.

Step 1 Derive the initial solution of VRPSPD by the modified saving method.
Step 2 Improve the above solutions by the modified tabu search.

Though the algorithm of the modified saving method itself is same as the original one, we need to note the following two special ideas in cost accounting. The first one is to derive the saving value on the Ton-Kilo basis. Referring to Fig. 2, the saving value between customer i and j is given by Eqs. (16) and (17) for Weber model and generalized Weber model, respectively. Apparently, the generalized model reduces to the ordinary when $\alpha = \beta = \gamma = 1$.

$$s_{ij}/c_v = d_{0i}(p_i - q_j + w_v) + d_{0j}(-p_i + q_j + w_v) - d_{ij}(p_i + q_j + w_v) \tag{16}$$

$$\begin{aligned} s_{ij}/(c_v\gamma) = {} & (q_i + w_v)^\alpha d_{0i}^\beta + (p_i + w_v)^\alpha d_{i0}^\beta + (q_j + w_v)^\alpha d_{0j}^\beta + (p_j + w_v)^\alpha d_{j0}^\beta \\ & -(q_i + q_j + w_v)^\alpha d_{0i}^\beta - (p_i + q_j + w_v) - (p_i + p_j + w_v)^\alpha d_{j0}^\beta \end{aligned} \tag{17}$$

Here, let the suffix be 0 for depot, $s_{ii} = 0$, and $d_{ij} = d_{ji}$.

The second one comes to the assertion that it is more economical to visit the new customer even if its saving value would become negative as long as its absolute value stays within the fixed charge of vehicle.

Moreover, for the simplicity of the algorithm, we assume the pickup load is less equal to the delivery one, i.e. $p_i \leq q_i$, $\forall i \in I$. This is a mild condition when looking at the cases of the bottled drinks and so on mentioned already. Then, Eqs. (19) and (20) are certainly satisfied under Eq. (18). In other word, it is enough to notify only the condition Eq. (18) as the loading constraint of vehicle.

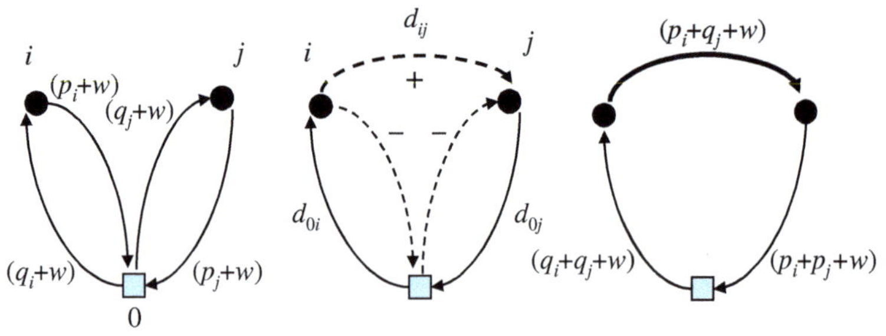

Fig. 2 Scheme to derive Ton-Kilo basis saving value

$$q_i + q_j \leq W_v \tag{18}$$

$$p_i + p_j \leq W_v \tag{19}$$

$$p_i + q_j \leq W_v \tag{20}$$

Since the (modified) saving method derives only an approximated solution, we try to improve it by applying the modified tabu search. There, we employed a few neighborhood operations known as insert, exchange and 2-opt both within each loop and between the loops. Actually, one of those operations is randomly selected under the control of tabu list whose length varies along with the problem size. Moreover, in the original algorithm of tabu search, only the improved neighbourhood solution can survive as long as it would not be involved in the tabu lists. In the modified method, however, even a degraded solution can be allowed to be a new tentative solution. This decision is made based on the probability whose function is known as Maxwell-Boltzmann and used in simulated annealing (Kirkpatrick et al. 1983). It is described as follow.

$$p = \begin{cases} 1 & if\ \Delta e \leq 0 \\ \exp(-\Delta e/T) & if\ \Delta e \leq \varepsilon \\ 0 & if\ \Delta e > \varepsilon \end{cases} \tag{21}$$

where

Δe and ε denote the amount of degradation of objective function value and a certain small value, respectively. Moreover, temperature T will be decreased geometrically along with the iteration.

Regardless of using the ordinal or the generalized models, the proposed method can work with in the same framework. It only requires to replace the saving value given by Eq. (16) with Eq. (17) in Step 1. In contrast, the foregoing mixed-integer bi-linear formulation turns to the non-linear one for the generalized model. This will expand the difficultly of solution greatly.

Table 2 Outline of parameter setting

Distance between customers	**[2, 50]**
Customer demand	$[2,100]/|K|^{0.25}$
Unladen weight	10
Payload	1,000
Fixed operating cost	50,000

5 Numerical Experiment

Numerical experiments were carried out to validate the effectiveness of the proposed method. In my best knowledge, there have been never existed the studies concerned with Ton-Kilo basis anywhere. Hence, we prepared the benchmark problems by ourselves. First, we randomly generated the prescribed numbers of customers within the rectangular region. Then, a depot is located in its center. The distances between the depot and customers and also those between every customer are given by Euclidian basis. Moreover, demands of each customer for delivery are randomly given within the prescribed ranges. Then, the pickup demands are set like $p_i = (0.3 + 0.7\ \mathrm{rand}())q_i$ so that $p_i \leqq q_i$, $\forall i \in I$ will be satisfied. The outline of those parameter setting is summarized in Table 2. We used PC with CPU: Intel(R) Celeron(R) 430 1.8 GHz, and RAM: 1 GB. The following discussions are made by averaging the results over 10 samples.

As an alternative to satisfy the requirement both on delivery and pickup of every customer, we can attain such final goal by doing them separately. In this case, we need to solve the delivery (D) and pickup (P) problems independently, and to sum those results to evaluate the total performance. Actually, we compared the results between SPD and such independent dealing (IP&D) through the procedure outlined in Fig. 3.

We summarize the results in Tables 3 and 4. They involve such large problems that have never been solved elsewhere. In the tables, the numbers in *Rate* column denote the improved rates of the final solution from the initial solution derived from the modified saving method. Due to the high non-linearity, results of the generalized model are poor (less than half) compared with those of the ordinary one. Due to the elastic coefficients [$\alpha = 0.894$, $\beta = 0.750$; ($\gamma = 1.726$)], however, the cost itself is less than that of the Weber model except for one exception.

The values in *IMP* column are given by 100(cost(IP&D)-cost(SPD))/cost (IP&D). They show how much we can save the cost by the simultaneous transport from the independent case. In any cases, we know it is able to decrease more than 60 % for the Weber model and till nearly 60 % for the generalized model. We should notice that the number of routes or necessary vehicles is also around 60 %. These results reveal the great advantage of the simultaneous transportation over the independent one.

In comparison, CPU time and iteration number of SPD also stay a bit below 50 % of those of IP&D. As a common property of the combinatorial optimization, CPU time itself expands rapidly according to the increase in problem size. Even

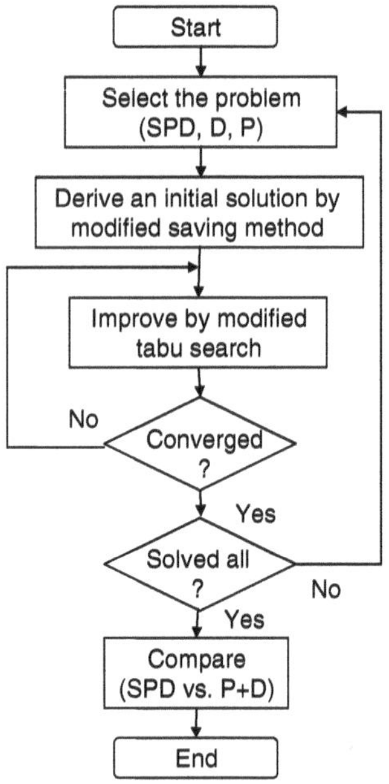

Fig. 3 Flow of the evaluation process

Table 3 Result for Weber model

	Size (−)	Cost (cost unit)	CPU (s)	Route# (−)	Iteration# (−)	Rate (−)	IMP (%)
Simultaneous	10	5.43E + 04	0.012	1.0	8.11E + 02	0.207	65.6
(SPD)	100	2.12E + 05	0.292	3.9	6.33E + 04	0.241	66.0
	300	4.84E + 05	3.502	9.0	5.64E + 05	0.214	61.3
	500	6.66E + 05	11.899	12.5	1.52E + 06	0.214	63.8
	1000	1.08E + 06	64.769	20.6	5.95E + 06	0.220	65.0
Independent	10	1.58E + 05	0.024	2.0	1.66E + 03	–	–
(IP&D)	100	6.24E + 05	0.563	6.8	1.31E + 05	–	–
	300	1.25E + 06	7.229	13.5	1.20E + 06	–	–
	500	1.84E + 06	24.551	20.3	3.40E + 06	–	–
	1000	3.09E + 06	138.634	32.9	1.38E + 07	–	–

for the largest problems, however, we can solve them within a reasonable time. Moreover, we show the convergence profiles in Fig. 4. Due to the high non-linearity, it seems the generalized case (open circle key) will not improve sufficiently compared with the original case, especially for the larger size problems.

To improve this poor performance in advance, it had better introduce the additional neighborhood operations that can derive more distant neighbors in the modified tabu search. As a whole, however, we can claim the proposed method can derive near optimum solutions of VRPSPD within reasonable computation times.

Table 4 Result for generalized Weber model

	Size (–)	Cost (cost unit)	CPU (s)	Route# (–)	Iteration# (–)	Rate (–)	IMP (%)
Simultaneous (SPD)	10	5.31E + 04	0.018	1.0	8.22E + 02	0.083	58.8
	100	2.06E + 05	1.425	3.9	6.33E + 04	0.105	58.9
	300	4.62E + 05	16.509	8.8	5.59E + 05	0.096	58.4
	500	6.65E + 05	50.753	12.7	1.47E + 06	0.091	58.9
	1000	1.12E + 06	251.795	21.4	5.91E + 06	0.089	59.8
Independent (IP&D)	10	1.29E + 05	0.04	2.0	1.66E + 03	–	–
	100	5.01E + 05	3.32	6.6	1.31E + 05	–	–
	300	1.11E + 06	43.106	14.0	1.23E + 06	–	–
	500	1.62E + 06	136.652	20.6	3.40E + 06	–	–
	1000	2.79E + 06	702.3	33.1	1.37E + 07	–	–

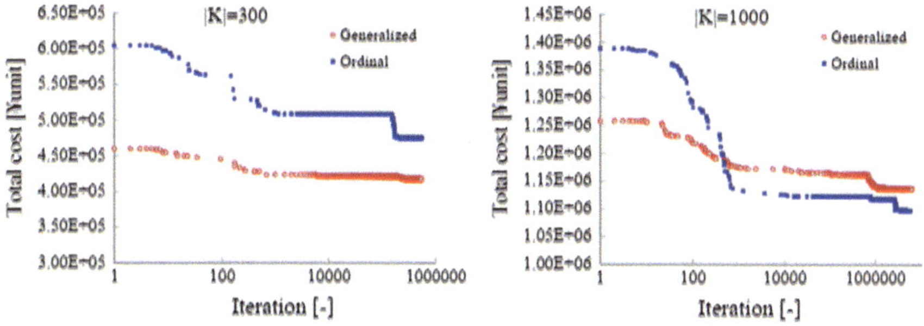

Fig. 4 Profiles of convergence

6 Conclusion

As a key technology for sustainable logistics under agile manufacturing and qualified service in distribution system, we have proposed a novel method for VRPSPD available for various real world applications. It is deployed in the framework proposed previously by making the best use of the components termed modified saving method and modified tabu search in a hybrid manner. Numerical experiments are carried out to validate the effectiveness of the proposed method.

In future studies, it is necessary to make earnest efforts for improving the solution ability especially for the generalized model. We also aim at extending the idea to cope with the problems more in general framework and consider various more practical conditions. For example, multi-depot, time window and duration distance are plausible conditions for such approach. Moreover, it is meaningful to turn our interest to multi-objective optimization that could associate with trade-off analysis among economics, risks, services, environment issues, etc.

Acknowledgments This research was partially supported by the Ministry of Education, Science, Sports and Culture, Grant-in-Aid, 25350442, Japan. We also thank our student Mr. Seiji Kato for his helpful support to prepare the manuscript.

Appendix A

The saving values on Ton-Kilo basis for the other types are summarized in Table A-1.

Table A-1. The Ton-Kilo basis saving values

Type	Weber model	Generalized Weber model
Delivery	$q_j(d_{0j} - d_{0i} - d_{ij}) + w_v(d_{0j} + d_{i0} - d_{ij})$	$\{(w_v + q_i)^\alpha - (w_v + q_i + q_j)^\alpha\}d_{0i}^\beta$ $+(w_v + q_j)^\alpha(d_{0j}^\beta - d_{ij}^\beta) + w_v^\alpha d_{i0}^\beta$
Pick up	$p_i(d_{i0} - d_{j0} - d_{ij}) + w_v(d_{0j} + d_{i0} - d_{ij})$	$\{(w_v + p_j)^\alpha - (w_v + p_i + p_j)^\alpha\}d_{j0}^\beta$ $+(w_v + p_i)^\alpha(d_{i0}^\beta - d_{ij}^\beta) + w_v^\alpha d_{0j}^\beta$

References

Ai TJ, Kachitvichyanukul V (2009) A particle swarm optimization for the vehicle routing problem with simultaneous pickup and delivery. Comput Oper Res 36:1693–1702

Cao E, Lai M (2007) Vehicle routing problem with simultaneous delivery and pick-up with time windows. In: International conference on transportation engineering 2007, pp 160–166

Catay B (2010) A new saving-based ant algorithm for the vehicle routing problem with simultaneous pickup and delivery. Expert Syst Appl 37:6809–6817

Chen S, Imai A, Zhao B (2005) A SA-based heuristic for the multi-depot vehicle routing problem. Japan Inst Navig 113–9:209–216 (in Japanese)

Goksal FP, Karaoglan I, Altiparmak F (2013) A hybrid discrete particle swarm optimization for vehicle routing problem with simultaneous pickup and delivery. Comput Ind Eng 65:39–53

Hezer S, Kara Y (2011) Solving vehicle routing problem with simultaneous delivery and pick-up using bacterial foraging optimization algorithm. In: Proceedings of the 41st international conference on computers and industrial engineering

Jozefowiez N, Semet F, Talbi EG (2008) Multiobjective vehicle routing problems. Eur J Oper Res 189:293–309

Jun Y, Kim B-I (2012) New best solutions to VRPSPD benchmark problems by a perturbation based algorithm. Expert Syst Appl 39:5641–5648

Kirkpatrick S, Gelatt CD, Vecch MP (1983) Optimization by simulated annealing. Science 220:671–680

Lai M, Cao E (2010) An improved differential evolution algorithm for vehicle routing problem with simultaneous pickups and deliveries and time windows. Eng Appl Artif Intell 23:188–195

Min H (1989) The multiple vehicle routing problem with simultaneous delivery and pick-up points. Transp Res Part A 23:377–386

Nagy G, Salhi S (2005) Heuristic algorithms for single and multiple depot vehicle routing problems with pickups and deliveries. Eur J Oper Res 162:126–141

Shimizu Y (2011a) Advanced saving method to evaluate economic concern. Trans Inst Syst Control Inf Eng 24:39–41 (in Japanese)

Shimizu Y (2011b) A Meta-heuristic approach for variants of VRP in terms of generalized saving method, Trans Inst Syst Control Inf Eng 24:287–295 (in Japanese)

Subramanian A, Uchoa E, Pessoa AA, Ochi LS (2011) Branch-and-cut with lazy separation for the vehicle routing problem with simultaneous pickup and delivery. Oper Res Lett 39:338–341

Subramanian A, Drummonda LMA, Bentes C, Ochi LS, Farias R (2010) A parallel heuristic for the vehicle routing problem with simultaneous pickup and delivery. Comput Oper Res 37:1899–1911

Xie RH, Qiu ZQ, Zhang YY (2007) A new heuristics for VRP with simultaneous delivery and pick-up. Int Conf Transp Eng 2007:2175–2180

Inventory Routing Problem with CO_2 Emissions Consideration

Nasir Alkawaleet, Yi-Fang Hsieh and Yanxiang Wang

Abstract Greenhouse gases emission is a major concern globally since they are key players in global warming. Some countries have signed the Kyoto Protocol, and set up some regulations to reduce their CO_2 emissions. Optimizing inventory and routing decisions can help in the reduction of CO_2 emissions if these emissions are taken into account by the decision makers. In the formulation, CO_2 emitted by transporting the product is modeled. The chapter investigates the effect of CO_2 emissions on the inventory and routing decisions determined over a given time horizon. The model is coded and solved in GAMS. The test results are used to indicate that emission costs should be considered when deciding the routing and inventory decisions.

Keywords Inventory routing · Green supply chain · Integer programming

1 Introduction

The traditional objective of minimizing the logistical costs in supply chain management has been upgraded to include the cost of greenhouse gases (GHG) emissions. Green logistics is the recent buzzword in supply chain management. It has received high attention from governments and business organizations. The

N. Alkawaleet (✉) · Y.-F. Hsieh · Y. Wang
Department of Engineering Systems and Management, Masdar Institute of Science and Technology, Abu Dhabi, UAE
e-mail: nalkawaleet@masdar.ac.ae

Y.-F. Hsieh
e-mail: yfhsieh@masdar.ac.ae

Y. Wang
e-mail: ywang@masdar.ac.ae

P. Golinska (ed.), *Logistics Operations, Supply Chain Management and Sustainability*, EcoProduction, DOI: 10.1007/978-3-319-07287-6_44,

environmentally sensitive logistical activities, especially transportation, have lead policy makers to change the transportation scheme to a more sustainable one with less negative effect on the environment (Lin et al. 2014).

Transportation is the largest end-use contributor toward global warming in the world. It relies on burning dwindling reserves of fossil fuels for energy, which emits greenhouse gases (GHGs) that contribute to climate change (Wakeland et al. 2012). Higher emissions can result not only from the energy needed for climate control, but also from the inherent yield losses that occur during storage. Protected cultivation is even more energy intensive.

Since most of developed nations agreed to legally binding targets for their emissions of the six major greenhouse gases according to the Kyoto Protocol in 1997, there has been an increasing interest in bringing environmentally factors into the routing problem (Know et al. 2013). To represent this situation, our model includes a cost-benefit assessment of acquiring carbon rights under a cap-and-trade regime.

The "*cap-and-trade*" approach to air pollution abatement was first demonstrated in a series of micro-economic computer simulation studies between 1967 and 1970 for the National Air Pollution Control Administration by Burton and William (1967). This approach offsets created through a baseline and credit approach, and a carbon tax are all market-based approaches that put a price on carbon and other greenhouse gases, and provide an economic incentive to reduce emissions, beginning with the lowest-cost opportunities. The main contribution of this research is adding the cap-and-trade regime into the model to give a more accurate decision for those countries which are environmentally sensitive.

A deterministic multi-period inventory routing problem for perishable goods (PIRP) is built by Richard (2013) to optimize the decision about when to deliver to each customer, how much to deliver to each customer in each time period, and how to route vehicles such that the sum of transportation and inventory costs is minimized while still meeting customer demands and perishability constraints.

The remainder of this chapter is organized as follows. Section 2 is the literature review. Section 3 is the problem description and formulation for product delivery under a cap-and-trade regime. In Sect. 4, computational analysis is presented. Section 5 will be the conclusion.

2 Literature Review

Inventory routing problems (IRPs) deal with not only routing but also inventory decisions for a single model. IRPs are considered as difficult problems, so many heuristic approaches were proposed in the previous literatures. Dror and Ball (1987) who were the first to consider a heuristic approach for solving multi-period IRPs. After several efforts for finding ways to solve different types of IRPs, Andersson et al. (2010) gave a classification and comprehensive literature review of the current state of the research from industrial aspects of combined inventory management and routing in maritime and road-based transportation. Afterwards, people started working on using column generation to solve IRPs. Richard (2013) modeled a specific inventory routing problem in which goods are perishable (PIRP), and they exploit the model structure to develop a column generation-based solution approach. The computational experiments were presented to demonstrate their effective methodology and the significant cost savings in the integration of routing and inventory.

For the delivery of perishable goods, there was rare consideration in the literatures of inventory cost. Federgruen et al. (1986) who were firstly integrate transportation and inventory models for perishable goods but only focused on the single period problem. They solved their model by introducing a heuristic algorithm based on interchanges. Tarantilis and Kiranoudis (2001) developed a threshold-based algorithm for solving the fresh milk distribution problem for one of the biggest diary companies in Greece. They modeled this specific problem as a heterogeneous fixed fleet vehicle routing problem. On the other hand, some studies concern perishable goods in the extended inventory models based on the economic order quantity (EOQ) policy. Giri and Chaudhuri (1998) proposed a deterministic models of perishable inventory with stock-dependent demand rate and nonlinear holding cost, and the holding cost were treated as two situations, the approximate optimal solution in both cases were derived.

The model introduced in supply chain design field focused on minimizing fixed and operating costs. With the raising of global warming concerns and sustainable environment issues, more and more studies focus on eco-friendly supply chain, considering carbon emission as one of the transportation costs. Nagurney et al. (2007) developed a new supply chain model in which the manufacturers can produce the homogeneous product in different manufacturing plants with associated distinct environmental emissions, and proved that the supply chain model with environmental concerns can be reformulated and solved as an elastic demand transportation network equilibrium problem. Cruz and Matsypura (2009) modeled and analyzed a supply chain networks with corporate social responsibility through integrated environmental decision-making, and especially they included the minimization of emissions (waste) to their model. Elhedhli (2012) took carbon emission into account in their supply chain network design problem, and the relationship between CO_2 emissions and vehicle weight is modeled using a concave function leading to a concave minimization problem. They utilized Lagrangian relaxation to decompose their problem into a capacitated facility location problem with single sourcing and concave knapsack problem which can be solved easily. The result showed that considering emission costs can change the optimal configuration of the supply chain and confirmed that emission costs should be taken into account when designing supply chain in jurisdictions with carbon costs. Recently some studies considered the consideration of CO_2 emissions in the design of supply chain, Abdallah et al. (2012); Diabat et al. (2013a); Diabat et al. (2014); Al Dhaheri et al. (2010); Diabat et al. (2013b); Govindan et al. (2013); Abdallah et al. (2013); Le et al. (2013); Diabat et al. (in press); Diabat et al. (2009); Diabat et al. (2013c); Al Zaabi et al. (2013). Some of these studies considered developing solution algorithms to tackle the developed mathematical programs such as Diabat et al. (2013a) and Diabat et al. (2014), or considered the cost of CO_2 emissions in the oil refinery industry such as Al Dhaheri et al. (2010). Govindan et al. (2012) evaluated of performance measures and supply chain profit behavior under buyback, revenue sharing, quantity flexibility and advanced purchase discount contracts versus no coordination and wholesale price systems.

3 Problem Description and Formulation

In our problem, we formulate a mixed integer programming for the inventory routing problem of a product distributed to a number of customers from a single distribution center (DC). The distribution center ships the product to its customers using heterogeneous fleet of trucks. In our formulation, we assume that in each time period, at most one truck will be used to transport goods to each customer and then returns back to the distribution center. The distribution center has a service level of 95 %, hence stock-outs at customers could take place which will lead the distribution center to pay a penalty for the corresponding customers. Moreover, we incorporate the cost benefit calculation associated with carbon trading. According to the third mechanism of the Kyoto protocol, carbon trading allows companies to buy and sell emissions credits. If a company produces below its cap, it will earn credits; otherwise it needs to purchase credit to make up the shortfall. We refer to (Facanha and Horvath 2007) for the values of the emission factors used to calculate CO_2 emissions for different truck classes (Fig. 1).

Indices

j customer
t time period
k truck type

Parameters

c_k cost per unit distance of truck of type k
d_j distance from distribution center to customer j
D_{jt} demand of customer j at time t
h_{jt} inventory holding cost of customer j in time period t
v_k load capacity of truck class k
p price of CO_2 per unit mass
n_k number of trucks of type k
cap limit of carbon emissions
r_k amount of carbon emissions per unit distance emitted by truck of type k
UB_j inventory upper bound of customer j
s_j penalty of having stock-outs at customer j
$\propto$ percentage of stock-out of demand

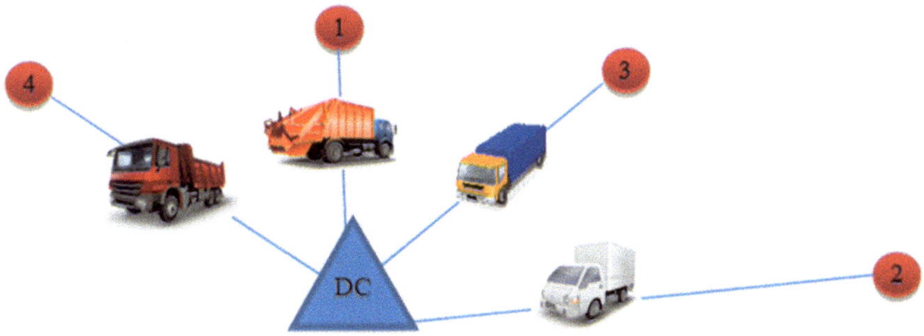

Fig. 1 Example of the problem with different truck classes

Decision Variables

$$x_{jtk} = \left\{ \begin{array}{ll} 1, & if\ customer\ j\ is\ visited\ by\ truck\ class\ k\ in\ time\ period\ t \\ 0, & otherwise \end{array} \right\}$$

y_{jt} amount delivered to customer j in time period t

I_{jt} inventory level of customer j at the end of time period t

$st_{jt.}$ stock-out at customer j in time period t

$$\text{Min} \sum_j \sum_t \sum_k c_k d_j x_{jtk} + \sum_j \sum_t h_{jt} I_{jt} + \sum_j \sum_t s_j st_{jt} + p\left[\sum_j \sum_t \sum_k r_k d_j x_{jtk} - cap\right] \tag{1}$$

Subject to

$$\sum_k x_{jtk} \leq 1, \quad \forall j, t \tag{2}$$

$$y_{jt} \leq \sum_k v_k x_{jtk}, \quad \forall j, t \tag{3}$$

$$I_{jt-1} + y_{jt} = D_{jt} + I_{jt} - st_{jt}, \quad \forall j, t \tag{4}$$

$$I_{jt} \leq UB_j, \ \forall j, t \tag{5}$$

$$\sum_j x_{jtk} \leq n_k, \ \forall t, k \tag{6}$$

$$st_{jt} \leq \alpha D_{jt}, \quad \forall j, t \tag{7}$$

$$x_{jtk} \in \{0, 1\}, \quad \forall j, t, k \tag{8}$$

$$y_{jt}, I_{jt} \geq 0, \quad \forall j, t \tag{9}$$

The objective function (1) represents the minimization of the sum of transportation cost, inventory cost, stock-out penalty and carbon emissions cost. Constraint (2) ensures that each customer is visited at most once in each time period. Constraint (3) implies that the load capacity of a truck must be respected. Constraint (4) defines the inventory balance. Constraint (5) defines the upper bound of the inventory level. Constraint (6) limits the number of truck units of each type. Constraint (7) defines the service level of the distribution center. Constraints (8) and (9) represent the condition of the decision variables.

4 Computational Analysis

In this section, we perform computational analysis on the model proposed above. We generated data randomly according to Richard (2013):

- Demand of a customer in any time period: random integer number in the interval [10, 100].
- Distance of customer j: random number in the interval [0, 500].
- Inventory holding cost of customer j: random number in the interval [4.6, 5.0].
- Beginning inventory: random integer in the interval [0, $D_{j1} + D_{j2}$].

In addition, we assumed the following:

- Inventory upper bound of customer j = max{Djt}, t = 1…T
- The total number of trucks equals the number of customers with the following breakdown:
  - 20 % Class 8b
  - 40 % Class 5
  - 40 % Class 2b

- The load capacities of truck classes are as follows:
  - Class 8b = max{D_{jt}}, $\forall j, t$
  - Class 5 = 50 % of class 8b capacity
  - Class 2b = 25 % of class 8b capacity

Data regarding emission factors of trucks are obtained from Facanha and Horvath (2007) and are shown below (Table 1):

In order to test the effect of the cap on the model, we added constraint to model the maximum allowable amount of CO_2 emissions.

Table 1 Emission factors

Truck type	CO_2 emissions (grams/mile)
Class 8b	2,337
Class 5	713
Class 2b	462

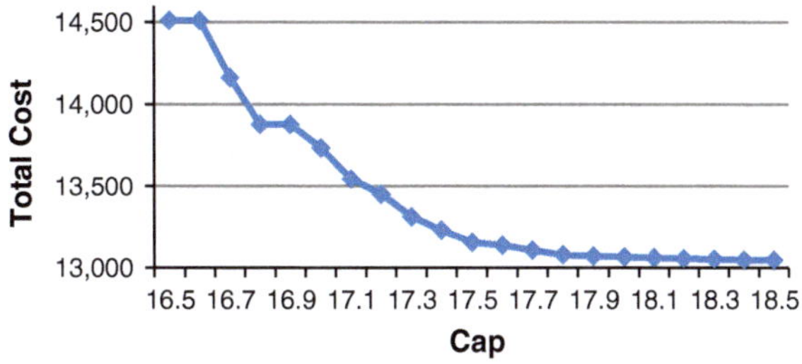

Fig. 2 Total cost versus cap

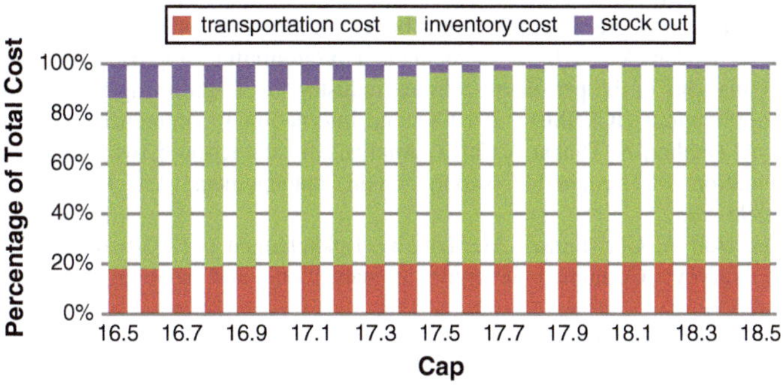

Fig. 3 Transportation cost, inventory cost and stock-out penalty versus cap

$$\sum_{j}\sum_{t}\sum_{k} r_k d_j x_{jtk} \leq cap \tag{10}$$

Twenty one scenarios for different cap values were tested, the scenarios were solved using ILOG CPLEX MIP solver in GAMS modeling language (Rosenthal 2012).

Figure 2 shows the results of running different scenarios for the value of the total cost with different values of cap. The trend shows that the total cost decreases with the increase of the emissions cap. This behaviour is sensible as the right hand side of Eq. (10) increases, this will expand the solution space and thus better objective value is expected. Figure 3 shows the breakdown of the total cost per cap value, the figure shows that transportation cost is almost the same, while the change is evident for the inventory cost and stock-out penalty.

5 Conclusion

In this chapter, we have introduced the problem of inventory routing with CO_2 emissions consideration. In the model, stock-outs were allowed with a maximum value of 5 % of the demand in that time period. In addition, three classes of trucks were used, class 8b, class 5 and class 2b. These classes have different load capacities, transportation costs and emission factors. The problem was formulated as a MIP and solved using the ILOG CPLEX MIP solver in GAMS. We conducted an experimental study on 21 instances and found out that the cost of emissions has an influence on the inventory and routing decisions.

References

Abdallah T, Diabat A, Simchi-Levi D (2012) Sustainable supply chain design: a closed-loop formulation and sensitivity analysis. Prod Plan Con 23(2–3):120–133

Abdallah T, Diabat A, Rigter J (2013) Investigating the option of installing small scale PVs on facility rooftops in a green supply chain. Int J Prod Econom 146(2):465–477

Al Dhaheri N, Diabat A (2010) A mathematical programming approach to reducing carbon dioxide emissions in the petroleum refining industry. In: Engineering Systems Management and Its Applications (ICESMA), 2010 Second International Conference on, IEEE

Al Zaabi S, Al Dhaheri N, Diabat A (2013) Analysis of interaction between the barriers for the implementation of sustainable supply chain management. Int J Adv Manuf Tech 68(1–4):895–905

Andersson H, Hoff A, Christiansen M, Hasle G, Løkketangen A (2010) Industrial aspects and literature survey: Combined inventory management and routing. Comput Oper Res 37:1515–1536

Bai R, Burke E.K., Kendall G (2008) Heuristic, meta-heuristic and hyper-heuristic approaches for fresh produce inventory control and shelf space allocation. J Oper Res Soc 59(10):1387–1397

Burton E, William S (1967) An Economic Analysis of the Control of Sulphur Oxides Air Pollution DHEW Program Analysis Report No. 1967–1969. Ernst and Ernst, Washington, DC

Cruz JM, Matsypura D (2009) Supply chain networks with corporate social responsibility through integrated environmental decision-making. Int J Prod Res 47:621–648

Diabat A, Aouam T, Ozsen L (2009) An evolutionary programming approach for solving the capacitated facility location problem with risk pooling. Int J App Dec Sci 2(4):389–405

Diabat A, Richard J. P., Codrington C. W (2013a) A Lagrangian relaxation approach to simultaneous strategic and tactical planning in supply chain design. Anna Oper Res 203(1):55–80

Diabat A, Abdallah T, Al-Refaie A, Svetinovic D, Govindan K (2013b) Strategic closed-loop facility location problem with carbon market trading. IEEE Trans Eng Manag 60(2):398–408

Diabat A, Abdallah T, Henschel A (2013c) A closed-loop location-inventory problem with spare parts consideration. Comput Oper Res

Diabat A, Battaïa O, Nazzal D (2014) An improved lagrangian relaxation-based heuristic for a joint location-inventory problem. Comput Oper Res

Diabat A, Abdallah T, Le T (2014) A hybrid tabu search based heuristic for the periodic distribution inventory problem with perishable goods. Ann Oper Res (in press)

Dror M, Ball M (1987) Inventory/routing: reduction from an annual to a short period problem. Naval Res Logistics 36(6):891–905

Elhedhli S, Merrick R (2012) Green supply chain network design to reduce carbon emissions. Transp Res Part D Transp Environ 17:370–379

Facanha C, Horvath A (2007) Evaluation of life-cycle air emission factors of freight transportation. Environ Sci Technol 41(20):7138–7144

Federgruen A, Prastacos G, Zipkin PH (1986) An Allocation and Distribution Model for Perishable Products. Oper Res 34:75–82

Giri B, Chaudhuri K (1998) Deterministic models of perishable inventory with stock-dependent demand rate and nonlinear holding cost. Eur J Oper Res 105:467–474

Govindan K, Diabat A, Popiuc M. N (2012) Contract analysis: A performance measures and profit evaluation within two-echelon supply chains. Comput Ind Eng 63(1):58–74

Govindan K, Popiuc M. N., Diabat A (2013) Overview of coordination contracts within forward and reverse supply chains. J Clean Prod 47:319–334

Le T, Diabat A, Richard, J-P, Yih Y (2013) A column generation-based heuristic algorithm for an inventory routing problem with perishable goods. Optim Lett 7:1481–1502

Lin C, Choy K, Ho G, Chung S, Lam H (2014) Survey of green vehicle routing problem: past and future trends. Expert Syst Appl 41(4):1118–1138

Kwon YJ, Choi YJ, Lee DH (2013) Hetrogenous fixed fleet vehicle routing considering carbon emission. Transp Res Part D 81–89

Nagurney A, Liu Z, Woolley T (2007) Sustainable supply chain and transportation networks. Int J Sustain Transp 1:29–51

Rosenthal RE (2012) A user's guide. GAMS Development Corporation, Washington, DC, USA

Tarantilis C, Kiranoudis C (2001) A meta-heuristic algorithm for the efficient distribution of perishable foods. J Food Eng 50:1–9

Wakeland W, Cholette S, Venkat J (2012) Food transportation issues and reducing carbon footprint. In Green technologies in food production and processing, Springer, London, US, pp 211–236

MIX
Papier aus verantwortungsvollen Quellen
Paper from responsible sources
FSC® C105338

If you have any concerns about our products,
you can contact us on
ProductSafety@springernature.com

In case Publisher is established outside the EU,
the EU authorized representative is:
Springer Nature Customer Service Center GmbH
Europaplatz 3, 69115 Heidelberg, Germany

Printed by Libri Plureos GmbH
in Hamburg, Germany